Quantum Molecular Dynamics

Quantum Molecular Dynamics

Energy Transfer and Reactivity

David A. Micha
University of Florida, Gainesville, FL, USA

WILEY

Published by John Wiley & Sons, Inc., Hoboken, New Jersey.

Library of Congress Cataloging-in-Publication Data

Names: A. Micha, David author | John Wiley & Sons publisher
Title: Quantum molecular dynamics : energy transfer and reactivity / David A. Micha, University of Florida, Gainesville, FL, USA.
Description: Hoboken, New Jersey : Wiley, [2026] | Includes index.
Identifiers: LCCN 2025022988 (print) | LCCN 2025022989 (ebook) | ISBN 9780470383735 hardback | ISBN 9781119319245 adobe pdf | ISBN 9781119319269 epub
Subjects: LCSH: Molecular dynamics | Quantum chemistry | Energy transfer
Classification: LCC QD461 .M5845 2026 (print) | LCC QD461 (ebook)
LC record available at https://lccn.loc.gov/2025022988
LC ebook record available at https://lccn.loc.gov/2025022989

Cover Design: Wiley
Cover Image: © Photobank.kiev.ua/Shutterstock

Set in 9.5/12.5pt STIXTwoText by Straive, Chennai, India

Contents

Preface

The interaction of colliding molecules leads to energy transfer and to reactivity. A full treatment of these phenomena, as they relate to experimental measurements or to the kinetics of molecular gases, requires quantum mechanics and statistical mechanics. Those phenomena are particularly complex when they involve electronic excitation or electron transfer, or when they occur in large many-atom systems. The understanding extracted from the interplay of theory and experiment impacts many areas relevant to engineering applications and to the environment. Related concepts and methods provide results for thermodynamic properties including chemical equilibrium constants, and kinetic properties including transport coefficients, energy relaxation times, and reaction rates.

The material covered in this book originated from a sequence of courses on chemical physics and quantum theory taught at the University of Florida for graduate students and undergraduate researchers in the Chemistry, Physics, and Engineering departments. The subjects were introduced assuming only basic knowledge of molecular interactions and dynamics, familiarity with quantum and statistical mechanics, and the mathematics of partial differential equations, special functions, and linear algebra in functional spaces. Starting from fundamental concepts and methods, the book proceeds to develop the needed quantum theory and related computational methods to allow for the calculation of physical properties. These are mainly rates and cross sections of energy transfer and reaction phenomena, as needed to interpret experimental measurements or to calculate kinetic properties.

Given all the available literature on molecular collisional phenomena and applications, it is relevant to point out why there is a need for yet another presentation of these subjects. It helps to introduce subjects in some detail, starting from fundamental concepts and proceeding to advanced methods, as is done in every chapter of this book. Also, many new developments have occurred in recent years on new computational methods prompted by the derivation of new software algorithms and by the availability of more powerful computer hardware. This has allowed for the calculation of new results, involving complex phenomena, and for larger molecular systems. Recent developments in information sciences and artificial intelligence provide large amounts of organized relevant data and ways to retrieve it, and their basic concepts are included in this book to provide tools for the treatment of desired data. Some of these new methods and results are integrated in what follows, after covering the traditional subjects on molecular collisions and gaseous kinetics. There is an enormous literature on the subjects of this book, including several thousand papers and reviews. References in the chapters are a small selection of all the relevant publications, chosen to help with the chapter's materials while avoiding overwhelming complexities. This book's author apologizes in advance to the contributors of the many relevant publications that could not be referenced.

The presentation in each chapter begins at a level accessible to graduate and advanced undergraduate students, and proceeds to present recent theoretical and computational developments. Treatments of elastic, inelastic, and reaction phenomena are first developed in simple cases, containing physical models and calculated properties comparable to that of measured values. The theory and mathematics are developed as needed to obtain numerical results. In cases involving a many-atom system or electronic excitation, the computational work may become very demanding and treatments must be carefully chosen to achieve results. Practical treatments of the molecular dynamics combine quantal and classical mechanics, and one is developed in several chapters with new and more details than presently available in the literature.

Chapter 1 provides a qualitative introduction to collisional phenomena, cross sections, and rates, with examples for energy transfer and reactivity in physical systems. Chapter 2 gives a detailed treatment of classical and quantum dynamics of elastic collisions, including differential and integral equation formulations, partial-wave analysis, and numerical methods for classical, quantal, and semiclassical dynamics, with examples for atom–atom and electron–atom scattering. Chapter 3 on inelastic (energy transfer) collision dynamics displays coupled-channel equations, matrix forms of partial-wave expansions, examples of two coupled states and resonance phenomena, distorted-wave, and optical model treatments. Chapter 4 describes electronically adiabatic energy transfer, with special attention to numerical methods, wavefunction expansions in basis sets of quantum states for atom–polyatomic and molecule–molecule collisions, and translational–rotational–vibrational energy transfer models and cross-section results.

Chapter 5 deals with the challenging subject of diabatic collisions, where atomic motions and electronic transitions are dynamically coupled, and treatments require choices of electronic representations, multiple electronic potential energy surfaces, and their couplings. This chapter covers conical intersections, seams, and geometric phase effects. Semiclassical descriptions are developed with new details, based on a general eikonal formulation of wavefunctions. Reactive collisions are discussed in Chapter 6, with a modern formulation based on coupled arrangement channels for both adiabatic and nonadiabatic reactions, including time-dependent formulations and some recent developments related to machine-learning procedures. It contains many examples of calculated cross sections. Finally, Chapter 7 is dedicated to general scattering operators and density operators. Partitioning of state-space is developed to construct effective Hamiltonians, multiple scattering is treated in a many-atom system, and coupled arrangement-channel operator equations for reactions are given here. Density operator developments are covered to include statistical averages from models and from experimental setups, and quantum-classical equations and related results are presented from eikonal and Wigner-transform treatments.

This book follows the author's previous work, Molecular Interactions, and the author reiterates the acknowledgment made in that book's preface. The author thanks his quantum theory science teacher at the Institute of Physics Balseiro, Bariloche (Argentina), Dr. Jose Balseiro, and Dr. Per-Olov Lowdin for his mentoring in quantum chemistry at the Quantum Chemistry Group, Uppsala University (Sweden). Much was learned from colleagues and researchers at the University of Wisconsin (Madison), Drs. Richard Bernstein and Joseph Hirschfelder, and at the University of California, San Diego, Dr. Keith A. Brueckner. He is grateful for the collaborations and hospitality at institutions visited during sabbaticals: Gothenburg University in Sweden, Harvard University, the Max Planck Institutes for Stroemungsforschung and for Astrophysics in Germany, Imperial College in London (England), the Institute of Theoretical Physics at the University of California Santa Barbara, JILA at the University of Colorado in Boulder, the Weizmann Institute in Rehovot (Israel), Florida State University, the École Normale Supérieure in Paris (France), and the Institute for Mathematics and its Applications at the University of Minnesota in Minneapolis. The author is especially grateful to the many graduate and undergraduate students who conducted research under his direction, as well as to postdoctoral associates and visiting scientists at the University of Florida. Due to their numbers, they are not listed here; their names can be found at the website https://people.clas.ufl.edu/Micha/.

The author is greatly appreciative of the organizations that provided essential financial support for his research and collaborations: the Swedish International Development Agency; the Alfred P. Sloan Foundation; the National Science Foundation of the USA, for many years of Principal Investigator research support, and funding of US-Latin American workshops; NASA; the Alexander von Humboldt Foundation Senior Scientists Awards twice; and the Dreyfus Foundation.

Thanks are also due to reviewers at the early stages when this book and the previous one were being planned, particularly Bernard Kirtman, George Schatz, and Victor Batista. Several of the author's research colleagues have been kind enough to read and comment on some of the book's chapters, in particular John Stanton, Adrian Roitberg, and Dmitri Kilin. Apologies are extended to the many colleagues and research visitors not mentioned due to text space limits!

The author appreciates the patience and support of the editors at John Wiley & Sons, Inc., over the years it took to have this book completed. Special thanks go to project senior editor Michael Leventhal for his guidance during production and publication. A special thanks is also due to the author's wife, Rebecca A. Micha, for her help in keeping the English grammar right in many paragraphs, and for her unwavering support.

October 2024

David A. Micha
Gainesville, Florida

1

Collisional Phenomena, Cross Sections, and Rates

CONTENTS

1.1 Electronic and Nuclear Motions in Collisional Phenomena

Matter consisting of particles such as atoms or molecules in a gas, where interparticle average distances are large compared to the size of the constituent particles, can be described in terms of pair interactions during collisions, insofar the probability of finding a third particle nearby is small and its effect on the interaction pair is negligible.

Collisions in a gas can be described in terms of the properties of constituent particles, their positions and velocities, and their interaction potential energies [1–4].

Quantum Molecular Dynamics: Energy Transfer and Reactivity, First Edition. David A. Micha.

This can sometimes be done with classical mechanics or more generally within quantum mechanics. For colliding particles A and B, it is convenient to describe their relative motion in terms of their interparticle distance and their relative velocities. The rate of encounters (or number of collisions per unit time) is proportional to the relative flux (or number of collisions per unit time and unit area traversed by the relative-motion trajectories), with the proportionality factor equal to a collisional cross section, a measure of the diameters of the particles dependent on their relative velocity.

If the gas has been in contact with a medium of a given temperature, it will eventually reach thermal equilibrium through energy exchange during collisions, and rate processes can be assumed to occur near thermal equilibrium. This remains stable insofar individual pair interactions occur in a diluted system with low particle densities.

There is extensive literature on concepts and applications of molecular collision phenomena and their relation to experimental methods such as molecular beam scattering and spectroscopy. At the introductory level, some (among many others) relevant books are in references [5–10]. At an intermediate level, a classic text is [11] among others [12–17], dealing with quantitative treatments. Advanced treatments with extensive use of the quantum formalism of scattering theory are found in [18–24]. Applications involving models, and calculations of cross sections of interest in the interpretation of experimental results obtained with molecular beam and spectroscopic techniques, are available in [25–35], and the use of cross sections as inputs in molecular reaction rates, gaseous transport, and kinetics, can be found in [36–45]. Mathematical and computational methods developed for the calculation of cross sections can be found in [46–51]. Applications to surface phenomena are also covered in several books, among them [52–54].

The interaction of molecules with photons leading to photoinduced phenomena such as dissociation and ionization and related theory can be found in [15, 55–57], and scattering of electrons by atoms and molecules are found in [11, 58, 59]. In addition, many recent relevant chapters about molecular collisions in gases and at surfaces and about molecular interactions with photons are found in volumes of several edited series on advances, such as *Advances in Chemical Physics*; *Advances in Quantum Chemistry*; *Advances in Atomic, Molecular and Optical Physics*; and *Annual Review of Physical Chemistry*.

Given all this literature on molecular collisional phenomena and applications, it is relevant to point out why there is a need for yet another presentation of these subjects. It helps to introduce subjects in some detail, starting from fundamental concepts and proceeding to advanced methods, as done in what follows here in every chapter. Also, many new developments have occurred in recent years in new computational methods prompted by the derivation of new algorithms and by the availability of more powerful computer hardware and software. Many new results have been generated involving complex phenomena and for larger molecular systems. Recent developments in information sciences and artificial intelligence provide large amounts of organized relevant data and ways to retrieve it. The language of concepts and methods in what follows provides tools to access desired data. They provide new insights and data useful in many applications. Some of these new methods and results are integrated into what follows after covering the traditional subjects on molecular collisions and gaseous kinetics. The presentation in each chapter begins at a level accessible to undergraduate students with knowledge of differential equations and special functions, who are familiar with introductions to quantum mechanics, and proceeds to present recent developments.

Insights on the quantum dynamics of molecular interactions involving a many-atom system can best be derived from the interplay of accurate experimental measurements, such as those obtained from crossed molecular beams or from time-resolved spectroscopy, and detailed theoretical treatments from quantum mechanics and statistics. Theory provides an interpretation of measurements and resulting data on molecular interactions and dynamics, while experimental measurements provide checks on the accuracy of models and calculations.

Figure 1.1 shows relationships mediated by quantum chemistry, molecular dynamics, and statistical mechanics, leading from electrons, nuclei, and photons

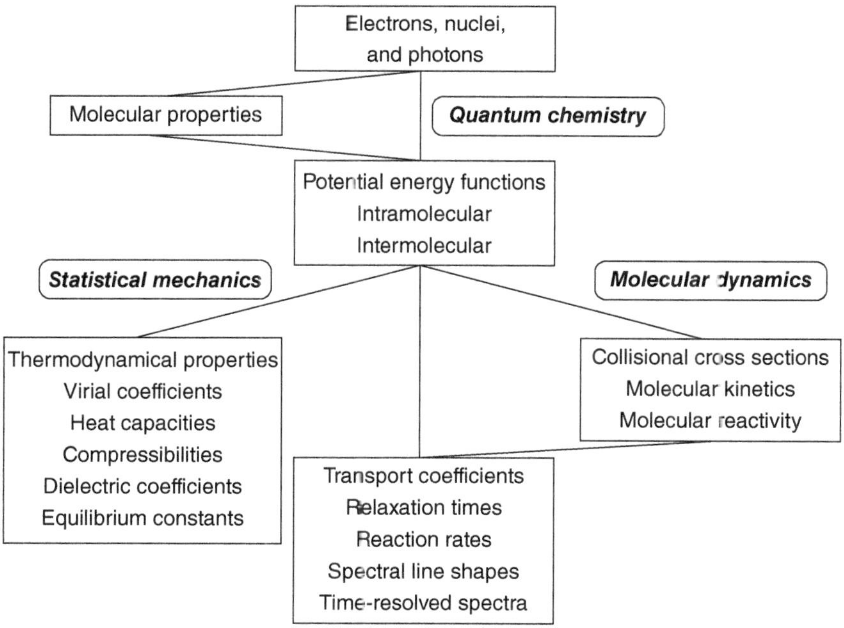

Figure 1.1 Relationships mediated by quantum chemistry, molecular dynamics, and statistical mechanics, leading from electrons, nuclei, and photons present in molecules and electromagnetic fields, to physical properties. Collisional cross sections, transport coefficients, and reaction rates follow from potential energy functions by means of molecular dynamics and statistical mechanics. Obtained from reference [4] / John Wiley & Sons.

present in molecules and electromagnetic fields, to physical properties. Collisional cross sections, transport coefficients, and reaction rates follow from potential energy functions by means of molecular dynamics and statistical mechanics. Obtained from reference [4].

1.2 Collisional Cross Sections

1.2.1 Definition of a Cross Section

Let us consider a collision process

$$\mathrm{A}(p) + \mathrm{B}(q) \rightarrow \mathrm{C}(r) + \mathrm{D}(s)$$

where the collision species A, B, C, and D may be electrons, atoms, molecules, ions, photons, or even a surface. The index p is a collection of quantum numbers specifying the internal (electronic and rovibrational) state of A, and similarly for the others. The pair (p, q) and the initial relative motion momentum $\vec{P}$ define the reactant-channel state $\alpha = (\vec{P}, p, q)$, and similarly β is used for products C and D and final momentum $\vec{P}'$.

For two stationary beams with velocities $\vec{v}_A$ and $\vec{v}_B$ colliding in a laboratory (LAB) frame, as shown in Figure 1.2, with an incoming flux $J_0^{(L)} = v\, n_A n_B\, \tau$ of A relative to B, equal to the number of A particles moving toward B per unit area and unit time, with n_A (*or* n_B) the number of particles A (or B) per unit volume, the increment $d\dot{N}_{\beta\alpha}$ of the $\alpha \rightarrow \beta$ reaction rate (pairs/unit time) may be expressed as

$$d\dot{N}_{\beta\alpha} = J_0^{(L)} d\sigma_{\beta\alpha}^{(L)}$$

$$d\sigma_{\beta\alpha}^{(L)} = I_{\beta\alpha}^{(L)}(\Omega_C, v) d\Omega_C.$$

Here $\vec{v} = \vec{v}_A - \vec{v}_B$ is the relative velocity of the particles, τ is the reaction volume, and $d\Omega_C$ is an increment of solid angle subtended by the detector of emerging product species C.

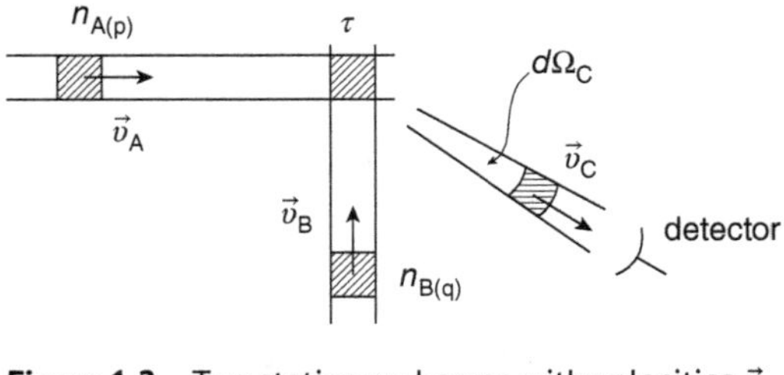

Figure 1.2 Two stationary beams with velocities $\vec{v}_A$ and $\vec{v}_B$ colliding in a laboratory (LAB) frame, leading to the formation of species C and D. Here τ is the reaction volume and $d\Omega_C$ is an increment of solid angle subtended by the detector of particle C.

Hence $n_B\,\tau$ is the number of particles B in the reaction volume and $v\,n_A$ is the flux of particles A relative to each B. The function $I^{(L)}_{\beta\alpha}$ is a differential cross section in the laboratory frame, with units of area. The integral cross section is defined by

$$\sigma^{(L)}_{\beta\alpha}(v) = \int d\Omega_C\, I^{(L)}_{\beta\alpha}(\Omega_C, v).$$

1.2.2 Conservation Laws

As long as each pair collision may be considered an isolated event, we must have conservation of the total mass, momentum, angular momentum, and energy of the system. Indicating the final values with primed symbols, we have conservation of

(a) Mass, $M_{tot} = M_A + M_B = M_C + M_D = M'_{tot}$
(b) Momentum, $\vec{P}_{tot} = \vec{P}_A + \vec{P}_B = \vec{P}_C + \vec{P}_D = \vec{P}'_{tot}$
(c) Angular Momentum

$$\vec{L}_{tot} = \vec{L}^A_{orb} + \vec{L}^A_{int} + \vec{L}^B_{orb} + \vec{L}^B_{int} = \vec{L}^C_{orb} + \vec{L}^C_{int} + \vec{L}^D_{orb} + \vec{L}^D_{int} = \vec{L}'_{tot}$$

(d) Energy

$$E_{tot} = E^A_{kin} + E^A_p + E^A_{bind} + E^B_{kin} + E^B_q + E^B_{bind}$$
$$= E^C_{kin} + E^C_r + E^C_{bind} + E^D_{kin} + E^D_s + E^D_{bind} = E'_{tot}$$

where M_A is the total mass of A, $\vec{P}_A$ is the momentum of the center of mass (CM) of A, $\vec{L}^A_{orb} = \vec{R}_A \times \vec{P}_A$ is the orbital angular momentum of the CM of A located at position $\vec{R}_A$, and $\vec{L}^A_{int}$ is the internal angular momentum of species A, which can be nuclear-rotational, electronic orbital or spin, or a combination of them.

Also, $E^A_{kin} = P^2_A/(2M_A)$ is the kinetic energy of A, E^A_p is the internal excitation energy of the state A(p), E^A_{bind} is the binding energy ($-I_A$ for ionization or $-D_A$ for dissociation) of A, of the "elementary particles" making up A, B, C, and D. An example is provided by the reaction $F + H_2 \rightarrow H + FH$, where the "elementary particles" are F, H, and H, and the binding energies are the dissociation energies of H_2 and FH with negative signs.

1.2.3 Collisions in the Center of Mass Frame

The CM frame is defined by introducing CM and relative positions $\vec{R}_{CM}$ and $\vec{R}$ as

$$\vec{R}_{CM} = (M_A\vec{R}_A + M_B\vec{R}_B)/M_{tot}$$
$$\vec{R} = \vec{R}_A - \vec{R}_B$$

and fixing the orientation of the frame axes with respect to the LAB frame. The positions of the CM of A and B are $\vec{R}_A = \vec{R}_{CM} + M_B\vec{R}/M_{tot}$ and $\vec{R}_B = \vec{R}_{CM} - M_A\vec{R}/M_{tot}$.

Letting the frame origin move with the CM velocity $\vec{V}_{CM}$ that satisfies (differentiating the above equations with respect to time)

$$M_A\vec{V}_A + M_B\vec{V}_B = M_{tot}\vec{V}_{CM} = \vec{P}_{CM}$$

it follows that

$$\vec{L}^A_{orb} + \vec{L}^B_{orb} = \vec{L}_{CM} + \vec{L}_{rel}$$

$$E^A_{kin} + E^B_{kin} = E_{CM} + E_{rel}$$

where, using the relative velocity $\vec{v} = d\vec{R}/dt$ and reduced mass $M = M_A M_B/M_{tot}$, one finds $E_{CM} = M_{tot}V^2_{CM}/2$, $E_{rel} = Mv^2/2$, $\vec{L}_{CM} = \vec{R}_{CM} \times \vec{P}_{CM}$, and $\vec{L}_{rel} = \vec{R} \times \vec{P}$, which is the relative angular momentum of A with respect to B in terms of the relative momentum $\vec{P} = M\vec{v}$.

Hence in the CM frame with its origin at $\vec{R}_{CM}$, one finds that $\vec{P}_{CM} = \vec{L}_{CM} = E_{CM} = 0$, and conservation laws apply with the modifications:

$$\vec{L}^A_{orb} + \vec{L}^B_{orb} = \vec{L}_{rel}$$

$$E^A_{kin} + E^B_{kin} = E_{rel} = P^2/(2M).$$

Relations between LAB and CM velocities are illustrated in the so-called Newton velocity diagram of Figure 1.2, where O and G are the origins of the LAB and CM frames, and

$$\overrightarrow{OG} = \vec{V}_{CM}, \quad \overrightarrow{BA} = \vec{v}$$

$$\overrightarrow{OA} = \vec{v}_A, \quad \overrightarrow{OB} = \vec{v}_B$$

$$\overrightarrow{GA} = \vec{u}_A, \quad \overrightarrow{GB} = \vec{u}_B$$

where the end points of velocities for particles A and B are labelled by A and B in Figure 1.3, and particle velocities in the CM frame are $\vec{u}_A$ and

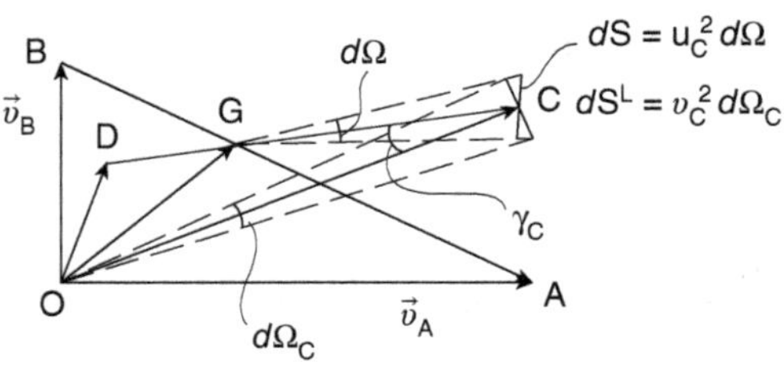

Figure 1.3 Relations between LAB and CM velocities illustrated in the so-called Newton velocity diagram. Here G is the location of the center-of-mass of the (A, B) and (C, D) pairs and detection of species C follows from its flux through the element of surface $dS = u_C^2\, d\Omega$ in the LAB frame or alternatively through $S^L = v_C^2\, d\Omega_C = dS\,\vec{u}_C \cdot \vec{v}_C/(u_C v_C) = dS\,\cos(\gamma_C)$.

$\vec{u}_B$, and V similarly for products C and D, written with primed symbols $\vec{V}'_{CM} = \vec{V}_{CM}$, $\vec{v}' = \vec{v}_C - \vec{v}_D$. Detection of species C follows from its flux through the element of surface $dS = u_C^2\, d\Omega$ in the CM frame or alternatively through $dS^L = v_C^2\, d\Omega_C = dS\, \vec{u}_C \cdot \vec{v}_C/(u_C v_C) = dS\, \cos(\gamma_C)$ in the LAB frame.

Reaction cross sections and rates for $\alpha \to \beta$ collisions of two molecules A and B can be described in both LAB or CM frames, in terms of their incoming molecular-pair flux, $J_0^{(L)}$ or J_0 [60]. In the LAB frame, the rate of pair-transitions per unit time is

$$d\dot{N}_{\beta\alpha}^{(L)}\left(\vec{v}_A, \vec{v}_B\right) = J_0^{(L)}\left(\vec{v}_A, \vec{v}_B\right) d\sigma_{\beta\alpha}^{(L)}\left(\vec{v}_A, \vec{v}_B\right).$$

This can be converted to a CM frame using that $\vec{v}_A = \vec{V}_{CM} + M_B\vec{v}/M_{tot}$ and $\vec{v}_B = \vec{V}_{CM} - M_A\vec{v}/M_{tot}$. It leads to $d\dot{N}_{\beta\alpha}(\vec{V}_{CM}, \vec{v}) = J_0(\vec{V}_{CM}, \vec{v})d\sigma_{\beta\alpha}(\vec{V}_{CM}, \vec{v})$ with the relation between fluxes given by

$$J_0^{(L)}\left(\vec{v}_A, \vec{v}_B\right) d\vec{v}_A d\vec{v}_B = J_0\left(\vec{V}_{CM}, \vec{v}\right) d\vec{V}_{CM} d\vec{v}.$$

In an experimental set up where the distribution of initial velocities $\vec{v}_I, I = A, B$, is thermal at a temperature T, given by the Maxwell–Boltzmann distribution

$$f_I(\vec{v}_I; T) = \left[M_I/(2k_BT)\right]^{1/2} exp\left[-M_I v_I^2/(2k_BT)\right],$$

the incoming volume densities are $n_I = N_I f_I$, and J_0 contains only V_{CM}^2. Integration over V_{CM} provides a cross-sectional increment $d\overline{\sigma}_{\beta\alpha}^{(L)}(v; T) = \overline{I}_{\beta\alpha}^{(L)}(\Omega_C, v; T)d\Omega_C$ for events where the product species C emerges within solid angles in $d\Omega_C$.

Corresponding to $d\Omega_C$, the solid angle in CM is $d\Omega$, and from invariance of the cross sections under LAB → CM transformations, it follows that for a single collision event,

$$I_{\beta\alpha}^{(L)}(\Omega_C, v)d\Omega_C = I_{\beta\alpha}(\Omega, v)d\Omega$$

which defines the CM differential cross section $I_{\beta\alpha}(\Omega, v)$. A similar relation for a thermal distribution gives in CM $\overline{I}_{\beta\alpha}(\Omega, v; T)$.

The relation between solid angle increments,

$$d\Omega_C = d\Omega\, (u_C/v_C)^2/\cos\gamma_C$$

$$\cos\gamma_C = \vec{u}_C \cdot \vec{v}_C/(u_C v_C)$$

obtained from the geometry of collision kinematics in velocity space, gives the relation between differential cross sections.

1.2.4 Classification of Collision Processes

For the generic reaction $A(p) + B(q) \to C(r) + D(s)$ and fixed reactant conditions (i.e. for fixed E_{tot}), a product channel is called open if $E'_{rel} > 0$ and closed if $E'_{rel} < 0$.

For open channels, we may have

(a) Elastic collisions

$$C = A, \quad D = B, \quad E_{rel} = E'_{rel}$$

(b) Inelastic collisions

$$C = A, \quad D = B, \quad E_{rel} \neq E'_{rel}$$

(c) Reactive collisions

$$C \neq A, \quad D \neq B, \quad E_{rel} \neq E'_{rel}$$

with

$$\Delta_R E = E^A_{bind} + E^B_{bind} - \left(E^C_{bind} + E^D_{bind}\right)$$

the reaction energy.

Elastic collisions are *coherent* if the states of A and B do not change and *incoherent* otherwise. Non-elastic collisions are *exoergic* if $E'_{rel} > E_{rel}$ and *endoergic* otherwise. Choosing A and C to indicate initial and final directions, respectively, collisions are *forward* if $\vec{v} \cdot \vec{v}' > 0$ and *backward* if $\vec{v} \cdot \vec{v}' < 0$.

An additional classification item can be introduced when an initial collision partner is a photon ϕ absorbed by a compound AB and leading to its dissociation, as in

(d) Photodissociation

$$\phi + AB \rightarrow A + B.$$

Also, one of the initial collisional partners can be a solid surface, with the particle–surface collision being elastic, inelastic, reactive, or dissociative.

1.3 Quantal Description of Collisions

1.3.1 Time-Dependent Quantal States

We consider again a general case $A(p) + B(q) \rightarrow C(r) + D(s)$. A quantal description is based on state wavefunctions $\Psi(t)$ that are solutions, evolving with time, of the Schrodinger equation for given initial and boundary conditions. The equation contains a Hamiltonian operator $\widehat{H}$ that can be constructed by adding the kinetic energies of all the atomic ion-cores and electrons, and their interaction energies, which can be Coulombic and other long- and short-range interparticle potential energies. The Hamiltonian may also contain spin–orbit coupling energies and time-dependent energies resulting from externally applied electromagnetic fields.

Starting from the time-dependent Schrodinger equation [18] and its boundary conditions (or normalization) when the collision partners A and B are initially at time t_{in} far apart,

$$\hat{H}\Psi(t) = i\hbar\partial\Psi/\partial t, \quad \langle\Psi|\Psi\rangle = 1$$

$$\Psi(t_{in}) = \Psi^{(in)} = \Psi^{(A)}\Psi^{(B)}.$$

Here $\hbar = h/(2\pi)$ is the reduced Planck constant, and an initial quantal state $\Psi^{(in)} = \Psi(t_{in})$ is given by the product of a pair of non-overlapping wavepackets $\Psi^{(w)}$, $w = A, B$, with velocities $\vec{v}_A$ and $\vec{v}_B$ and initial many-atom states for each reactant species. The bracket notation indicates integration over all the wavefunction variables. The total Hamiltonian operator $\hat{H}$ can be written as a sum of terms for relative and internal motions plus their coupling energy operator and also plus an external time-dependent energy from applied fields or due to an environment,

$$\hat{H} = \hat{H}_{rel} + \hat{H}_{int} + \hat{H}_{coupl} + \hat{H}_{ext}(t)$$

where $\hat{H}_{rel} = \hat{K}$ is the operator in the CM frame for the relative motion kinetic energy, $\hat{H}_{int} = \hat{K}_{int} + \hat{V}_{int} = \hat{H}_A + \hat{H}_B$ is the Hamiltonian for the internal motions of isolated A and B in a set of internal states $\{\phi_n\}$, and $\hat{H}_{coupl} = \hat{V}_{coupl}$ is an interaction potential energy between relative and internal motions. Similar expressions apply for C + D. The coupling $\hat{V}_{coupl}$ goes to zero when reactants or products are far apart. The wavefunction evolves in time as a wavepacket $\Psi(t)$ with asymptotic boundary conditions after collision at a final time t_{fn} given by

$$\Psi(t) \approx \Psi^{(0)}(t) + \Psi^{(sc)}(t), \quad t \geq t_{fn},$$

where $\Psi^{(0)}$ is the evolved initial wavepacket and $\Psi^{(sc)}$ is a scattered component. After a collision, the probability at times $t \geq t_{fn}$ for transition into a final state $\Psi^{(fn)} = \Psi^{(C)}\Psi^{(D)}$ is obtained from $P^{(fn)}(t) = |\langle\Psi^{(fn)} \mid \Psi(t)\rangle|^2$ and the transition rate is $dP^{(fn)}/dt$.

A usual experimental setup for measurement of cross sections involves scattering of a stationary beam of molecules A and B in a pair-state $\phi_n = \phi_a^{(A)}\phi_b^{(B)}$ with given translational velocities $\vec{v}_A$ and $\vec{v}_B$, respectively, and relative velocity $\vec{v} = \vec{v}_A - \vec{v}_B$ ending in a state β for the C + D pair. In this case, it is relevant to introduce an initial flux $\vec{J}_\alpha^{(0)} = \vec{v}\,\rho_\alpha$, the number of incoming colliding pairs per unit time and unit traversed area given as a product of velocity times a particle density ρ_α per unit volume, and to calculate a steady-state flux $\vec{J}_{\alpha\to\beta}$ for transitions between initial and final states [12, 15, 20, 61]. This can be done for the time-independent Hamiltonian of an isolated collision event, by first calculating steady states $\Psi_\nu^{(+)}$, with $\nu = (\vec{k}, n)$, as solutions of the equation $\hat{H}\Psi_\nu^{(+)} = E_\nu\Psi_\nu^{(+)}$, for total energies

$E_\nu = E_{rel} + E_p^A + + E_q^B$, and relative momentum $\vec{P} = \hbar\vec{k}$, with scattering boundary conditions, and by expressing the wavepacket $\Psi_\alpha(t)$ as a superposition

$$\Psi_\alpha(t) = \int d\nu A_\alpha(\nu)\Psi_\nu^{(+)} \exp(-iE_\nu t/\hbar)$$

shown as a symbolic addition and integration over states ν, with $A_\alpha(\nu)$ selected to form the incoming scattering wavepacket. The connection between time-dependent and time-independent treatments of collisions is described in detail in several books [12, 20] and reviews, and in applications with emphasis on molecular systems [15, 61].

1.3.2 Time-Independent Steady States

The total Hamiltonian operator $\hat{H}$ for a collision in an isolated space is

$$\hat{H} = \hat{K} + \hat{H}_{int} + \hat{H}_{coupl}.$$

For an isolated collision in a collision region without any external forces, the Hamiltonian is time-independent and the time variable can be separated in the equation of motion of the wavefunction, to work within a time-independent description, where

$$\Psi_\alpha(t) = \Psi_\alpha^{(+)} \exp(-iE_\alpha t/\hbar)$$

$$\hat{H}\Psi_\alpha^{(+)} = E_\alpha\Psi_\alpha^{(+)}, \quad \Psi_\alpha^{(+)} \approx \Psi_\alpha^{(0)} + \Psi_\alpha^{(sc)} \quad \text{for} \quad R \to \infty$$

with the state index $\alpha = (\vec{k}, n)$ for internal states $n = (p, q)$ for A and B (or pair I), and R is the relative distance between the CMs of the final species C and D. The total energy is conserved during a collision and given for reactants by $E_\alpha = E_{rel}(\vec{k}) + E_{int}(I) = E_{rel}(\vec{k}') + E_{int}(I') = E_\beta$ for the products. The stationary wave $\Psi_\alpha^{(+)}$ is an eigenfunction of the Hamiltonian as shown, and the previous normalization boundary condition for wavepackets must now be replaced by a scattering asymptotic condition at large distances R from the collision region. In regions where $\hat{V}_{coupl} = 0$, we can write the total wavefunction in the CM frame as a product of the eigenfunctions of $\hat{K}$ and of $\hat{H}_{int}$, which we assume are known. This allows us to separate the free motion and internal motions states in the total wavefunction at large distances.

At the large distances where the reaction products are detected, the wavefunction is given by a sum of the incoming unperturbed wave $\Psi_\alpha^{(0)} = (2\pi)^{-\frac{3}{2}} \exp(i\vec{k}_\alpha \cdot \vec{R})\phi_\alpha$ plus a scattering wave $\Psi^{(sc)} = \sum_\beta \psi_{\beta\alpha}^{(sc)}(\vec{R})\phi_\beta$ generated by interactions in the collision region, leading to products C and D in internal states β. Consistent with this, the flux of particle C with respect to D detected in state $\beta = (\vec{k}', n')$ is $\vec{J}_{\beta\alpha} \approx \vec{J}_{\beta\alpha}^{(sc)}$.

The flux operator along the direction $\vec{n} = \vec{R}/R$ can be obtained from the momentum operator $\hat{P} = -i\hbar\nabla_R$, where ∇_R is a gradient along the pair relative position $\vec{R}$, and from the density operator $\hat{\rho}(\vec{R})$ as [18]

$$\hat{\vec{J}} = (-i\hbar/2M)\left[\nabla_R \cdot \hat{\rho}(\vec{R}) - \hat{\rho}(\vec{R}) \cdot \nabla_R\right]$$

with an expectation value $\vec{J}_\alpha = \left\langle \Psi_\alpha^{(+)} \middle| \hat{\vec{J}} \middle| \Psi_\alpha^{(+)} \right\rangle$. Writing the density operator as $\hat{\rho}(\vec{R}) = \delta(\vec{R} - \hat{\vec{R}}) \sum_{n'} |\phi_{n'}\rangle\langle\phi_{n'}|$ in terms of the Dirac delta function for a position density and with a ket-bra notation for internal states, it is possible to identify the state-to-state scattered flux $\vec{J}_{\beta\alpha}^{(sc)}$. The differential cross sections $I_{\beta\alpha}(\Omega, v) = d\sigma_{\beta\alpha}/d\Omega$ for C, emerging through a detection area $dA = R^2 d\Omega$ within solid angle $d\Omega$ in the direction given by a unit vector $\vec{n}$, can be obtained from the incoming and outgoing flux terms as

$$d\sigma_{\beta\alpha}/d\Omega = \vec{J}_{\beta\alpha}^{(sc)} \cdot \vec{n}\, R^2 / J_\alpha^{(0)}.$$

With the denominator given by the incoming unperturbed flux. This cross section can be obtained from the asymptotic wavefunction $\psi_{\beta\alpha}^{(sc)}$ components in $\Psi_\alpha^{(sc)}$.

1.4 Examples of Physical Systems and Phenomena

1.4.1 Overview of Phenomena

An overview of phenomena for frequently encountered physical systems can be organized as

(A) Electronically adiabatic heavy-particle collisions.
(B) Electronically diabatic heavy-particle collisions.
(C) Electron and photon scattering by molecules.

Their description can be done in terms of stationary eigenfunctions Ψ_E for fixed energy, if the Hamiltonian of the system is time-independent, or using time-dependent wavefunctions $\Psi(t)$ if there are external time-dependent fields or if it is convenient to construct initial localized wavepackets that evolve in time. These wavefunctions depend on all the electronic variables (positions and spins) $\boldsymbol{X}$ and all the nuclear (or ion core) positions $\boldsymbol{Q} = (\vec{R}, \boldsymbol{Q}')$.

An electronically adiabatic collision involves a single electronic state and its related potential energy dependent on chosen nuclear positions; electronic motions are separated from nuclear motions by the Born–Oppenheimer decoupling approximation [4, 18], which fixes nuclear positions, and the collision involves only atomic CM degrees of freedom and motion on a single potential energy surface (or PES) function of atomic positions. Electronic rearrangement due to excitation or electron transfer instead involves several electronic states and the related PESs as well as the couplings of electronic states mediated by nuclear motions, in diabatic (or non-adiabatic) phenomena. The following examples show involved degrees of freedom and Hamiltonians needed for quantal descriptions.

The scattering of electrons or of photons by a single molecule requires introduction of a Hamiltonian that includes the scattering particle (electron or photon), molecule, and their interaction energies to begin with. Electronic excitations however may lead to dissociation into two or more interacting fragments, which can be treated as in the final scattering stage in a collision of heavy particles. These subjects are treated in detail in several published books on electron scattering [11, 58, 59] and photon scattering [55, 56, 62–64] and in review chapters [65, 66]. They are briefly described in what follows but are not otherwise developed in this book.

The following examples are a selection of results briefly illustrating calculated and measured cross sections. The subjects are expanded in the following chapters.

1.4.2 Electronically Adiabatic Heavy-Particle Collisions

(a) Two structureless atoms (e.g. Na + Hg, He + He)
Indicating with $\vec{R}$ the relative position vector between their CMs, the Hamiltonian contains

$$\widehat{K} = -\hbar^2\nabla_R^2/(2M), \quad \widehat{H}_{int} = 0, \quad \widehat{V} = \varepsilon f(R/R_m)$$

written in terms of the gradient operator ∇_R, where f is a functional form of the interaction potential energy, with a long-range attraction, short-range strong repulsion and a well in between, and ε for the energy of the well depth at position R_m [4]. The wavefunction is $\Psi_{\vec{k}}(\vec{R})$ for relative momentum $\vec{P} = \hbar\,\vec{k} = M\vec{v}$ and can be conveniently expanded in spherical harmonic functions of the relative position vector angles.

Differential and integral cross sections can be calculated from accurate potential energy functions obtained from electronic structure calculations or by parametrization. Elastic scattering differential cross sections in the CM frame with relative collision energy $E = Mv^2/2$ and into solid angles $\Omega = (\vartheta, \varphi)$ are given by $d\sigma/d\Omega = I(\vartheta, v)$ due to axial symmetry, as shown in Figure 1.4 for Na + Hg. They display broad oscillations for varying angles called rainbow fixtures and fast oscillations resulting from quantal wave superposition.

Cross sectional features are affected by the size and masses of colliding partners, as shown in Figure 1.5, from LAB measurements for pairs of noble gas atoms from He to Xe. In addition to rainbow and quantal oscillations, this figure shows large

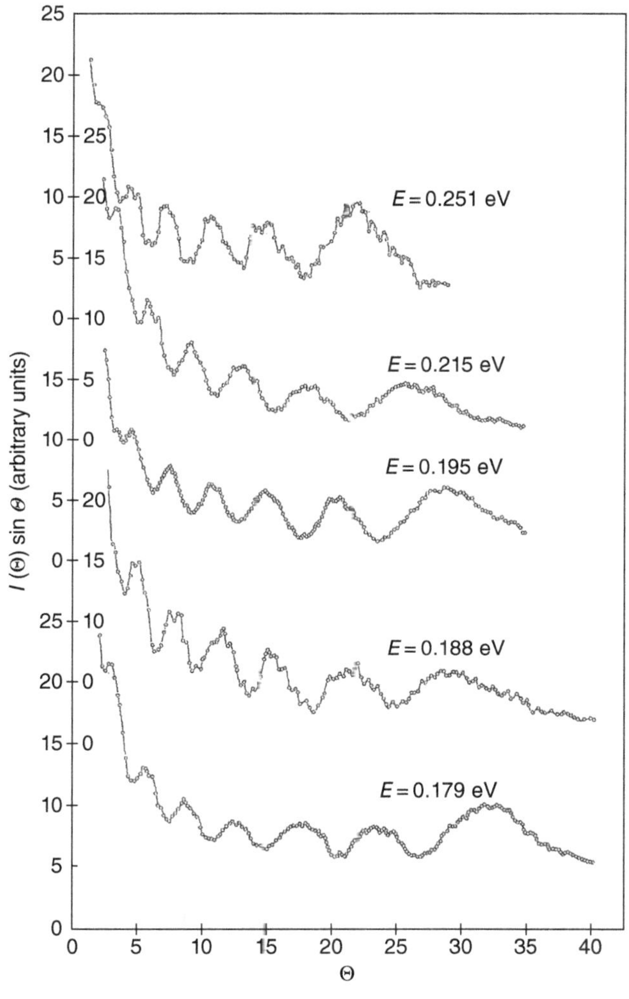

Figure 1.4 Angular distribution $I(\Theta)\sin(\Theta)$ cross section vs scattering angle Θ for Na + Hg for changing CM collisional energies E. The broad oscillations are rainbow fixtures, and the fast oscillations result from quantal wave superposition. Adapted from [67].

oscillations due to quantal identical-particle exchange symmetry that are more prominent in the lighter atom pairs and disappear for the heavier pairs.

(b) Nonreactive atom–diatomic pair (e.g. He + N_2)
Let $\vec{R}$ be the position vector of the atom relative to the CM of the diatomic, and $\vec{r}$ be the relative position vector of the atoms in the diatomic with a reduced mass m

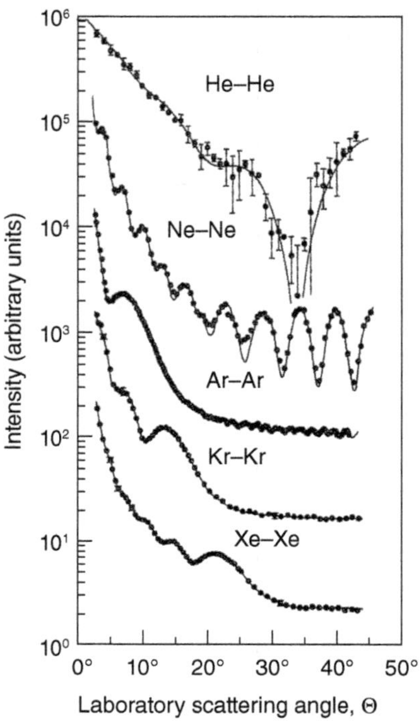

Figure 1.5 Laboratory elastic differential cross sections vs LAB angle Θ for identical pairs of noble gas atoms. In addition to rainbow and quantal oscillations, this figure shows large oscillations due to quantal identical-particle exchange symmetry that are more prominent in the lighter atom pairs and disappear for the heavier ones. [68] / AIP Publishing.

of internal motion and an internal potential energy $v(r)$. Then

$$\hat{K} = -\hbar^2\nabla_R^2/(2M)$$

$$\hat{H}_{int} = -\hbar^2\nabla_r^2/(2m) + v(r)$$

where $v(r)$ can be an expansion in the displacement $r - r_e$ from the equilibrium diatomic distance, or a Morse potential, and the coupling potential can be expanded in terms of Legendre polynomials $P_l(\cos\gamma)$ as [4]

$$V(R, r, \gamma) = \sum_{l=0}^{l_{max}} V_l(R, r)P_l(\cos\gamma)$$

where $\gamma = \cos^{-1}(\vec{R}\cdot\vec{r}/Rr)$. The wavefunction is of form $\Psi_\alpha(\vec{R},\vec{r})$ and can be expanded in spherical harmonic functions of the angles of $\vec{R}$ and $\vec{r}$.

Cross sections can be obtained vs scattering angles and collisional energies, for vibrational–rotational transitions $(v, j) \to (v', j')$ between diatomic states of quantum vibrational and rotational numbers (v, j) and energy levels $E_{v,j}$. In a CM reference frame, with relative collision energy $E = Mv^2/2$ and relative momentum change $\vec{P} \to \vec{P}'$ into solid angles $\Omega = (\vartheta, \varphi)$, cross sections are

$$d\sigma^{(CM)}_{v'j',vj}/d\Omega = I^{(CM)}_{v'j',vj}(\Omega, v).$$

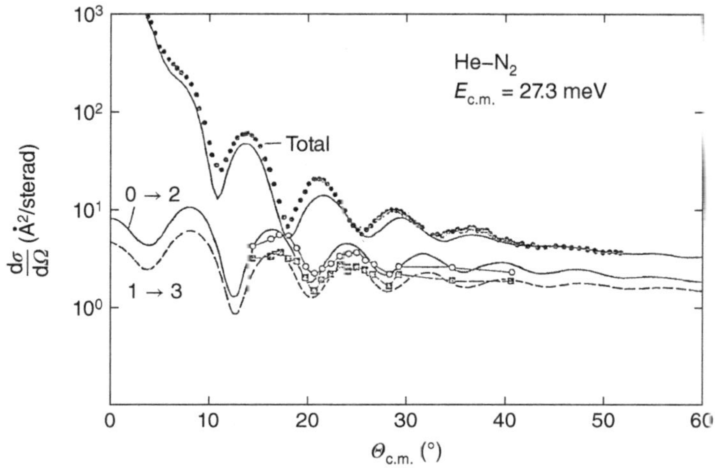

Figure 1.6 Measured rotationally inelastic differential cross section vs scattering angle for He−N_2 collisions in the CM frame, for the collision-induced rotational transitions $j = 0 \to 2$ and $j = 1 \to 3$ in N_2. Source: Adapted from [69].

Comparisons of measured and calculated cross sections vs scattering angles for selected collision energies provide information on the shapes of intermolecular potentials. Results for He + N_2 are shown in Figure 1.6.

(c) Reactive atom–diatomic pair (e.g. H + H_2)

We can use the same coordinates as in (b), but there are three equivalent sets for the three possible rearrangements A + BC, B + CA, and C + AB, defining the arrangements j = 1, 2, and 3. The total potential energy E is conveniently expressed as a function of the distances r_{AB}, r_{BC}, and r_{CA} of each of the three atom–atom pairs, or alternatively in terms of Jacobi vectors $(\vec{R}_{A,BC}, \vec{r}_{BC})$. Isocontours of the potential energy (such as for H + H_2) in a linear conformation show valleys for reactants and products and an activation barrier between them [4]. The total Hamiltonian can be written in three equivalent ways for the three arrangements, as

$$\hat{H} = \hat{K}^{(j)} + \hat{H}^{(j)}_{int} + \hat{V}^{(j)}.$$

The coupling $\hat{V}^{(j)}$ in arrangement j is a function of the Jacobi vectors $(\vec{R}_j, \vec{r}_j)$. The wavefunction is a superposition of coupled reaction-channel states that are functions of the channel variables, so that $\Psi_\alpha = \sum_j \Psi^{(j)}(\vec{R}_j, \vec{r}_j)$. Calculations can be done expanding the arrangement components in basis sets of diatomic vibrational–rotational functions [12, 13, 15], or as a sequence of coupled atom–diatomic scattering amplitudes using transition operators [70]

Cross sections can be calculated and measured for transitions between specific arrangement channels 1 and j, for given collisional energies E in CM, and as functions of the scattering solid angle for product species detected in arrangement j. Indicating with symbols α and ν the collection of state quantum numbers for arrangements 1 and j, differential reaction cross sections are of form $d\sigma_{j\nu,1\alpha}/d\Omega_j$ for products emerging in arrangement j. Quantal treatments are generally needed to identify reaction mechanisms and to clarify whether long-lived states (or scattering resonance states) are created during reactions. These aspects can be clarified generating partial state-to-state cross sections for each total angular momentum J of the triatomic system [71, 72].

Results for reactive $D + H_2\ (v, j) \rightarrow HD\ (v', j') + H$ at the CM collision energy $E = 0.78$ eV are shown in panels (a), (b), and (c) of Figure 1.7 part (A) vs CM scattering angles, for a sequence of transitions $(v = 0,\ j = 0) \rightarrow (v' = 0,\ j' = 0 - 11)$ and $(v = 0, j = 0) \rightarrow (v' = 1,\ j' = 0 - 6)$. Full lines are calculated values and dashed lines are fits to measured values. Theoretical values obtained from accurate potential energy surfaces are compared with experimental values and show good agreements in demanding tests of the PES accuracy and quantal dynamics treatment.

Figure 1.7 part (B) shows an alternative presentation of experimental measurements of the reaction $F + p\text{-}H_2 \rightarrow HF\ (v') + H$ at a LAB collision energy of 1.84 kcal/mol, with product HF ($v' = 1, 2, 3$) flux intensities vs molecular velocity in isocontour plots. For an incoming F atom beam from right to left, the figures show that product intensities peak primarily in the back direction and that secondly some $v' = 3$ intensity appears in the forward direction. Reproduction of these experimental plots with quantal dynamics calculations requires very accurate potential energy surfaces, as shown in [72].

(d) Atom–polyatomic and molecule–polyatomic pairs

Collisions involving polyatomic molecules can be described starting from the total potential energy function of all the atomic positions. For scattering of molecule A with atoms at positions $\vec{R}_a$ and molecule B with positions $\vec{R}_b$, the total potential energy is of form $E(\mathbf{R}) = \sum_{a,b} v_{ab} + \sum_{a,b,b'} v_{abb'} + \cdots$, a sum over pairs of atoms, triplets, and so on. Introducing the molecular CMs $\vec{R}_A$ and $\vec{R}_B$ and relative atomic positions such as $\vec{r}_a = \vec{R}_a - \vec{R}_A$, the interaction potential energy is $V(\vec{R}, \mathbf{R}^A, \mathbf{R}^B) = E(\vec{R}, \mathbf{R}^A, \mathbf{R}^B) - E_A(\mathbf{R}^A) - E_B(\mathbf{R}^B)$, with $\mathbf{R}^A = \{\vec{r}_a\}$ and $\mathbf{R}^B = \{\vec{r}_b\}$ [4].

A general treatment of the quantal dynamics of scattering can be done introducing dynamical atom–atom correlation functions for the whole system and by relating cross sections to the correlation functions [75]. Measured cross sections give the intensities of projectile energy distributions for initial collision energy E vs scattering angle Ω following target energy transfer ε. For impulsive collisions

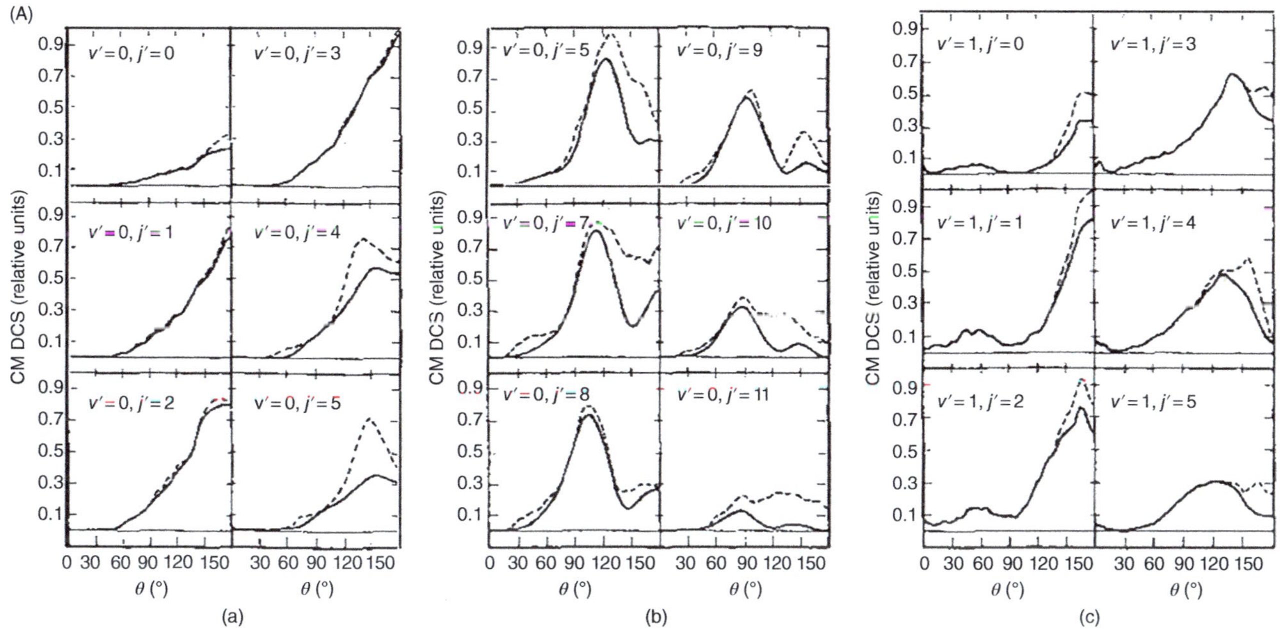

Figure 1.7 (A) Measured (dashed lines) and calculated (full lines) differential cross sections for reactive $D + H_2$ $(v, j) \rightarrow$ HD (v', j') + H at the CM collision energy $E = 0.78$ eV. (B) Detailed measured cross sections for F + p-H_2 $\rightarrow$ HF (v') + H at a LAB collision energy of 1.84 kcal/mol, shown as velocity flux isocontours for emerging HF molecules in vibrational states $v' = 1-3$. Source: (A) Adapted from [73]; (B) [74] / AIP Publishing.

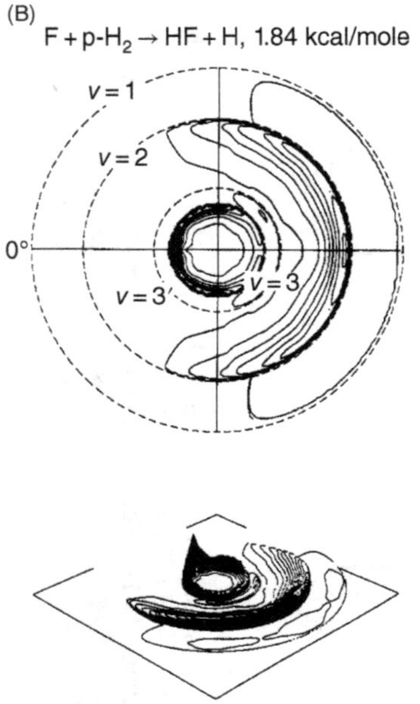

Figure 1.7 (*Continued*)

at hyperthermal energies of a projectile atom A with a target B containing atoms b, the double differential inelastic cross section is

$$\frac{d^2\sigma}{d\varepsilon\, d\Omega} = \sum_b \frac{d\sigma_b}{d\Omega} \frac{dP_b}{d\varepsilon}$$

where $d\sigma_b/d\Omega$ is an atomic cross-sectional component of the target and $dP_b/d\varepsilon$ is a rate of energy transfer obtained from each atom correlation function in the target [76].

An example is $Li^+ + CO_2$, which can be described in terms of the position $\vec{R}$ of the atom relative to the center of mass of the triatomic, angles Ω_M giving the orientation of the triatomic in the CM frame, and three internal vibrational coordinates $\boldsymbol{q}_V$ for its two bond distances and one interbond angle. For impulsive collisions at energy $E = 2.81$ eV and a vibrational temperature $T_V = 300$ K leading to fast vibrations and slow rotations of the target, energy loss measurements provide cross sections for vibrational normal mode transitions with changing quantum numbers $n_j = n_{j,fn} - n_{j,in}$ for symmetric stretch, asymmetric stretch, and bending modes $j = 1, 2, 3$, vs CM scattering angles, as shown in Figure 1.8.

(e) Atom–surface collisions

The mechanisms of atom–surface collisions are similar to those in atom–polyatomic collisions, except that surfaces can display atomic periodicities leading to collisional intensity variations related to quantal interference of scattered waves [4, 53].

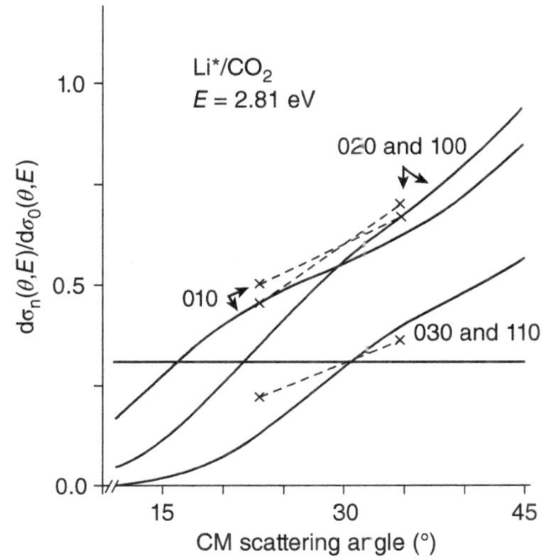

Figure 1.8 Calculated cross section for $Li^+ + CO_2$ at collisional energy $E = 2.81$ eV and a vibrational temperature $T_V = 300$ K with fast vibrations and slow rotations of the target, leading to vibrational normal mode transitions with changing quantum numbers $n_j, j = 1, 2, 3$. The crosses are experimental results. Source: [76] / with permission of Elsevier.

An example such as Rg + LiF(001) is described by the position $\vec{R}$ of the rare-gas atom over the (001) crystal surface of solid LiF and by the positions $\boldsymbol{Q}_S$ of all the atoms on the surface and below it, which interact with He, in a reference frame attached to the surface. The interaction potential energy function $V(X, Y, Z)$ for this system displays the translational symmetry of the (001) surface containing Li^+ and F^- ions. The Z-coordinate is perpendicular to the LiF(001) *fcc* surface, and the X-coordinate runs through F^- ions at the surface.

Cross sections for Rg + LiF(001), Rg = Ne, Ar, Kr, Xe, vs final scattering angle θ_F for fixed initial angle $\theta_S = 40°$ from the surface axis are shown in Figure 1.9 [77]. Here, the temperatures are $T_S = 300$ K for the surface and $T_G = 573$ K for the Rg gas. The scattering of lighter atoms display a double maximum corresponding to rainbow scattering angles.

1.4.3 Electronically Diabatic Heavy-Particle Collisions

(a) Electronically diabatic atom–atom collisions

Electronic rearrangement during collisions may involve excitation and charge transfer or ionization of product species when excited reactants are present. These phenomena may be simultaneously present. Examples are:

$H^+ + H(1s) \rightarrow H^+ + H(nlm)$ (e^- excitation)
$H^+ + H(1s) \rightarrow H(nlm) + H^+$ (e^- transfer)
He^* $(1s2s\ ^1S) + Ar(3p^6\ ^1S) \rightarrow He(1s_2\ ^1S) + Ar^+(3p^5\ ^3P) + e^-$ (ionization)

where (nlm) are the quantum numbers of a hydrogen atomic state.

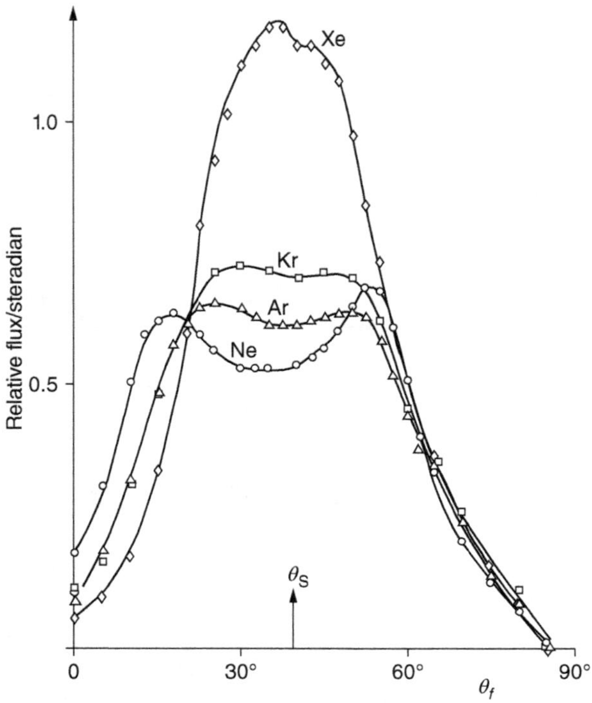

Figure 1.9 Cross sections for Rg + LiF(001), Rg = Ne, Ar, Kr, Xe, vs final scattering angle θ_F for fixed initial angle $\theta_S = 40°$ from the surface axis. Source: [77] / AIP Publishing.

The Hamiltonian contains now both nuclear ($\boldsymbol{Q}$) and electronic ($\boldsymbol{X}$) variables and must be written as a sum over energies of interacting nuclei and electrons or a sum over many-electron atoms A and B Hamiltonians and their interaction energy,

$$\hat{H} = \hat{K}_n + \hat{K}_e + \hat{V}_{nn} + \hat{V}_{ne} + \hat{V}_{ee} = \hat{H}_A + \hat{H}_B + \hat{H}_{AB}$$

In a CM reference frame with relative interatomic position $\vec{R}$, the interaction potential is $\hat{H}_{AB} = V(\vec{R}, \boldsymbol{X})$. An eigenfunction $\Psi_\alpha(\vec{R}, \boldsymbol{X})$ can be constructed as sums of molecular AB many-electron functions in a configuration interaction with its terms multiplying scattering functions of relative nuclear positions, or as products of atomic functions of electron variables for A and B, times relative motion functions, in a valence-bond description. Many-electron functions must be made invariant under electron exchange permutations between the collision partners [4]. Their expansion in a basis set of electronic states Γ introduce a matrix of diagonal and non-diagonal potential energy functions, and scattering leads to momentum-induced transitions $\Gamma \rightarrow \Gamma'$ among the electronic states.

The treatment simplifies when only one electron is active in the collision. Figure 1.10 shows a collection of theoretical and experimental total cross sections $\sigma^{(tr)}(E_L)$ for electron transfer in $H^+ + H(1s)$ (or $D(1s)$) vs LAB collision energy, with

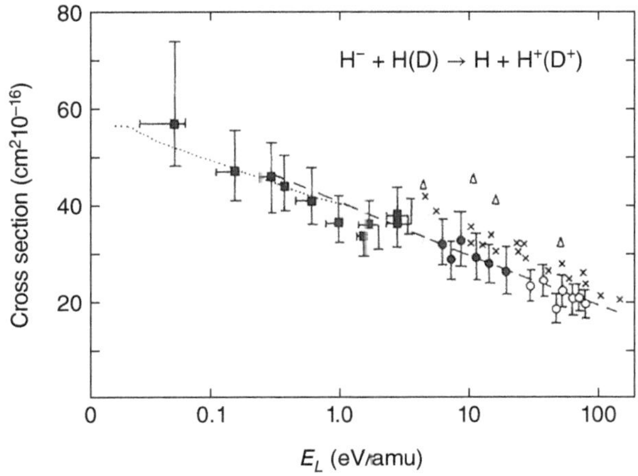

Figure 1.10 Measured and calculated cross sections for electron transfer $H^+ + H(1s)$ (or D(1s)) → H(1s) + H^+ (or D^+) vs LAB collision energy E_L, with theory results shown as full and dashed lines and experimental measurements as dots and crosses. Source: [44] / Oxford University Press.

theory results shown as full and dashed lines and experimental measurements as dots and crosses [44].

(b) Electronically diabatic atom–diatom collisions
An example is given by the following set of coupled arrangements,

$$H^+ + H_2\left(1\sigma_g^2\ {}^1\Sigma_g^+\right) \to H(1s) + {H_2}^+\left(1\sigma_g\ {}^2\Sigma_g^+\right)\ (e^-\ \text{transfer})$$
$$H^+ + H_2\left(1\sigma_g^2\ {}^1\Sigma_g^+\right) \to H_2\left(1\sigma_g^2\ {}^1\Sigma_g^+\right) + H^+\ (\text{H displacement}),$$
$$H^+ + H_2\left(1\sigma_g^2\ {}^1\Sigma_g^+\right) \to {H_2}^+\left(1\sigma_g\ {}^2\Sigma_g^+\right) + H(1s)\ (\text{H} - \text{transfer})$$

which illustrates the complexity resulting from coupling of several electronic arrangements. In a treatment where the three arrangement channels are allowed to couple, the total wavefunction for the three atoms is of form $\Psi_\alpha = \sum_j \Psi^{(j)}(\vec{R}_j, \vec{r}_j, \boldsymbol{X})$, and all terms must be made invariant under electron exchange between collision partners [4]. Here again transitions between electronic states are mediated by nuclear momentum couplings.

Figure 1.11 shows scattering beam intensities of ions H^+ in collisions at an initial CM energy $E = 20$ eV with a target of $H_2(v_i = 0)$, leading to (a) inelastic $H^+ + H_2(v_f)$ and (b) charge transfer $H + {H_2}^+(v_f)$, vs energy loss [78]. Peaks correspond to excitation of vibrational states of the target, with their locations slightly changed due to rotational excitation of the diatomic. The endoergicity of $\Delta E = 1.84$ eV in charge transfer is clearly seen in the late appearance of vibrational peaks for (b). The locations of peaks are in good agreement with the vibrational excitation energies of $H_2(v_f)$ and ${H_2}^+(v_f)$.

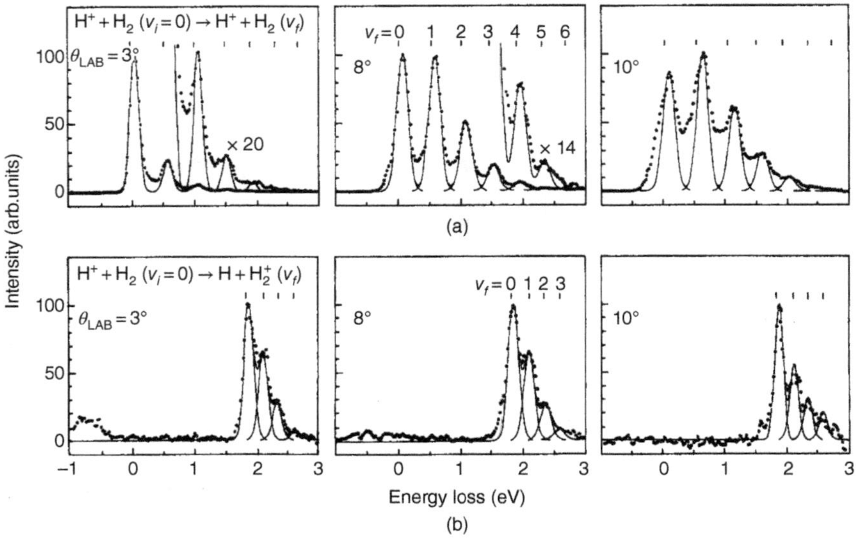

Figure 1.11 Intensities of scattered H^+ beams vs energy loss for (a) inelastic and (b) charge transfer measured, for $H^+ + H_2$. Source: [78] / AIP Publishing.

1.4.4 Electron and Photon Scattering by Molecules

(a) Electron–atom and electron–molecule systems

The total Hamiltonian here is a sum of terms for the electron kinetic energy, internal molecular motions, and the electron–molecule interaction, given by

$$\hat{H} = \hat{H}_{el} + \hat{H}_{mol} + \hat{H}_{el,mol}$$

with the system described by a wavefunction $\Psi_\alpha(\boldsymbol{x}_0, \boldsymbol{x}_1, \ldots, \boldsymbol{x}_{N_{el}}, \boldsymbol{Q})$ for a projectile electron of position and spin variables $\boldsymbol{x}_0$ and wavevector $\vec{k}_0$, scattered by a target with N_{el} electrons and nuclei (or atomic ions) at locations $\boldsymbol{Q}$ in an initial state Γ_0. This wavefunction can be expanded in a basis set to electronic states Γ of the target, with expansion coefficients describing the scattering of the electron.

Generally for a molecule with N_{el} electrons, taking the CM frame origin at the position of the molecular CM, let $\vec{r}_0$ be the position of the incoming electron and $\vec{r}_i, i = 1$ to N_{el}, be those of the molecular electrons. The Hamiltonian terms involve electrons (0 or e) and nuclei (n), or atomic ion cores, interacting through Coulomb potential as

$$\hat{K} = \hat{H}_{el} = -\hbar^2\nabla_0^2/(2m_e)$$

$$\hat{H}_{mol} = \hat{H}_{mol}^{(ee)} + \hat{H}_{mol}^{(en)} + \hat{H}_{mol}^{(nn)}, \quad \hat{H}_{el,mol} = \hat{H}_{el,mol}^{(0e)} + \hat{H}_{el,mol}^{(0n)}$$

The corresponding wavefunction for electronic position and spin variables $\boldsymbol{x} = (\vec{r}, \zeta)$ can be expanded in a basis set of molecular electronic states leading to a coupled-states (or close-coupling) treatment of the scattering, with coefficients dependent on the position of the projectile electron. Their asymptotic form provides scattering amplitudes and cross sections.

A simple example and thoroughly studied system is $\mathrm{e} + \mathrm{He}(1\mathrm{s}^2)$, leading to elastic and inelastic scattering [59],

$$\mathrm{e} + \mathrm{He}(1\mathrm{s}^2) \rightarrow \mathrm{e} + \mathrm{He}(1\mathrm{s}\, nl)$$

which can be described as the excitation of a two-electron system including polarization of the target. The corresponding wavefunction is of form $\Psi_\alpha(\boldsymbol{x}_0, \boldsymbol{x}_1, \boldsymbol{x}_2)$, for electronic position and spin variables $\boldsymbol{x} = (\vec{r}, \zeta)$ and $\vec{r}_0$ the position of the incoming electron, with the He nucleus at the origin of the reference frame. However, accurate calculations require a more detailed treatment as a three-electron system with full consideration of polarization and electron exchange effects.

In some cases, it is possible to describe elastic collisions with a one-electron model by representing the atom by means of a static potential energy derived from the atomic charge density $n_A(r_0)$, so that $\widehat{H}_{at} \approx E_{at}$ and

$$V(r_0) = -e\,\phi_A(r_0), \quad \nabla_0^2 \phi_A(r_0) = -4\pi n_A(r_0)$$

where ϕ_A is the electric potential function obtained from the Poisson equation, given the charge density. Alternatively, one can introduce an effective atomic charge function $eZ_A(r_0)$ and write $V = -Z(r_0)e^2 r_0^{-1}$, where $Z_A(r_0)$ reproduces the atomic selfconsistent field at radius r_0 and generally satisfies $Z_A(0) = Z_A$ (the nuclear charge number) and $Z_A(\infty) = Z_A - N$, the net asymptotic charge number for the atom A.

An example of electron–diatom systems is $\mathrm{e} + \mathrm{H}_2$, with the diatomic initially in its ground electronic and vibrational–rotational state $\Gamma_g = (1\sigma_g^2\ {}^1\Sigma_g^+, v = 0, l = 0, m_l = 0)$, which may lead to elastic scattering where the diatomic stays in its initial state or to inelastic scattering with the diatomic undergoing electronic and (or) vibrational–rotational excitation. Figure 1.12 shows cross sections for excitation of $\mathrm{H}_2\left(1\sigma_g^2\ {}^1\Sigma_g^+\right)$ into the excited state $1\sigma_g 2\sigma_g a\ {}^3\Sigma_g$, with a differential cross section vs scattering angle in (a) for collision energy of 20 eV and an integral cross section vs collision energy in (b). From [79], their Figure 25. The collision can be treated as a three-electron system undergoing state changes for nuclei at different orientations and bond-stretch values.

In the case where there is no electronic excitation, it is again possible to introduce an effective potential energy dependent only on the position of the scattered electron but with additional dependence on diatomic vibrational and orientation variables r and ω.

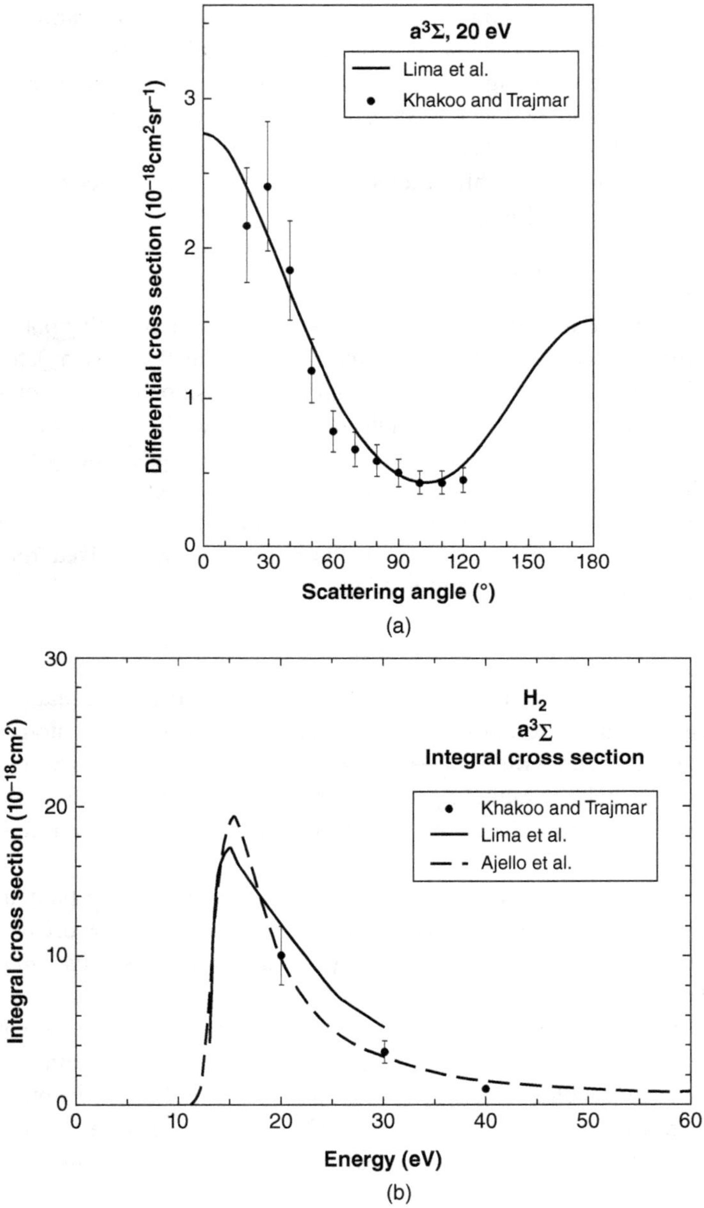

Figure 1.12 Excitation of $H_2\left(1\sigma_g^2\ {}^1\Sigma_g^+\right)$ into the excited state $a^3\Sigma$ induced by electron collisions. (a) Differential cross section for collision energy of 20 eV and (b) the integral cross section. Source: [79] / with permission of Elsevier.

Electronic or vibrational excitation may lead to fragmentation of the molecule, with the fragments emerging from the interaction region along measurable angles, and with given translational energies.

(b) Photon–atom and photon–molecule systems
Here the photon is described as a radiation particle (a quantum of electromagnetic field $\vec{\mathcal{E}}$), and its interaction with a molecule is treated with Hamiltonian terms for the radiation and for its coupling to the molecule [56]. The field $\vec{\mathcal{E}}$ is a superposition of waves with wavevector $\vec{q}$ and wavelength $\lambda = 2\pi/|\vec{q}|$.

In the dipole approximation for the coupling, valid for photon field wavelengths large compared to the molecular size, and using creation and annihilation operators for photons of given wavevector $\vec{q}$ and polarization σ, the total Hamiltonian is a sum of terms for radiation, molecular motions, and their interaction, given by

$$\hat{H} = \hat{H}_{rad} + \hat{H}_{mol} + \hat{H}_{rad,mol}$$

$$\hat{H}_{rad} = \sum_{\vec{q},\sigma} (\hbar qc)\hat{a}^{\dagger}_{\vec{q}\sigma}\,\hat{a}_{\vec{q}\sigma}$$

$$\hat{H}_{rad,mol} = -\vec{\mathcal{E}} \cdot \vec{D}$$

with an electric field operator

$$\vec{\mathcal{E}} = \sum_{\vec{q},\sigma} \mathcal{E}_{\vec{q}\sigma}\hat{a}^{\dagger}_{\vec{q}\sigma} + \mathcal{E}_{\vec{q}\sigma}{}^{*}\,\hat{a}_{\vec{q}\sigma}$$

and a molecular electric dipole operator $\vec{D} = \sum_I c_I \vec{r}_I$, where $\hat{a}_{\vec{q}\sigma}$ is a photon-annihilation operator and the dipole is given in terms of molecular constituent charges c_I (of electrons and nuclei) located at positions $\vec{r}_I$.

The wavefunction for photons initially incoming in a state with wavevector $\vec{q}_{in}$ and polarization σ_{in} and for the molecule in the ground state Γ_g is of the form $\Psi_\alpha(\boldsymbol{X}, \boldsymbol{Q}, \vec{q}, \sigma)$ with total energy $E_c = E_{\vec{q}_{in},\sigma_{in}} + E_{\Gamma_g} = E_{\vec{q},\sigma} + E_\Gamma$ and with the initial scattering states $\alpha = (\vec{q}_{in}, \sigma_{in}, \Gamma_g)$ specified as a collection of quantum numbers for the states of the photon and molecule. The final state $(\vec{q},\ \sigma,\ \Gamma)$ after scattering may be the result of an elastic or inelastic collision, where the internal molecular state and the wavelength of the photon change, or it may be a photoinduced dissociation state of the molecule.

An example is the photolysis of the BrCl($v'' = 0, 1, 2$) molecule in three parent vibrational states, induced by absorption of a photon with wavelengths in the range 330–570 nm, shown in Figure 1.13. Photofragmentation leads to Br($4p^5$ $P_{3/2}$) + Cl($3p^5$ $P_{3/2}$) and also to vibrationally excited Cl* in Br($4p^5$ $P_{3/2}$) + Cl*($3p^5$ $P_{1/2}$). Dissociation cross sections σ_d for the two products have been measured, and the ratio $\Gamma = \sigma_d(\mathrm{Br}+\mathrm{Cl}^*)/[\sigma_d(\mathrm{Br}+\mathrm{Cl}) + \sigma_d(\mathrm{Br}+\mathrm{Cl}^*)]$ vs photon energies is shown in Figure 1.13, as well as calculated ratios from wavepacket modeling of the photolysis, showing excellent agreement. This comparison is detailed in [80].

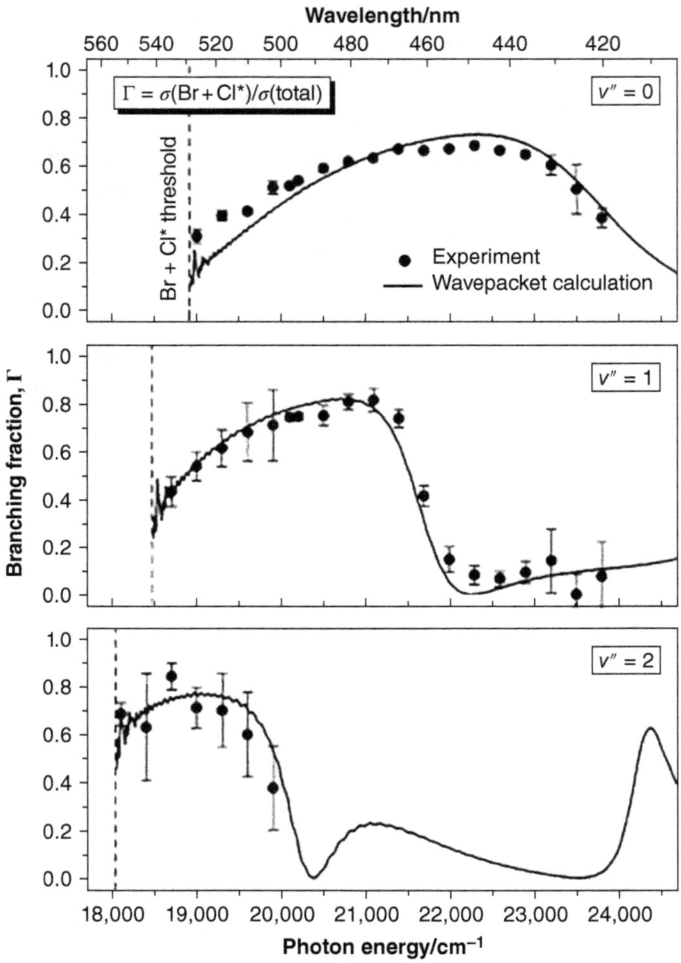

Figure 1.13 Fragmentation ratio $\Gamma = \sigma_d(\mathrm{Br + Cl^*})/[\sigma_d(\mathrm{Br + Cl}) + \sigma_d(\mathrm{Br + Cl^*})]$ vs photon energies of the BrCl($v'' = 0, 1, 2$) molecule induced by absorption of a photon. Experimental points are compared to calculations from a wavepacket treatment of the photolysis. Source: [80] / AIP Publishing.

1.5 Transport, Energy Relaxation, and Reaction Rates in Gases

Given state-to-state transition probabilities resulting from molecular interactions, it is possible to combine them with statistical mechanical distributions of the molecules in the gas phase at or near thermal equilibrium, to generate macroscopic properties [1, 81]. Among these, transport properties deal with diffusion (mass transport), viscosity (momentum transport), and thermal conductivity (kinetic energy transport). The relation between collisional cross sections and transport coefficients can be described in detail for dilute gases and temperatures not far from ambient values [2].

Similarly, statistical thermodynamics provides ways to calculate energy relaxation rates from state-to-state molecular transitions and chemical reaction rates from cross sections, for dilute gases [1, 81, 82].

1.6 Concepts and Methods in the Quantal Modelling

The above examples are a very limited survey of the many physical systems and phenomena relating to molecular collisional dynamics. They can be nevertheless studied with the tools provided by classical and quantum mechanics of systems of interacting particles and statistical treatments of particle fluxes in colliding beams and gases near thermal equilibrium.

Experimental measurements of cross sections and their theoretical interpretation or prediction show the need for detailed quantal treatments. Introduction of the concepts of quantum theory for scattering waves is essential as a guide to detailed quantitative treatments, which can provide numerical results for comparisons and tests. In the present cases of molecular dynamics an important initial simplification makes use of the separation of electronic and nuclear motions based on the very different masses of electrons and nuclei. This leads to a progression of models starting with the simple molecular interactions where the electronic system remains in its ground state in electronically adiabatic phenomena, followed by non-adiabatic processes. Another progression in modelling goes from simple two-atom systems to polyatomic interactions and interactions with solid surfaces. An additional challenge is presented by reactive phenomena, where bonds are broken and formed in chemical reactions.

The following chapters have been planned to proceed from simple concepts to advanced quantal treatments, which can be implemented numerically to obtain values of collisional properties. They proceed from simple systems to larger ones and from simple adiabatic collisions to complex non-adiabatic phenomena involving electronic rearrangements and chemical reactions.

References

1 J. O. Hirschfelder, C. F. Curtiss and R. Byron Bird, *Molecular Theory of Gases and Liquids*, New York: John Wiley & Sons, Inc., 1954.

2 G. C. Maitland, M. Rigby, E. Brian Smith and W. A. Wakeham, *Intermolecular Forces*, Oxford, England: Oxford University Press, 1981.

3 A. J. Stone, *The Theory of Intermolecular Forces*, Oxford, England: Clarendon Press, 1996.
4 D. A. Micha, *Molecular Interactions. Concepts and Methods*, Hoboken, NJ, USA: Wiley Publ. Co., 2019.
5 R. E. Weston and S. H. A., *Chemical Kinetics*, Englewood Cliffs, NJ, USA: Prentice-Hall, 1972.
6 E. E. Nikitin, *Theory of Elementary Atomic and Molecular Processes in Gases*, Oxford, England: Clarendon Press, 1974.
7 I. W. M. Smith, *Kinetics and Dynamics of Elementary Gas Reactions*, London, England: Butterworth Publ., 1980.
8 R. E. Johnson, *Introduction to Atomic and Molecular Collisions*, New York: Plenum Press, 1982.
9 R. D. Levine and R. B. Bernstein, *Molecular Reaction Dynamics and Chemical Reactivity*, Oxford, England: Oxford University Press, 1987.
10 J. N. Murrell and S. D. Bosanac, *Introduction to the Theory of Atomic and Molecular Collisions*, New York: Wiley Publ. Co., 1989.
11 N. F. Mott and H. S. W. Massey, *Theory of Atomic Collisions*, Oxford, England: Oxford University Press, 1965.
12 L. S. Rodberg and R. M. Thaler, *Introduction to the Quantum Theory of Scattering*, New York: Academic Press, 1967.
13 M. S. Child, *Molecular Collision Theory*, New York: Academic Press, 1974.
14 G. C. Schatz and M. Ratner, *Quantum Mechanics in Chemistry*, Englewood, NJ: Prentice-Hall, 1993.
15 J. Z.-H. Zhang, *Theory and Application of Quantum Molecular Dynamics*, Singapore: World Scientific, 1999.
16 G. D. Billing and K. V. Mikkelsen, *Advances Molecular Dynamics and Chemical Kinetics*, New York: Wiley-Interscience, 1997.
17 G. G. Balint-Kurti and A. P. Palov, *Theory of Molecular Collisions*, London: Royal Society of Chemistry, 2015.
18 A. Messiah, *Quantum Mechanics*, Volumes 1 and 2, Amsterdam: North-Holland, 1962.
19 M. L. Goldberger and K. M. Watson, *Collision Theory*, New York: Wiley, 1964.
20 R. G. Newton, *Scattering Theory of Waves and Particles*, New York: McGraw-Hill, 1966.
21 R. D. Levine, *Quantum Mechanics of Molecular Rate Processes*, Oxford, England: Oxford University Press, 1969.
22 J. R. Taylor, *Scattering Theory*, New York: Wiley, 1972.
23 C. J. Joachain, *Quantum Collision Theory*, Amsterdam, Holland: Elsevier, 1975.
24 S. K. Adhikari and K. L. Kowalski, *Dynamical Collision Theory and Its Applications*, New York: Academic Press, 1991.

25 D. R. Bates and I. Estermann (Editors), *Advances in Atomic and Molecular Physics*, Volume 1, New York: Academic Press, 1965.

26 J. Ross (Editor), *Advances in Chemical Physics: Molecular Beams*, Volume 10, New York: Interscience, 1966.

27 H. Hartmann (Editor), *Chemische Elementarprocesse*, Berlin: Springer-Verlag, 1968.

28 C. Schlier (Editor), *Molecular Beams and Reaction Kinetics*, New York: Academic Press, 1970.

29 E. W. McDaniels, V. Cermak, A. Dalgarno, E. E. Ferguson and L. Friedman, *Ion-Molecule Reactions*, New York: John Wiley & Sons, Inc, 1970.

30 M. A. D. Fluendy and K. P. Lawley, *Chemical Applications of Molecular Beam Scattering*, Boca Raton, FL: Chapman and Hall/CRC, 1973.

31 K. P. Lawley (Editor), *Advances in Chemical Physics: Molecular Scattering*, Volume 30, New York: Wiley, 1975.

32 J. P. Toennies, "The calculation and measurement of cross section for rotational and vibrational excitation," *Annu. Rev. Phys. Chem.*, vol. 27, p. 225, 1976.

33 R. B. Bernstein (Editor), *Atom-Molecule Collision Thweory: A Guide to the Experimentalist*, New York: Plenum Press, 1979.

34 R. B. Bernstein, *Chemical Dynamics Via Molecular Beams and Laser Techniques*, Oxford, England: Oxford University Press, 1982.

35 D. G. Truhlar (Editor), *Potential Energy Surfaces and Dynamics Calculations*, New York: Plenum Press, 1981.

36 W. H. Miller (Editor), *Dynamics of Molecular Collisions Parts A and B*, New York: Plenum Press, 1976

37 F. A. Gianturco, *The Transfer of Energy by Collisions*, Berlin: Springer-Verlag, 1979.

38 D. Henderson (Editor), *Theoretical Chemistry: Theory of Scattering Parts A and B*, Volume 6, New York: Academic Press, 1981.

39 J. M. Bowman (Editor), *Molecular Collision Dynamics*, Berlin: Springer-Verlag, 1983.

40 E. E. Nikitin and S. Y. Umanskii, *Theory of Slow Atomic Collisions*, Berlin: Springer-Verlag, 1984.

41 M. Baer (Editor), *Theory of Chemical Reaction Dynamics*, Volumes I, II, III, IV, Boca Raton, FL, USA: CRC Press, 1985.

42 D. C. Clary (Editor), *The Theory of Chemical Reaction Dynamics*, Dordrecht, The Netherlands: D. Reidel, 1986.

43 F. A. Gianturco, *Collision Theory for Atoms and Molecules*, New York: Plenum Press, 1989.

44 B. H. Bransden and M. R. C. McDowell, *Charge Exchange in the Theory of Ion-Atom Collisions*, Oxford, England: Clarendon Press, 1992.

45 J. M. Bowman (Editor), *Advances in Molecular Vibration and Collision Dynamics*, Volumes 2A and 2B, London, England: JAI Press, 1994.

46 B. Alder, S. Fernbach and M. Rotenberg, *Methods in Computational Physics vol. 10: Atomic and Molecular Scattering*, New York: Academic Press, 1971.

47 K. Smith, *The Calculation of Atomic Collision Processes*, New York: Wiley, 1971.

48 M. S. Child (Editor), *Semiclassical Methods in Molecular Scattering and Spectroscopy*, Dordrecht, The Netherlands: Reidel, 1980.

49 B. C. Eu, *Semiclassical Theories of Molecular Scattering*, Berlin: Springer-Verlag, 1984.

50 S. E. Koonin, *Computational Physics*, San Francisco, CA, USA: Benjamin/Cummings, 1986.

51 R. E. Wyatt, *Quantum Dynamics with Trajectories*, New York: Springer Science, 2005.

52 F. O. Goodman (Editor), *Dynamics Aspects of Surface Physics. E. Fermi School Course LVIII*, Bologna, Italy: Compositori, 1974.

53 F. O. Goodman and H. Y. Wachman, *Dynamics of Gas-Surface Scattering*, New York: Academic Press, 1976.

54 G. D. Billing, *Dynamics of Molecule-Surface Collisions*, New York: Wiley-Interscience, 2000.

55 T. Rescigno, V. McVoy and B. Schneider, *Electron-Molecule and Photon-Molecule Collisions*, New York: Plenum Press, 1978.

56 C. Cohen-Tannoudji, J. Dupont-Roc and G. Grynberg, *Atom-Photon Interactions*, New York: Wiley-Interscience, 1992.

57 R. Schinke, *Photodissociation Dynamics*, Cambridge, England: Cambridge University Press, 1993.

58 S. Geltman, *Topics in Atomic Collision Theory*, New York: Academic Press, 1969.

59 B. H. Bransden, *Atomic Collision Theory*, Reading, MA, USA: Benjamin-Cummings, 1983.

60 L. M. Raff and D. L. Thompson, "The classical trajectory approach to reactive scattering," in *Theory of Chemical Reaction Dynamics*, Volume III, (Editor: M. Baer), Boca Raton, FL, USA: CRC Press, 1985, pp. 1–121.

61 D. A. Micha, "Effective Hamiltonian methods for molecular collisions," in *Advances in Quantum Chemistry*, Volume 8, New York: Academic Press, 1974, pp. 231–288.

62 S. Mukamel, *Nonlinear Optical Spectroscopy*, Oxford, England: Oxford University Press, 1995.

63 V. May and O. Kuhn, *Charge and Energy Transfer Dynamics in Molecular Systems* (2nd edition), Weinheim, Germany: Wiley-VCH Verlag, 2004.

64 A. Nitzan, *Chemical Dynamics in Condensed Phases*, Oxford, England: Oxford University Press, 2006.

65 P. G. Burke and C. J. Joachain (Editors), *Photon and Electron Collisions with Atoms and Molecules*, New York: Plenum Press, 1997.

66 D. A. Micha, "Generalized response theory for a photoexcited many-atom system," in *Advances in Quantum Chemistry*, 71, (Editors: J. R. Sabin and R. Cabrera-Trujillo), Amsterdam, Holland: Elsevier, 2015, pp. 195–220.

67 U. Buck and H. Pauli, "Determination of intermolecular potentials by the inversion of molecular beam scattering data II. High resolution measurements for Na-Hg," *J. Chem. Phys.*, vol. 51, p. 1929, 1971.

68 J. M. Farrar, T. P. Schafer and Y. T. Lee, "Intermolecular potentials of symmetric rare gas pairs from elastic differential cross section measurements," *AIP Conf. Proc.*, vol. 11, pp. 279–295, 1973.

69 M. Faubel, K.-H. Kohl, J. P. Toennies, K. T. Tang and Y. Y. Yung, "The He–N_2 anisotropic Van der Waals potential. Test of a simple model using state-to-state differential scattering cross-sections," *Faraday Discuss. Chem. Soc.*, vol. 73, pp. 205–220, 1982.

70 D. A. Micha, "General theory of reactive scattering: a many-atom approach," in *Theory of Chemical Reaction Dynamics*, Volume II, *Chapter 3*, (Editor: M. Baer), Boca Raton, FL, USA: CRC Press, 1985, pp. 182–218.

71 W. H. Miller, "Recent advances in quantum mechanical reactive scattering theory," *Annu. Rev. Phys. Chem.*, vol. 41, p. 245, 1990.

72 F. J. Aoiz, L. Banares, B. Martinez-Haya, J. F. J. F. Castillo, D. E. Manolopoulos, K. Stark and H.-J. Werner, "Ab initio simulation of molecular beam experiments for the F + H_2 to HF + H reaction," *J. Phys. Chem. A*, vol. 101, pp. 6403–6414, 1997.

73 J. Z. H. Zhang and W. H. Miller, "Quantum reactive scattering via the S-matrix version of the Kohn variational principle: differential and integral cross sections for D + H_2 --> HD + H," *J. Chem. Phys.*, vol. 91, p. 1528, 1989.

74 D. M. Neumark, A. M. Wodtke, G. N. Robinson, C. C. Hayden and Y. T. Lee, "Molecular beam studies of the F + H_2 reaction," *J. Chem. Phys.*, vol. 82, p. 3045, 1985.

75 D. A. Micha and E. F. Vilallonga, "The collisional time-correlation function approach to molecular energy transfer," in *Advances in Chemical Physics*, Volume 84, (Editors: I. Progogine and S. A. Rice), New York: John Wiley & Sons, 1993, pp. 1–72.

76 E. F. Vilallonga and D. A. Micha, "The calculation of time-correlation functions for molecular collisions," *Phys. Rep.*, vol. 212, pp. 329–390, 1992.

77 D. R. O'Keefe, R. L. Palmer and J. N. Smith, "Scattering of noble gas atoms by LiF(001)," *J. Chem. Phys.*, vol. 55, p. 4572, 1971.

78 G. Niedner, M. Noll, J. P. Toennies and C. Schlier, "Observation of vibrationally resolved charge transfer in H^+ + H_2 at E_{CM}=20 eV," *J. Chem. Phys.*, vol. 87, p. 2685, 1987.

79 M. J. Brunger and S. J. Bruckman, "Electron-molecule scattering cross sections I: experimental techniques and data for diatomic molecules," *Phys. Rep.*, vol. 357, p. 215, 2002.

80 M. Beckert, E. R. Wouters, M. N. R. Ashfold and E. Wrede, "High resolution ion imaging study of BrCl photolysis in the wavelength range 330–570 nm," *J. Chem. Phys.*, vol. 119, p. 9576, 2003.

81 D. A. McQuarrie, *Statistical Mechanics*, New York: Harper and Row Publ., 1973.

82 P. W. Atkins and R. S. Friedman, *Molecular Quantum Mechanics* (3rd edition), New York: Oxford University Press, 1997.

2

Elastic Collisions

CONTENTS

Quantum Molecular Dynamics: Energy Transfer and Reactivity, First Edition. David A. Micha.

2.1 Elastic Collision Cross Sections

2.1.1 Classical Mechanics Treatment

We consider the relative motion of species A with respect to B, with reduced mass $M = M_A M_B/(M_A + M_B)$, for an interaction potential energy $V(R)$, a function of the interparticle distance R. More generally, as described in the first chapter, elastic collisions occur between two species when there is no change in the internal states of the two species resulting from their interaction. The potential energy for such situations involves not only their CM relative position but also depends on all internal coordinates. This more general case is developed in the chapter on inelastic collisions and the following ones.

We describe the collision in terms of classical trajectories, as that of scattered particles in a CM–SF reference frame with space-fixed axes orientations, involving trajectory relative position $\vec{R}(t)$ and momentum $\vec{P}(t)$. Relative positions and momenta evolve in time as given by their equations of motion. Using Hamilton's equations, they satisfy

$$\frac{d\vec{R}}{dt} = \frac{\partial H}{\partial \vec{P}}, \quad \frac{d\vec{P}}{dt} = -\frac{\partial H}{\partial \vec{R}}$$

with a Hamiltonian function

$$H(\vec{P}, \vec{R}) = \vec{P} \cdot \vec{P}/(2M) + V(R)$$

which gives $dH/dt = 0$. For initial conditions $(\vec{P}_{in}, \vec{R}_{in})$ and a constant total energy $H(\vec{P}_{in}, \vec{R}_{in}) = E$, the initial kinetic energy is $P_{in}^2/(2M) = E$ for a collision starting where $V(R) = 0$ outside the potential energy region. We take the relative motion velocity $\vec{v} = \vec{P}/M$ to be initially parallel to the Z-axis and the relative angular momentum $\vec{L} = \vec{R} \times \vec{P}$ parallel to the X-axis, choosing the trajectory to be on the plane (Y, Z), with an initial impact parameter $Y_{in} = b$, and azimuthal angle φ around the Z-axis, which remains constant for a radial potential energy. After the collision and in a region of vanishing potential energy, at times $t \geq t_{out}$, the final momentum is $\vec{P}_{out}$ pointing in the scattering direction Θ relative to the Z-axis.

An increment $d\sigma$ of a cross sections in a SF frame is given by

$$d\dot{N} = J_0 d\sigma$$

with $\dot{N}$ the scattering rate of colliding pairs per unit time and $J_0 = \upsilon\, n_A n_B\, \tau$ the incoming flux or number of colliding pairs in the volume τ for molecular densities n_A and n_B, per unit time and unit area moving across a plane perpendicular to the relative motion for an initial relative velocity v and scattering of A relative to B into solid angle Ω. This rate can be transformed to a CM–SF frame in terms of CM and relative velocities, to obtain a differential cross section $I(\Omega, \upsilon)$ given by $d\sigma = I(\Omega, \upsilon)d\Omega$. The scattering rate can be obtained in calculations starting with a bundle of trajectories of given initial scattering flux, following the trajectories to collect a statistical distribution of final states with the needed angular distribution.

Starting with Cartesian coordinates (X, Y, Z), we transform to spherical coordinates (R, Θ, Φ). In an isolated collision with total relative energy E starting at large negative Z and initial impact parameter $Y_{in} = b$, it is convenient to distinguish among the orbital angle $\vartheta(t; b, E)$, which is a function of time and dependent parametrically on the impact parameter and energy, the deflection function $\chi(b, E) = \lim \vartheta(t; b, E)$, $t \to \infty$, that is the orbit angle at times after the collision, and the scattering angle $\Theta = |\chi(mod\ 2\pi)|$ with $0 \le \Theta \le \pi$, that gives the direction of the scattered particle detector.

The collision dynamics of two particles can be treated as mentioned, following their relative position and momentum over time for given initial conditions. However, when the total energy E is constant during an isolated collision, in a stationary process, it is also possible to directly obtain the orbital angle ϑ as a function of the relative position R and, from this, the scattering angle Θ vs impact parameters b giving a differential cross section.

From conservation of the relative collision energy $E = Mv^2/2$ and the orbital angular momentum $L = Mvb$, we can write [1, 2]

$$E = P_R^2/2M + V_L(R)$$

$$P_R = M\frac{dR}{dt} = M\frac{dR}{d\vartheta}\frac{d\vartheta}{dt} = -M\frac{dR}{d\vartheta}\frac{L}{MR^2}$$

$$V_L(R) = V(R) + L^2/(2MR^2)$$

where the effective radial potential $V_L(R)$ is shown in Figure 2.1 for a typical atom–atom interaction energy with a minimum energy ϵ at a distance R_m or at $x_m = R/R_m = 1$. As the centrifugal potential energy increases, the effective potential displays a maximum at positive energies and distances larger than R_m. This disappears at even larger centrifugal energies.

The potential energy function $V(R)$ for two interacting atoms is in general strongly repulsive at short distances, where closed electronic shells overlap, and weakly attractive at very long distances due to dispersion (or Van der Waals) forces, with a minimum energy $-\varepsilon$ at the location R_m. The potential energy function can be analytically parametrized or piecewise fitted to reproduce experimental results [4]. Generally, the relative motion trajectory in the potential $V_L(R)$ displays turning points as the relative velocity goes to zero for $V_L(R) = E$ at $R = R_{trn}(E, b)$, $trn = 1, 2, \ldots$.

Solving for $d\vartheta/dR$, one finds the differential equation

$$\frac{d\vartheta}{dR} = -\frac{L}{R^2\{2M[E - V_L(R)]\}^{1/2}}$$

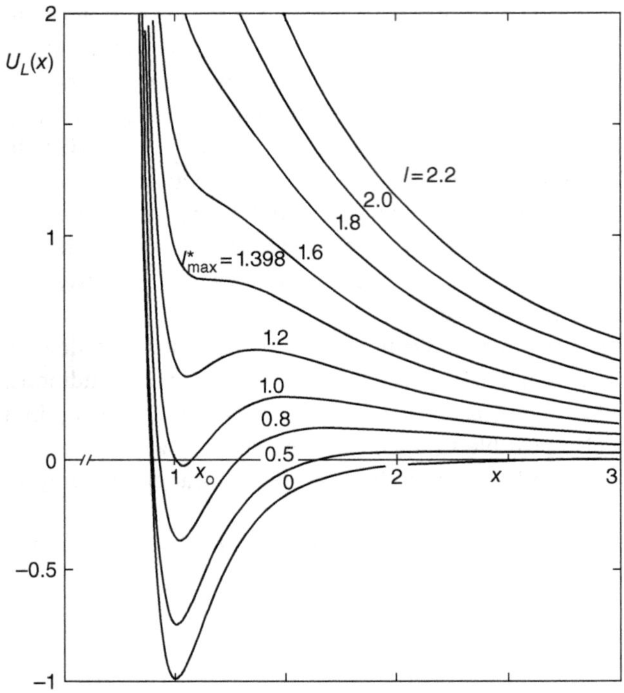

Figure 2.1 Effective radial potential $U_L(R) = V_L(R)/\varepsilon$ vs reduced radial distance $x = R/R_m$, and for values of the reduced angular momentum parameter $l^* = Mvb/(2M\varepsilon R_m^2)^{1/2}$, for a typical atom–atom potential $V(R)$ repulsive at short distances and attractive at large distances. Calculated for a Lennard-Jones (L-J 12–6) potential function. Source: [3] / Springer Nature.

and integrating from the initial distance $R \to \infty$ far from the origin along the incoming trajectory to the outer distance of closest approach, $R_b(E)$, at which $dR/dt = 0$, one finds for the orbital angle at closest approach

$$\vartheta_b(E) = \pi - b\int_{R_b}^{\infty} dR\left\{R^2\left[1 - \frac{V(R)}{E} - \frac{b^2}{R^2}\right]^{\frac{1}{2}}\right\}^{-1}$$

and, considering that the trajectory is symmetric around this angle direction, adding angles shows that the deflection angle is $\chi(b, E) = 2\vartheta_b(E) - \pi$, from which it follows the value of the scattering angle as $\Theta = |\chi(mod\ 2\pi)|$. These relations are shown in Figure 2.2 adapted from [3], for a bundle of trajectories with varying reduced impact parameter $\beta = b/R_m$, also giving the deflection angle $\chi(b, E)$ vs β, which displays a minimum at the so-called rainbow angle χ_r.

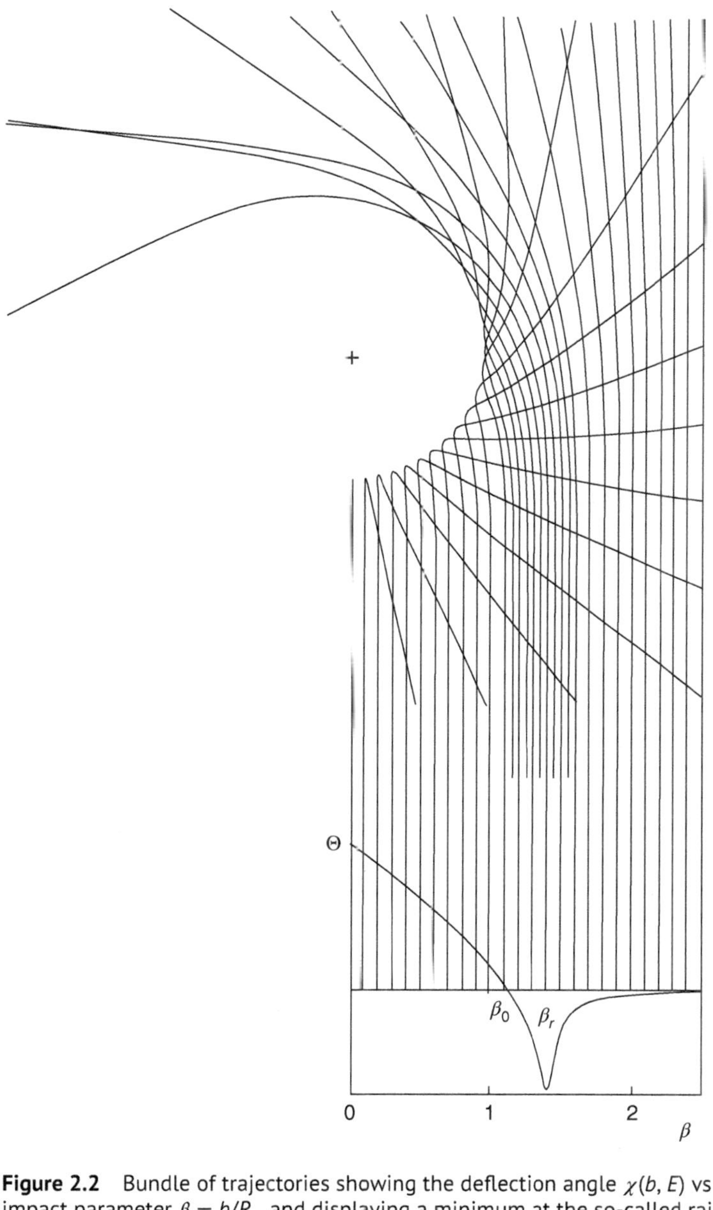

Figure 2.2 Bundle of trajectories showing the deflection angle $\chi(b, E)$ vs the reduced impact parameter $\beta = b/R_m$ and displaying a minimum at the so-called rainbow angle χ_r. Source: [3] / Springer Nature.

Cross sections follow from the statistics of a bundle of trajectories with an initial distribution of impact parameters and conservation of their flux (number of trajectories per unit time and unit cross area), insofar rates of incoming N_{inc} and scattered N_{sct} trajectories satisfy $dN_{inc}/dt = dN_{sct}/dt$. Equating increments of initial and final scattering rates for an initial flux $J^{(0)}$ and range db of impact parameters gives $J^{(0)} 2\pi b\, db = J^{(0)} d\sigma$ and therefore the differential of scattering cross section $d\sigma = 2\pi b\, db$.

The inverse function $b(\Theta, E)$ is found in general to be a multivalued function of $\Theta = |\chi(mod\ 2\pi)|$. Labelling its branches with an index r, the differential cross section into an element of solid angle $d\Omega = 2\pi \sin(\Theta) d\Theta$ is then

$$\frac{d\sigma}{d\Omega} = \sum_r \left| \frac{b\, db}{\sin(\Theta) d\Theta} \right|_r$$

and the integral cross section $\sigma(E)$ follows by integration over the solid angles. The differential cross section (Figure 2.3) shows, for this classical mechanics treatment, a divergence at the so-called rainbow angle (by analogy with the atmospheric rainbow), which is replaced by a maximum in a quantal treatment, as shown in what follows.

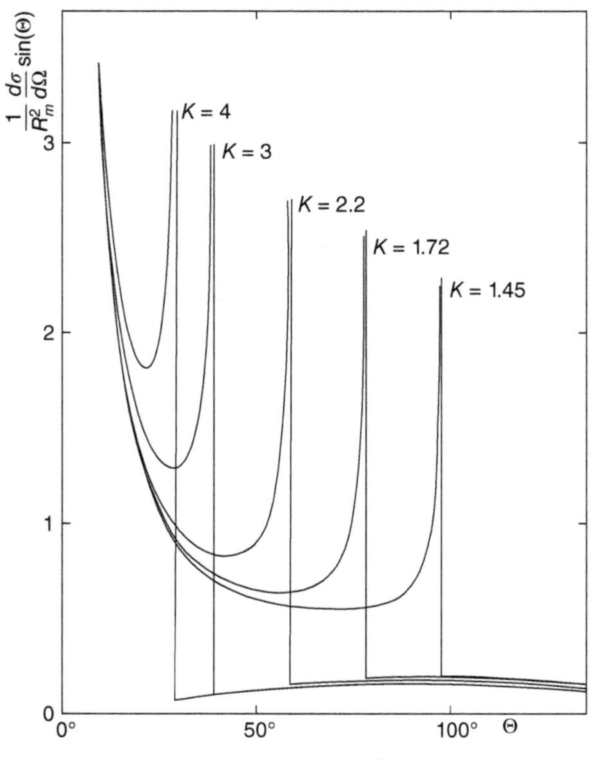

Figure 2.3 Reduced differential cross section times $\sin(\Theta)/R_m^2$ vs scattering angle, showing its divergence at the rainbow angle, for the reduced energy parameter $K = E/\varepsilon$. Source: [3] / Springer Nature for the (L-J 12–6) potential function.

2.1.2 Quantal Treatment

In a wavemechanical description, the bundle of classical trajectories is instead treated in terms of a wavepacket function $\Psi(\vec{R}, t)$ of position and time, with an initial spatial distribution $\Psi^{(in)}(\vec{R}) = \Psi(\vec{R}, t_{in})$ encompassing all initial positions of the bundle and containing the initial distribution of relative momenta $\vec{P}$, derived from the Fourier transform $\widetilde{\Psi}^{(in)}(\vec{P})$ of the wavepacket function [1]. For a localized wavepacket much larger than the scattering target and moving toward it, waves are scattered over time and finally emerge at $t \geq t_{fn}$ as $\Psi(\vec{R}, t) \approx \Psi^{(0)}(\vec{R}, t) + \Psi^{(sc)}(\vec{R}, t)$, where $\Psi^{(0)}$ is a component unaffected by scattering and $\Psi^{(sc)}$ is a scattered component moving outwardly.

The wavefunction is obtained from solutions of the time-dependent Schrodinger equation [1]

$$\widehat{H}\,\Psi(\vec{R}, t) = i\hbar\partial\Psi(\vec{R}, t)/\partial t$$

with the initial value $\Psi^{(in)}(\vec{R})$ and normalized to a total probability $\int d^3R\, |\Psi(\vec{R}, t)|^2 = 1$. Insofar collision events are isolated, the Hamiltonian operator $\widehat{H}$ is given by a sum of kinetic and potential energies as $\widehat{H} = \widehat{K} + V(R)$ and the potential energy is independent of time. The wavepacket can be expanded in a basis set of stationary scattering states $\psi^{(+)}_{\vec{k}}(\vec{R})$, with $\vec{P} = \hbar\vec{k}$ and energy $E(k) = (\hbar k)^2/(2M)$, and outgoing scattering boundary conditions as

$$\Psi(\vec{R}, t) = \int d^3k\, A(\vec{k})\psi^{(+)}_{\vec{k}}(\vec{R}) \exp[-iE(k)t/\hbar]$$

with the amplitude $A(\vec{k})$ chosen from initial conditions. The following treatment focuses on the calculation of stationary scattering states.

The relative momentum $\vec{P}$ between the collision particles is related to the wavelength λ by $P = Mv = h/\lambda = \hbar k$, where h is the Planck constant and $\hbar = h/(2\pi)$. The time-independent Schrodinger equation for the wavefunction ψ describing the collision of two structureless particles is

$$\widehat{H}\psi^{(+)}_{\vec{k}}(\vec{R}) = E\psi^{(+)}_{\vec{k}}(\vec{R})$$

where

$$\widehat{H} = -\frac{\hbar^2\nabla^2_R}{2M} + V(R)$$

and $\psi^{(+)}_{\vec{k}}$ is a state with incoming relative momentum $\vec{P} = \hbar\vec{k} = M\vec{v}$ and outgoing (scattered) spherical waves as asymptotic conditions. For $V = 0$, this state is a free-motion wave

$$\psi^{(0)}_{\vec{k}}(\vec{R}) = (2\pi)^{-\frac{3}{2}} \exp(i\vec{k} \cdot \vec{R})$$

normalized here with a Dirac delta function so that

$$\left\langle \psi_{\vec{k}}^{(0)} \middle| \psi_{\vec{k}'}^{(0)} \right\rangle = \delta(\vec{k} - \vec{k}')$$

and with density $\rho^{(0)} = \left|\psi_{\vec{k}}^{(0)}\right|^2 = (2\pi)^{-3}$. For scattering by the potential $V(R)$, one can write the state's asymptotic form for $R \to \infty$ as $\psi_{\vec{k}}^{(+)} \approx \psi_{\vec{k}}^{(0)} + \psi_{\vec{k}}^{(sc)}$ with

$$\psi_{\vec{k}}^{(sc)}(\vec{R}) \approx (2\pi)^{-\frac{3}{2}} \frac{\exp(i\,kR)}{R} f_{\vec{k}}^{(+)}(\Omega)$$

where $f_{\vec{k}}^{(+)}(\Omega)$ is the scattering amplitude into solid angle Ω, with dimension of length.

For a single pair of colliding particles, the scattering rate into the solid angle $d\Omega = dS/R^2$ through an element of surface dS in the direction $\vec{n} = \vec{R}/R$, is

$$d\dot{N}_{sc} = |J^{(0)}|d\sigma = \vec{J}^{(sc)} \cdot \vec{n}\, dS = \vec{J}^{(sc)} \cdot \vec{n} R^2 d\Omega$$

where $|J^{(0)}| = v/(2\pi)^3$ is the incoming flux, $d\sigma$ is an increment of elastic scattering cross section, and $\vec{J}^{(sc)}$ is an scattering flux vector. This can be obtained from the expectation value of the flux operator $\hat{\vec{J}}_{\vec{R}} = [\hat{\vec{P}}_{\vec{R}}\, \delta(\vec{R} - \vec{R}') + \delta(\vec{R} - \vec{R}')\hat{\vec{P}}_{\vec{R}}]/(2M)$ as $\vec{J}^{(sc)}(\vec{R}) = \left\langle \psi_{\vec{k}}^{(sc)} \middle| \hat{\vec{J}}_{\vec{R}} \middle| \psi_{\vec{k}}^{(sc)} \right\rangle$ with the bracket indicating an integral over $\vec{R}'$ variables. The flux-average expression, using that the momentum operator is $\hat{\vec{P}}_{\vec{R}} = (\hbar/i)\nabla_{\vec{R}}$, becomes

$$\vec{J}^{(sc)}(\vec{R}) = \frac{\hbar}{2Mi} \left\{ \psi_{\vec{k}}^{(sc)}(\vec{R})^* \nabla_{\vec{R}} \psi_{\vec{k}}^{(sc)}(\vec{R}) - \left[\nabla_{\vec{R}} \psi_{\vec{k}}^{(sc)}(\vec{R})^*\right] \psi_{\vec{k}}^{(sc)}(\vec{R}) \right\}$$

and, writing the gradient operator in spherical coordinates, as

$$\nabla_{\vec{R}} = \frac{1}{R} \left[\vec{n}_R \frac{\partial}{\partial R} R + \vec{n}_\Theta \frac{\partial}{\partial \Theta} + \vec{n}_\Phi \frac{1}{\sin(\Theta)} \frac{\partial}{\partial \Phi} \right]$$

the flux expectation value from the scattered wavefunction is found to be

$$\vec{J}^{(sc)}(\vec{R}) \cdot \vec{n} = \frac{v}{(2\pi)^3} \frac{1}{R^2} \left| f_{\vec{k}}^{(sc)}(\Omega) \right|^2 + \mathcal{O}\left(\frac{1}{R^3}\right)$$

with the last term of order R^{-3} and negligible at very large distances. Replacing this in the expression for the scattering rate above gives for the differential cross section

$$\frac{d\sigma}{d\Omega} = (\vec{J}^{(sc)} \cdot \vec{n}/|J^{(0)}|)R^2 = \left| f_{\vec{k}}^{(+)}(\Omega) \right|^2 = I(\Omega, E)$$

Therefore, this can be obtained from the asymptotic amplitude of the scattered wave.

The previous treatment of stationary states can be used to reconstruct the time-dependent wavepacket $\Psi(\vec{R}, t)$, and it shows that at large distances, the total state is a superposition of a free wave plus outgoing scattered waves containing the exponential factor $\exp\{i\ [kR - E(k)t/\hbar]\}$. An alternative description, equally general, gives the total state as a sum of a free wave plus incoming scattering waves with asymptotic factors $\exp\{i\ [-kR - E(k)t/\hbar]\}$. The corresponding stationary states are $\psi_{\vec{k}}^{(-)} \approx \psi_{\vec{k}}^{(0)} + \psi_{\vec{k}}^{(sc-)}$ with $\psi_{\vec{k}}^{(sc-)}(\vec{R}) \approx (2\pi)^{-\frac{3}{2}}[\exp(-i\,kR)/R]f_{\vec{k}}^{(-)}(\Omega)$ and differential cross sections are given by $\left|f_{\vec{k}}^{(-)}(\Omega)\right|^2$. The relation between the two stationary forms is $\psi_{\vec{k}}^{(+)}(\vec{R})^* = \psi_{-\vec{k}}^{(-)}(\vec{R})$.

Accurate calculations of differential cross sections vs scattering angle for a range of collision energies produce very detailed results with functional structures that can be compared with high-resolution experimental measurements. As an example, Figure 2.4 shows experimentally measured differential cross sections $I(\Theta, \mathrm{E})\sin(\Theta)$ for Na–Hg vs scattering angle Θ and for chosen collision energies E [5]. They display broad oscillations for each energy, at a principal rainbow angle and supernumerary smaller ones, and also display rapid oscillations due to quantal interference of partial waves with different angular momenta. All these features can be reproduced with numerical results for cross sections, provided that accurate potential energy functions have been first calculated or have been fit to the experimental results [5, 6].

2.1.3 Scattering Resonances and Quantal Exchange Symmetry

Quantal treatments are essential for a complete description of molecular scattering. In addition to the fundamental treatment of quantal state coherence, two other quantal phenomena are tunneling through an effective potential barrier and the exchange symmetry effect due to coherence between states of identical interacting species.

The wavefunction of a particle of mass M with a given energy E below the maximum energy barrier in the effective potential energy V_L has components at both shorter and longer distances than a maximum location at R_0, related by tunneling. As the energy E and relative angular momenta change, the wavefunction amplitude in the inner region can go through a large increase at an energy E_r, indicating that the particle has acquired a new localized state around the minimum of the potential. This is called a quantal resonance state with a resonance energy E_r and angular momentum quantum number l_r. The wavefunction also has a component at large distance with the same energy, related to the resonance state by quantal tunneling through the barrier. This is a well-known quantal effect [1], which can be detected in the shape of differential and integral cross sections [7]. Detailed experimental measurements showing the effects of resonances in integral cross sections have been available for atom–atom and atom–molecule systems [8]. It will be treated in more detail in the following section on radial wavefunctions, and more generally in the chapters on inelastic and reactive collisions.

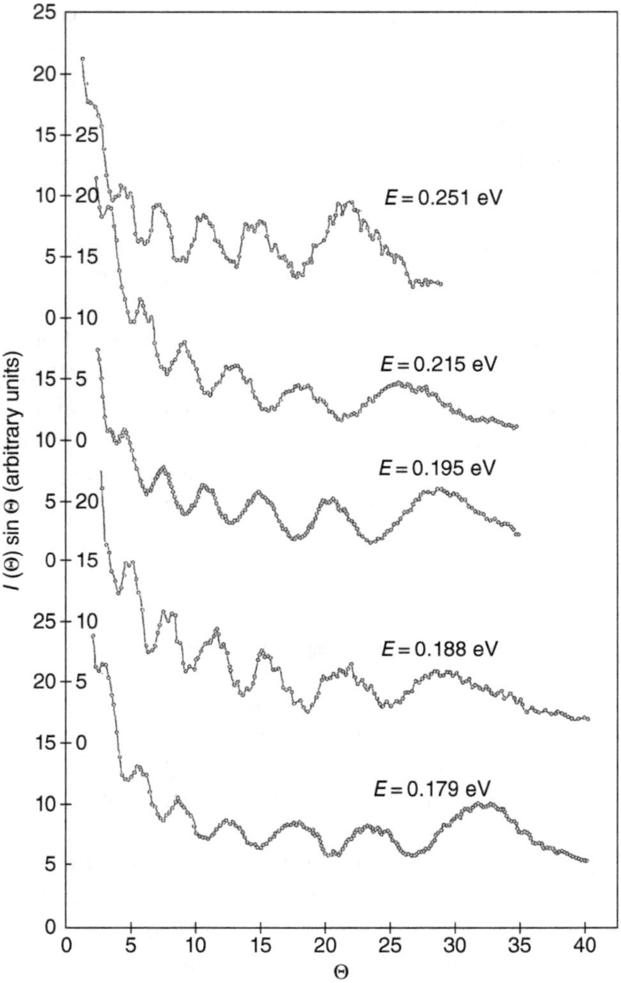

Figure 2.4 Experimentally measured differential cross sections $I(\Theta, E)\sin(\Theta)$ for Na–Hg vs scattering angle Θ and for chosen collision energies E. Features can be reproduced with numerical results from accurate potential energy functions. Source: [5] / John Wiley & Sons.

Another purely quantal effect relates to exchange symmetry of identical particles. It is detected in the collisions involving identical species, like Rg–Rg rage–gas collisions with Rg = He, Ar, Kr, Xe [5]. More generally, composite species A and B containing several electrons and nuclei can be classified as boson or fermion composites, and their quantal states must be constructed to be symmetric for bosons or anti-symmetric for fermions, under an exchange symmetry $A \leftrightarrow B$.

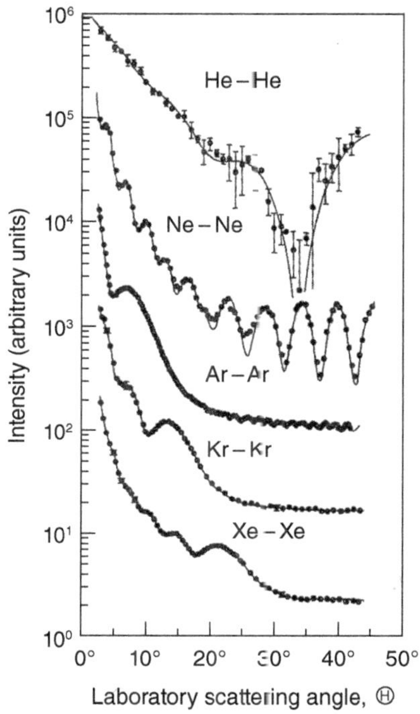

Figure 2.5 Differential cross sections for pairs of identical rare-gas atoms, where the exchange symmetry effect is seen to be very prominent for He and also for Ne. Source: [5] / John Wiley & Sons.

In the simple case of elastic scattering, this means that the scattering amplitude must be of form

$$\psi_{\vec{k}}^{(sc\pm)}(\vec{R}) \approx (2\pi)^{-\frac{3}{2}} \frac{\exp(i\,kR)}{R} \left[f_{\vec{k}}^{(sc)}(\Omega) \pm f_{\vec{k}}^{(sc)}(\Omega')\right)$$

where Ω' is the solid angle resulting from the coordinates exchange in $\vec{R} \leftrightarrow \vec{R}' = -\vec{R}$. The amplitude superposition effect can be detected in cross sections, as displayed in Figure 2.5, where the effect is seen to be very prominent for He and also for Ne. It is less noticeable for the heavier atoms due to extensive averaging over their many scattered partial waves.

2.2 Integral Equation and Approximations

2.2.1 Green Functions and Integral Equations

The equation for the scattering wavefunction and its boundary condition can be combined in an integral equation, introducing the free scattering Green function (or free propagator) for outgoing waves $G_0^{(+)}(\vec{R},\vec{R}')$, which satisfies

$$\frac{\hbar^2}{2M}\left(k^2+\nabla_R^2\right)G_0^{(+)}(\vec{R},\vec{R}')=\delta(\vec{R}-\vec{R}')$$

and is given by [1]

$$G_0^{(+)}(\vec{R},\vec{R}')=-\frac{1}{4\pi}\frac{2M}{\hbar^2}\frac{\exp\left(ik|\vec{R}-\vec{R}'|\right)}{|\vec{R}-\vec{R}'|}$$

Taking the limit for $\mathcal{R}$ much larger than the range of $V(R)$, while R' is within its range, so that $|\vec{R}-\vec{R}'|\approx R(1-\vec{R}'\cdot\vec{R}/R)$, replacing this in the exponential, and keeping the leading term in the denominator, one finds

$$G_0^{(+)}\approx-\frac{1}{4\pi}\frac{2M}{\hbar^2}\exp(ikR)R^{-1}\exp\left(-ik\vec{n}\cdot\vec{R}'\right)$$

$$\approx-\frac{1}{4\pi}\frac{2M}{\hbar^2}\exp(ikR)R^{-1}(2\pi)^{3/2}\psi_{\vec{k}'}^{(0)}(\vec{R}')^*$$

where $\vec{k}'=k\vec{n}$ and $\vec{n}=\vec{R}/R$. The solution of the differential equation for the wavefunction, in integral form, is

$$\psi_{\vec{k}}^{(+)}(\vec{R})=\psi_{\vec{k}}^{(0)}(\vec{R})+\int d^3R'\,G_0^{(+)}(\vec{R},\vec{R}')V(R')\psi_{\vec{k}}^{(+)}(\vec{R}')$$

as can be verified by differentiation of both sides. At large distances and using the previous asymptotic form of the free propagator, the amplitude multiplying an outgoing spherical wave gives

$$f_{\vec{k}}^{(+)}(\Omega)=-\frac{1}{4\pi}\frac{2M}{\hbar^2}(2\pi)^3T_{\vec{k}\vec{k}'}$$

$$T_{\vec{k}'\vec{k}}=\left\langle\psi_{\vec{k}'}^{(0)}|V|\psi_{\vec{k}}^{(+)}\right\rangle=\int d^3R\,\psi_{\vec{k}'}^{(0)}(\vec{R})V(R)\psi_{\vec{k}}^{(+)}(\vec{R})$$

where $T_{\vec{k}'\vec{k}}$ is a scattering transition matrix element between two states with the shown initial and final wavevectors.

Stationary states $\psi_{\vec{k}}^{(-)}(\vec{R})$ corresponding to incoming waves are given by

$$\psi_{\vec{k}}^{(-)}(\vec{R})=\psi_{\vec{k}}^{(0)}(\vec{R})+\int d^3R'\,G_0^{(-)}(\vec{R},\vec{R}')V(R')\psi_{\vec{k}}^{(-)}(\vec{R}')$$

with a new free propagator $G_0^{(-)}(\vec{R},\vec{R}')=-(1/4\pi)(2M/\hbar^2)[\exp(-ik|\vec{R}-\vec{R}'|)/|\vec{R}-\vec{R}'|]$, and transition elements $T_{\vec{k}'\vec{k}}=\left\langle\psi_{\vec{k}'}^{(-)}\middle|V\middle|\psi_{\vec{k}}^{(0)}\right\rangle$ follow from its asymptotic form.

The integral equation can be written in a compact operator form introducing the Dirac bracket notation and coordinate representation for states and operators, for both outgoing and incoming stationary states, with $\psi_{\vec{k}}^{(\pm)}(\vec{R}) = \left\langle \vec{R} \middle| \psi_{\vec{k}}^{(\pm)} \right\rangle$ and $G_0^{(\pm)}(\vec{R}, \vec{R}') = \left\langle \vec{R} \middle| \hat{G}_0^{(\pm)} \middle| \vec{R} \right\rangle$, to obtain

$$\left| \psi_{\vec{k}}^{(\pm)} \right\rangle = \left| \psi_{\vec{k}}^{(0)} \right\rangle + \hat{G}_0^{(\pm)} \hat{V} \left| \psi_{\vec{k}}^{(\pm)} \right\rangle$$

which can be formally solved as

$$\left| \psi_{\vec{k}}^{(\pm)} \right\rangle = \left| \psi_{\vec{k}}^{(0)} \right\rangle + \hat{G}_0^{(+)} \hat{T}^{(\pm)} \left| \psi_{\vec{k}}^{(0)} \right\rangle$$

in terms of the transition operator

$$\hat{T}^{(\pm)} = \hat{V} + \hat{V}\, \hat{G}_0^{(\pm)}\, \hat{T}^{(\pm)}$$

as can be verified by substitution in the previous equation. Its matrix elements are $T_{\vec{k}'\vec{k}} = \left\langle \psi_{\vec{k}'}^{(0)} \middle| \hat{T}^{(\pm)} \middle| \psi_{\vec{k}}^{(0)} \right\rangle$ in the basis set of free motion states.

2.2.2 Born Expansion

The above integral equation for $\psi_{\vec{k}}^{(+)}$ can be solved by iteration, to obtain the Born perturbation expansion in powers of the potential energy strength, as

$$\left| \psi_{\vec{k}}^{(+)} \right\rangle = \sum_{n \geq 0} \left| \psi_{\vec{k}}^{(n)} \right\rangle$$

$$\left| \psi_{\vec{k}}^{(n+1)} \right\rangle = \left| \psi_{\vec{k}}^{(0)} \right\rangle + \hat{G}_0^{(+)} \hat{V} \left| \psi_{\vec{k}}^{(n)} \right\rangle, \quad n \geq 0$$

This gives the $(n+1)$-th order term for the transition amplitude as $T_{\vec{k}'\vec{k}}^{(n+1)} = \left\langle \psi_{\vec{k}'}^{(0)} \middle| V \middle| \psi_{\vec{k}}^{(n)} \right\rangle$. The first-order Born approximation is then $T_{\vec{k}'\vec{k}}^{(B1)} = \left\langle \psi_{\vec{k}'}^{(0)} \middle| V \middle| \psi_{\vec{k}}^{(0)} \right\rangle$. This simplifies for a radial potential energy into

$$T_{\vec{k}'\vec{k}}^{(B1)} = \int d^3R\, V(R) \exp[i(\vec{k} - \vec{k}') \cdot \vec{R}]/(2\pi)^3$$

$$= 4\pi \int_0^\infty \frac{R^2 dR}{(2\pi)^3} V(R) \sin(qR)/(qR)$$

which is a simple sine Fourier transform and depends only on the momentum transfer $\hbar\vec{q} = \hbar\vec{k} - \hbar\vec{k}'$ with a wavevector of magnitude $q = 2k\sin(\Theta/2)$ involving the angle Θ between the initial and final momenta. This approximation works best for small interaction potential energies and small momentum transfer and is limited to potential energy functions with a finite sine Fourier transform for $RV(R)$.

2.2.3 Eikonal Wavefunctions and Semiclassical Treatments

The connection between classical and quantal treatments of scattering has been extensively considered in the literature, to provide physical insights and computational procedures. The connection has been done introducing the propagation of scattering wavepackets [9–11], path integrals [12–16], and Bohmian trajectories [17–20], among others. Here we adopt a straightforward treatment in terms of eikonal wavefunctions [21], which can also be extended to inelastic and reactive collisions and can be implemented computationally.

We can write quite generally the scattering function for an initial relative motion momentum $\hbar\vec{k}$ in an eikonal representation [1] as

$$\psi_{\vec{k}}^{(+)}(\vec{R}) = \chi_{\vec{k}}(\vec{R}) \exp[iS_{\vec{k}}(\vec{R})/\hbar]$$

where we take $S_{\vec{k}}(\vec{R})$ to be real-valued. This expression provides a starting treatment for semiclassical approximations, where $S_{\vec{k}}(\vec{R})$ can be related to an action integral in classical mechanics [22], while $\chi_{\vec{k}}(\vec{R})$ provides a spatial probability amplitude going asymptotically to a constant value consistent with the asymptotic scattering conditions.

An equation for the function $S(\vec{R})$ at each fixed wavevector $\vec{k}$ can be obtained first introducing an effective potential energy function $V(\vec{R})$ to describe a classical-like motion with a local momentum function $\vec{P}(\vec{R}) = \partial S/\partial\vec{R}$. The effective potential can be chosen on physical considerations and can include a statistical average over asymptotic or propagation conditions. The action function is chosen to satisfy a Hamilton–Jacobi (or H–J) partial differential equation for stationary scattering with a total energy E, as

$$V(\vec{R}) + \frac{1}{2M}\frac{\partial S}{\partial \vec{R}} \cdot \frac{\partial S}{\partial \vec{R}} - E = 0$$

Replacing the scattering function into the time-independent Schrodinger equation for stationary scattering states with Hamiltonian $\hat{H} = -\hbar^2\nabla_R^2/(2M) + V(R)$ provides the differential equations for $\chi(\vec{R})$,

$$\left[\frac{1}{2M}\left(\frac{\hbar}{i}\frac{\partial}{\partial\vec{R}} + \frac{\partial S}{\partial\vec{R}}\right)^2 + V(\vec{R}) - E\right]\chi(\vec{R}) = 0$$

to be solved at each initial relative motion momentum $\vec{P}_{in} = \hbar\vec{k}$, with a spatial amplitude going asymptotically to a constant value as $R \to \infty$. The squared term here indicates a scalar product of the vector argument in parenthesis, times itself. The equations for $S(\vec{R})$ and $\chi(\vec{R})$ are coupled and must be solved together, with their coupling dependent on the choice of the effective potential energy $V(\vec{R})$. The H–J equation must be solved with asymptotic boundary conditions $S(\vec{R}) \approx \vec{k}\cdot\vec{R}$ for $R \to \infty$, and from its solution, one finds $\vec{P}(\vec{R}) = \partial S/\partial\vec{R}$ with magnitude $P = [2M(E - V)]^{1/2}$.

A self-consistent effective potential energy function $V(\vec{R})$ can be constructed based on physical considerations, to allow for changes of $\vec{P}(\vec{R})$ along each path with a given amplitude $\chi(\vec{R})$ [21]. Differentiating $\psi^{(+)}(\vec{R})$ in its equation for a given $\vec{k}$ value, multiplying times $\chi(\vec{R})^*$ to the left and taking the real part, the action function is found to satisfy a Hamilton–Jacobi partial differential equation with a quantum potential energy $V(\vec{R}) = V_{qu}(\vec{R})$, in

$$V_{qu}(\vec{R}) + \frac{1}{2M}\frac{\partial S}{\partial \vec{R}} \cdot \frac{\partial S}{\partial \vec{R}} - E = 0$$

$$V_{qu}(\vec{R}) = V(\vec{R}) + V'(\vec{R}) + V''(\vec{R})$$

$$V'(\vec{R}) = \frac{i\hbar}{2M}\frac{\partial S}{\partial \vec{R}} \cdot \left[\left(\frac{\partial \chi}{\partial \vec{R}}\right)^{*} \chi(\vec{R}) - \chi(\vec{R})^{*}\frac{\partial \chi}{\partial \vec{R}}\right] \Big/ |\chi(\vec{R})|^2$$

$$V''(\vec{R}) = -\frac{\hbar^2}{2M}\frac{1}{2}\left[\left(\frac{\partial^2 \chi}{\partial \vec{R}^2}\right)^{*} \chi(\vec{R}) + \chi(\vec{R})^{*}\frac{\partial^2 \chi}{\partial \vec{R}^2}\right] \Big/ |\chi(\vec{R})|^2$$

with the magnitude of the additional potential energies $V'(\vec{R})$ and $V''(\vec{R})$ dependent on how large are the changes of $\chi(\vec{R})$ over scattering distances, and both vanishing as $R \to \infty$.

For a general effective potential function, the action function can be obtained for a set $\{\vec{R}_s\}$ of position vectors and related initial momenta $\{\vec{P}_s = \hbar\vec{k}\}$ from a path integral as

$$S_s(\vec{R}) = S(\vec{R}_s) + \int_{\vec{R}_s}^{\vec{R}} d\vec{R} \cdot \vec{P}(\vec{R}) = \vec{k} \cdot \vec{R} + \int_{\vec{R}_s}^{\vec{R}} d\vec{R} \cdot [\vec{P}(\vec{R}) - \vec{k}]$$

while the amplitudes follow from the equation

$$\left[-\frac{\hbar^2}{2M}\left(\frac{\partial}{\partial \vec{R}}\right)^2 + \frac{\vec{P}}{M} \cdot \frac{\hbar}{i}\frac{\partial}{\partial \vec{R}} + \frac{\hbar}{i}\left(\frac{\partial}{\partial \vec{R}} \cdot \frac{\vec{P}}{M}\right) + \frac{P^2}{2M} + V(\vec{R}) - E\right] \chi(\vec{R}) = 0$$

with $P^2/(2M) + V(\vec{R}) - E = V(\vec{R}) - V(\vec{R}) \approx 0$ for $R \to \infty$.

As written, the equation for $S(\vec{R})$ may be difficult to solve numerically for a given $V(\vec{R})$. For example, $V_{qu}(\vec{R})$ contains $|\chi(\vec{R})|^2$ in denominators, which can go to zero. However, solutions can be constructed which satisfy the differential equations and scattering boundary conditions, with a general choice $\chi(\vec{R}) = \exp[F(\vec{R})]$, where $F(\vec{R}) = F'(\vec{R}) + iF''(\vec{R})$, and both $F'(\vec{R})$ and $F''(\vec{R})$ are real-valued functions.

The differential equation for $\chi(\vec{R})$ leads to

$$-\frac{\hbar^2}{2M}[(\nabla F)^2 + \nabla^2 F] + \frac{\hbar}{2iM}\nabla \cdot \vec{P} + \frac{\hbar}{iM}\vec{P} \cdot \nabla F + \frac{P^2}{2M} + V(R) - E = 0$$

where the grad (or Del) symbol $\nabla. = \partial./\partial\vec{R}$ has been used, and this equation can in principle be solved for $F(\vec{R})$ to obtain $\chi(\vec{R})$. The quantum potential $V_{qu}(\vec{R})$ can be obtained from $F(\vec{R})$ as

$$V'(\vec{R}) = \vec{P} \cdot \vec{w}(\vec{R}), \quad \vec{w}(\vec{R}) = (\hbar/M)\nabla F''$$

$$V''(\vec{R}) = -(\hbar^2/M)\,[(\nabla F')^2 + \nabla^2 F']$$

where $\vec{w}(\vec{R})$ is a velocity increment.

In practice, the eikonal treatment is applied for short wavelengths $\lambda = 2\pi/k$ with $\lambda/R_m \ll 1$ and $\nabla^2\chi/k^2 \ll 1$, to situations where $F(\vec{R})$ and $S(\vec{R})$ show only small changes over distances R_m typical of the potential energy range. This can be quantified testing that $|(\nabla F)^2/\nabla F| \ll \lambda^{-1}$ and $|\nabla^2 F/\nabla F| \ll \lambda^{-1}$. An expansion of equation terms in powers of λ^n can be done keeping terms to the lowest order in λ^1. This implies omitting $(\nabla F)^2$ and $\nabla^2 F$ in the above equations. Writing the resulting differential equation for $F(\vec{R})$ as $\hat{D}[\vec{R}; F(\vec{R})] = 0$, taking real and imaginary parts, we then find that to this order $Re[\hat{D}] = 0$ reproduces the H–J equation, with $\vec{P} \cdot (\hbar/M)\nabla F'' = V(\vec{R}) - V(\vec{R})$ satisfied by $F'' = 0$, and that $Im[\hat{D}] = 0$ gives $\vec{P} \cdot \nabla F' = -\nabla \cdot \vec{P}/2$ to this order. Solutions for $F'(\vec{R})$ can be obtained in special cases, as described below.

Solutions to the equations can be obtained at high collision energies compared to the magnitude of the potential energy $V(R)$ traversed during collision so that $E \gg V(R)$ and one deals with a short-wavelength limit. Introducing cylindrical coordinates (ρ, φ, Z) in a CM–SF frame, with ρ an axial distance and φ an axial angle around Z, consider collisions starting at $Z_{in} \to -\infty$ and initial momentum $\vec{P}_{in} = \hbar\vec{k}$, $\vec{k} = k\vec{n}_z$ parallel to the Z-axis, with $E = (\hbar k)^2/2M$. The functions $S(\vec{R})$ and $F(\vec{R})$ and the gradient operator ∇ can be re-expressed in the cylindrical coordinates. They are independent of φ for a radial potential energy, by symmetry. Furthermore, insofar they change little over a wavelength for a smooth potential energy, it can be assumed that gradients in ρ are negligible, so that here $\nabla. \approx \vec{n}_z \partial./\partial Z$. The action $S(\vec{R}) = S_{\vec{k}}(Z; \rho)$ follows from

$$\frac{1}{2M}\left(\frac{\partial S_{\vec{k}}}{\partial Z}\right)^2 = E - V(\vec{R})$$

integrating between Z_{in} and Z, rearranging terms and letting $Z_{in} \to \infty$ to get

$$S_{\vec{k}}(Z; \rho) = \hbar k Z + \int_{-\infty}^{Z} dZ' \left(\{2M[E - V(Z'; \rho)]\}^{1/2} - \hbar k\right)$$

and $\vec{P} = P_Z(Z; \rho)\vec{n}_z, P_Z = \partial S_{\vec{k}}/\partial Z$. The differential equation for $\chi(Z; \rho)$ is now

$$\left[\frac{P_Z}{M} \cdot \frac{\hbar}{i}\frac{\partial}{\partial Z} + \frac{\hbar}{iM}\frac{\partial P_Z}{\partial Z} + \frac{P_Z{}^2}{2M} + V(Z; \rho) - E\right] \chi(Z; \rho) = 0$$

In the present limit of short wavelengths, $V \cong V$ and $P_Z = \{2M[E - V(Z; \rho)]\}^{1/2}$. Letting $\chi(Z; \rho) = \exp[F(Z; \rho)]$ to be solved for $F'(Z; \rho)$, the resulting equation is simply

$$P_Z \partial F'/\partial Z = -(\partial P_Z/\partial Z)/2$$

It gives $F'(Z; \rho) = -\ln[P_Z/(\hbar k)]/2$ and

$$\chi(Z; \rho) = 2\pi^{-3/2}[P_Z(Z; \rho)/(\hbar k)]^{-1/2}$$

in the full wavefunction. Here we can assume that the momentum component $P_Z(Z;\ \rho)$ is larger than zero along the trajectory for all the impact parameters, insofar we are dealing with collisions where all trajectories eventually emerge from the potential energy region and we can exclude the backward scattering with $\Theta = \pi$.

The present wavefunction expressions are similar to the ones derived in the JWKB approximation for one-dimensional radial scattering functions [2, 7]. However, they differ from that treatment for one-dimensional radial motion, developed in great detail in the JWKB treatment [1, 2], where the radial momentum P_R goes to zero at a turning point when $V(R) = E$. In that case, it is necessary to separately treat classically allowed and forbidden radial regions and the region around the turning point. The present treatment for three-dimensional motion avoids this situation. Connection formulae valid for tunneling along the Z coordinate can be similarly given for our eikonal wavefunctions in one-dimensional cases.

The transition matrix elements are given by $T_{\vec{k}'\vec{k}} = \int d^3R\, \psi_{\vec{k}'}^{(0)}(\vec{R})V(R)\psi_{\vec{k}}^{(+)}(\vec{R})$, with $\psi_{\vec{k}}^{(+)}(\vec{R}) \cong \chi_{\vec{k}}(Z;\rho)\exp[iS_{\vec{k}}(Z;\rho)/\hbar]$ so that

$$T_{\vec{k}'\vec{k}} \cong \frac{1}{2\pi^3}\int d\rho d\varphi dZ\, J\begin{pmatrix} X & Y \\ \rho & \varphi \end{pmatrix} \exp(-i\vec{k}'\cdot\vec{R})V(Z;\rho)\left[\frac{\hbar k}{P_Z(Z;\rho)}\right]^{\frac{1}{2}}$$
$$\times \exp[iS_{\vec{k}}(Z;\rho)/\hbar]$$

where

$$J\begin{pmatrix} X & Y \\ \rho & \varphi \end{pmatrix} = \begin{vmatrix} \partial X/\partial\rho & \partial X/\partial\varphi \\ \partial Y/\partial\rho & \partial Y/\partial\varphi \end{vmatrix}$$

is the Jacobian determinant for the transformation between components of $\vec{R}$ and the impact parameter length and its azimuthal angle around the Z-axis, in the element of volume $d^3R = dXdYdZ$. Letting $X = \rho\cdot\cos(\varphi)$, $Y = \rho\cdot\sin(\varphi)$ gives $J = \rho$. Also, given that $R = (\rho^2 + Z^2)^{1/2}$, $V(R) = V(Z;\ \rho)$ for a fixed impact parameter. The integrand contains $\exp(-i\vec{q}\cdot\vec{R})$ with $\vec{q} = \vec{k}' - \vec{k}$, and assuming small scattering angles, it follows that $\vec{q}$ is effectively on the (xy)-plane, so that $\vec{q}\cdot\vec{R} \cong \vec{q}\cdot\vec{\rho}$. This gives the first-order eikonal approximation

$$f^{(E1)} = -\frac{1}{4\pi}\frac{2M}{\hbar^2}(2\pi)^3 T_{\vec{k}\vec{k}'} = -\frac{M}{2\pi\hbar^2}\int_0^\infty \rho\, d\rho \int_0^{2\pi} d\varphi \int_{-\infty}^\infty dZ$$
$$\times \exp(-i\,\vec{q}\cdot\vec{\rho})V(Z;\rho)\left[\frac{\hbar k}{P_Z(Z;\rho)}\right]^{\frac{1}{2}} \exp\left[i\Delta S_{\vec{k}}(Z;\rho)\right]$$

$$\Delta S_{\vec{k}}(Z;\rho) = \int_{-\infty}^{Z} dZ'\,(\{2M[E - V(Z';\rho)]\}^{1/2} - \hbar k)$$

Since the integrand in $f^{(E1)}$ is independent of the angle φ, the integral over this variable can be done and gives the zero-th order Bessel function [23]

$$J_0(q\rho) = (2\pi)^{-1} \int_0^{2\pi} d\varphi \, \exp(-iq\rho\cos\varphi)$$

which leads to

$$f^{(E1)}(\Theta; k) = \int_0^{\infty} \rho \, d\rho J_0(q\rho) A(\rho; k)$$

where $q = 2k\sin(\Theta/2)$ and

$$A(\rho; k) = -\frac{M}{\hbar^2} \int_{-\infty}^{\infty} dZ \, V(Z;\rho) \left[\frac{\hbar k}{P_Z(Z;\rho)}\right]^{\frac{1}{2}} \exp\left[i\Delta S_{\vec{k}}(Z;\rho)\right].$$

Here the product $V(Z;\rho)[\hbar k/P_Z(Z;\rho)]^{\frac{1}{2}}$ can be replaced by an expression from the equation for $\chi_{\vec{k}}$ which eliminates reference to the potential energy function.

A slightly more accurate description is obtained by constructing the eikonal form of the wavefunction $\psi_{\vec{k}}^{(+)}$ for an incoming momentum $\vec{P}_{in} = \hbar\vec{k}_A$ with the average wavevector $\vec{k}_A = (\vec{k} + \vec{k}')/2$, which is perpendicular to $\vec{q} = \vec{k} - \vec{k}'$, and by defining new cylindrical coordinates with the new Z-axis along $\vec{k}_A$ and impact parameter $\vec{\rho}$ so that $\vec{q} \cdot \vec{R} = \vec{q} \cdot \vec{\rho}$ without approximation. The final formula is the same, but it does not require assuming small scattering angles. The momentum transfer is replaced by $\vec{q}_A = \vec{k}_A - \vec{k}' = \vec{q}/2$.

The differential cross-section follows from $d\sigma/d\Omega = |f^{(E1)}(\Theta;\, k)|^2$, and the integral cross section can be obtained from the optical theorem result $\sigma(v) = 4\pi k^{-1}\, Im\,[f^{(+)}(0;\, k)]$ (to be derived in the following section) using that $J_0(0) = 1$. This leads to

$$\sigma^{(E1)}(v) = 4\pi \int_0^{\infty} \rho d\rho Im[A(\rho; k)]$$

$$Im[A(\rho; k)] = -\frac{M}{\hbar^2} \int_{-\infty}^{\infty} dZ \, V(Z;\rho) \left[\frac{\hbar k}{P_Z(Z;\rho)}\right]^{\frac{1}{2}} \sin[\Delta S_{\vec{k}}(Z;\rho)]$$

Therefore, the cross-section calculations reduce to two consecutive quadratures, first over Z and then over ρ. The integral over ρ involves a rapidly oscillating integrand and may require special integration techniques. It can also be approximated with stationary phase methods or by piecewise integrations [24].

The H–J partial differential equation, which is nonlinear and of first order in the unknown $S(\vec{R})$, can be solved introducing characteristic path solutions [22, 23, 25] of relative momentum $\vec{P} = \partial S/\partial\vec{R}$ and position $\vec{R}$ variables as functions of a parameter that can be identified as the time variable t. These variables can be obtained for a fixed energy E introducing position $\vec{R}(t)$ and momentum $\vec{P}(t)$ functions of the time, satisfying the Hamiltonian equations

$$\frac{d\vec{R}}{dt} = \frac{\partial \mathcal{H}}{\partial \vec{P}}, \quad \frac{d\vec{P}}{dt} = -\frac{\partial \mathcal{H}}{\partial \vec{R}}$$

with

$$\mathcal{H}(\vec{P}, \vec{R}) = \vec{P} \cdot \vec{P}/(2M) + V(\vec{P}, \vec{R})$$

which gives $d\mathcal{H}/dt = 0$, by solving them for a set of initial conditions with $\vec{P}_{in} = \vec{P}(t_{in}), \vec{R}_{in} = \vec{R}(t_{in})$. This provides $\vec{R}(t; \vec{P}_{in}, \vec{R}_{in})$ and $\vec{P}(t; \vec{P}_{in}, \vec{R}_{in})$ constructed to satisfy the initial conditions, for a constant total energy $\mathcal{H}(\vec{P}, \vec{R}) = E$.

For a collision with relative momentum $\vec{P}(t_{in}) = \hbar\vec{k}$ pointing along the Z-axis, with ρ and φ the axial distance and angle around the Z-axis, the trajectories are given by $\vec{R}(t, \rho, \varphi; \vec{k})$ and $\vec{P}(t, \rho, \varphi; \vec{k})$. The eikonal wavefunction phase and amplitude can be obtained as $S(\vec{R}) = S(t, \rho, \varphi)$ and $\chi(\vec{R}) = \chi(t, \rho, \varphi) = \exp[F(t, \rho, \varphi)]$, by transforming their differential equations from the variables in $\vec{R}$ to new variables (t, ρ, φ).

The trajectories give, over time, the momentum $\vec{P}(t, \rho, \varphi)$ starting from $\vec{P}_{in} = \hbar\vec{k}$ and from this the line integral in

$$S_{\vec{k}}(t, \rho, \varphi) = \vec{k} \cdot \vec{R}(t, \rho, \varphi; \vec{k}) + \Delta S_{\vec{k}}, \quad \Delta S_{\vec{k}} = \int_{t_{in}}^{t} dt' \frac{d\vec{R}}{dt'} \cdot [\vec{P}(t', \rho, \varphi; \vec{k}) - \vec{k}]$$

The equation for $\chi_{\vec{k}}(t, \rho, \varphi)$ can be transformed changing there the gradient operator in the variables (X, Y, Z) into derivatives in the new variables (t, ρ, φ) using Jacobian determinants [24, 26, 27]. The approximation of short wavelengths gives in particular $\vec{P}/M = d\vec{R}/dt, P^2/(2M) + V(\vec{R}) - E = V(\vec{R}) - V(\vec{R}) \cong 0$, and $(\vec{P}/M) \cdot \nabla\chi_{\vec{k}} = d\chi_{\vec{k}}/dt$, leading to the equation

$$\left[\frac{\hbar}{i}\frac{d}{dt} - i\, U_{\vec{k}}(\vec{R})\right] \chi_{\vec{k}}(\vec{R}) = 0$$

where $U_{\vec{k}}(\vec{R}) = (\hbar/M)\nabla \cdot \vec{P}$ is a density-loss potential function accounting for the change of flux density along a trajectory, with $\vec{R} = \vec{R}(t, \rho, \varphi)$. This equation gives a

$$\chi_{\vec{k}}(t, \rho, \varphi) = \chi_{\vec{k}}(t_{in}, \rho, \varphi) \exp\left[-\int_{t_{in}}^{t} dt'\, U_{\vec{k}}(t', \rho, \varphi)/\hbar\right] = (2\pi)^{-\frac{3}{2}} \exp[F_{\vec{k}}(t, \rho, \varphi)]$$

incorporating scattering initial conditions for the eikonal amplitude.

The eikonal wavefunction is then, for a general trajectory generated in the potential energy function $V(R)$,

$$\psi_{\vec{k}}^{(+)}(t, \rho, \varphi) = (2\pi)^{-\frac{3}{2}} \exp[F_{\vec{k}}(t, \rho, \varphi)] \exp[iS_{\vec{k}}(t, \rho, \varphi)/\hbar]$$

and the transition amplitude can be obtained from $T_{\vec{k}'\vec{k}} = \left\langle \psi^{(0)}_{\vec{k}'} \middle| V \middle| \psi^{(+)}_{\vec{k}} \right\rangle$ transforming integration variables from (X, Y, Z) to (t, ρ, φ). This gives

$$T_{\vec{k}'\vec{k}} = (2\pi)^{-3} \int dt d\rho d\varphi \, J \begin{pmatrix} X, Y, Z \\ t, \rho, \varphi \end{pmatrix} e^{i(\vec{k}'-\vec{k})\cdot\vec{R}} V(R) \exp(F_{\vec{k}} + i\Delta S_{\vec{k}})$$

where all the trajectories are started with $\vec{P}_{in} = \hbar\vec{k}$, and $J \begin{pmatrix} X, Y, Z \\ t, \rho, \varphi \end{pmatrix}$ is the 3×3 Jacobian determinant [27] for the transformation of variables from (X, Y, Z) to (t, ρ, φ) involving the derivatives for $X(t, \rho, \varphi)$, $Y(t, \rho, \varphi)$, and $Z(t, \rho, \varphi)$. The previously described transition amplitude for a straight-line trajectory follows here as a special case when $\vec{R} = \vec{R}_{in} + (\hbar\vec{k}/M)(t - t_{in})$.

Cross sections can also be obtained in a simpler but more approximate treatment, from a bundle of trajectories $\{\vec{P}_{in}, \vec{R}_{in}\}$, with a given initial flux J_0, to calculate a rate increment $d\dot{N} = J_0 d\sigma$, now however with trajectories generated by an effective potential energy $V(\vec{P}, \vec{R})$. This provides statistical values for differential cross sections $I(\Omega, v) = d\sigma/d\Omega$ that account for changes of $V(R)$ into $V(\vec{R}) = V_{qu}(\vec{R})$, due to quantum effects.

2.3 Partial-Wave Analysis

2.3.1 Radial Wavefunctions and Phase Shifts

For a radial potential $V(R)$, both the free motion and total Hamiltonians, $\hat{H}_0$ and $\hat{H}$, commute with the angular momentum operators $\hat{L}_\xi, \xi = x, y, z$, and one can use the eigenfunctions of angular momentum to expand the scattering wavefunctions. Since the scattering problem is symmetric around the relative momentum $\vec{P}$ direction of the incoming wave, it is convenient to choose the z-axis along it and to introduce spherical coordinates (R, Θ, Φ). We expand the free motion and scattering functions in a basis set of Legendre polynomials $P_l(\cos\Theta)$ [23] satisfying

$$\hat{L}^2 P_l = \hbar^2 l(l+1) P_l$$

$$\int_{-1}^{1} d\mu \, P_l(\mu) P_{l'}(\mu) = 2\,\delta_{ll'}/(2l+1)$$

For $V = 0$, we have free motion, with a Hamiltonian

$$\hat{H}_0 = -\frac{\hbar^2 \nabla_R^2}{2M} = -\frac{\hbar^2}{2M} \frac{1}{R} \frac{\partial^2}{\partial R^2} R + \frac{\hat{L}^2}{2MR^2}$$

and the plane wave solutions may be expressed in terms of Bessel functions $J_\nu(x)$ [23] with ν a half-integer [23]. In effect,

$$\exp(i\vec{k} \cdot \vec{R}) = \sum_{l=0}^{\infty} (2l+1) i^l P_l(\cos\Theta) u_l(kR)/(kR)$$

where we have used the regular Riccati–Bessel function

$$u_l(x) = \sqrt{\frac{\pi x}{2}} J_{l+\frac{1}{2}}(x) \approx \sin\left(x - \frac{l\pi}{2}\right)$$

with its limit as shown for large x. It satisfies the differential equation

$$\left[\frac{d^2}{dx^2} - \frac{(l+1/2)^2 - 1/4}{x^2} - 1\right] u_l(x) = 0$$

and is small for $x = kR \leq \sqrt{l(l+1)}$. In addition, we shall use the Riccati–Neumann functions, which are singular at the origin $R = 0$,

$$v_l(x) = \sqrt{\frac{\pi x}{2}} J_{-l-\frac{1}{2}}(x) \approx \cos\left(x - \frac{l\pi}{2}\right)$$

and the Riccati–Hankel functions

$$w_l^{(\pm)}(x) = v_l(x) \pm iu_l(x) \approx \exp\left[\pm i\left(x - \frac{l\pi}{2}\right)\right]$$

with the shown asymptotic forms for large x.

Writing quite generally

$$\psi_{\vec{k}}(\vec{R}) = \sum_{l=0}^{\infty} \frac{(2l+1)i^l P_l(\cos\Theta)\psi_l(R;k)}{kR},$$

involving a radial wavefunction $\psi_l(R;k)$, its form for free motion is the regular solution $\psi_l(R;k) = u_l(kR)$. For $V(R) \neq 0$, and scattering with outgoing waves, we write $\psi_l(R;k) = \psi_l^{(+)}(R;k)$ and find, substituting the expansion in the Schrodinger equation, that the radial scattering function satisfies

$$\left[\frac{d^2}{dR^2} + k^2 - U(R) - \frac{l(l+1)}{R^2}\right] \psi_l^{(+)}(R;k) = 0$$

with $k = P/\hbar$ and $U(R) = 2MV(R)/\hbar^2$. Provided $R^2V(R) \to 0$ for $R \to 0$ and $R^2V(R) \approx 0$ for $R \to \infty$, we can satisfy the scattering boundary conditions choosing $\psi_l^{(+)}(0;k) = 0$ and

$$\psi_l^{(+)}(R;k) \approx u_l(kR) + f_l(k)w_l^{(+)}(kR) \approx \sin\left(kR - \frac{l\pi}{2}\right) + f_l(k)\exp\left[i\left(kR - \frac{l\pi}{2}\right)\right]$$

as found comparing the radial wavefunction equation at the boundaries with the equation for free motion. Other potential energy functions, with different behaviors at the origin of the radial coordinate or at large distances, must be treated as special cases, as will be shown in what follows.

Replacing this asymptotic form in the scattering function, one finds the scattering amplitude

$$f^{(+)}(\Theta;k) = k^{-1}\sum_{l=0}^{\infty}(2l+1)P_l(\cos\Theta)f_l(k)$$

or, for an arbitrary direction $\Omega_{\vec{k}}$ of the momentum $\vec{P} = \hbar\vec{k}$, using that $\cos\Theta = \vec{k}\cdot\vec{R}/kR$ and that the addition theorem [23] for the orthonormal spherical harmonics $Y_{lm}(\Omega), -l \leq m \leq l$ gives $(2l+1)P_l(\vec{k}\cdot\vec{R}/kR) = 4\pi\sum_{l,m} Y_{lm}(\Omega_{\vec{k}})^* Y_{lm}(\Omega)$,

$$f_k^{(+)} = 4\pi k^{-1}\sum_{l,m} f_l(k) Y_{lm}(\Omega_{\vec{k}})^* Y_{lm}(\Omega)$$

The integral cross section follows from

$$\sigma(v) = 2\pi\int_0^{\pi} d\Theta\ \sin(\Theta)|f^{(+)}(\Theta;k)|^2$$
$$= 4\pi k^{-2}\sum_{l=0}^{\infty}(2l+1)|f_l(k)|^2$$

In addition, from the inequality

$$\left|\sum_{l=0}^{\infty}(2l+1)f_l(k)\right|^2 \geq \sum_{l=0}^{\infty}(2l+1)|f_l(k)|^2$$

it follows that the forward differential cross section, with $P_l(\cos 0) = 1$, gives the bound $(d\sigma/d\Omega)_{frw} \geq \sigma(v)/(4\pi)$.

An alternative asymptotic form for $\psi_l^{(+)}$ follows by first calculating the real-valued radial eigenfunction with asymptotic form

$$\psi_l^{(K)}(R;k) \approx u_l(kR) + K_l(k)v_l(kR)$$

where K_l is a real-valued coefficient convenient for calculations, and using that $v_l = w_l^{(+)} - iu_l$, to obtain

$$\psi_l^{(+)}(R;k) \approx C_l^{(K)}\psi_l^{(K)}(R;k), Cl_{(K)} = 1 - iK_l$$

from which it follows that $f_l = K_l/(1 - iK_l)$. This gives

$$Im(f_l) = \frac{K_l^2}{1 + K_l^2} = |f_l|^2$$

and the optical theorem

$$\sigma(v) = 4\pi k^{-1} Im[f^{(+)}(0;k)]$$

relating the integral cross section to the imaginary part of the forward scattering amplitude.

Another form of the radial wavefunction symmetric in terms of incoming and outgoing radial waves is given by a radial function $\psi_l^{(S)}(R;k)$ constructed so that its asymptotic form is

$$\psi_l^{(S)}(R;k) \approx w_l^{(-)}(kR) - S_l(k)w_l^{(+)}(kR)$$

which compared to $\psi_l^{(K)}(R;k)$ gives $S_l = (1+iK_l)/(1-iK_l)$ so that $|S_l(k)| = 1$ and is a special case of the scattering matrix (to be defined in a following chapter), here for elastic scattering in a radial potential energy.

An enlightening equivalent expression for $f^{(+)}(\Theta;k)$ as a superposition of partial waves follows by defining partial phase shifts $\delta_l(k)$ in

$$K_l = \tan(\delta_l)$$

Writng, for fixed k, the two radial wavefunctions

$$\psi_l^{(0)}(R;k) \approx \sin\left(kR - \frac{l\pi}{2}\right)$$

$$\psi_l^{(+)}(R;k) \approx C_l \sin\left(kR - \frac{l\pi}{2} + \delta_l\right)$$

the term δ_l is found to be a phase shift, obtained comparing the corresponding n-th root $R_{l,n}^{(0)}$ and $R_{l,n}$ of the oscillating functions $\psi_l^{(0)}$ and $\psi_l^{(+)}$, respectively, giving

$$\delta_l = \lim_{n \gg 1} k\left(R_{l,n}^{(0)} - R_{l,n}\right)$$

with $n = 0$ near the origin, as shown in Figure 2.6.

It follows that

$$\delta_l < 0 \text{ for } U(R) > 0, \text{ repulsive at all } R$$

$$\delta_l > 0 \text{ for } U(R) < 0, \text{ attractive at all } R$$

and $\delta_l \cong 0$ for $U(R) \ll l(l+1)/R^2$.

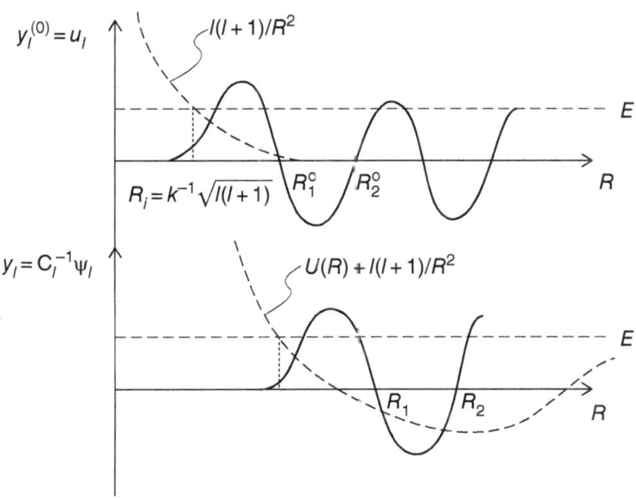

Figure 2.6 Zeroes of the radial wavefunctions $y_l^{(0)} = C_l^{-1}\psi_l^{(0)}$ and $y_l^{(+)} = C_l^{-1}\psi_l^{(+)}$ for no potential and in an effective radial potential vs radial distance.

It is also found that $S_l = \exp(2i\delta_l)$ and $f_l = \exp(i\delta_l)\sin(\delta_l)$ from which

$$f^{(+)}(\Theta; k) = k^{-1} \sum_{l=0}^{\infty} (2l+1) P_l(\cos\Theta) \exp(i\delta_l)\sin(\delta_l)$$

$$\sigma(v) = 2\pi \int_0^{\pi} d\Theta \,\sin(\Theta) |f^{(+)}(\Theta; k)|^2 = \frac{4\pi}{k^2} \sum_{l=0}^{\infty} (2l+1)\sin^2(\delta_l)$$

providing the integral cross section $\sigma(v)$ from phase shifts.

2.3.2 Exceptional Cases

Two exceptional potential energy functions, which do not satisfy the boundary conditions at the origin and large R, as mentioned above for the solutions of the radial equation, are: (a) the Coulomb potential energy between two electric charges and (b) the hard-core potential. These can be treated as special cases as follows.

(a) The Coulomb interaction energy $V(R) = q_1 q_1/(4\pi\epsilon_0 R)$ between charges q_1 and q_1 in vacuum:

The radial solution may be directly found in terms of the confluent hypergeometric function $F(\alpha|\beta|z)$ and is given by [1, 23]

$$\begin{aligned}\psi_l^{(+)}(R; k) &= e^{i\sigma_l} A_l e^{ix} x^{l+1} F(l+1+i\gamma|2l+2|-2ix) \\ &\approx e^{i\sigma_l} \sin\left[x - \gamma\ln(2x) - \frac{l\pi}{2} + \sigma_l\right)\end{aligned}$$

where $x = kR$, $\gamma = q_1 q_2 M/(4\pi\epsilon_0 \hbar^2 k)$, $\sigma_l = \arg[\Gamma(l+1+i\gamma)]$, and $A_l = 2^l \exp(-\pi\gamma/2)$ $|\Gamma(l+1+i\gamma)|/(2l+1)!$, in terms of the gamma function Γ.

The solution to the partial differential Schrodinger equation in spherical coordinates can also be found as [1]

$$\psi_{\vec{k}}^{(+)}(\vec{R}) = C\, \exp(ikZ) F[-iq_1 q_2/(4\pi\epsilon_0 2k)|1|ikR(1-\cos\Theta)]$$

where F is again the confluent hypergeometric function. This provides a scattering amplitude and a differential cross section

$$\frac{d\sigma}{d\Omega} = \left(\frac{q_1 q_2}{4\pi\epsilon_0 4E}\right)^2 \operatorname{cosec}^4\left(\frac{\Theta}{2}\right)$$

with $E = \hbar^2 k^2/2M$, which is the same expression as obtained in a classical treatment. However, the phase of the complex-valued scattering amplitude must be obtained from the quantal treatment [1].

(b) The hard-core potential defined by $V(R) = \infty$, $R < a$; $V(R) = 0$, $R > a$:

In this case, the condition at the origin must be replaced by $\psi_l^{(+)}(a; k) = 0$ and a real-valued radial solution is

$$y_l^{(c)}(R; k) = u_l(kR) - \frac{u_l(ka)}{v_l(ka)} v_l(kR)$$

valid for all $R > a$. The related $K_l(k)$ amplitude and core phase-shift follow as

$$K_l^{(c)}(k) = -\frac{u_l(ka)}{v_l(ka)} = \tan\left(\delta_l^{(c)}\right)$$

The scattering amplitude $f^{(c)}(\Theta; k)$ is obtained from this scattering phase shift as a sum over partial waves. For small Θ and $ka \gg 1$, requiring a sum over many partial waves, we can treat a large $l = \lambda$ as a continuous variable, and use $P_l(\cos\Theta) \cong J_0(2\lambda y)$, with $y = \sin(\Theta/2)$, [23], to write

$$f^{(c)}(\Theta; k) = \frac{1}{ik} \int_0^{\infty} d\lambda\, \lambda\, J_0(2\lambda y) \left[\exp\left(2i\delta_l^{(c)}\right) - 1\right]$$

with the differential cross section given by $d\sigma^{(c)}/d\Omega = |f^{(c)}(\Theta; k)|^2$. The integral cross section follows from the optical theorem as

$$\sigma^{(c)}(v) = \frac{4\pi}{k} Im[f^{(c)}(0; k)] = \frac{4\pi}{k^2} \int_0^{\infty} d\lambda\, \lambda \left[1 - \cos\left(2\delta_\lambda^{(c)}\right)\right]$$

which can also be obtained from $K_l^{(c)}(k)$ as a function of the relative collision velocity.

2.3.3 Radial Integral Equations

It is computationally convenient to look for real-valued solutions $\psi_l^{(K)}(R; k)$ of the differential equation with convenient asymptotic conditions or its equivalent integral equation, to construct the scattering radial function. A regular solution behaving as u_l at large distances may be expressed in terms of the radial Green function (or propagator) $G_{0,l}(R, R')$ that satisfies

$$\left[k^2 + \frac{d^2}{dR^2} - \frac{l(l+1)}{R^2}\right] G_{0,l}(R, R') = \delta(R - R')$$

Its form is

$$G_{0,l}(R, R') = -k^{-1} u_l(kR_<) v_l(kR_>)$$

with $R_<$ (or $R_>$) the smaller (or larger) between R and R'. This propagator can be verified replacing it in its differential equation, and its normalization follows integrating the equation over R between limits $R = R' \pm \epsilon$. The wavefunction solution is

$$\psi_l^{(K)}(R; k) = u_l(kR) + \int_0^{\infty} dR' G_{0,l}(R, R') U(R') \psi_l^{(K)}(R'; k)$$

from which asymptotically, the K-scattering amplitudes is

$$K_l = -k^{-1} \int_0^{\infty} dR' u_l(kR') U(R') \psi_l^{(K)}(R'; k).$$

The above integral equation can be solved by iteration. But, a more convenient procedure follows from conversion first to a Volterra integral equation [28] This is done breaking up the integral for $0 \le R' < \infty$ into integrals for $0 \le R' < R$ and $R \le R' < \infty$ and introducing the Volterra integral equation

$$y_l(R; k) = u_l(kR) + \int_0^{R} dR' G_{0,l}(R, R') U(R') y_l(R'; k).$$

Insofar both $\psi_l^{(K)}$ and y_l are solutions of the same differential equation and are both regular at $R = 0$; they must be linearly related, so that $\psi_l^{(K)}(R; k) = A_l y_l(R; k)$ with the constant A_l obtained from the two integral equations above. This gives

$$A_l = 1 \Big/ \left[1 + k^{-1} \int_0^{\infty} dR\, v_l(kR) U(R) y_l(R; k) \right]$$

which provides a solution $\psi_l^{(K)}(R; k)$ propagating $y_l(R; k)$ from its integral equation values for $R' < R$.

A similar procedure can be followed to obtain $\psi_l^{(+)}(R; k)$ in the Volterra integral equation for a complex-valued $y_l^{(+)}(R; k)$ from the above equations with $v_l(kR)$ replaced by the Riccati–Hankel function $w_l^{(+)}(kR)$. It can be written as $y_l^{(+)}(R; k) = \alpha_l(R; k) u_l(kR) + \beta_l(R; k) v_l(kR)$ with complex-valued coefficients shaped by the interaction potential and goes asymptotically to $y_l^{(+)}(R; k) \approx a_l(k) u_l(kR) + b_l(k) v_l(kR)$ as $R \to \infty$. Again, since both $y_l^{(+)}(R; k)$ and $\psi_l^{(+)}(R; k)$ are regular (non-divergent) scattering solutions, they must be proportional. Hence asymptotically

$$y_l^{(+)}(R; k) \approx D_l(k) \left[u_l(kR) + f_l(k) w_l^{(+)}(kR) \right]$$

$$D_l(k) = 1 + k^{-1} \int_0^{\infty} dR\, w_1^{(+)}(kR)\, U(R) y_l^{(+)}(R; k)$$

$$f_l(k) = -k^{-1} \int_0^{\infty} dR\, u_l(kR)\, U(R) y_l^{(+)}(R; k) / D_l(k).$$

The so-called Jost function $D_l(k)$ is an analytic function of the complex variable k under very general conditions. This can be used for the study of scattering resonance states, whose complex energies are obtained from $D_l(k) = 0$ and are of the form $k_r = \kappa_r - i\lambda_r$ for a resonance r.

The integral equation for $y_l(R; k)$ can be solved by iteration to generate a Born–von Neumann series expansion. In terms of operators, using the bracket notation and the coordinate representation, with $y_l(R) = \langle R \mid y_l \rangle$ and $G_{0,l}(R, R') = \langle R|\hat{G}_{0,l}|R'\rangle$, we can write the integral equation and its iterative expansion as

$$y_l = u_l + \hat{G}_{0,l} U y_l$$

$$y_l = \left[1 + \sum_{n=1}^{\infty} (\hat{G}_{0,l} U)^n\right] u_l$$

which is an expansion of the solution in powers of the partial-wave strength $\eta_l = ||\hat{G}_{0,l} U||$ and gives at large R an expansion of the coefficients $a_l(k)$ and $b_l(k)$ in powers of the strength. This strength can be estimated for a given potential energy to ascertain the accuracy of the power expansion. Its magnitude depends on the combined effects of the potential energy value at each distance and the value of the free-motion wavefunctions u_l and v_l at that distance.

The radial wavefunction for a general potential with a hardcore and variable energy at larger distances may be treated as before, but with the radial component $u_l(kR)$ replaced by $y_l^{(c)}(R; k)$ as given above.

2.3.4 Resonance Energies and Angular Momenta

Details of resonance scattering can be described by means of the Jost function $D_l(k)$ [7, 29]. Fixing l and varying k, one can locate the so-called Breit–Wigner resonance energies, while fixing k and varying l gives the Regge resonances [7, 29]. Rewriting the asymptotic radial function as

$$y_l^{(+)}(R; k) \approx [a_l(k) - ib_l(k)]u_l(kR) + b_l(k)w_l^{(+)}(kR) \text{ as } R \to \infty$$

with $D_l(k) = a_l(k) - ib_l(k)$ and insofar $w_l^{(+)}(x) \approx \exp\left[+i\left(x - \frac{l\pi}{2}\right)\right]$, we can identify states that grow asymptotically with distance by looking for roots of the equation $D_l(k) = 0$ with values $k_\rho = \kappa_\rho - i\lambda_\rho$, $\lambda_\rho \geq 0$, for resonance quantum numbers $\rho = (l, r)$. Expanding $y_l^{(+)}(R; k)$ as a function of k around k_ρ, this gives

$$y_l^{(+)}(R; k_\rho) \approx b_\rho w_l^{(+)}(k_\rho R) \approx b_\rho i^{-l} \exp(i\kappa_\rho R) \exp(\lambda_\rho R)$$

showing an oscillating function with exponentially increasing amplitude.

To understand the physical meaning of this radial function, it helps to introduce its time dependence by forming a wavepacket of energy $E = \hbar^2 k^2/2M$ as

$$\chi_l(R, t; k) = y_l^{(+)}(R; k) \exp(-iEt/\hbar)$$

and to expand the energy function of k around k_ρ to obtain

$$E = E_\rho + \mathcal{O}(|k - k_\rho|) = E'_\rho - i\Gamma_\rho/2$$

where E'_ρ and $\Gamma_\rho = (\hbar^2/2M)4\kappa_\rho\lambda_\rho$ are real valued. This gives asymptotically

$$\chi_l(R,t;k_\rho) \approx b_\rho i^{-l} \exp\left[i\left(\kappa_\rho R - \frac{E'_\rho t}{\hbar}\right)\right] \exp\left(\lambda_\rho R - \frac{\Gamma_\rho t}{2\hbar}\right)$$

and shows a wavefunction amplitude changing with position and time so that, at its maximum, increments satisfy $\lambda_\rho dR = (\Gamma_\rho/2\hbar)dt$. This in turn means that the wavepacket moves with its maximum at a constant speed $dR/dt = \Gamma_\rho/(2\hbar\lambda_\rho)$. Furthermore, from $|\chi_l|^2 \approx \exp(-\Gamma_\rho t/\hbar)$ as $R \to \infty$, it is clear that the state probability decreases with a decay time $\tau_\rho = \hbar/\Gamma_\rho$, which identifies Γ_ρ as an energy width for decay and $\Gamma_\rho/\hbar$ as a decay rate.

The implications for properties of differential cross sections can be found returning to the partial scattering amplitude. It is also given by $f_l(k) = b_l(k)/D_l(k)$ with $b_l(k) = [D_l(k) - D_l(k)^*]/(2i) = -Im[D_l(k)]$. Expanding numerator and denominator around $k = k_\rho$ and keeping lowest order terms give, in the partial and angular scattering amplitudes, the resonance plus background terms are

$$f_l(k) = f_l^{(res)}(k) + f'_l(k),\ \ f_l^{(res)}(k) = -\lambda_\rho/(k - \kappa_\rho + i\lambda_\rho)$$

$$f_k(\Theta) = f_k^{(res)}(\Theta) + f'_k(\Theta),\ \ f_k^{(res)}(\Theta) = k^{-1}\sum_{l\geq 0}(2l+1)P_l(\cos\Theta)f_l^{(res)}(k).$$

The differential cross section $d\sigma/d\Omega = (d\sigma/d\Omega)^{(res)} + (d\sigma/d\Omega)' + (d\sigma/d\Omega)^{(int)}$ follows from this as a sum over partial waves. However, for values of κ_ρ very close to k, and small λ_ρ (meaning narrow energy width Γ_ρ at the energy E'_ρ), there is one resonance partial amplitude dominant in the summation for an orbital quantum number l_ρ. This shows in the angular distribution, which displays a maximum or a minimum in angles around $\Theta = \pi/2$, respectively, for even or odd l_ρ.

Phase shifts can also be obtained around a resonance from the Jost function insofar

$$S_l(k) = \exp(2i\delta_l) = D_l(k)^*/D_l(k)$$

for all k, and using again an expansion around $k = k_\rho$ in numerator and denominator, solving for the phase shift gives now

$$\delta_l(k) = \delta_l^{(res)}(k) + \delta_l^{\prime(k)}, \delta_l^{(res)}(k) = \tan^{-1}[\lambda_\rho/(k - \kappa_\rho)].$$

This shows that for a chosen energy or wavenumber $k = \kappa_\rho$, the phase-shift tangent goes through infinity at an orbital quantum number l_ρ, for which the phase shift jumps by a magnitude of π.

Other aspects of resonance scattering can be found in the literature [7, 29] and also in related treatments of so-called Siegert states [30] or using a wave-operator formalism [31] and also using analytical properties of the wavefunction in complex-valued coordinates.

2.4 Numerical Methods for Scattering

2.4.1 Numerical Procedures for Differential Equations

Scattering radial wavefunctions $y_l(R; k)$ satisfy a linear second-order differential equation which can be solved numerically by a variety of procedures. This must be done while satisfying boundary conditions at the origin $R = 0$ and at large distances. For strong repulsive atomic cores, the potential energy function $V(R)$ has large positive values at short distances, and the wavefunction decreases exponentially at distances smaller than the inner classical turning point $R_{L,t}$, where $V_L(R) = E$. At large distance, past any centrifugal energy barriers, the wavefunction oscillates with an amplitude which is constant (for steady scattering) or exponentially increasing (for quasibound decay) amplitude and provides phase shifts from which cross sections can be calculated. Numerical procedures for solving the scattering equations rely on relative distance step-by-step algorithms or on potential energy piecewise expansions and are briefly described as follows.

For a typical potential energy with parameters R_m, ϵ, a differential equation with dimensionless quantities is

$$\left[\frac{d^2}{dx^2} + A^2 - Bg(x) - \frac{l(l+1)}{x^2}\right] y_l(x) = 0$$

with $x = R/R_m$, $g(x) = V(R)/\epsilon$, $A^2 = k^2 R_m{}^2$, and with $B = 2M\epsilon R_m{}^2/\hbar^2$, being a measure of the potential-well strength. To impose boundary conditions, an inner classical turning point x_t is identified, and calculations are started at a distance $x_{in} = x_t - 0.5$ or, where $Bg(x_{in}) + l(l+1)/x_{in}{}^2 \geq 10A^2$, with $y_l(x_{in}) = 0$ and $(dy_l/dx)_{in} = 10^{-10}$ or with a value obtained from a short wavelength approximation.

Step-by-step procedures are helpful when the wavelength λ of a wavefunction $y_l(x)$ is large compared with a typical potential energy distance R_m so that relatively few increments of x are needed for propagation from initial to asymptotic values. This is the case for collisions in gases at very low temperatures (like mK-s) for near threshold photodissociation and for collisions of electrons.

Outward integration can be done by steps at $x_{n+1} = x_n + h$ with increments $h \leq (\lambda/R_{m)})/16$ obtained from the wavelength λ, that can be decreased to test accuracy. A variety of algorithms have been used to advance the solution, with an efficient procedure for the scattering differential equation provided by the de Vogelaere algorithm, advantageous for differential equations without a first derivative, of form $d^2y/dx^2 = f(x, y)$ [32]. This proceeds from (y_n, y_n') to $(y_{n+1}, y_{n+1}{}')$ through $y_{n+1/2}$, with $y' = dy/dx$, as follows. Writing

$$y_s = y(x + sh),\ \ y_s' = (dy/dx)_{x+sh},\ \ f_s = f(x + sh, y),$$

the algorithm to advance from y_0 to y_1 is, within an error of order h^4,

$$y_{1/2} = y_0 + \frac{h}{2}y'_0 + \frac{h^2}{24}(4f_0 - f_{-1/2})$$

$$y_1 = y_0 + hy'_0 + \frac{h^2}{6}(f_0 + f_{1/2})$$

$$y'_1 = y'_0 + \frac{h^2}{6}(f_0 + 4f_{1/2} + f_1).$$

To start at any x_0, one constructs $y_{-1/2} = y_0 - (h/2)y'_0 + (h^2/8)f_0$. Other details and tests may be found in reference [32]. An alternative algorithm is the Numerov method, frequently used for electronic and atomic collisions, also with an error of order h^4 [33].

Phase shifts can be extracted at a large final distance x_{fn} such that $Bg(x_{fn}) \ll [A^2 - l(l+1)/x_{fn}{}^2]$. They can be extracted from the zeros $x_{l,p}$ of $y_l(x)$ past x_{fn}. From $y_l^{(K)}(x;A) \approx u_l(Ax) + K_l(k)v_l(Ax)$ in terms of the Riccati functions and with $K_l(k) = \tan[\delta_l(k)]$, it follows that

$$\delta_l(k) = -\tan^{-1}\left[\frac{u_l(Ax_{l,p})}{v_l(Ax_{l,p})}\right].$$

Otherwise, the radial function written also as $y_l(x;A) \approx a_l\, u_l(Ax) + b_l v_l(Ax)$ can be obtained at x_{fn} extracting its constant coefficients using Wronskians $W[f,g;x] = f(x)g'(x) - f'(x)g(x)$, with $W[u_l, v_l; x] = constant$, from y_l and $y'_l = dy_l/dx$, as

$$a_l = \frac{W[y_l, v_l; x_{fn}]}{W[u_l, v_l; x_{fn}]}, \quad b_l = \frac{W[y_l, u_l; x_{fn}]}{W[u_l, v_l; x_{fn}]}.$$

An alternative to step-by-step calculations is a procedure which constructs the radial solution in a piecewise fashion [34]. It avoids the need for very small step increments, found to be computationally very time consuming in propagation when wavefunction wavelengths are short. It instead constructs local solutions to local potential energy functions starting from

$$\left[\frac{d^2}{dR^2} + k^2 - U_l(R)\right]\psi_l^{(+)}(R;k) = 0$$

with $U_l(R) = 2MV(R)/\hbar^2 - l(l+1)/R^2$. This effective potential function, for each chosen l, is divided into pieces within left and right boundaries $[R_l, R_r]$ and is approximated locally by a $U_0(R)$ function, omitting l here for convenience, which can be constant, linear, or quadratic in R for convenience and accuracy. Two local linearly independent solutions $A(R)$ and $B(R)$ are introduced with Wronskian $W[A, B; R] = A(R)\,(dB/dR) - (dA/dR)B(R)$ satisfying $dW/dR = 0$ as follows from the differential equation:

$$\left[\frac{d^2}{dR^2} + k^2 - U_0(R)\right]F(R) = 0, F(R) = A(R) \text{ or } B(R).$$

Within the chosen interval, one can write $\psi_l^{(+)}(R;k) = \psi(R)$ as

$$\psi(R) = a(R)A(R) + b(R)B(R)$$

with slowly varying coefficients $a(R)$ and $b(R)$ chosen to satisfy $a'(R)A(R)+b'(R)B(R) = 0$ and with primes indicating R-derivatives. From the above ψ and from $\psi'(R) = a(R)A'(R)+b(R)B'(R)$, one can solve for $a(R)$ and $b(R)$ in terms of ψ and ψ'. Further differentiating $a(R)$ and $b(R)$ and using the differential equations for $A(R)$ and $B(R)$, one arrives at the coupled equations

$$a' = -W^{-1}B(U - U_0)(Aa + Bb)$$

$$b' = W^{-1}A(U - U_0)(Aa + Bb)$$

providing two coupled first-order equations for the varying coefficients. Insofar as these are slowly varying with $(U - U_0)$, it is possible to obtain accurate solutions by perturbation expansions in powers of $(U - U_0)/k^2$. Integrating both sides over $R_l \leq R' \leq R_r$ and assuming that the coefficients are known at R_l, the answers at R_r to first order are

$$a_r = a_l - I_{BA}a_l - I_{BB}b_l$$

$$b_r = b_l + I_{AA}a_l + I_{AB}b_l$$

where

$$I_{FG} = \int_{R_l}^{R_r} dR'[W^{-1}F(U - U_0)G]_{R'}.$$

When the sector potential U_0 is chosen to be constant over R, the functions F and G are simply trigonometric, and the integrals are readily performed.

Other treatments of the differential equation for scattering have been described and implemented, in particular the variable phase method which generates a function $\delta_l(R)$ [35] the log-derivative method to directly calculate $z_l(x) = (dy_l/dx)/y_l(x)$ [36] and the **R**-matrix method which relates a function $y_l(x_{l,r})$ and its derivative $(dy_l/dx)_{l,r}$ at left and right positions $x_{l,r}$ of a distance sector i and propagates the matrix $\mathbf{R}(i)$ from a sector to the next one [37, 38]. These three are efficient improvements over the step-by-step procedures, especially as they avoid the need for small step increments in the generation of scattering functions with short wavelength, and they also avoid numerical stability issues that arise from the presence of exponentially increasing amplitudes in classically forbidden regions of the potential energy vs x.

2.4.2 Numerical Procedures for Integral Equations

2.4.2.1 Stepwise Solutions of the Volterra Integral Equation

Parallel developments have been made using integral equations, which are usually more stable for propagation treatments. The real-valued solution for the radial scattering function $\psi_l^{(K)}(R;k)$ given by the Volterra integral equation

$$y_l(R;k) = u_l(kR) + \int_0^R dR' G_{0,l}(R,R')U(R')y_l(R';k)$$

provides a straightforward numerical procedure for numerical solutions using stepwise propagation of $y_l(R;k)$ as R increases. The integral can be replaced by a quadrature sum on a set of points R_n, $n = 1$ to N, with quadrature weights w_n, chosen so that $R_0 = 0$ and $R_N = R$. It would appear as if $y_l(R_N;k)$ to the left would also be present to the right in $R' = R_N$, which would require solving for its value at each needed $R_N = R$.

However, it has been pointed out that in this Volterra integral equation with a factored Green function $G_{0,l}(R_n, R'_{n'})$, the upper-most quadrature terms with R_N to the right are found to cancel, so that the summation ends at $n = N - 1$ and $y_l(R_N;k)$ to the left is given without further treatment [28]. This provides a very simple and stable procedure for generation of $y_l(R;k)$, which can also be used in regions where the function changes exponentially. It takes the form

$$y_l(R_N;k) = u_l(kR_N) + \sum_{n=1}^{n=N-1} G_{0,l}(R_N, R'_n)U(R'_n)y_l(R'_n;k)w_n$$

and requires generation of the Riccati functions $u_l(kR_n)$ and $v_l(kR_{n'})$ at the quadrature points, with small steps so as to capture the radial function oscillations with R.

2.4.2.2 Piecewise Solutions

An alternative treatment also based on this integral equation again replaces stepwise summations with piecewise components [39, 40]. As done for the differential equation piecewise treatment, the effective potential function is divided into pieces within left and right boundaries $[R_l, R_r]$ and is approximated locally by a $U_0(R)$ function which can be constant, linear, or quadratic in R, for convenience and accuracy. Two local linearly independent solutions $F(R) = A(R)$ and $B(R)$ are introduced, omitting here the subindex l for now, which satisfy

$$\left[\frac{d^2}{dR^2} + k^2 - U_0(R)\right] F(R) = 0, \quad F(R) = A(R) \text{ or } B(R)$$

and are used to construct a local Green function as

$$G_{0,loc}(R,R') = k^{-1}[A(R)B(R') - A(R')B(R)]$$

For a constant U_0 in $[R_l, R_r]$, these are simply sin and cos functions. The integral with $y(R')$ for $0 \leq R' \leq R$ can be partitioned into sectors, where $R_l \leq R' \leq R_r$, and can be written in each sector as

$$y(R') = aA(R') + bB(R')$$

with coefficients a and b obtained from $y(R)$ and dy/dR at R_l. Replacing this and the local Green function in the local integral equation,

$$y(R) = y(R_l) + \int_{R_l}^{R} dR' G_{0,loc}(R, R') U_0(R') y(R')$$

and letting $R = R_r$ to obtain $y(R_r)$, each sector is found to contain integrals of type

$$I_{FG} = \int_{R_l}^{R_r} dR' F(R') U_0(R') G(R')$$

with $F(R')$ and $G(R')$ either $A(R)$ or $B(R)$ [39]. The solution for $y(R)$ is a sum over sectors from an initial $R_l = R_{in}$ up to a large $R_r = R$. The integrals are simple analytical expressions for constant piecewise sectors. This treatment decreases the number of terms appearing in integrals to the right and also avoids having to generate Riccati functions at quadrature points. It also shows stability when exponential functions appear locally due to the presence of classically forbidden regions of the potential energy vs R at some total energies E [40].

2.4.2.3 Expansion of Amplitude Densities

An alternative to the piecewise expansion in the integral equation for the wavefunction $y(R)$ is to instead introduce an integral equation for the amplitude density $\xi_l(R) = U(R)y_l(R; k)$ and solve for it first [41]. This new function goes asymptotically to zero as $U(R)$ does and also is zero at some initial R_{in}. It is usually square-integrable and can be expanded in an orthonormal basis set $\{\eta_n(R), n = 1 \text{ to } N_B\}$ of L^2 functions. The integral equation

$$\xi_l(R) = U(R)u_l(kR) + \int_0^R dR' G_{0,l}(R, R') \xi_l(R') = \sum_n C_n^{(l)} \eta_n(R)$$

then leads to an algebraic equation $C_n^{(l)} = A_n^{(l)} + \sum_{n'} B_{nn'}^{(l)} C_{n'}^{(l)}$ where

$$A_n^{(l)} = \int_0^{\infty} dR\, \eta_n(R) U(R) u_l(kR)$$

$$B_{nn'}^{(l)} = \int_0^{\infty} dR\, \eta_n(R) \int_0^R dR' G_{0,l}(R, R') \eta_{n'}(R')$$

Once the expansion coefficients $C_n^{(l)}$ have been obtained, they can be used to construct scattering amplitudes from the previous Section 2.3.3 as

$$K_l = -k^{-1} A_l \int_0^{\infty} dR' u_l(kR') \xi_l(R')$$

$$A_l = 1 \Big/ \left[1 + k^{-1} \int_0^{\infty} dR\, v_l(kR) \xi_l(R) \right]$$

for standing waves or

$$f_l(k) = -k^{-1} \int_0^\infty dR\, u_l(kR)\xi_l(R)/D_l(k)$$

$$D_l(k) = 1 + k^{-1} \int_0^\infty dR\, w_l^{(+)}(kR)\xi_l(R)$$

for outgoing waves. The basis set can be chosen, given $U(R)$, to minimize the number of expansion coefficients or as a set of trigonometric functions to simplify the above integrals.

2.4.2.4 Improvements by a Variational Procedure

The Kohn variational method is based on a variational principle of scattering theory, which provides the scattering amplitude as a stationary functional of a trial scattering wavefunction, and gives a way to improve on a chosen scattering solution [1, 42]. The variational principle can be stated for the transition amplitude $T_{\vec{k}'\vec{k}}$ and alternatively for the K-matrix or S-matrix amplitudes.

For numerical calculations, it is convenient to implement it as in the Kohn variational method, with a functional $\mathcal{T}_{\vec{k}'\vec{k}}[\tilde{\psi}]$ of a $\tilde{\psi}$ trial function, providing a stationary value for the transition amplitude, to be found variationally. Its form is

$$\mathcal{T}_{\vec{k}'\vec{k}}[\tilde{\psi}] = \left\langle \tilde{\psi}_{\vec{k}'}^{(-)} | (E - \hat{H}) | \tilde{\psi}_{\vec{k}}^{(+)} \right\rangle$$

and its variation, obtained from $\delta\mathcal{T}_{\vec{k}'\vec{k}} = \mathcal{T}_{\vec{k}'\vec{k}}[\tilde{\psi} + \delta\tilde{\psi}] - \mathcal{T}_{\vec{k}'\vec{k}}[\tilde{\psi}]$, satisfies [43]

$$\delta\left(\mathcal{T}_{\vec{k}'\vec{k}} - T_{\vec{k}'\vec{k}}\right) = \mathcal{O}[\delta\tilde{\psi}]^2$$

so that a variation $\delta\tilde{\psi}$ of the trial function in the functional gives the transition amplitude with an error of second order in the variation. This variational principle and related ones are derived in the following chapter on the formalism of scattering phenomena. The above variational result can be used to calculate an improved scattering function starting with a good trial function so that $\psi = \tilde{\psi} + \delta\tilde{\psi}$ and $T_{\vec{k}'\vec{k}} = \tilde{T}_{\vec{k}'\vec{k}} + \delta T_{\vec{k}'\vec{k}}$ to first order in the trial function accuracy. Then insofar $\delta\mathcal{T}_{\vec{k}'\vec{k}} = \mathcal{T}_{\vec{k}'\vec{k}}[\psi] - \mathcal{T}_{\vec{k}'\vec{k}}[\tilde{\psi}] = -\mathcal{T}_{\vec{k}'\vec{k}}[\tilde{\psi}]$, we find that $\delta T_{\vec{k}'\vec{k}} = T_{\vec{k}'\vec{k}} - \tilde{T}_{\vec{k}'\vec{k}} = -\mathcal{T}_{\vec{k}'\vec{k}}[\tilde{\psi}]$ and

$$T_{\vec{k}'\vec{k}} = \tilde{T}_{\vec{k}'\vec{k}} - \mathcal{T}_{\vec{k}'\vec{k}}[\tilde{\psi}] + \mathcal{O}[\psi - \tilde{\psi}]^2.$$

A practical way to construct a good trial function is to parametrize it with a set of variables $\{\lambda_j, j = 1\ to\ N\}$ plus a variable amplitude $\tilde{T}$ so that $\mathcal{T}_{\vec{k}'\vec{k}}[\tilde{\psi}] = F_{\vec{k}'\vec{k}}(\{\lambda_j\}, \tilde{T})$. One procedure then requires that $F_{\vec{k}'\vec{k}}(\{\lambda_j\}, \tilde{T}) = 0$ is solved to give a transition amplitude $\tilde{T}_{\vec{k}'\vec{k}}(\{\lambda_j\})$ to $\mathcal{O}[\delta\tilde{\psi}(\{\lambda_j\})]^2$. A second procedure, somewhat simpler numerically, imposes that the first derivatives of the function $\mathcal{T}_{\vec{k}'\vec{k}}[\tilde{\psi}] = F_{\vec{k}'\vec{k}}(\{\lambda_j\}, \tilde{T})$ vanish with respect to variation of its parameters. The $N+1$ equations $\partial F_{\vec{k}'\vec{k}}/\partial\lambda_j = 0$ and $\partial F_{\vec{k}'\vec{k}}/\partial\tilde{T}=0$ give solutions for the $N+1$ parameters and an improved $T_{\vec{k}'\vec{k}} = \tilde{T}_{\vec{k}'\vec{k}} - F_{\vec{k}'\vec{k}}(\{\lambda_j\}, \tilde{T})$ accurate to first order in parameter variations.

These developments can also be adapted to calculations of improved radial scattering functions $\psi_l(R;k)$ and related phase shifts $\delta_l(k)$ after partial wave analysis [7]. Following the notation in Section 2.3.1, a possible choice is

$$\tilde{\psi}_l^{(\pm)}(R;k) = u_l(kR) + \tilde{f}_l(R;k)w_l^{(\pm)}(kR) + \sum_{n=1}^{N} \tilde{c}_{l,n}(k)\eta_n(R)$$

where the shape of the trial wavefunction in an inner region is expanded in an orthonormal basis set $\{\eta_n(R)\}$, with coefficients $\tilde{c}_{l,n}$ to be calculated, and the basis set and $\tilde{f}_l(R;k)$ are chosen so that $\tilde{\psi}_l^{(\pm)}(0;k) = 0$. Replacing these trial expressions in the transition functional for each partial wave and integrating over R gives an expression $\mathcal{T}[\tilde{\psi}_l] = F_l(\{\tilde{c}_{l,n}\}, \tilde{f}_l)$ bilinear in the set $\{\tilde{c}_{l,n}\}$. Solving for this set, with one of the two procedures described above, gives an improved transition amplitude.

2.4.3 Numerical Procedures for Semiclassical Scattering

In the case of scattering with a general potential energy function $V(\vec{R})$, trajectories must be numerically generated solving for the classical dynamics. A numerical procedure for a semiclassical treatment starts with the generation of bundles of trajectories by integration of equations of motion for the six variables $\vec{R}(t)$ and $\vec{P}(t)$, with a chosen set of initial conditions in phase space [44], given by $(\{\vec{P}(t_{in}), \vec{R}(t_{in})\}_s = \{(\vec{P}_s, \vec{R}_s), s = 1 \text{ to } N_{in}\}$, a set of N_{in} points in phase space chosen to reconstruct the initial wavefunction. The set of all the evolved trajectories over time form a moving bundle in phase space, with $\vec{R}(t;\vec{P}_s,\vec{R}_s)$ and $\vec{P}(t;\vec{P}_s,\vec{R}_s)$ constructed to satisfy the initial conditions, for a constant total energy $H_{qu}(\vec{P},\vec{R}) = E$. Changing variables from $\vec{R}$ to a set (t, b, φ) containing time and two space variables for an atom–atom collision and choosing initial conditions to define a bundle of trajectories to be propagated in time for each initial phase space point $(\vec{P}_s, \vec{R}_s)$, the eikonal wavefunction phase and amplitude can be obtained solving their differential equations in the variables $(t,\ \rho,\ \varphi)$. This must be done in principle solving for the Hamiltonian equations with the quantal potential V_{qu}. But, in practice, semiclassical calculations are done in the limit of short wavelengths as described in Section 2.2.3, with $V_{qu} \cong V$.

The trajectory Hamiltonian equations for all six variables $(\vec{P}, \vec{R})$ in a set $\mathbf{Y} = (Y_1, \ldots, Y_6)$ are formally given by

$$d\mathbf{Y}/dt = \mathbf{F}[\mathbf{Y}(t)]$$

with $\mathbf{F} = (F_1, \ldots, F_6)$ the set of Hamiltonian gradients and can be solved by a numerical algorithm starting with initial conditions $\mathbf{Y}_{in} = \mathbf{Y}(t_{in})$. The algorithm can be a single-step procedure of the Runge–Kutta type, with $\mathbf{Y}(t + \Delta t)$ at a time increment Δt, obtained from $\mathbf{Y}(t)$ and its time derivatives. This is a self-starting procedure, which however requires generation of derivatives of $\mathbf{F}$ with respect to $\mathbf{Y}$, a computationally inefficient step. It can be avoided using instead a finite quadrature formula such as

$$\mathbf{Y}(t_n + \Delta t) = \mathbf{Y}(t_n) + \Delta t \sum_{i=0}^{r-1} w_i \, \mathbf{F}[\mathbf{Y}(t_n + \zeta_i \Delta t)]$$

where w_i and ζ_i are weights and locations for a quadrature of order r in Δt [44] to advance a time step after t_n. A choice of parameters for a single-step procedure is $\zeta_i = i/(r-1)$, with the weight w_i obtained from the time-derivative expression to the desired order.

An alternative is a multistep procedure, where $\mathbf{Y}(t_n + \Delta t)$ is obtained from $\mathbf{Y}(t_{n-p})$ for $p = 0$ to m. A frequently used procedure is a Verlet algorithm, which employs relative position, velocity, and acceleration $\vec{a} = -M^{-1}\partial V/\partial \vec{R}$ of functions $\vec{R}(t), \vec{v}(t)$, and $\vec{a}(t)$ in a sequence

$$\begin{aligned}
\vec{R}(t + \Delta t) &= \vec{R}(t) + \Delta t \, \vec{v}(t) + (\Delta t)^2 \vec{a}(t)/2 \\
\vec{v}(t + \Delta t/2) &= \vec{v}(t) + \Delta t \, \vec{a}(t)/2 \\
\vec{v}(t + \Delta t) &= \vec{v}(t + \Delta t/2) + \Delta t \, \vec{a}(t + \Delta t)/2,
\end{aligned}$$

which allows for testing of energy conservation at $t + \Delta t$. This procedure and its improvements are usually followed in the molecular dynamics simulation of liquids [45]. The force $-\partial V/\partial \vec{R}$ can be obtained by analytical differentiation of a potential energy function or it can be obtained numerically at time $t_n + \Delta t$ from its values along a trajectory at times t_{n-p} for $p = 0$ to m.

An alternative general treatment of multistep procedures has been developed for the Hamiltonian equations using a Liouvillian propagator and the Trotter factorization of its exponential, which leads to reversible multiple time-scale dynamics [46]. Multistep treatments can further be improved with predictor–corrector procedures. These procedures must be chosen to provide a desired accuracy and to be computationally efficient.

Choosing desired initial conditions and a self-starting integration procedure, the trajectories with $\vec{R}(t,\rho,\varphi), \vec{P}(t,\rho,\varphi)$ can be generated and from them the phase and amplitude of the eikonal wavefunction $\psi^{(+)}_{\vec{k}}(t,\rho,\varphi) = (2\pi)^{-3/2} \exp [F_{\vec{k}}(t,\rho,\varphi) + iS_{\vec{k}}(t,\rho,\varphi)/\hbar]$ in the limit of short wavelengths. Transition amplitudes then follow from

$$T_{\vec{k}'\vec{k}} = (2\pi)^{-3} \int \rho d\rho d\varphi dt \, J\begin{pmatrix} R, \Theta, \Phi \\ t, \rho, \varphi \end{pmatrix} \mathrm{e}^{i\vec{k}'\cdot\vec{R}\,(t,\rho,\varphi)} \, V[\vec{R}(t,\rho,\varphi)] e^{F_{\vec{k}}(t,\rho,\varphi)+iS_{\vec{k}}(t,\rho,\varphi)/\hbar}$$

by numerical integration. This has been done for elastic scattering in a Lennard–Jones (12-6) potential energy function with its potential well value ε parametrized for the systems Hg + H_2 and Hg + Li and with results obtained for differential cross sections $d\sigma/d\Theta$ at a reduced energy $E/\varepsilon = 2.6$. Comparisons with accurate results, generated by adding partial waves from numerical integration of radial differential equations, show very good agreement over a wide range of scattering angles [24].

2.4.4 Extraction of Potential Functions by Inversion

Accurate calculations of differential cross sections vs scattering angle for a range of collision energies produce very detailed results when using parametrized potential energy functions, that can be compared with high-resolution experimental measurements. Numerical values depend on the quality of the interaction potential energy, and cross section vs angles and velocities are very sensitive to the potential energy values over the whole range of distances. This allows extraction of the potential functions by fitting of potential function parameters in an inversion procedure [3].

As an example, the high-resolution measured differential cross sections $I(\Theta, E)$ $\sin(\Theta)$ for Na–Hg vs scattering angle Θ and for chosen collision energies E, previously shown in Figure 2.4, have been reproduced with an accurate potential energy function. Details are fully obtained with a potential energy function shown in Figure 2.7, which has been constructed in piecewise form fitting separately short-range, intermediate, and long-range components [5]. This illustrates a numerical inversion procedure from scattering cross sections to the potential energy function.

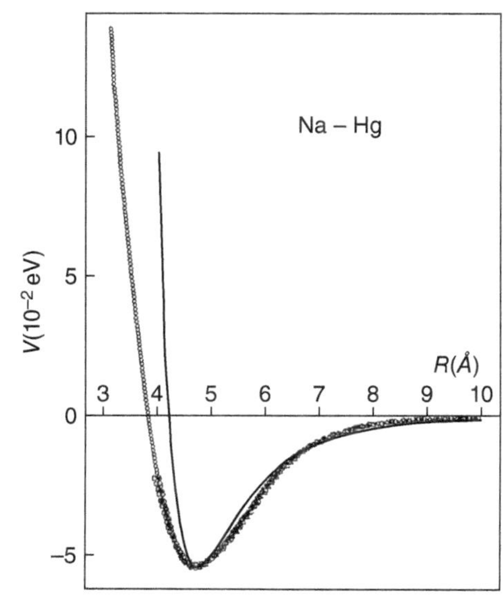

Figure 2.7 Potential energy function for Na–Hg vs radial distances constructed in piecewise form fitting separately short-range, intermediate, and long-range components. Source: [5] / John Wiley & Sons.

2.5 Examples

2.5.1 Atom–Atom Collisions

For two closed shell atoms such as Ar–He, we have a strongly repulsive short-range interaction of closed electronic shells and a long-range attraction due to induced dipole-induced dipole attraction (or Van der Waals attraction), with an energy minimum in between. We can choose its analytical form as the Lennard–Jones (12–6) function

$$V(R) = \epsilon \left[\left(\frac{R}{R_m} \right)^{-12} - 2 \left(\frac{R}{R_m} \right)^{-6} \right]$$

where typically the potential energy well depth is $\epsilon \cong 0.1$ eV, at its minimum location $R_m \cong 0.5$ nm, and the kinetic energy $\hbar^2 k^2/(2M)$ involves a reduced mass $M \cong 10$ amu. At a thermal collision energy with a wavenumber k_{th} and $l_{max} \approx k_{th} 2R_m \gg 1$, gives δ_l as an almost continuous function of the orbital quantum number l.

The phase shifts can be obtained from the asymptotic form of the radial wavefunction $\psi_l^{(+)}(R;k)$ generated by numerically solving its differential or integral equation for all relevant l. It is shown in sketches of Figure 2.8 for the phase shift vs orbital quantum number l and values of the parameter $A = kR_m$ [3, 5, 6], typical of results for He–Ar elastic collisions under a L–J (12-6) potential. At large A = 40, the phase shifts show a maximum at the glory l_g and at the rainbow l_r values. At small A = 10, the phase shift discontinuity indicates that the L–J potential

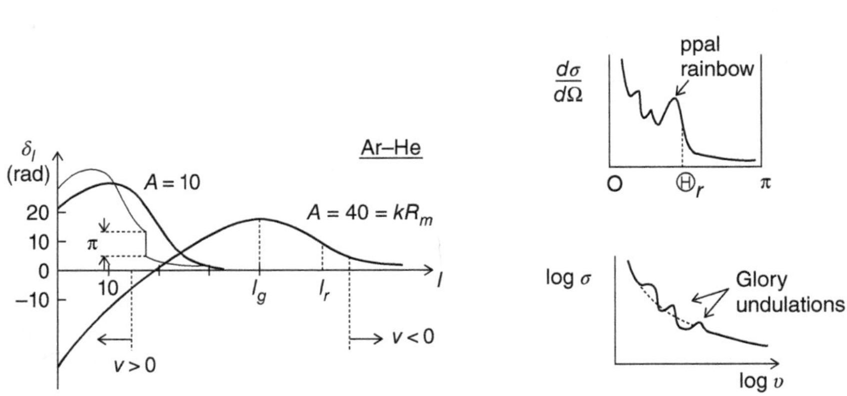

Figure 2.8 Sketches of the phase shift δ_l vs orbital quantum number l and chosen reduced wavevectors $A = kR_m$, for a pair such as Ar + He, generated for a LJ(12-6) potential energy. Panels to the right give a typical differential cross section $d\sigma/d\Omega$ vs scattering angle, and an integral cross section $\sigma(\upsilon)$ vs collision velocity.

holds a quasibound (resonance or long-lived) state. The region around the maximum l_g relates to the "glory" oscillations in the integral cross section $\sigma(v)$, and the region around the inflection point l_r relates to the "rainbow" structure of the differential cross section $I(\Theta, E)$. Panels to the right sketch the differential cross section $I(\Theta, E) = d\sigma/d\Omega$ vs scattering angle Θ showing principal and supernumerary rainbow oscillations and also the integral cross section vs scattering velocity in a log–log plot displaying glory oscillations.

2.5.2 Electron Scattering by Atoms

The scattering potential $v(r)$ can be obtained from the radial electronic charge density $\rho(r)$ of the atom for an electron at distance r from the nucleus, satisfying Poisson's equation in atomic units,

$$\nabla^2 v(r) = -4\pi\rho(r).$$

Take for example electron scattering by He(ls^2) in its Hartree–Fock state that we assume remains unchanged. Using for the electron orbital the simple function $\varphi_{1s}(r) = [(\zeta/a_B)^3/\pi]^{1/2} \exp(-\zeta r)$ with an exponential value $\zeta = 1.688$ au and density $\rho(r) = \varphi_{1s}(r)^2$, the averaged electrostatic interaction is

$$v(r) = -\left(\frac{2}{r} + \zeta\right) \exp(-2\zeta r)$$

which gives rise to only a few phase shifts at low velocities ($k \leq 2$ au), as shown in Figure 2.9 vs the wavevector k [7, 47].

These results from a static potential $v(r)$ however are only qualitatively correct and must be improved by incorporating into the effective potential energy contributions from electron exchange between projectile and target electrons and from electronic polarization of the target [47].

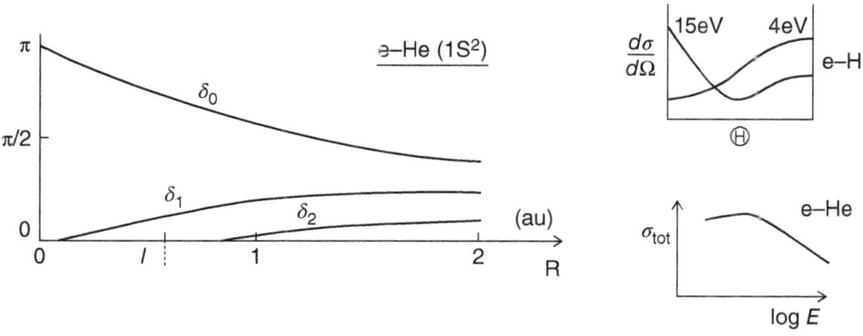

Figure 2.9 Sketches of phase shifts δ_l, $l = 0, 1, 2$, vs the relative scattering wavevector k for e$^-$ + He(ls^2). Panels to the right show typical differential cross sections vs angle for two collision energies and the integral cross section vs collision energy.

References

1 A. Messiah, *Quantum Mechanics*, Volume 1, Amsterdam: North-Holland, 1962.

2 M. S. Child, *Molecular Collision Theory*, New York: Academic Press, 1974.

3 H. Pauly, "Elastic scattering cross sections I: spherical potentials," in *Atom-Molecule Collision Theory*, (Editor: R. B. Bernstein), New York: Plenum Press, 1979, pp. 111–191.

4 D. A. Micha, *Molecular Interactions: Concepts and Methods*, Hoboken, NJ, USA: Wiley, Inc., 2020.

5 U. Buck, "Elastic scattering," *Adv. Chem. Phys.*, vol. 30, p. 313, 1975.

6 R. B. Bernstein, "Quantum effects in elastic molecular scattering," in *Advances in Chemical Physics: Molecular Beams*, Volume 10, (Editor: J. Ross), New York: Wiley, 1966, pp. 75–134.

7 B. H. Bransden, *Atomic Collision Theory*, Reading, MA, USA: Benjamin-Cummings, 1983.

8 J. P. Toennies, W. Welz and G. Wolf, "Molecular beam scattering studies of orbiting resonances and the determination of van der Waals potentials for H-Ne, Ar, Kr, and Xe and for H2-Ar, Kr, and Xe," *J. Chem. Phys*, vol. 71, p. 614, 1979.

9 E. J. Heller, "Time-dependent approach to semiclassical dynamics," *J. Chem. Phys.*, vol. 62, p. 1544, 1975.

10 D. J. Tannor, *Introduction to Quantum Mechanics: A Time-dependent Perspective*, New York: University Science Books, 2004.

11 E. Deumens and Y. N. Ohrn, "Electron-nuclear dynamics with diabatic and adiabatic wave packets," *J. Phys. Chem.*, vol. 92, p. 3181, 1988.

12 R. P. Feynman and A. R. Hibbs, *Quantum Mechanics and Path Integrals, Readings*, Mass: Addison-Wesley, 1965.

13 W. H. Miller, "Classical-limit quantum mechanics and the theory of molecular collisions," in *Advances in Chemical Physics*, vol. 25, New York: Wiley, 1974, pp. 69–177.

14 N. Makri and W. H. Miller, "Monte Carlo path integration for the real time propagator," *J. Chem. Phys.*, vol. 89, p. 2170, 1988.

15 K. G. Kay, "Semiclassical initial value treatments of atoms and molecules," *Annu. Rev. Phys. Chem.*, vol. 56, p. 255, 2005.

16 A. Donoso, Y. Zheng and C. C. Martens, "Simulation of quantum processes using entangled trajectory molecular dynamics," *J. Chem. Phys.*, vol. 119, p. 5010, 2003.

17 R. E. Wyatt, *Quantun Dynamics with Trajectories*, New York: Springer Science, 2005.

18 E. R. Bittner, "Quantum inirial value representationusing approximate brownian trajectories," *J. Chem. Phys.*, vol. 119, p. 1358, 2003.

19 I. Burghardt and L. S. Cederbaum, "Hydrodynamic equations for mixed quantum states. I. general formulation," *J. Chem. Phys.*, vol. 115, p. 10303, 2001.

20 D. A. Micha and I. Burghardt (Editors), *Quantum Dynamics of Complex Molecular Systems*, Heidelberg, Germany: Springer-Verlag, 2007.

21 D. A. Micha, "A self-consistent eikonal treatment of electronic transitions in molecular collisions," *J. Chem. Phys.*, vol. 78, p. 7138, 1983.

22 H. Goldstein, C. P. Poole Jr., and J. L. Safko, *Classical Mechanics* (3rd edition), San Francisco, CA: Addison-Wesley, 2002.

23 G. Arfken, *Mathematical Methods for Physicists* (2nd edition), New York: Academic Press, 1970.

24 J. M. Cohen and D. A. Micha, "Angular distribution in electronically adiabatic hyperthermal collisions. An eikonal approach," *J. Chem. Phys.*, vol. 98, p. 2023, 1993.

25 S. J. Farlow, *Partial Differential Equations for Scientists and Engineers*, New York: Dover Publ., 1993.

26 J. C. Y. Chen and K. M. Watson, "Electronic transitions in slow collisions of atoms and molecules," *Phys. Rev.*, vol. 174, p. 152, 1968.

27 J. W. Dettman, *Mathematical Methods of Physics and Engineering* (2nd edition), New York: McGraw-Hill, 1969.

28 W. N. Sams and D. J. Kouri, "Noniterative solutions of integral equations for scattering. I. Single channels," *J. Chem. Phys.*, vol. 51, p. 4809, 1969.

29 J. R. Taylor, *Scattering Theory*, New York: John Wiley & Sons, 1972.

30 D. A. Micha, "Expansion of the T matrix for resonance collisions," *J. Math. Phys.*, vol. 8, p. 1716, 1967.

31 D. A. Micha and E. Brandas, "Variational methods in the wave-operator formalism. A unified treatment of bound and quasi-bound electronic and molecular states," *J. Chem. Phys.*, vol. 55, p. 4792, 1971.

32 W. A. Lester and R. B. Bernstein, "Computational procedure for the close-coupled rotational excitation problem: Scattering of diatomic molecules by atoms," *J. Chem. Phys.*, vol. 48, p. 4896, 1968.

33 K. Smith, *Calculation of Atomic Collision Processes*, New York: Wiley Publ. Co., 1971.

34 R. G. Gordon, "New method for constructing wavefunctions for bound states and scattering," *J. Chem. Phys.*, vol. 51, p. 14, 1969.

35 F. Calogero, *The Variable Phase Approach to Potential Scattering*, New York: Academic Press, 1967.

36 B. R. Johnson, "The multichannel log-derivative method for scattering calculations," *J. Comput. Phys.*, vol. 13, p. 445, 1973.

37 J. C. Light and R. B. Walker, "An R-matrix apporach to the solution of coupled equations for atom-molecule reactive scattering," *J. Chem. Phys.*, vol. 65, p. 4272, 1976.

38 B. I. Schneider and R. B. Walker, "The coupled-channel R-matrix propagation method," *J. Chem. Phys.*, vol. 70, p. 2466, 1979.

39 M. J. Redmon and D. A. Micha, "A computational method fro multichannel scattering: Applications to rotational excitation and long-lived states of He-N_2," *Chem. Phys. Lett.*, vol. 28, p. 341, 1974.

40 M. J. Redmon and D. A. Micha, "Interaction potential and dynamics for Li-F collisions," *Int. J. Quantum Chem. Symp.*, vol. 8, no. Symp. 8, p. 253, 1974.

41 D. J. Kouri, "General theory of reactive scattering: The integral equation approach (Chapter 4)," in *Theory of Chemical Reaction Dynamics*, Volume I, (Editor: M. Baer), Boca Raton, FL, USA: CRC Press, 1985, pp. 163–226.

42 B. Lipman and J. Schwinger, "Variational Principles for Scattering Processes. I," *Phys. Rev.*, vol. 79, p. 469, 1949.

43 W. Kohn, "Variational methods in nuclear collision problems," *Phys. Rev.*, vol. 74, p. 1763, 1948.

44 R. N. Porter and L. M. Raff, "Classical trajectory methods in molecular collisions," in *Dynamics of Molecular Collisions Part B* (Editor: W. H. Miller), New York, Plenum Press, 1976, pp. 1–52.

45 M. P. Allen and D. J. Tildesley, *Computer Simulation of Liquids*, New York: Oxford University Press, 1989.

46 M. Tuckerman and B. J. Berne, "Reversible multiple time scale molecular dynamics," *J. Chem. Phys.*, vol. 97, p. 1990, 1992.

47 N. F. Mott and H. S. W. Massey, *Theory of Atomic Collisions, Oxford*, England: Oxford University Press, 1965.

3

Inelastic Collisions: Dynamics

CONTENTS

3.1 Inelastic Collision Cross Sections

3.1.1 Kinematics of Inelastic Collisions

We consider inelastic collision processes between molecules A and B in internal states p and q, changing as

$$A(p) + B(q) \rightarrow A(p') + B(q')$$

at fixed total energy $E = E^A_{kin} + E^{(A)}_p + E^B_{kin} + E^{(B)}_q = E^{A'}_{kin} + E^{(A)}_{p'} + E^{B'}_{kin} + E^{(B)}_{q'}$ in a Laboratory (LAB) reference frame where E^A_{kin} and E^B_{kin} are the kinetic energies of molecules A and B, while $E^{(A)}_p$ and $E^{(B)}_q$ are their state internal energies, respectively.

Introducing the center-of-mass (CM) positions $\vec{R}_A$ and $\vec{R}_B$ of the molecules in a space-fixed (SF) LAB frame, the pair CM and relative positions are $\vec{R}_{CM} = (M_A\vec{R}_A + M_B\vec{R}_B)/M_{tot}$ and $\vec{R} = \vec{R}_A - \vec{R}_B$. Reciprocally, the CM positions of the molecules are $\vec{R}_A = \vec{R}_{CM} + M_B\vec{R}/M_{tot}$ and $\vec{R}_B = \vec{R}_{CM} - M_A\vec{R}/M_{tot}$, with $M_{tot} = M_A + M_B$. Each molecular orientation is described by its orientation angles in its local frame axes. Kinetic energies satisfy $E^A_{kin} + E^B_{kin} = E_{CM} + E_{rel}$ and the addition of orbital angular momenta gives $\vec{L}^A_{orb} + \vec{L}^B_{orb} = \vec{L}_{CM} + \vec{L}_{rel}$, with $\vec{L}_{CM} = \vec{R}_{CM} \times \vec{P}_{CM}$ and with the relative angular momentum $\vec{L}_{rel} = \vec{R} \times \vec{P}$ of A with respect to B in terms of the relative momentum $\vec{P} = M\vec{v}$, with $\vec{v} = \vec{v}_A - \vec{v}_B$. In the CM frame, with its origin at $\vec{R}_{CM} = 0$, one finds that $\vec{P}_{CM} = \vec{L}_{CM} = E_{CM} = 0$. It follows that $\vec{L}^A_{orb} + \vec{L}^B_{orb} = \vec{L}_{rel}$ and $E^A_{kin} + E^B_{kin} = E_{rel} = P^2/(2M)$ for the initial collisional event, with $\vec{L}'_{rel}$ and E'_{rel} the final relative values. The amount of energy transfer is $\varepsilon = E'_{rel} - E_{rel} = E_n - E_{n'}$ for initial internal state $n = (p, q)$ and final $n' = (p', q')$.

Inelastic collisions are *coherent* if the states of A and B do not change and *incoherent* otherwise. Nonelastic collisions are *exoergic* if $E'_{rel} > E_{rel}$ and *endoergic* otherwise. Choosing A to indicate initial and final directions, collisions are *forward* if $\vec{v} \cdot \vec{v}' > 0$ and *backward* if $\vec{v} \cdot \vec{v}' < 0$. A detailed inelastic collision event specifies the initial and final internal states $\alpha = (\vec{k}, n)$ and $\alpha' = (\vec{k}', n')$ and the relative velocities $\vec{v}$ and $\vec{v}'$, and the related detailed differential cross section in the CM frame is then $d^6\sigma_{\alpha'\alpha}/(d^3v'd^3v)$, constrained by total energy conservation. A more usual description involves only the angular and energy transfer distribution given by the double-differential cross section $d^3\sigma_{\alpha'\alpha}/(d^2\Omega d\varepsilon)$, with Ω the solid angle direction of $\vec{v}'$ with respect to $\vec{v}$. A further simplification arises when one is interested only in the initial and final internal energies $E^{(AB)}_{int} = E^{(A)}_p + E^{(B)}_q$ and $E^{(AB)'}_{int} = E^{(AB)}_{int} - \varepsilon$, for given initial relative kinetic energy, with a cross section $d^3\sigma\left(\Omega, \varepsilon; E^{(AB)}_{int}, E_{rel}\right)/(d^2\Omega d\varepsilon)$, which can be obtained in classical [1] or quantal descriptions [2, 3].

Molecules contain electronic and nuclear component particles, with a set electronic variables (spatial and spin) $\mathbf{X}$ and a set of nuclear (or atomic ion core) positions $\mathbf{Q} = (\vec{R}, \boldsymbol{q}_A, \boldsymbol{q}_B)$, with $\boldsymbol{q}_A = \{\vec{r}_a, a = 1, \ldots, N_A\}, \boldsymbol{q}_B = \{\vec{r}_b, b = 1, \ldots, N_B\}$ containing the internal position vectors of atoms in species A and B and $\vec{R}$ giving the positions of the CM of A and B. The total wavefunction of a molecular pair can be expanded in a basis set of N_{el} molecular electronic states $\Phi_\Gamma(\mathbf{X}; \mathbf{Q})$, each a function of electronic variables for fixed nuclei, in the space-fixed (SF) reference frame with axes (X, Y, Z) defined in a LAB setup. The label Γ contains all the electronic orbital and spin quantum numbers. Here we consider molecular collisions involving atoms and molecules with closed electronic shells, at relative velocities low compared with electron velocities. In these cases, a system with $\Gamma = g$ evolving in its ground electronic state $\Phi_g(\mathbf{X}; \mathbf{Q})$ undergoes a so-called adiabatic process, with the potential energy of the atomic ion cores given by the electronic average $E_g(\mathbf{Q}) = \langle \Phi_g | \hat{H}_\mathbf{Q} | \Phi_g \rangle_\mathbf{Q}$ of a Hamiltonian $\hat{H}_\mathbf{Q}$ containing electronic energies and the interaction of all atomic ion charges for fixed core positions [4].

Electronic states in a scattering event must be chosen to go asymptotically to products of molecular states, so that $\Phi_\Gamma(\mathbf{X}; \mathbf{Q}) \approx \Phi_\gamma^{A,B}(\mathbf{X}; \boldsymbol{q}_A, \boldsymbol{q}_B) = \Phi_{\Gamma_A}^A(\mathbf{X}_A; \boldsymbol{q}_A)$ $\Phi_{\Gamma_B}^B(\mathbf{X}_B; \boldsymbol{q}_B)$ as the CM intermolecular distance $R \to \infty$. The interaction potential energy is then $V(\mathbf{Q}) = E_g(\mathbf{Q}) - E_A(\boldsymbol{q}_A) - E_B(\boldsymbol{q}_B)$.

In more detail, we consider energy transfer involving relative translational (T) and internal rotational (R) and vibrational (V) motions of the molecules, induced by collisional events. The total Hamiltonian of the collision system is

$$\hat{H} = \hat{H}_0 + V = \hat{K}_R + \hat{H}_{\boldsymbol{q}}(\vec{R})$$

$$\hat{H}_{\boldsymbol{q}}(\vec{R}) = \hat{H}_{int}(\boldsymbol{q}) + V(\vec{R}, \boldsymbol{q})$$

$$\hat{H}_{int}(\boldsymbol{q}) = \hat{H}_A + \hat{H}_B$$

where $\hat{H}_I$ is the Hamiltonian operator of species $I = A, B$, including the kinetic energy of its moving atoms, and their interactions in its molecular ground electronic state. The potential energy function $V(\vec{R}, \boldsymbol{q}), \boldsymbol{q} = (\boldsymbol{q}_A, \boldsymbol{q}_B)$, can be constructed in a number of ways. One useful decomposition is in terms of short-(SR) and long-range (LR) potential energy functions of the relative intermolecular distance R, with SR strongly repulsive and a weak LR interaction cut-off at short R distances [4], so that

$$V(\vec{R}, \boldsymbol{q}) = V_{SR}(\vec{R}, \boldsymbol{q}) + V_{LR}(\vec{R}, \boldsymbol{q})$$

in which case the weak LR interaction can be treated as a perturbation in the collision dynamics. An alternative representation of the potential energy is as a function of all the atomic position vectors, so that

$$V(\vec{R}, \boldsymbol{q}) = \sum_{a,b} v_{ab}(\vec{r}_a, \vec{r}_b) + \sum_{a,b,b'} v_{abb'}(\vec{r}_a, \vec{r}_b, \vec{r}_{b'}) + \cdots$$

as a sum over pair, triplet, and so on, groups of atoms. This representation is helpful in the treatment of collisions involving many atoms in polyatomic systems and can be used in multi-scattering formulations of collisions.

As an example, consider a three-atom A + BC system in a CM reference frame with fixed-axes orientations. In a SF frame, the Jacobi positions are $\vec{R}_o$, $\vec{r}$, and $\vec{R}$, with corresponding CM, internal and relative momenta $\vec{P}_o$, $\vec{p}$, and $\vec{P}$, and corresponding masses M_o, m, and M. In a CM frame, $\vec{P}_o = 0$, $\vec{R}_o = 0$, and the treatment involves only six position variables instead of nine. Here again we consider an electronically adiabatic collision where the interaction potential is a function only of the atomic positions for a given electronic state. A typical PES is shown in Figure 3.1 as isocontours of potential energy vs internal $r = r_{BC}$ and relative $R = r_{A,BC}$ distances, at a fixed bond angle.

Simplifying by working in the CM frame, the Hamiltonian is $H_{CM} = P^2/(2M) + p^2/(2m) + V(\vec{r}, \vec{R})$, with $V(\vec{r}, \vec{R}) = v(\vec{r}) + V^{(rel)}(\vec{r}, \vec{R})$, where $v(\vec{r})$ is the internal diatomic potential energy and $V^{(rel)}$ is the remaining coupling potential energy between internal and relative motions. Equations of classical motion follow by differentiating this Hamiltonian with respect to position and momentum variables, to obtain forces and generalized momenta given initial conditions $(\vec{r}, \vec{R})_{in}$ and $(\vec{p}, \vec{P})_{in}$ at a starting time t_{in} when the collision partners are far apart and their coupling potential energy $V^{(rel)}$ is vanishing. Details have been covered in several places [5, 6].

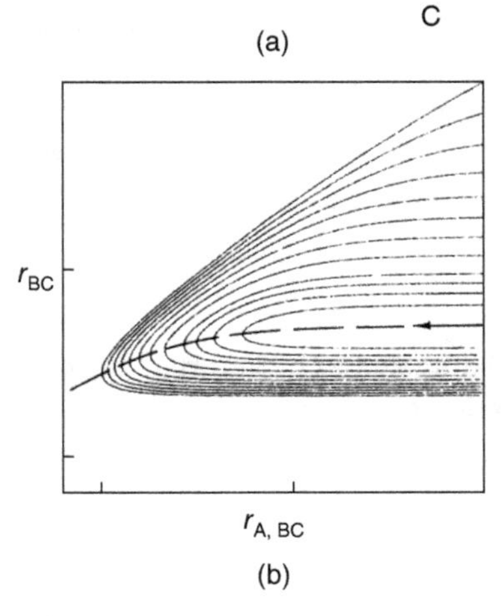

Figure 3.1 (a) Collisional coordinates (R, r, θ) for atoms A, B, and C; (b) isocontours of potential energy vs internal $r = r_{BC}$ and relative $R = r_{A,BC}$ distances, at a bond angle $\vartheta_{ABC} = \pi$, for a typical potential energy surface of A + BC.

Continuing with the example of A + BC for three atoms with ion cores at CM locations $\vec{r}_a, \vec{r}_b, \vec{r}_c$, a connection can be established between the shapes of potential energy surfaces (PESs) and the distributions of final T, R, and V energies, given the initial ones. Considering the PES $V(r_{ab}, r_{bc}, \vartheta_{abc})$ for ABC, with a fixed interbond angle $\vartheta_{abc} = \pi$, as a function of bond distances r_{ab}, r_{bc}, isocontours of the interaction potential surface in Figure 3.1 show a valley with a minimum along $r_{a,bc}$ and a contracting bond distance r_{bc} as the atom approaches (as shown), or alternatively an expanding bond, for fixed interbond angle. Internal rotational motions are typically slower than vibrational motions.

The description of scattering events leading to energy transfer can be done in terms of classical dynamics, quantal dynamics, and semiclassical dynamics depending on the properties being calculated and what experimental results are studied. Treatments involve formulations in terms of ordinary and differential equations and boundary conditions needed to obtain cross sections and rates [3, 5, 7] and also treatments using operator methods and collisional time-correlation functions relating to rate phenomena [8].

A general theory of inelastic collisions, that contain as special cases the mentioned alternative dynamics treatments, is a time-dependent formulation where scattering states $\Psi_\alpha(\mathbf{Q}, t; \vec{k}_\alpha)$ for an initial internal state n and relative momentum $\vec{P}_\alpha = \hbar\vec{k}_\alpha$ are wavepackets evolving in time as solutions of the Schrodinger time-dependent equation of motion for given initial conditions [9, 10]. They can be constructed as superpositions of steady-state wavefunctions $\Psi^{(+)}_{\alpha'}(\mathbf{Q}), \alpha' = (\vec{k}', n')$, obtained as eigenfunctions of the time-independent Schrodinger equation of energies $E_{\alpha'}$, with asymptotic conditions for scattering, as

$$\Psi_\alpha(\mathbf{Q}, t; \vec{k}_\alpha) = \sum_{n'} \int d^3k' A_{\alpha'\alpha} \Psi^{(+)}_{\alpha'}(\mathbf{Q}) \exp\left(-iE_{\alpha'}t/\hbar\right)$$

with coefficients $A_{\alpha'\alpha}$ chosen to construct the initial wavepacket. A calculation of cross sections can be done in terms of wavepackets or using steady-state functions. Computational methods have been extensively developed for stationary states, and the following treatments in this chapter focus on that time-independent formulation.

3.1.2 Classical Dynamics Treatment

Classical treatments have been presented in great detail in several chapters of monographs [6, 11–13], and only the main aspects are briefly described here. The connection between classical and quantal treatments can be done taking the limit of short wavelengths in the wavemechanical version of quantum mechanics, as shown in what follows, or generally in terms of path integrals [14].

Generally, a set of $F = 3N$ generalized coordinates $\boldsymbol{Q} = (Q_1, Q_2, \dots, Q_F)$ can be introduced for a system with N atoms, with their atomic (Cartesian, or Jacobi, or curvilinear) coordinates dependent on the time t. In a CM frame, $F = 3N - 3$.

A collision can be described introducing the relative position vector $\vec{R}$ of the CM of molecules A and B and their internal coordinates $\boldsymbol{q} = (\boldsymbol{q}_A, \boldsymbol{q}_B)$. These contain rotational coordinates, in general given for each top-like molecular structure by three Euler orientation angles [15], and vibrational coordinates derived from the bond distances, bond angles, and torsion angles of the rigid structure. Considering their time-dependence and related relative translational, rotational, and vibrational momenta $\boldsymbol{P} = (P_1, P_2, \dots, P_F)$ for motion in a potential energy surface $V[\boldsymbol{Q}(t)]$, the classical dynamics of the whole system leads to a total energy $E[\boldsymbol{P}(t), \boldsymbol{Q}(t)] = E_T + E_{R,V}$ containing translational and rovibrational energies. In some cases, for systems with light atoms, it is also possible to separate rotational and vibrational energies, so that $E_{R,V} = E_R + E_V$, however with vibrational motions affected by rotational motions. Those energies change from their initial values by amounts $\Delta E = \Delta E_T + \Delta E_{R,V}$ during a collision. Qualitatively, for bound rotational and vibrational motions, these changes occur with rates κ_R and κ_V induced by torque forces for rotational changes and by compression forces for vibrational changes.

For typical rotational and vibrational forces in molecules, one finds that $\kappa_R \ll \kappa_V$ and that these motions can be approximately treated at different time scales [16]. Rotations can be assumed to occur over long times, with vibrational coordinates averaged at their equilibrium values, and vibrational motions can be modelled assuming that molecular orientations are fixed. More detailed classical treatments follow from differential equations of motion for given initial conditions.

To describe a detailed dynamics with generalized coordinates $(Q_1, Q_2, \dots, Q_F)$, we first introduce corresponding generalized velocities $\dot{Q}_n = dQ_n/dt$, $n = 1$ to F. The potential energy function, taken here to depend only on position coordinates, is $\mathcal{V}(Q_1, Q_2, \dots, Q_F)$, while the kinetic energy function is generally of form $\mathcal{K}(Q_1, Q_2, \dots, Q_F, \dot{Q}_1, \dot{Q}_2, \dots, \dot{Q}_F)$. The Lagrangian function of the system is $\mathcal{L} = \mathcal{K} - \mathcal{V}$ and the generalized momenta are given by $P_n = \partial\mathcal{L}/\partial\dot{Q}_n$, which are alternatively used instead of velocities in a treatment with a Hamiltonian function

$$\mathcal{H} = \mathcal{K}(Q_1, Q_2, \dots, Q_F, P_1, P_2, \dots, P_F) + \mathcal{V}(Q_1, Q_2, \dots, Q_F).$$

Equations of motion for positions and momenta follow from the set of Hamiltonian equations

$$\dot{P}_n = -\partial\mathcal{H}/\partial Q_n \text{ and } \dot{Q}_n = \partial\mathcal{H}/\partial P_n = \partial\mathcal{K}/\partial P_n$$

to be supplemented with initial conditions in the sets $\boldsymbol{Q}_{in} = \boldsymbol{Q}(t_{in})$ and $\boldsymbol{P}_{in} = \boldsymbol{P}(t_{in})$.

These sets of initial conditions in the $2F$-dimensional phase space $(\boldsymbol{P}, \boldsymbol{Q})$ must be constructed in accordance with experimental conditions, coming from a molecular-beams arrangement or from a thermodynamical environment. Classical treatments however do not account for experimental quantal effects and can only give results comparable to experimental measurements in an average sense, obtained from bundles of trajectories from a distribution of initial position and velocity values and from final statistical distributions.

In a molecular beams experiment, for example, for an atom–diatomic collision A + BC, an initial intermolecular relative translational energy E_T and initial molecular internal rotational and vibrational energies $E_{R,V}$ can be selected. These provide the relative velocity v_T from $E_T = M_1 v_T^2/2$ and internal angular momentum $L_{int} = L_{BC}$ from $E_{R,int} = {L_{BC}}^2/(2I_{BC})$, with I_{BC} the diatomic moment of inertia. A statistical distribution of initial conditions in the phase space of variables $(\boldsymbol{P}, \boldsymbol{Q})$ includes in addition the impact parameter b of the collision, assuming that incoming trajectories are along a Z-axis in CM, and the relative angular momentum component $L_\perp$ (perpendicular to the initial plane of collision) or alternatively the angle η between the initial collision plane and the XZ-plane. A calculation of scattering probabilities must include a sampling of initial intermolecular distances R_1, molecular orientation angles $(\vartheta, \varphi)_1$, planar angle η_1, and internal bond distances r_1.

Given a distribution of initial parameters, $W_{in}(\Lambda_{in})$, with $\Lambda_{in} = (R_1, \eta_1, \vartheta_1, \varphi_1$ r_1; v_T, $E_{R,V}$, $b)$, and propagating over time the bundle of classical trajectories originating on its set of values, for chosen variables v_T, $E_{R,V}$, b, leads to a final distribution $W_{fn}(\boldsymbol{P}, \boldsymbol{Q}; v_T, E_{R,V}, b)$ of points in the phase space of variables $(\boldsymbol{P}, \boldsymbol{Q})$ after collisions. This distribution can be analyzed to extract the fraction of trajectories leading to final energy distribution of $v'_T, E'_{R,V}$ values. For the phase space with final parameters Λ_{fn} after a collision with final values $v'_T, E'_{R,V}$, the partial inelastic collision probability is

$$P^{(inel)}\left(v'_T, E_{R,V}{}'; v_T, E_{R,V}, b\right) = \int_{\Lambda_{fn}} d\boldsymbol{P}\, d\boldsymbol{Q}\, W_{fn}(\boldsymbol{P}, \boldsymbol{Q}; v_T, E_{R,V}, b)$$

from which it is possible to construct differential cross sections $d\sigma^{(inel)}/d\Omega$ into scattering solid angle $d\Omega = 2\pi b\, db/R^2$, and integral cross sections $\sigma_{1\to 1'}$ as

$$\sigma^{(inel)} = 2\pi \int b\, db\, P^{(inel)}\left(v'_T, E'_{R,V}; v_T, E_{R,V}, b\right).$$

Detailed calculations can be performed integrating the differential equations of motion with time changing in step-by-step methods for given initial conditions. General qualitative relations can be deduced between a distribution of an initial energy fraction $f_X = E_X/(E_T + E_R + E_V)$, $X = T, R, V$, and a final one f'_X for each type of PES [6].

3.1.3 Quantal Treatment

In general, the collision dynamics of two many-atom molecules A and B, remaining in their ground electronic states, is governed by an interaction potential energy $V(\boldsymbol{Q})$, which vanishes as the intermolecular distance $R \to \infty$. A steady-state scattering wavefunction $\Psi_\alpha^{(+)}(\boldsymbol{Q})$, depending on all the atomic position variables of the two molecules, specifies an initial state $\alpha = (\vec{k}_\alpha, n)$ of the pair and satisfies scattering boundary conditions for outgoing waves [5, 7, 17]. In more detail, we introduce the molecular internal (vibrational–rotational) coordinates $\boldsymbol{q}_A$ and $\boldsymbol{q}_B$ and the relative position $\vec{R}$, with $\mathbf{Q} = (\vec{R}, \boldsymbol{q})$ and $\boldsymbol{q} = (\boldsymbol{q}_A, \boldsymbol{q}_B)$, and scattering states $\Psi_\alpha^{(+)}(\vec{R}, \boldsymbol{q}; \vec{k}_\alpha)$. The relative motion wavevector $\vec{k}_\alpha$ points in the relative motion direction $\vec{n} = \vec{P}/P$ of a CM–SF reference frame and has its magnitude given by $k_\alpha^2 = (2M/\hbar^2)(E - E_n)$.

Internal motions are described by the equation

$$\hat{H}_{int}\phi_n(\boldsymbol{q}) = E_n\phi_n(\boldsymbol{q})$$

where $\hat{H}_{int} = \hat{H}_A + \hat{H}_B$ is the internal motion Hamiltonian including the kinetic energy operators for atoms in A and B. Also, $\phi_n(\boldsymbol{q}) = \varphi_p^{(A)}(\boldsymbol{q}_A)\varphi_q^{(B)}(\boldsymbol{q}_B)$ is the internal state of the pair with quantum numbers $n = (p, q)$ and $E_n = E_p^{(A)} + E_q^{(B)}$ is the total internal energy. The molecules move with relative position $\vec{R}$ between their CM under the influence of an interaction potential energy $V(\vec{R}, \boldsymbol{q}) = \mathrm{V}(\vec{R}, \boldsymbol{q}_A, \boldsymbol{q}_B)$. For free motion at large intermolecular distances R where $V \approx 0$, we have an unperturbed Hamiltonian $\hat{H}_0 = \hat{K}_R + \hat{H}_{int}$, including the relative kinetic energy operator $\hat{K}_R = -(\hbar^2/2M)\nabla_R^2$, and

$$\hat{H}_0\Psi_\alpha^0\left(\vec{R}, \boldsymbol{q}; \vec{k}_\alpha\right) = E\Psi_\alpha^0\left(\vec{R}, \boldsymbol{q}; \vec{k}_\alpha\right)$$

with a free-motion eigenfunction given by $\Psi_\alpha^0(\vec{R}, \boldsymbol{q}; \vec{k}_\alpha) = \psi^0(\vec{R}; \vec{k}_\alpha)\phi_a(\boldsymbol{q})$, where the collisional relative momentum is $\hbar\vec{k}_\alpha = \vec{P}_\alpha$, and

$$\psi^0\left(\vec{R}; \vec{k}_\alpha\right) = (2\pi)^{-\frac{3}{2}} \exp\left(i\vec{k}_\alpha \cdot \vec{R}\right)$$

is a relative free-motion state normalized so that the relative spatial density is $|\psi^0|^2 = (2\pi)^{-3}$.

The total energy for a reduced mass $M = M_A M_B/(M_A + M_B)$ of the molecular pair is $E = P_\alpha^2/2M + E_n$ and a corresponding flux of particles in the state α is given by $\vec{J}_\alpha^0 = (2\pi)^{-3}\vec{P}_\alpha/M$. Open internal channels (detected at large distances) are defined as those with energy $E_n \le E$ and closed channels as those with $E_n > E$. For closed channels, the relative momenta have imaginary values and we write $k_\alpha = i\kappa_\alpha$, a purely imaginary value with $\kappa_\alpha > 0$ by choice, which generates a bound state decreasing exponentially for large relative distance.

For $V \neq 0$, and $\hat{H} = \hat{H}_0 + V = \hat{K}_R + \hat{H}_{\boldsymbol{q}}(\vec{R})$, with $\hat{H}_{\boldsymbol{q}}(\vec{R}) = \hat{H}_{int}(\boldsymbol{q}) + V(\vec{R}, \boldsymbol{q})$, the scattering wavefunction is in general a superposition of internal states, as

$$\Psi_\alpha^{(+)}(\vec{R}, \boldsymbol{q}) = \sum_{n'} \psi_{\alpha'\alpha}^{(+)}(\vec{R})\, \phi_{n'}(\boldsymbol{q})$$

with $\alpha' = \left(\vec{k}_{\alpha'}, n'\right)$.

This contains, at large relative distances, outgoing waves in all the open α' channels and vanishing waves in the closed α' channels and satisfies

$$\hat{H}\Psi_{\alpha}^{(+)}\left(\vec{R},\boldsymbol{q};\vec{k}_{\alpha}\right) = E\Psi_{\alpha}^{(+)}\left(\vec{R},\boldsymbol{q};\vec{k}_{\alpha}\right)$$

$$\Psi_{\alpha}^{(+)} \approx \psi^{0}\left(\vec{R};\vec{k}_{\alpha}\right) + \sum_{n'\ \text{open}} \psi_{\alpha'\alpha}^{(sc)}(\vec{R})\,\phi_{n'}(\boldsymbol{q}),\quad R \to \infty$$

$$\psi_{\alpha'\alpha}^{(sc)}(\vec{R}) = (2\pi)^{-\frac{3}{2}} f_{\alpha'\alpha}^{(+)}(\Omega) R^{-1} e^{ik_{\alpha'}R}$$

where $f_{\alpha'\alpha}^{(+)}$ is the scattering amplitude for the $\alpha \to \alpha'$ transition.

Proceeding by generalization of elastic scattering, the rate of inelastic $\alpha \to \alpha'$ scattering at a relative kinetic energy $P^2/2M$ into the element of solid angle $d\Omega$ in the direction of unit vector $\vec{n}$ through the surface $dS = R^2 d\Omega$ is

$$d\dot{N}_{\alpha'\alpha} = \left|\vec{J}_{\alpha}^{0}\right| d\sigma_{\alpha'\alpha} = \vec{J}_{\alpha'\alpha} \cdot \vec{n}\, dS$$

which defines the differential of inelastic cross section $d\sigma_{\alpha'\alpha}$ into solid angle $d\Omega = dS/R^2$ in terms of the incoming flux $\vec{J}_{\alpha}^{0}$ and the inelastic state-to-state scattered flux $\vec{J}_{\alpha'\alpha}$. This inelastic flux follows from the gradient differentiation of $\psi_{\alpha'\alpha}^{(sc)}(\vec{R})$ and $\vec{J}_{\alpha'\alpha}^{(sc)}(\vec{R}) = \left\langle \psi_{\alpha'\alpha}^{(sc)} \middle| \widehat{\vec{J}_{\vec{R}}} \middle| \psi_{\alpha'\alpha}^{(sc)} \right\rangle$, with

$$\widehat{\vec{J}_{\vec{R}}} = \left[\widehat{\vec{P}_{\vec{R}}}\,\delta\left(\vec{R} - \widehat{\vec{R}'}\right) + \delta\left(\vec{R} - \widehat{\vec{R}'}\right)\widehat{\vec{P}_{\vec{R}}}\right] \Big/ (2M)$$

giving the result

$$\vec{J}_{\alpha'\alpha} \cdot \vec{n} = \left(k_{\alpha'}/M\right)(2\pi)^{-3} R^{-2} \left|f_{\alpha'\alpha}^{(+)}(\Omega)\right|^2$$

and the differential inelastic cross section

$$\frac{d\sigma_{\alpha'\alpha}}{d\Omega} = \frac{k_{\alpha'}}{k_{\alpha}} \left|f_{\alpha'\alpha}^{(+)}(\Omega)\right|^2.$$

The integral cross section $\sigma_{\alpha'\alpha}$ follows by integration over the solid angle Ω.

3.1.4 Semiclassical Treatment

3.1.4.1 Eikonal Description

It is sometimes convenient to treat selected space and momentum variables of the atoms in the collision dynamics as classical degrees of freedom, dependent on time and varying in accordance with classical equations of motion. Variables within the set $\boldsymbol{Q} = (Q_1, Q_2, \dots, Q_F) = (\vec{R}, \boldsymbol{q})$ and the corresponding generalized momenta in the set $\boldsymbol{P} = (P_1, P_2, \dots, P_F) = (\vec{P}, \boldsymbol{p})$ can be selected on physical considerations to be treated as classical-like while the remining ones are treated as quantal variables. Classical-like motions are associated with large momenta $P_n = \hbar k_n = h/\lambda_n$ and short associated wavelengths $\lambda_n \ll R_m$, a typical dimension for potential energy variations. Short wavelengths involve numerical integration of quantal scattering differential equations over small steps of the related space coordinates, in computationally inefficient procedures, which can be avoided in a semiclassical treatment. A quantal-classical, or semiclassical, treatment allows derivation of equations of motion coupling quantal and classical-like dynamics, leading to more efficient computational procedures.

To describe the treatment in some detail, it is convenient to choose here the relative position and momentum vectors $\vec{R}$ and $\vec{P}$ as classical-like. The internal rotational motions of the colliding molecules can sometimes be also treated as classical-like if they satisfy the short-wavelength criteria. The classical-like variables are coupled to quantal degrees of freedom $\boldsymbol{q}$ in a quantal-classical or semiclassical treatment, which involves expansion functions $\phi_n(\boldsymbol{q})$. Here we describe a semiclassical approach introducing relative position and momentum $\vec{R}(t)$ and $\vec{P}(t)$ as classical functions of a time t, along lines described in the chapter on elastic collisions, now expanded for inelastic collisions. A general semiclassical treatment can be based on an eikonal description as we follow here [18] or on path integral methods [14, 19]. The present treatment incorporates features of several previous treatments of inelastic and reactive collisions based on the eikonal representation [20, 21].

We start our treatment with the time-independent Schrodinger equation for total energy E and a formally exact decomposition for the stationary total (molecular) wavefunction $\Psi(\vec{R}, \boldsymbol{q})$ of the system with N atoms in a CM–SF frame, dependent on position variables $\boldsymbol{q} = \{Q_n, n = 1 \text{ to } 3N-3\}$ describing quantal rotational and vibrational motions. Our first step in the description of the collisional dynamics is to write the wavefunction of the system as a superposition of eikonal forms $\chi(\vec{R}, \boldsymbol{q})\exp[iS(\vec{R})/\hbar]$ [22] and to follow a treatment developed in [18], where $S(\vec{R})$ is a real-valued phase function to be identified with a mechanical action [1] in an effective potential energy $\mathcal{V}(\vec{R})$ constructed from physical considerations. A general wavefunction is written as

$$\Psi(\vec{R}, \boldsymbol{q}) = \int d\Lambda\, \mathrm{a}(\Lambda)\chi(\vec{R}, \boldsymbol{q}; \Lambda)\exp[iS(\vec{R}; \Lambda)/\hbar]$$

in an eikonal representation, given as a sum of eikonal functions with amplitudes $\mathrm{a}(\Lambda)$ and parameters Λ chosen to satisfy boundary conditions. The action function satisfies a partial differential equation and must be supplied with boundary conditions for scattering, leading to a local relative momentum $\vec{P} = \partial S/\partial\vec{R}$.

The effective potential energy can be constructed from the potential energy $V(\vec{R}, \boldsymbol{q})$ quite generally by introducing a density operator $\widehat{w}(\vec{P}, \vec{R})$ in the space of molecular internal variables $\boldsymbol{q}$ to form the average $\mathcal{V}(\vec{P}, \vec{R}) = tr_{\boldsymbol{q}}[\widehat{w}(\vec{P}, \vec{R})\widehat{V}(\vec{R})]$ where $\widehat{V}(\vec{R}) = \int d\boldsymbol{q}\ |\boldsymbol{q}\rangle\, V(\vec{R}, \boldsymbol{q})\,\langle\boldsymbol{q}|$ is an operator in the ket-bra notation, and the trace refers to internal variables.

Examples of averages follow from $\widehat{w}(\vec{R}) = |\phi_n\rangle\langle\phi_n|$ giving $\mathcal{V}(\vec{R}) = \langle\phi_n|\widehat{V}(\vec{R})|\phi_n\rangle = V_{nn}(\vec{R})$, with the effective potential equal to the potential energy in a single internal state n, or from $\widehat{w}(\vec{R}) = \sum_{n,n'} w_{nn'}(\vec{R})|\phi_n\rangle\langle\phi_{n'}|$ with $\mathcal{V}(\vec{R}) = \sum_{n,n'} w_{nn'}(\vec{R}) V_{nn'}(\vec{R})$, an effective potential energy containing diagonal and off-diagonal (coupling) matrix elements of the interaction energy. The weights $w_{nn'}(\vec{R})$ can be chosen to change the interaction potential from a $V_{nn}(\vec{R})$ for internal state n to another potential energy $V_{n'n'}(\vec{R})$ as $\vec{R}$ varies, with the change induced by a dynamical $n \to n'$ coupling. Alternatively, a density distribution $w(\vec{R}_s, \vec{P}_s, \boldsymbol{q}), s = 1$ to N_B, can be constructed to introduce an average of $\widehat{V}(\vec{R}, \boldsymbol{q})$ over a bundle of initial conditions with a discrete distribution $\{\vec{R}_s, \vec{P}_s, s = 1 \text{ to } N_B\}$ of relative motion positions and momenta, to form

$$\overline{V}(\vec{R}) = \sum_s \int d\boldsymbol{q}\, w\left(\vec{R}_s, \vec{P}_s, \boldsymbol{q}\right) V\left(\vec{R}, \boldsymbol{q}\right)$$

Given a chosen effective potential energy $\mathcal{V}$, the action function $S(\vec{R})$ is obtained from the Hamilton–Jacobi (or H–J) equation

$$\mathcal{V}(\vec{P}, \vec{R}) + \frac{1}{2M}\frac{\partial S}{\partial \vec{R}} \cdot \frac{\partial S}{\partial \vec{R}} - E = 0$$

a nonlinear first-order differential equation to be solved for $S(\vec{R})$ with the asymptotic conditions $S_\alpha(\vec{R}) \approx \hbar\vec{k}_\alpha \cdot \vec{R}, R \to \infty$, for an initial state α. Introducing the momentum vector $\vec{P}(\vec{R}) = \partial S/\partial\vec{R}$, the action function can be obtained from a path integral in

$$S_\alpha(\vec{R}) = S_\alpha(\vec{R}_{in}) + \int_{\vec{R}_{in}}^{\vec{R}} d\vec{R} \cdot \vec{P}(\vec{R}) = \hbar\vec{k}_\alpha \cdot \vec{R} + \int_{\vec{R}_{in}}^{\vec{R}} d\vec{R} \cdot \left[\vec{P}(\vec{R}) - \hbar\vec{k}_\alpha\right]$$

The pre-exponential amplitude function in the eikonal expression can be obtained replacing $\Psi(\boldsymbol{Q}) = \chi(\vec{R}, \boldsymbol{q}) \exp[iS(\vec{R})/\hbar]$, with $\boldsymbol{Q} = (\vec{R}, \boldsymbol{q})$, in the time-independent Schrodinger equation $\left[\widehat{K}_{\vec{R}} + \widehat{H}_{\boldsymbol{q}}(\vec{R}) - E\right]\Psi(\boldsymbol{Q}) = 0$ containing the Hamiltonian operator $\widehat{H}_{\boldsymbol{q}}(\vec{R}) = \widehat{H}_{int}(\boldsymbol{q}) + V(\vec{R}, \boldsymbol{q})$ for internal and interaction energies at a given relative position $\vec{R}$, and the relative motion kinetic energy operator (for mass-scaled coordinates) in $\widehat{K}_{\vec{R}}\Psi(\mathbf{Q}) = -(\hbar^2/2M)\partial^2\Psi(\mathbf{Q})/\partial\vec{R}^2$. Using the bracket notation for internal variables, the pre-exponential function $\chi(\vec{R}, \boldsymbol{q}) = \langle\boldsymbol{q}|\chi(\vec{R})\rangle$ satisfies the related differential equation

$$\left[\widehat{L}_{\vec{R}} + \widehat{H}_{\boldsymbol{q}}(\vec{R}) - E\right]\left|\chi(\vec{R})\right\rangle = 0$$

$$\widehat{L}_{\vec{R}}(\vec{P}) = -\frac{\hbar^2}{2M}\left[\frac{\partial^2}{\partial\vec{R}^2} + 2\frac{i}{\hbar}\vec{P} \cdot \frac{\partial}{\partial\vec{R}} + \frac{i}{\hbar}\left(\frac{\partial}{\partial\vec{R}} \cdot \vec{P}\right) - \frac{1}{\hbar^2}\vec{P} \cdot \vec{P}\right]$$

$$= \frac{1}{2M}\left[\frac{\hbar}{i}\frac{\partial}{\partial\vec{R}} + \vec{P}(\vec{R})\right]^2$$

with the square meaning a scalar product of the vector argument times itself, and solutions can be obtained generalizing the previous treatment for elastic scattering by using here an expansion in internal states for an initial state α so that

$$\chi_\alpha(\vec{R}, \boldsymbol{q}) = \sum_{n'} \xi_{\alpha'\alpha}(\vec{R})\, \phi_{n'}(\boldsymbol{q})$$

with $\xi_{\alpha'\alpha}(\vec{R}) = \exp\left[F_{\alpha'\alpha}(\vec{R})\right]$. Replacing this in the above equation for χ_α gives a set of coupled differential equations for the complex exponential functions $F_{\alpha'\alpha}(\vec{R})$ to be constructed to satisfy asymptotic conditions for scattering. These require that asymptotically $\xi_{\alpha\alpha}(\vec{R})$ becomes a constant independent of distance and $\xi_{\alpha'\alpha}(\vec{R}) \approx \mathcal{O}(R^{-1})$, for $R \to \infty$.

As described for elastic scattering, equations simplify for short wavelengths $\lambda = 2\pi/k$ when state amplitudes $|\chi(\vec{R})\rangle$ are slowly varying over relative distances of order $R_m \gg \lambda$. Letting $\partial\chi/\partial\vec{R} = \nabla\chi$, the short-wavelength limit implies that $\|\nabla^2\chi\|/k^2 \ll 1$, $\|(\nabla\chi)^2\|/\|\nabla\chi\| \ll \lambda^{-1}$, and $\|\nabla^2\chi\|/\|\nabla\chi\| \ll \lambda^{-1}$. In this limit, $\mathcal{V}(\vec{P}, \vec{R}) \cong \overline{V}(\vec{R})$ and $P^2/2M = E - \mathcal{V}(\vec{R})$. The amplitude function satisfies the simpler first-order partial differential equation

$$\left[-i\frac{\hbar}{M}\vec{P}\cdot\frac{\partial}{\partial\vec{R}} - i\frac{\hbar}{M}\left(\frac{\partial}{\partial\vec{R}}\cdot\vec{P}\right) + \hat{H}_{Q'}(\vec{R}) - \overline{V}(\vec{R})\right]|\chi(\vec{R})\rangle = 0.$$

Replacing here the expansion in internal states, for an initial channel α, and projecting on $\langle\phi_{n''}|$ gives coupled equations for the set of $\xi_{\alpha''\alpha}(\vec{R})$ functions,

$$-i\frac{\hbar}{M}\vec{P}\cdot\frac{\partial\xi_{\alpha''\alpha}}{\partial\vec{R}} - \left[\overline{V}(\vec{R}) + i\,U_{\vec{k}}(\vec{R})\right]\xi_{\alpha''\alpha}(\vec{R}) + \sum_{\alpha'} H_{n''n'}(\vec{R})\xi_{\alpha'\alpha}(\vec{R}) = 0$$

with $H_{n''n'}(\vec{R}) = \langle\phi_{n''}|\hat{H}_{Q'}(\vec{R})|\phi_{n'}\rangle = E_{n'}\delta_{n''n'} + V_{n''n'}(\vec{R})$, and $U_{\vec{k}}(\vec{R}) = (\hbar/M)\nabla\cdot\vec{P}$ in an imaginary-valued potential energy giving flux loss.

An explicit form for the effective potential, leading to a self-consistent relation between the action and amplitude functions for inelastic and reactive collisions, can be found [18] requiring that the action function $S(\vec{R})$ should be real-valued. Its differential equation can be obtained multiplying $[\hat{L}_{\vec{R}} + \hat{H}_{qQ'}(\vec{R}) - E]|\chi(\vec{R})\rangle = 0$ on the left the times $\chi(\vec{R}, \boldsymbol{q})^*$ and integrating over $\boldsymbol{q}$ to obtain, for a given set of parameters Λ,

$$V_{qu}(\vec{P}, \vec{R}) + \frac{1}{2M}\frac{\partial S}{\partial\vec{R}}\cdot\frac{\partial S}{\partial\vec{R}} - E = 0$$

which is a Hamilton–Jacobi (or H–J) partial differential equation for the action function $S(\vec{R})$ [1]. This contains a quantum potential $V_{qu}(\vec{P}, \vec{R}) = \overline{\mathcal{V}}(\vec{R}) + \mathcal{V}'(\vec{R}) + \mathcal{V}''(\vec{R})$, as a functional of χ, and its first and second derivatives with respect to nuclear displacements, given by

$$\overline{\mathcal{V}}(\vec{R}) = \left\langle\chi\middle|\hat{H}_{Q'}\middle|\chi\right\rangle / \langle\chi|\chi\rangle$$

$$\mathcal{V}'(\vec{R}) = \frac{i\hbar}{2M}\vec{P}\cdot\left[\left\langle\frac{\partial\chi}{\partial\vec{R}}\middle|\chi\right\rangle - \left\langle\chi\middle|\frac{\partial\chi}{\partial\vec{R}}\right\rangle\right]\Big/\langle\chi|\chi\rangle$$

$$\mathcal{V}''(\vec{R}) = -\frac{\hbar^2}{2M}\frac{1}{2}\left[\left\langle\frac{\partial^2\chi}{\partial\vec{R}^2}\middle|\chi\right\rangle + \left\langle\chi\middle|\frac{\partial^2\chi}{\partial\vec{R}^2}\right\rangle\right]\Big/\langle\chi|\chi\rangle$$

which looks similar to the one derived for elastic scattering, but now with brackets indicating integration over $\boldsymbol{q}$, for fixed $\vec{R}$. In the limit of short wavelengths, $V_{qu}(\vec{P},\vec{R}) \cong \overline{\mathcal{V}}(\vec{R}) + \mathcal{V}'(\vec{R})$ and the H–J equation can be solved with an exponential representation of $\xi_{a'a}(\vec{R})$.

3.1.4.2 Trajectories Description

Alternatively, similarly to the treatment of elastic collisions, solutions to the partial differential equation for $S(\vec{R})$ can be obtained from characteristic paths, as solutions of coupled ordinary differential equations of a parameter t, which can be physically identified with the time variable [1, 23]. These are Hamiltonian equations of motion for momentum $\vec{P}(t)$ and position $\vec{R}(t)$ variables at a fixed energy E, with initial-value conditions for trajectories (or paths) labelled by an index s for each initial set of phase-space points $(\vec{P},\vec{R})_s$, $s = 1$ to N_{in}, at an initial time t_{in}. The paths are in principle entangled over time, but their interference frequently average out during a collision and lead to cancellation of interference effects on scattering probabilities. A simple treatment can be first developed assuming that each path is independent, and this can be extended to allow for path interference when needed [24]. The eikonal phase is given by a path-integral as $S(t;\vec{P}_s,\vec{R}_s)$. The pre-exponential function $\chi(\vec{R},\boldsymbol{q})$ can be obtained from its differential equation transforming variables and the space gradient to time and initial variables for relative motion, to generate $\chi(t;\vec{P}_s,\vec{R}_s,\boldsymbol{q})$.

The set of relative momentum $\vec{P} = \partial S/\partial\vec{R}$ and position $\vec{R}$ variables in the H–J first-order partial differential equation for a fixed energy E can be identified with position $\vec{R}(t)$ and momentum $\vec{P}(t)$ functions, which are the characteristic path functions. They satisfy the Hamiltonian equations

$$\frac{d\vec{R}}{dt} = \frac{\partial\mathcal{H}}{\partial\vec{P}}, \quad \frac{d\vec{P}}{dt} = -\frac{\partial\mathcal{H}}{\partial\vec{R}}$$

with

$$\mathcal{H}(\vec{P},\vec{R}) = \vec{P}\cdot\vec{P}/(2M) + \mathcal{V}(\vec{P},\vec{R})$$

to be solved for each initial set of conditions $(\vec{P}_s,\vec{R}_s)$, $s = 1$ to N_{in}, starting at a time t_{in}. The set of all these trajectories from $\vec{R}(t;\vec{P}_s,\vec{R}_s)$ and $\vec{P}(t;\vec{P}_s,\vec{R}_s)$ form over time a moving bundle in phase space, constructed to satisfy the initial conditions, and constant total energy $\mathcal{H}(\vec{P},\vec{R}) = E$. The action and amplitude functions can be written as functions of time and initial variables, given by $S(\vec{R}) = S(t; \vec{P}_s, \vec{R}_s)$ and $\chi(\vec{R}, \boldsymbol{q}) = \chi(t, \boldsymbol{q}; \vec{P}_s, \vec{R}_s)$.

The trajectories give $S(t; \vec{P}_s, \vec{R}_s)$ from the line integral in

$$S(t) = S(\vec{R}_s) + \int_{\vec{R}_s}^{\vec{R}(t)} d\vec{R} \cdot \vec{P} = \hbar\vec{k} \cdot \vec{R}(t) + \int_{t_{in}}^{t} dt \frac{d\vec{R}}{dt} \cdot (\vec{P} - \hbar\vec{k}).$$

For each trajectory, the related amplitudes $|\chi(\vec{R})\rangle$ satisfying the differential equation $[\hat{L}_{\vec{R}} + \hat{H}_{Q'} - E]|\chi(\vec{R})\rangle = 0$ can be re-expressed in the new variables involving time and initial positions. This is straightforward after introduction of the short-wavelength approximation, insofar in this case $\mathcal{V}(\vec{P}, \vec{R}) \cong \overline{\mathcal{V}}(\vec{R}) + \mathcal{V}'(\vec{R})$ and $\vec{P}(t) = M\, d\vec{R}/dt$, so that $-i(\hbar/M)\vec{P} \cdot \nabla\chi = -i\hbar d\chi/dt$, and

$$\left[-i\hbar\frac{d}{dt} + \hat{H}_{\boldsymbol{q}}(\vec{R}) - \mathcal{V}(\vec{R}) - iU_{\vec{k}}(\vec{R})\right] |\chi(t)\rangle = 0$$

with $U_{\vec{k}}(\vec{R}) = (\hbar/M)\nabla \cdot \vec{P}$ and a $\vec{R}(t; \vec{P}_s, \vec{R}_s)$. The scalar functions $\overline{\mathcal{V}}(\vec{R})$ and $U_{\vec{k}}(\vec{R})$ inside the bracket depend on time through $\vec{R}(t)$ and can be transformed away introducing a time-dependent phase factor so that $|\chi(t)\rangle = |\overline{\chi}(t)\rangle \exp[\gamma(t)]$ with

$$\gamma(t) = -\frac{i}{\hbar}\int_{t_{in}}^{t} dt' \left[\mathcal{V}(\vec{R}') - iU_{\vec{k}}(\vec{R}')\right]$$

and $\vec{R}' = \vec{R}(t'; \vec{P}_s, \vec{R}_s)$, giving the simpler equation

$$\left(-i\hbar\frac{d}{dt} + \hat{H}_{\boldsymbol{q}}[\vec{R}(t)]\right)\overline{\chi}(\boldsymbol{q}, t) = 0$$

with $\overline{\chi}(\boldsymbol{q}, t) = \langle \boldsymbol{q}|\overline{\chi}(t)\rangle$.

Further expanding $\overline{\chi}_\alpha(\boldsymbol{q}, t) = \sum_{n'} A_{\alpha'\alpha}(t)\, \phi_{n'}(\boldsymbol{q})$ gives a set of coupled ordinary differential equations for the channel amplitudes $A_{\alpha'\alpha}(t)$,

$$-i\hbar\frac{dA_{\alpha'\alpha}}{dt} + \sum_{n''} H_{n'n''}(t)A_{\alpha''\alpha}(t) = 0$$

with $H_{n'n''}(t) = \langle \phi_{n'}|\hat{H}_{\boldsymbol{q}Q'}[\vec{R}(t)]|\phi_{n''}\rangle = E_{n'}\delta_{n'n''} + V_{n'n''}[\vec{R}(t)]$. The initial conditions for $t = t_{in}$ follow from $\vec{R}(t_{in}) = R_{in} \to \infty$ and $\vec{P}_{in} = \hbar\vec{k}_\alpha$.

Reintroducing a bundle of trajectories with labels s and boundary-condition parameters Λ_s, the scattering wavefunction is given, for an initial state α, by

$$\Psi(\vec{R}, \boldsymbol{q}) = \sum_s \Lambda_s \chi^{(s)}(\boldsymbol{q}, t) \exp[i\, S_s(t)/\hbar] = \Psi^{(Eik)}(\boldsymbol{q}, t)$$

with $\chi^{(s)}(\boldsymbol{q}, t) = \overline{\chi}(t, \boldsymbol{q}; \vec{P}_s, \vec{R}_s) \exp[\gamma(t, \vec{P}_s, \vec{R}_s)]$, in a semiclassical (or eikonal) approximation. Further expanding $\chi^{(s)}$ in the basis set of internal states $\phi_{\alpha'}(\boldsymbol{q})$ and comparing $\Psi_\alpha^{(Eik)}(\boldsymbol{q}, t)$ for $t \to \infty$ with the full quantal wavefunction $\Psi_\alpha^{(+)}(\vec{R}, \boldsymbol{q}; \vec{k}_\alpha)$ with $R \to \infty$ in the previous subsection, for an initial state α and after a collision where relative motion is along emerging angle Ω, it is possible to extract a scattering amplitude $f^{(Eik)}_{\alpha'\alpha}(\Omega)$. A more detailed treatment of this procedure is given in the chapter on electronically diabatic collisions.

An alternative procedure, more direct but less accurate, is to calculate state-to-state cross sections from trajectories based on the use of initial bundles $\{\vec{P}_s, \vec{R}_s\}$ of trajectories chosen to construct an initial incoming flux $J_{0,\alpha}$ of colliding molecular pairs.

This allows calculation of a rate increment $\Delta\dot{N}_{\alpha'\alpha}(t; \vec{P}_s, \vec{R}_s) = J_{0,\alpha}\Delta\sigma_{\alpha'\alpha}(t; \vec{P}_s, \vec{R}_s)$, now however with trajectories generated by an effective potential energy $\mathcal{V}(\vec{P}, \vec{R})$. It provides statistical values at a final time from $\sum_s \Delta\sigma_{\alpha'\alpha}(t_{fn}; \vec{P}_s, \vec{R}_s)/N_{in} = \overline{\Delta\sigma_{\alpha'\alpha}}$ as differential cross sections $I_{\alpha'\alpha}(\Omega, v) = \overline{\Delta\sigma_{\alpha'\alpha}}/\Delta\Omega$ that account for changes of the dynamics when the averaged potential energy $\overline{\mathcal{V}}(\vec{R})$ changes into $\mathcal{V}(\vec{P}, \vec{R}) = V_{qu}(\vec{P}, \vec{R})$, containing quantum effects from the eikonal amplitude $\chi(t, \boldsymbol{q}; \vec{P}_s, \vec{R}_s)$.

3.2 Coupled-Channel Equations

3.2.1 Differential Equations and Boundary Conditions

To solve for the scattering states, we expand $\Psi^{(+)}_\alpha$ in a basis set of orthonormalized target states $\phi_{\alpha'}$ including both open and closed channels. Hence

$$\Psi^{(+)}_\alpha(\vec{R}, \boldsymbol{q}) = \sum_{\text{all } n'} \psi^{(+)}_{\alpha'\alpha}(\vec{R})\, \phi_{n'}(\boldsymbol{q})$$

$$\psi^{(+)}_{\alpha'\alpha} \approx \psi^0\left(\vec{R}; \vec{k}_\alpha\right)\delta_{\alpha'\alpha} + (2\pi)^{-\frac{3}{2}} f^{(+)}_{\alpha'\alpha}(\Omega) R^{-1} e^{ik_{\alpha'}R}, \quad \alpha' \text{ open}$$

$$\psi^{(+)}_{\alpha'\alpha} \approx C_{\alpha'\alpha}(\Omega) R^{-1} e^{-\kappa_{\alpha'}R}, \quad \alpha' \text{ closed}$$

Replacing this expansion in the differential equation for $\Psi^{(+)}_\alpha$, with $\widehat{H} = \widehat{K}_R + \widehat{H}_{int}(\boldsymbol{Q}') + V(\vec{R}, \boldsymbol{Q}')$, and projecting on $\langle\phi_{n'}|$, one gets the set of coupled differential equations for scattering amplitudes, with $E - E_{n'} = (\hbar^2 k^2_{\alpha'})/2M$, as

$$\frac{\hbar^2}{2M}\left(k^2_{\alpha'} + \nabla^2_R\right)\psi^{(+)}_{\alpha'\alpha}(\vec{R}) = \sum_{n''} V_{n'n''}(\vec{R})\psi^{(+)}_{\alpha''\alpha}(\vec{R})$$

where $V_{n'n''}(\vec{R}) = \langle\phi_{n'}|V|\phi_{n''}\rangle$ and the bracket here signifies integration over $\boldsymbol{q}$.

3.2.2 Integral Equations

This may be solved for open channel states in terms of the Green function $G^{(+)}_{0,\alpha'}(\vec{R}, \vec{R}')$ for each uncoupled channel, a solution of

$$\frac{\hbar^2}{2M}\left(k^2_{\alpha'} + \nabla^2_R\right) G^{(-)}_{0,\alpha'}(\vec{R}, \vec{R}') = \delta(\vec{R} - \vec{R}')$$

given by

$$G^{(+)}_{0,\alpha'}(\vec{R},\vec{R}') = -\frac{1}{4\pi}\frac{2M}{\hbar^2}\frac{\exp\left(ik_{\alpha'}|\vec{R}-\vec{R}'|\right)}{\left|\vec{R}-\vec{R}'\right|}$$

$$G^{(+)}_{0,\alpha'} \approx -\frac{1}{4\pi}\frac{2M}{\hbar^2}\exp\left(ik_{\alpha'}R\right)R^{-1}\exp\left(-ik_{\alpha'}\vec{n}\cdot\vec{R}'\right),\quad \text{for } R\to\infty,\ \ \vec{n}=\vec{R}/R$$

to obtain the set of coupled integral equations

$$\psi^{(+)}_{\alpha'\alpha}(\vec{R}) = \psi^{0}(\vec{R};\vec{k}_\alpha)\delta_{\alpha'\alpha} + \sum_{\alpha''}\int d^3R'\,G^{(+)}_{0,\alpha'}(\vec{R},\vec{R}')V_{\alpha'\alpha''}(\vec{R}')\psi^{(+)}_{\alpha''\alpha}(\vec{R}')$$

A similar expression is valid for closed α' channels, obtained by omitting the first term to the right and replacing the wavenumber $k_{\alpha'}$ in the exponent of $G^{(+)}_{0,\alpha'}$ with an imaginary valued $i\kappa_{\alpha'}$ to construct an exponentially decaying propagator $G^{(d)}_{0,\alpha'}$ as $|\vec{R}-\vec{R}'|\to\infty$.

Letting $R\to\infty$ in the integral equation and using the asymptotic form of the Green function, one finds, with $\vec{k}_{\alpha'} = k_{\alpha'}\vec{n}$ in the direction of scattering, that the scattering amplitude is given by

$$f^{(+)}_{\alpha'\alpha}(\Omega) = -\frac{1}{4\pi}\frac{2M}{\hbar^2}(2\pi)^3\sum_{\alpha''}\int d^3R\,\psi^{0}\left(\vec{R};\vec{k}_{\alpha'}\right)^{*}V_{\alpha'\alpha''}(\vec{R})\psi^{(+)}_{\alpha''\alpha}\left(\vec{R};\vec{k}_\alpha\right)$$

or alternatively

$$f^{(+)}_{\alpha'\alpha}(\Omega) = -(2\pi)^2\frac{M}{\hbar^2}T_{\alpha'\alpha}\left(\vec{k}_{\alpha'},\vec{k}_\alpha\right),\quad T_{\alpha'\alpha}\left(\vec{k}_{\alpha'},\vec{k}_\alpha\right) = \left\langle\Psi^0_\alpha\left(\vec{k}_{\alpha'}\right)\middle|V\middle|\Psi^{(+)}_\alpha\left(\vec{k}_\alpha\right)\right\rangle$$

where $T_{\alpha'\alpha}(\vec{k}_{\alpha'},\vec{k}_\alpha)$ is a state-to-state inelastic transition amplitude, and the bracket here signifies integration over $\vec{R}$.

Similarly as for elastic collisions, a first iteration of the coupled integral equations gives the first Born approximation

$$T^{(B1)}_{\alpha'\alpha}\left(\vec{k}_{\alpha'},\vec{k}_\alpha\right) = \left\langle\Psi^0_\alpha(\vec{k}_{\alpha'})\middle|V\middle|\Psi^0_\alpha\left(\vec{k}_\alpha\right)\right\rangle = \int\frac{d^3R}{(2\pi)^3}\exp\left(-i\vec{q}_{\alpha'\alpha}\cdot\vec{R}\right)V_{\alpha'\alpha}(\vec{R})$$

where $\hbar\vec{q}_{\alpha'\alpha} = \hbar\vec{k}_{\alpha'} - \hbar\vec{k}_\alpha$ is the collisional momentum transfer. This approximation can be used for small momentum transfer where the collision is dominated by the long-range potential energy function and where short-range repulsions can be ignored.

The integral equation for $\psi^{(+)}_{\alpha'\alpha}(\vec{R})$ can be written in a compact way introducing a Dirac bra-ket notation with states $|\vec{R}\rangle$ in the relative coordinate satisfying $\langle\vec{R}'|\vec{R}\rangle = \delta(\vec{R}'-\vec{R})$, for relative motion wavefunctions given as $\psi^{(+)}_{\alpha'}(\vec{R}) = \langle\vec{R}|\psi^{(+)}_{\alpha'}\rangle$, a component in the space of relative motion states. The integral equation is then

$$\left|\psi^{(+)}_{\alpha'}\right\rangle = \left|\psi^0\right\rangle\delta_{\alpha'\alpha} + \sum_{\alpha''}\widehat{G}^{(+)}_{0,\alpha'}\widehat{V}_{\alpha'\alpha''}\left|\psi^{(+)}_{\alpha''}\right\rangle$$

involving operators $\widehat{G}^{(+)}_{0,\alpha}$ and $\widehat{V}_{\alpha'\alpha''}$ with matrix elements $\langle\vec{R}'|\widehat{G}^{(+)}_{0,\alpha}|\vec{R}\rangle = G^{(+)}_{0,\alpha}(\vec{R}',\vec{R})$ and $\langle\vec{R}'|\widehat{V}_{\alpha'\alpha}|\vec{R}\rangle = \delta(\vec{R}'-\vec{R})V_{\alpha'\alpha}(\vec{R})$. This equation can be expanded as a series in powers of $\widehat{G}^{(+)}_0\widehat{V}$ by iteration of the right-hand side unknown $|\psi^{(+)}_{\alpha''}\rangle$.

3.2.3 Partial-Wave Expansion

The set of coupled partial differential equations for inelastic scattering may be transformed into a (larger) set of coupled ordinary differential equations by means of a partial-wave expansion using spherical coordinates for the relative motion and its orbital angular momentum operator $\hat{L}$. We split the Hamiltonian operator $\hat{H}$ into

$$\hat{H} = -\frac{\hbar^2}{2M}\frac{1}{R}\frac{\partial^2}{\partial R^2}R + \hat{H}_L + V$$

$$\hat{H}_L = \hbar^2\hat{L}^2/2MR^2 + \hat{H}_{int}$$

and introduce the eigenfunctions of $\hat{H}_L$ with

$$\hat{H}_L\Phi_\lambda(\Omega, \boldsymbol{q}) = E_\lambda\Phi_\lambda(\Omega, \boldsymbol{q})$$

where

$$\Phi_\lambda(\Omega, \boldsymbol{q}) = Y_{lm}(\Omega)\phi_n(\boldsymbol{q})$$

with (lm) the orbital angular momentum quantum numbers, and $\lambda = (l, m, n)$.

For free motion, when $V = 0$, and the relative momentum $\vec{P}_\alpha = \hbar\vec{k}_\alpha$, the wavefunction is

$$\Psi_\alpha^0\left(\vec{R}, \boldsymbol{q}; \vec{k}_\alpha\right) = \psi^0\left(\vec{R}; \vec{k}_\alpha\right)\phi_n(\boldsymbol{q})$$

$$\psi^0\left(\vec{R}; \vec{k}_\alpha\right) = (2\pi)^{-\frac{3}{2}}e^{i\vec{k}_\alpha.\vec{R}} = (2\pi)^{-\frac{3}{2}}\sum_{l,m}4\pi i^l(k_\alpha R)^{-1}u_l(k_\alpha R)Y_{lm}(\Omega)Y_{lm}(\Omega_\alpha)^*$$

where $u_l(x)$ is the regular Riccati–Bessel function, satisfying $u_l(x) = \sqrt{\pi x/2}\,J_{l+\frac{1}{2}}(x) \approx \sin(x - l\pi/2)$ in terms of the half-integer Bessel function, and Ω_α contains the orientation angles of $\vec{k}_\alpha$. Alternatively,

$$\Psi_\alpha^0\left(\vec{R}, \boldsymbol{q}; \vec{k}_\alpha\right) = (2\pi)^{-\frac{3}{2}}\sum_{l,m}4\pi i^l(k_\alpha R)^{-1}u_l(k_\alpha R)Y_{lm}(\Omega_\alpha)^*\Phi_\lambda(\Omega, \boldsymbol{q})$$

with $\Phi_\lambda(\Omega, \boldsymbol{q}) = Y_{lm}(\Omega)\phi_n(\boldsymbol{q})$.

We perform next a double expansion of the full scattering wavefunction $\Psi_\alpha^{(+)}(\vec{R}, \boldsymbol{q}; \vec{k}_\alpha)$ during a collision with $V \neq 0$, treated as a function of both space variables and wavevector variables, so that

$$\Psi_\alpha^{(+)}\left(\vec{R}, \boldsymbol{q}; \vec{k}_\alpha\right) = \sum_{\lambda'}\psi_{\lambda'\alpha}^{(+)}\left(R; \vec{k}_\alpha\right)\Phi_{\lambda'}(\Omega, \boldsymbol{q})$$

$$\psi_{\lambda'\alpha}^{(+)}\left(R; \vec{k}_\alpha\right) = (2\pi)^{-\frac{3}{2}}\sum_{l,m}4\pi i^l(k_\alpha R)^{-1}\psi_{\lambda'\lambda}(R; k_\alpha)Y_{lm}(\Omega_\alpha)^*$$

containing the radial scattering function $\psi_{\lambda'\lambda}(R; k_\alpha) \approx \sin(k_\alpha R - l\pi/2)\delta_{\lambda'\lambda}$ when λ' is an open channel and with $\psi_{\lambda'\lambda}(R; k_\alpha) = u_l(k_\alpha R)\delta_{\lambda'\lambda}$ in the absence of coupling.

This expansion must contain both open channel states that appear asymptotically at large distance and closed channel states that decrease exponentially but are also important for shaping the total wavefunction at short distances.

Replacing the function $\Psi_\alpha^{(+)}(\vec{R}, \boldsymbol{q}; \vec{k}_\alpha)$, with all its variables, into the scattering equation and taking scalar products by successively multiplying on the left with $\Phi_{\lambda'}(\Omega, \boldsymbol{q})^*$ and with $Y_{l'm'}(\Omega_\alpha)$ and integrating over Ω, Ω_α, and $\boldsymbol{q}$ for fixed R, one arrives at the set of equations

$$\left[k_{\alpha'}^2 + \frac{d^2}{dR^2} - \frac{l'(l'+1)}{R^2}\right] \psi_{\lambda'\lambda}(R; k_\alpha) = \sum_{\lambda''} U_{\lambda'\lambda''}(R)\psi_{\lambda''\lambda'}(R; k_\alpha)$$

$$U_{\lambda'\lambda''}(R) = (2M/\hbar^2)\langle\Phi_{\lambda'}|V|\Phi_{\lambda''}\rangle_R$$

which are coupled ordinary differential equations to be solved with scattering asymptotic conditions.

Generalizing the procedure used for elastic scattering, we choose as boundary conditions

$$k_\alpha^{-1/2}\psi_{\lambda'\lambda}(R; k_\alpha) \approx k_\alpha^{-1/2} u_l(k_\alpha R)\delta_{\lambda'\lambda} + k_{\alpha'}^{-1/2} w_l^{(+)}\left(k_{\alpha'}R\right) f_{\lambda'\lambda}(k_\alpha), \quad \lambda' \text{ open}$$

$$k_\alpha^{-1/2}\psi_{\lambda'\lambda}(R; k_\alpha) \approx k_{\alpha'}^{-1/2} e^{\frac{il'\pi}{2}} w_l^{(+)}\left(i\kappa_{\alpha'}R\right) f_{\lambda'\lambda}(k_\alpha), \quad \lambda' \text{ closed}$$

where $w_l^{(+)}(x) \approx \exp[i(x - l\pi/2)]$ is a Riccati–Hankel function. The $w_l^{(+)}\left(i\kappa_{\alpha'}R\right)$ function decays asymptotically as $\exp\left(-\kappa_{\alpha'}R\right)$, and the factors $k_\alpha^{-1/2}$ have been introduced so that all the asymptotic functions give the same flux value, while flux changes are included in the amplitude coefficients $f_{\lambda'\lambda}$. Then $\psi_{\alpha'\alpha}^{(+)}(\vec{R})$ contains asymptotically a plane wave plus outgoing waves with amplitudes

$$f_{\alpha'\alpha}^{(+)}(\Omega; \vec{k}_\alpha) = \frac{1}{\sqrt{k_\alpha k_{\alpha'}}} \sum_{l,m} \sum_{l',m'} 4\pi^{l-l'} f_{\lambda'\lambda}(k_\alpha) Y_{l'm'}(\Omega) Y_{lm}(\Omega_\alpha)^*$$

where Ω and Ω_α give the orientation angles of the scattered wave momentum along $\vec{n} = \vec{R}/R$ and of the initial wavevector $\vec{k}_\alpha$, respectively.

Integral cross sections follow by choosing $\vec{k}_\alpha$ along the z-axis, in which case $Y_{lm}(\Omega_\alpha) = (2l+1)\delta m_0/(4\pi)$. Integrating the differential cross section over solid angle increments $d\Omega$ and using that $\langle Y_{lm}|Y_{l'm'}\rangle = \delta_{ll'}\delta_{mm'}$, we obtain the inelastic channel-to-channel integral cross sections

$$\sigma_{\alpha'\alpha}(k_\alpha) = \frac{4\pi}{k_\alpha^2} \sum_{l',m'} \sum_{l} \sum_{\bar{l}} i^{l-\bar{l}} (2l+1)^{1/2} (2\bar{l}+1)^{1/2} f_{\lambda'\lambda_0}{}^* f_{\lambda'\bar{\lambda}_0}$$

or

$$\sigma_{\alpha'\alpha}(k_\alpha) = \frac{4\pi}{k_\alpha^2} \sum_{l',m'} \left| \sum_{l} (2l+1)^{1/2} i^l f_{\lambda'\lambda_0} \right|^2$$

where we used the notation $\lambda_0 = (l0\alpha)$ and $\bar{\lambda}_0 = (\bar{l}0\alpha)$. This can be rewritten as

$$\sigma_{\alpha'\alpha}(k_\alpha) = \frac{4\pi}{k_\alpha^2}\sum_{l'}(2l'+1)P_{\alpha'\alpha}^{(l')}$$

$$P_{\alpha'\alpha}^{(l')} = \sum_{m'}\left|\sum_{l}\left(\frac{2l+1}{2l'+1}\right)^{1/2} i^l f_{\lambda'\lambda_0}\right|$$

in terms of a partial transition probability $P_{\alpha'\alpha}^{(l')}$. Adding over all final open channels, we obtain the total integral inelastic cross section $\sigma_{tot,\alpha}(k_\alpha) = \sum_{\alpha'}\sigma_{\alpha'\alpha}(k_\alpha)$.

3.3 Matrix Form of Partial-Wave Equations

3.3.1 Differential Equations and Boundary Conditions

The previous equations may be rewritten in a matrix notation that simplifies further developments. The index λ, called here a partial-wave channel index or simply a channel index, runs over N_o values for open channels and N_c values for closed channels, with matrices of total order $N = N_o + N_c$. Defining $N \times N$ matrices with operator-valued elements

$$\hat{\boldsymbol{D}}_R = \left[\delta_{\lambda'\lambda}\left(k_\alpha^2 + \frac{d^2}{dR^2} - \frac{l(l+1)}{R^2}\right)\right],$$

$$\boldsymbol{U}(R) = \left[U_{\lambda'\lambda}(R)\right], \quad \boldsymbol{\psi}^{(+)}(R) = \left[\psi_{\lambda'\lambda}(R)\right]$$

the set of coupled differential equations is then

$$\hat{\boldsymbol{D}}_R\boldsymbol{\psi}^{(+)}(R) = \boldsymbol{U}(R)\boldsymbol{\psi}^{(+)}(R).$$

Boundary conditions may also be summarized in matrix form, classifying states into open (o) and closed (c) classes with N_o and N_c states, respectively, writing $\boldsymbol{\psi}^{(+)}$ in the block form

$$\boldsymbol{\psi}^{(+)} = \begin{pmatrix}\boldsymbol{\psi}_{oo}\\ \boldsymbol{\psi}_{co}\end{pmatrix}$$

that is a $N \times N_o$ matrix with $N_o \times N_o$ and $N_c \times N_o$ blocks for open and closed components. Using the previous regular Riccati–Bessel functions $u_l(x)$ and also the Riccati–Neumann functions $\upsilon_l(x)$ which are singular at the origin $R = 0$ and are asymptotically

$$\upsilon_l(x) = \sqrt{\frac{\pi x}{2}}J_{-l-\frac{1}{2}}(x) \approx \cos\left(x - \frac{l\pi}{2}\right),$$

it is convenient to define the $N_o \times N_o$ matrices

$$\boldsymbol{I}^{(s)}(R) = \left[\delta_{\lambda'\lambda}k_\alpha^{-\frac{1}{2}}u_l(k_\alpha R)\right]$$

$$\boldsymbol{I}^{(c)}(R) = \left[\delta_{\lambda'\lambda} k_\alpha^{-\frac{1}{2}} v_l(k_\alpha R)\right]$$

$$\boldsymbol{I}^{(\pm)}(R) = \boldsymbol{I}^{(c)}(R) \pm i\boldsymbol{I}^{(s)}(R)$$

which go asymptotically to sin, cos, and exponentially oscillating functions. Then the asymptotic conditions for $\boldsymbol{\psi}_{oo}$ imply

$$\boldsymbol{\psi}_{oo}(R) \approx [\boldsymbol{I}^{(s)}(R) + \boldsymbol{I}^{(+)}(R)\boldsymbol{T}_{oo}]\boldsymbol{k}_o^{1/2}$$

where $\boldsymbol{T}_{oo} = \left[t_{\lambda'\lambda}\right]$ and $\boldsymbol{k}_o = \left[\delta_{\lambda'\lambda} k_\alpha\right]$, with indices running over open channels.

Other useful asymptotic forms are the real-valued form

$$\boldsymbol{\psi}_{oo}^{(K)}(R) \approx [\boldsymbol{I}^{(s)}(R) + \boldsymbol{I}^{(c)}(R)\boldsymbol{K}_{oo}]\boldsymbol{k}_o^{1/2}$$

and the in-out symmetric form

$$\boldsymbol{\psi}_{oo}^{(S)}(R) \approx [\boldsymbol{I}^{(-)}(R) - \boldsymbol{I}^{(+)}(R)\boldsymbol{S}_{oo}]\boldsymbol{k}_o^{1/2}$$

from which, comparing asymptotic forms in $\boldsymbol{\psi}_{oo}(R) \approx \boldsymbol{\psi}_{oo}^{(K)}(R)\boldsymbol{A}^{(K)} \approx \boldsymbol{\psi}_{oo}^{(S)}(R)\boldsymbol{A}^{(S)}$,

$$\boldsymbol{T}_{oo} = \boldsymbol{K}_{oo}(\boldsymbol{1}_{oo} - i\boldsymbol{K}_{oo})^{-1} = (2i)^{-1}(\boldsymbol{S}_{oo} - \boldsymbol{1}_{oo}),$$

$$\boldsymbol{S}_{oo} = (\boldsymbol{1}_{oo} + i\boldsymbol{K}_{oo})(\boldsymbol{1}_{oo} - i\boldsymbol{K}_{oo})^{-1}.$$

The matrix $\boldsymbol{K}_{oo}$ is real-valued and symmetric when the potential matrix $\boldsymbol{U}(R)$ is Hermitian and symmetric. The so-called scattering matrix $\boldsymbol{S}_{oo}$ may then be shown to be symmetric and unitary. Unitarity and reciprocity (transposal symmetry) follow generally from

$$\left\langle \Psi_\beta^{(+)} \middle| \widehat{K}_R \Psi_\alpha^{(+)} \right\rangle_R - \left\langle \widehat{K}_R \Psi_\beta^{(+)} \middle| \Psi_\alpha^{(+)} \right\rangle_R = 0$$

$$\left\langle \Psi_\beta^{(+)*} \middle| \widehat{K}_R \Psi_\alpha^{(+)} \right\rangle_R - \left\langle \widehat{K}_R \Psi_\beta^{(+)*} \middle| \Psi_\alpha^{(+)} \right\rangle_R = 0$$

where brackets are calculated for fixed R, and the limit of large R is taken. This gives

$$S_{\lambda'\lambda} = S_{\lambda\lambda'}$$

$$\sum_{\lambda'} S_{\lambda\lambda'} S_{\lambda'\lambda''}{}^* = \delta_{\lambda\lambda''}$$

for a potential energy satisfying the symmetry condition $U_{\lambda'\lambda} = U_{\lambda\lambda'}$. These equations may also be derived from time-reversal invariance and flux conservation, respectively.

The quantity $|S_{\lambda'\lambda}|^2$ can be interpreted as the probability for finding the molecular pair in the final channel λ' after a collision event that started in the initial channel λ.

The closed-channel boundary conditions may be expressed in terms of the exponentially growing and decaying $N_c \times N_c$ matrices

$$\boldsymbol{I}^{(g)}(R) = \left[\delta_{\lambda'\lambda} k_\alpha^{-\frac{1}{2}} e^{-i(l+1)\pi/2}\, u_l(i|k_\alpha|R)\right]$$

$$\approx \left[\delta_{\lambda'\lambda} k_\alpha^{-\frac{1}{2}} \exp(\kappa_\alpha R)/2\right]$$

$$\boldsymbol{I}^{(d)}(R) = \left[\delta_{\lambda'\lambda} k_\alpha^{-\frac{1}{2}} e^{il\pi/2} w_l^{(+)}(i|k_\alpha|R)\right]$$

$$\approx \left[\delta_{\lambda'\lambda} k_\alpha^{-\frac{1}{2}} \exp(-\kappa_\alpha R)\right]$$

by means of which the block of decaying relative motion wavefunctions take the asymptotic form

$$\boldsymbol{\psi}_{co}(R) \approx \boldsymbol{I}^{(d)}(R)\boldsymbol{T}_{co}\boldsymbol{k}_o^{1/2}.$$

The $(N_o + N_c) \times N_o$ matrix of wavefunction solutions for all the initial open channels gives asymptotically

$$\boldsymbol{\psi}^{(+)} = \begin{pmatrix} \boldsymbol{\psi}_{oo}(R) \\ \boldsymbol{\psi}_{co}(R) \end{pmatrix} \approx \begin{pmatrix} \boldsymbol{I}^{(s)}(R)\boldsymbol{k}_o^{1/2} \\ \boldsymbol{0} \end{pmatrix} + \begin{pmatrix} \boldsymbol{I}^{(+)}(R)\boldsymbol{T}_{oo}\boldsymbol{k}_o^{1/2} \\ \boldsymbol{I}^{(d)}(R)\boldsymbol{T}_{co}\boldsymbol{k}_o^{1/2} \end{pmatrix}.$$

The set of $N = N_o + N_c$ coupled second-order differential equations, together with regular boundary conditions at the origin $R = 0$ and the above asymptotic conditions for open and closed channels as $R \to \infty$, define a unique mathematical solution of the scattering problem.

3.3.2 Integral Equations

Differential equations and their asymptotic conditions can be combined in a treatment with coupled integral equations. The $\boldsymbol{\psi}^{(+)}(R) = \left[\psi_{\lambda'\lambda}(R)\right]$ physical solutions for initial channel λ and final channel λ' can be expressed for each initial channel λ as linear combinations of $N_o + N_c$ numerical linear independent regular solutions $\boldsymbol{y}_\lambda(R)$ of $\widehat{\boldsymbol{D}}_R \boldsymbol{y}_\lambda(R) = \boldsymbol{U}(R)\boldsymbol{y}_\lambda(R)$ organized as a column matrix $\boldsymbol{y}_\lambda(R) = \left[y_{\lambda'\lambda}(R)\right]$. The row of such numerical solutions for $N = N_c + N_c$ initial numerical conditions provide a $N \times N$ square matrix of functions as

$$\boldsymbol{y} = (\boldsymbol{y}_1, \boldsymbol{y}_2, \ldots, \boldsymbol{y}_N) = \begin{pmatrix} \boldsymbol{y}_{oo} & \boldsymbol{y}_{oc} \\ \boldsymbol{y}_{co} & \boldsymbol{y}_{cc} \end{pmatrix}$$

satisfying $\widehat{\boldsymbol{D}}_R \boldsymbol{y}(R) = \boldsymbol{U}(R)\boldsymbol{y}(R)$ and all asymptotic conditions for both open and closed channels. The physical solutions are of block form

$$\boldsymbol{\psi}^{(+)} = \begin{pmatrix} \boldsymbol{\psi}_{oo} \\ \boldsymbol{\psi}_{co} \end{pmatrix} = \begin{pmatrix} \boldsymbol{y}_{oo} & \boldsymbol{y}_{oc} \\ \boldsymbol{y}_{co} & \boldsymbol{y}_{cc} \end{pmatrix} \begin{pmatrix} \boldsymbol{C}_{oo} \\ \boldsymbol{C}_{co} \end{pmatrix}$$

with blocks $\boldsymbol{C}_{oo}$ and $\boldsymbol{C}_{co}$ which can be related to scattering matrices $\boldsymbol{T}_{oo}$ and $\boldsymbol{T}_{co}$ at large distances.

Generalizing the procedure for elastic scattering, we can construct diagonal matrices of free motion radial Green functions satisfying $\widehat{\boldsymbol{D}}_R \boldsymbol{G}^{(0)}(R,R') = \delta(R-R')\mathbf{I}$, with $\mathbf{I}$ the $N \times N$ identity matrix, for the sets of open and closed channels, with diagonal elements obtained from the diagonal blocks in

$$\boldsymbol{G}^{(0)}(R,R') = \begin{pmatrix} \boldsymbol{G}_{oo}(R,R') & \mathbf{0} \\ \mathbf{0} & \boldsymbol{G}_{cc}(R,R') \end{pmatrix}$$
$$\boldsymbol{G}_{oo}(R,R') = \boldsymbol{I}^{(s)}(R_>)\boldsymbol{I}^{(c)}(R_<) - \boldsymbol{I}^{(c)}(R_>)\boldsymbol{I}^{(s)}(R_<)$$
$$\boldsymbol{G}_{cc}(R,R') = \boldsymbol{I}^{(g)}(R_>)\boldsymbol{I}^{(d)}(R_<) - \boldsymbol{I}^{(d)}(R_>)\boldsymbol{I}^{(g)}(R_<)$$

of dimensions $N_o \times N_o$ and $N_c \times N_c$, respectively, with $R_>$ (or $R_<$) the larger (or smaller) between R and R'. Then the physical solution is

$$\boldsymbol{\psi}^{(+)}(R) = \boldsymbol{\psi}^{(0)}(R) + \int_0^\infty dR'\, \boldsymbol{G}^{(0)}(R,R')\boldsymbol{U}(R')\boldsymbol{\psi}^{(+)}(R')$$

a Fredholm-type integral equation with a kernel over an infinite domain, with

$$\boldsymbol{\psi}^{(0)}(R) = \begin{pmatrix} \boldsymbol{I}^{(s)}(R) & \mathbf{0} \\ \mathbf{0} & \boldsymbol{I}^{(d)}(R) \end{pmatrix} \begin{pmatrix} \mathbf{I}_{oo} \\ \mathbf{0}_{co} \end{pmatrix}.$$

However, for computational purposes, it is more convenient to first construct solutions $\boldsymbol{y}_\lambda(R)$ of Volterra-type integral equations [25]. This can be accomplished with free Green functions $\mathbf{g}_{oo}$ and $\mathbf{g}_{cc}$ satisfying again $\widehat{\boldsymbol{D}}_R(R,R')\mathbf{g}_{aa} = \delta(R-R')\mathbf{I}_{aa}$, now given by

$$\mathbf{g}_{oo}(R,R') = \boldsymbol{I}^{(s)}(R)\boldsymbol{I}^{(c)}(R') - \boldsymbol{I}^{(c)}(R)\boldsymbol{I}^{(s)}(R'),\quad R > R',\quad \mathbf{g}_{oo}(R,R') = 0,\quad R < R'$$
$$\mathbf{g}_{cc}(R,R') = \boldsymbol{I}^{(g)}(R)\boldsymbol{I}^{(d)}(R') - \boldsymbol{I}^{(d)}(R)\boldsymbol{I}^{(g)}(R'),\quad R > R',\quad \mathbf{g}_{cc}(R,R') = 0,\quad R < R'.$$

The total free radial Green function matrix is

$$\mathbf{g}_0(R,R') = \begin{pmatrix} \mathbf{g}_{oo}(R,R') & \mathbf{0} \\ \mathbf{0} & \mathbf{g}_{cc}(R,R') \end{pmatrix}$$

and numerical solutions are given by

$$\boldsymbol{y}(R) = \boldsymbol{y}^{(0)}(R) + \int_0^R dR'\, \mathbf{g}_0(R,R')\boldsymbol{U}(R')\boldsymbol{y}(R')$$
$$\boldsymbol{y}^{(0)}(R) = \begin{pmatrix} \boldsymbol{I}^{(s)}(R) & \mathbf{0} \\ \mathbf{0} & \boldsymbol{I}^{(d)}(R) \end{pmatrix}$$

with free motion radial Green functions satisfying $\widehat{\boldsymbol{D}}_R \mathbf{g}_0(R,R') = \delta(R-R')\mathbf{I}$, which contain within $\mathbf{g}_0$ oscillating functions for open channels and both exponentially increasing and decreasing amplitudes in the closed channels, for large R.

However, the physical solutions $\psi^{(+)}$ contain asymptotically only exponentially decreasing amplitudes in the closed channels and must be constructed choosing the coefficient matrices C_{oo} and C_{co} so that the exponentially increasing components are cancelled.

At large distances,

$$\begin{pmatrix} y_{oo} & y_{co} \\ y_{oc} & y_{cc} \end{pmatrix} \approx \begin{pmatrix} I^{(s)} & 0 \\ 0 & I^{(g)} \end{pmatrix} a + \begin{pmatrix} I^{(c)} & 0 \\ 0 & I^{(d)} \end{pmatrix} b, \quad R \to \infty$$

where a and b are constant $N \times N$ matrices obtained from the above integral equation for $y(R)$ as $R \to \infty$, which give

$$a = \mathbf{I} + \int_0^\infty dR' \begin{pmatrix} I^{(c)} & 0 \\ 0 & I^{(d)} \end{pmatrix}_{R'} U(R')y(R'), \quad b = -\int_0^\infty dR' \begin{pmatrix} I^{(s)} & 0 \\ 0 & I^{(g)} \end{pmatrix}_{R'} U(R')y(R').$$

Scattering amplitudes from initial open channels follow from matching to the asymptotic conditions for real-valued outgoing waves, in

$$\psi^{(K)} \approx \begin{pmatrix} y_{oo} & y_{oc} \\ y_{co} & y_{cc} \end{pmatrix} \begin{pmatrix} C_{oo} \\ C_{co} \end{pmatrix} \approx \begin{pmatrix} I^{(s)}(R)k_o^{1/2} \\ 0 \end{pmatrix} + \begin{pmatrix} I^{(c)}(R)K_{oo}k_o^{1/2} \\ I^{(d)}(R)K_{co}k_o^{1/2} \end{pmatrix}$$

which provide the relations

$$a \begin{pmatrix} C_{oo} \\ C_{co} \end{pmatrix} = \begin{pmatrix} \mathbf{I}_{oo}k_o^{-1/2} \\ 0_{co} \end{pmatrix}, \quad b \begin{pmatrix} C_{oo} \\ C_{co} \end{pmatrix} = \begin{pmatrix} K_{oo}k_o^{-1/2} \\ K_{co}k_o^{-1/2} \end{pmatrix}$$

and a set of $2 \times N \times N_o$ algebraic equations for the same number of unknowns in C_{oo}, C_{co}, K_{oo}, and K_{co}. They give

$$\begin{pmatrix} K_{oo} \\ K_{co} \end{pmatrix} = ba^{-1} \begin{pmatrix} \mathbf{I}_{oo} \\ 0_{co} \end{pmatrix}$$

and also $\psi^{(K)}(R)$ from the known C_{oo}, C_{co} coefficients, and from $y(R)$. The scattering function $\psi^{(+)}(R)$ follows from the relations between the $N \times N_o$ matrices K and T.

It can in principle be obtained numerically by iteration of its integral equation, but more practically it requires numerical procedures which avoid the propagation of exponentially increasing terms as the distance R increases.

An alternative expression for the numerical solutions is

$$y(R) = I_a(R)A(R) - I_b(R)B(R)$$

$$I_a = \begin{pmatrix} I^{(s)} & 0 \\ 0 & I^{(g)} \end{pmatrix}, \quad I_b(R) = \begin{pmatrix} I^{(c)}(R) & 0 \\ 0 & I^{(d)}(R) \end{pmatrix}$$

where $A(R)$ and $B(R)$ are $N \times N$ matrices of coefficient functions of R, satisfying coupled integral equations. At large distances $R > R_{out}$ where interaction potential energies are null and the collision is over, $\boldsymbol{A}(R) = \boldsymbol{a}$ and $\boldsymbol{B}(R) = \boldsymbol{b}$ are matrices of constants and provide scattering matrices as follows.

Outward integration of the equation for $\boldsymbol{y}(R)$ past $R = R_{out}$ give also its derivative $\boldsymbol{y}'(R) = d\boldsymbol{y}/dR$ and the relations

$$\boldsymbol{y}(R) = \boldsymbol{I}_a(R)\boldsymbol{a} + \boldsymbol{I}_b(R)\boldsymbol{b}$$

$$\boldsymbol{y}'(R) = \boldsymbol{I}_a{}'(R)\boldsymbol{a} + \boldsymbol{I}_b{}'(R)\boldsymbol{b}$$

where $\boldsymbol{a}$ and $\boldsymbol{b}$ can be obtained from $\boldsymbol{y}$ and $\boldsymbol{y}'$ in terms of Wronskians $W[\boldsymbol{f},\boldsymbol{g}] = \boldsymbol{f}\boldsymbol{g}' - \boldsymbol{g}\boldsymbol{f}'$ as

$$\boldsymbol{a} = W[\boldsymbol{y},\boldsymbol{I}_b]/W[\boldsymbol{I}_a,\boldsymbol{I}_b] \text{ and } \boldsymbol{b} = W[\boldsymbol{I}_a,\boldsymbol{y}]/W[\boldsymbol{I}_a,\boldsymbol{I}_b].$$

Furthermore, $\boldsymbol{\psi}^{(K)}(R) \approx \boldsymbol{y}(R)\boldsymbol{c}$, where the scattering function is written as real-valued to extract the $\boldsymbol{K}$ matrix, gives the relation

$$[\boldsymbol{I}_a(R)\boldsymbol{a} + \boldsymbol{I}_b(R)\boldsymbol{b}]\boldsymbol{c} = [\boldsymbol{I}_a(R) + \boldsymbol{I}_b(R)\,\boldsymbol{K}]\boldsymbol{k}^{1/2}$$

and $\boldsymbol{K} = \boldsymbol{b}\boldsymbol{a}^{-1} = W[\boldsymbol{I}_a,\boldsymbol{y}]/W[\boldsymbol{y},\boldsymbol{I}_b]$, from which $\boldsymbol{K}_{oo}$ can be extracted, and from it $\boldsymbol{T}_{oo}$ and $\boldsymbol{S}_{oo}$.

However, the generation of matrices $\boldsymbol{K}$ and $\boldsymbol{T}$ requires exponentially increasing functions in $\boldsymbol{I}_a$ at large distances and leads to loss of accuracy in coupled-channel calculations. Practical numerical procedures must be implemented to avoid propagation of channel functions coupled to exponentially increasing terms [25].

3.4 Collisions Involving Two Coupled Channels

3.4.1 Two Open Channels

As a simple example, consider a collision with total energy E involving two internal states $\phi_\alpha(Q)$ labeled by $\alpha = 1, 2$ with internal energies $E_\alpha < E$, coupled due to transfer of energy between R and Q motions, both in open collision channels and for the partial wave $l = 0$. The potential energy is $V(R, Q)$ with matrix elements $V_{\alpha\beta}(R)$ in the two-state basis set and asymptotic values $V_{11}(\infty) < E$, $V_{22}(\infty) < E$, and $V_{12}(\infty) = 0$.

The coupled radial equations for scattering wavefunctions $\psi_{11}(R)$ and $\psi_{21}(R)$, starting in channel 1, are

$$\left(k_1^2 + \frac{d^2}{dR^2}\right)\psi_{11}(R) = U_{11}(R)\psi_{11}(R) + U_{12}(R)\psi_{21}(R)$$

$$\left(k_2^2 + \frac{d^2}{dR^2}\right)\psi_{21}(R) = U_{21}(R)\psi_{11}(R) + U_{22}(R)\psi_{21}(R)$$

with $(\hbar^2 k_\alpha^2)/2M = \mathrm{E} - E_\alpha$ and similar equations for $\psi_{12}(R)$ and $\psi_{22}(R)$ starting in channel 2. These equations can be numerically solved with regular-origin values and asymptotic boundary conditions for open channels. Cross sections are obtained from the asymptotic transition amplitudes $T_{\beta\alpha}$, related to scattering amplitudes by $f^{(+)}_{\beta\alpha} = (k_\beta k_\alpha)^{-1/2} T_{\beta\alpha}$ so that, insofar there is spherical symmetry for $\ell = 0$,

$$\frac{d\sigma_{\beta\alpha}}{d\Omega} = \frac{k_\beta}{k_\alpha} \left| f^{(+)}_{\beta\alpha} \right|^2$$

that gives for the integral cross section the relation $\sigma_{\beta\alpha} = 4\pi \, |T_{\beta\alpha}|^2 / k_\alpha^2$.

Numerical integration is best done in real variables to generate numerical solutions in $\boldsymbol{y}(R) = [y_{nm}(R)]$, $(m, n = 1, 2)$, which go asymptotically as *sin* and *cos* functions. They can be combined to obtain real-valued physical solutions with asymptotic forms

$$\psi_{1\alpha}(R) k_\alpha^{-1/2} \approx k_1^{-1/2} \sin(k_1 R) + k_1^{-1/2} \cos(k_1 R) K_{1\alpha}$$

$$\psi_{2\alpha}(R) k_\alpha^{-1/2} \approx k_2^{-1/2} \cos(k_2 R) K_{2\alpha}$$

in terms of the 2×2 matrix $\boldsymbol{K}$, giving the related scattering matrices $\boldsymbol{T} = \boldsymbol{K}(\boldsymbol{1} - \boldsymbol{K})^{-1} = (\boldsymbol{S} - \boldsymbol{1})/(2i)$.

Collecting the numerical functions in the 2×2 matrix $\boldsymbol{y}$, the 2×1 matrices $\boldsymbol{\psi}_\alpha$ with elements $\psi_{\beta\alpha}$ can be expressed as combinations $\boldsymbol{\psi}_\alpha(R) = \boldsymbol{y}(R)\boldsymbol{C}_\alpha$, with a 2×1 matrix of coefficients $\boldsymbol{C}_\alpha$. The equality of those functions and their derivatives in $d\boldsymbol{\psi}_\alpha/dR = (d\boldsymbol{y}/dR)\boldsymbol{C}_\alpha$ at large distances gives four linear algebraic equations to be solved for the four unknowns in the matrices $\boldsymbol{C}_\alpha$ and $\boldsymbol{K}_\alpha$. This may be repeated for $\alpha = 1, 2$.

The resulting 2×2 scattering $\boldsymbol{S}$ matrix contains eight real parameters, but the symmetry condition for the complex-valued $S_{12} = S_{21}$ gives two relations, and the unitarity condition gives three relations, so that only three parameters are needed to define the matrix. These can be chosen to be A, η_1 and η_2 such that

$$S_{\alpha\alpha} = (1 - A)^{\frac{1}{2}} \exp(2i\eta_\alpha), \quad \alpha = 1, 2$$

$$S_{12} = S_{21} = iA \exp[i(\eta_1 + \eta_2)]$$

giving a general parametrization for any two-state scattering process.

3.4.2 Resonances for One Open- and One Closed-Channel

Another instructive simple example is a collision with total energy E involving again two internal states $\phi_\alpha(Q)$ labeled by $\alpha = 1, 2$ but now with state ϕ_1 in an open channel of internal energies $E_1 < E$, coupled to state ϕ_2 with energy $E_2 > E$ and therefore in a closed channel at this total energy [2, 5]. The model potential energy, for the partial wave $l = 0$, is again $V(R, Q)$ but now with matrix elements of asymptotic values $V_{11}(\infty) < E$, $V_{22}(\infty) > E$, and $V_{12}(\infty) = 0$. The coupled radial equations are the same as above, but now $E - E_2 = \left(\hbar^2 k_2^2\right)/2M \leq 0$ shows that the wavenumber in channel 2 must be chosen as an imaginary-valued $k_2 = i\kappa_2$, with a real-valued κ_2. The two states start at small distances as regular (finite) values, but they differ asymptotically, with ϕ_1 an oscillatory scattering wave at large distances while ϕ_2 must decay exponentially.

Introducing a Green function $G^{(2)}(R, R')$ for the closed channel 2, satisfying

$$\left(k_2^2 + \frac{d^2}{dR^2} - U_{22}(R)\right) G^{(2)}(R, R') = \delta(R - R'),$$

a formal solution for $\psi_{21}(R)$ is

$$\psi_{21}(R) = \int_0^R dR'\; G^{(2)}(R, R')U_{21}(R')\psi_{11}(R')$$

which can be replaced in the equation for $\psi_{11}(R)$ to obtain its implicit differential-integral equation. Here $G^{(2)}$ contains both exponentially increasing and decreasing functions $\exp(\pm\kappa_2 R)$ for large R, insofar at large distances in closed channel 2, we have that $U_{22}(\infty) > (2M/\hbar^2)E_2 - {\kappa_2}^2$.

Solutions for the equation

$$\left(k_1^2 + \frac{d^2}{dR^2}\right)\psi_{11}(R) = U_{11}(R)\psi_{11}(R) + U_{12}(R)\int_0^R dR'\; G^{(2)}(R, R')U_{21}(R')\psi_{11}(R')$$

can be obtained noticing that $U_{12}(R)\psi_{21}(R)$ is zero near the origin $R = 0$, where $\psi_{21}(R)$ vanishes, and also at large distances insofar $U_{12}(R) \approx 0$ for $R \to \infty$, and the product is finite in between. It can be therefore expanded in an orthonormal basis set $\{\eta_n(R), n = 1 \text{ to } N^{(2)}\}$ of L^2 functions.

A convenient way of doing this is to introduce the double expansion

$$G^{(2)}(R, R') = \sum_n \sum_{n'} \eta_n(R) G^{(2)}_{n,n'} \eta_{n'}(R')^* = \sum_n \sum_{n'} \langle R|\eta_n\rangle G^{(2)}_{n,n'} \langle \eta_{n'}|R'\rangle$$

where the Dirac bracket notation is used, giving

$$U_{12}(R)\psi_{21}(R) = U_{12}(R)\sum_n \sum_{n'} \langle R|\eta_n\rangle G^{(2)}_{n,n'} \langle \eta_{n'}|U_{21}\psi_{11}\rangle$$

with the bracket here implying integration over R'. This shows that the effect of coupling of the open channel to the closed one can be described by an integrodifferential equation for $\psi_{11}(R)$,

$$\left(k_1^2 + \frac{d^2}{dR^2}\right)\psi_{11}(R) = U_{11}(R)\psi_{11}(R) + \int_0^R dR'\; W_1^{(2)}(R, R')\psi_{11}(R')$$

where a new nonlocal potential energy function

$$W_1^{(2)}(R, R') = U_{12}(R)\sum_n \sum_{n'} \langle R|\eta_n\rangle G^{(2)}_{n,n'} \left\langle \eta_{n'}|R'\right\rangle U_{21}(R')$$

has been added to the potential $U_{11}(R)$. Both U_{11} and $W_1^{(2)}$ decrease at large distances, allowing solutions with scattering boundary conditions,

$\psi_{11}(R) \approx \sin(k_1 R) + \cos(k_1 R) K_{11}$, and related forms with the transition amplitude T_{11} or the scattering amplitude S_{11}, both obtained from K_{11}.

Special choices for the expansion basis set allow for detailed treatment of the closed channel effects, including resonance scattering [26, 27]. In particular, we consider cases where the potential energy function $U_{22}(R)$ leads to formation of bound states along the R variable, with self-energies $\varepsilon_{2,n}$, $n = 1$ to $N^{(2)}_{bnd}$, above E_2 such that $E - E_2 - \varepsilon_{2,n} = -(\hbar^2/2M)\kappa_{2,n}{}^2$, and with orthonormalized states $\eta^{(2)}_n(R)$ satisfying

$$\left[-\kappa_{2,n}{}^2 + \frac{d^2}{dR^2} - U_{22}(R)\right] \eta^{(2)}_n(R) = 0.$$

The Green function in channel 2 can be expanded in this set of bound states plus a set of scattering functions for $U_{22}(R)$. As the total energy goes through $E = E_2 + \varepsilon_{2,n}$, the nonlocal potential energy $W^{(2)}_1 = W^{(2)}_{1,r} + W^{(2)}_{1,nr}$ displays a resonance term $W^{(2)}_{1,r}$ with sharp large variations near the locations of the $\varepsilon_{2,n}$ bound energies, as

$$W^{(2)}_{1,r}(R, R') = U_{12}(R)(\hbar^2/2M) \sum_n \left\langle R \middle| \eta^{(2)}_n \right\rangle (E - E_2 - \varepsilon_{2,n} + i\epsilon)^{-1} \times \left\langle \eta^{(2)}_n \middle| R' \right\rangle U_{21}(R')$$

with $\epsilon \to 0$, showing resonance energy structures, plus a nonresonance term $W^{(2)}_{1,nr}$. Energy widths for these resonances can be obtained solving the differential equation for $\psi_{11}(R)$. A solution for $\psi_{11}(R)$, including this resonance potential energy term in its equation, provides shifts and widths for resonance energies near $\varepsilon_{2,n}$. The transition amplitude then takes a form $T_{11} = T^{(nr)}_{11} + T^{(r)}_{11}$ with $T^{(r)}_{11}(E)$ displaying resonance peaks. This is shown in more detail in a following section on effective Hamiltonian treatments.

3.5 Distorted-Waves Treatment

3.5.1 General Distortion Potential

It is convenient, in a variety of inelastic collisions, to break-up their treatment into two stages, with the first one involving a portion $V^{(1)}$ of the interaction potential energy $V(\vec{R}, \boldsymbol{q})$ suitable for accurate calculations or for description of a physically relevant potential term such as a strongly repulsive short-range interaction. This provides distorted-wave solutions $\Psi^{(+)}_{1,\alpha}(\boldsymbol{Q})$ to be used in a second stage that accounts for the effects of the remaining potential energy function $V^{(2)} = V - V^{(1)}$ [3, 5, 10]. A distorted-waves treatment can be used in a perturbation expansion of the wavefunction $\Psi^{(+)}_{\alpha}(\boldsymbol{Q})$ in powers of $V^{(2)}$ with terms containing only the known wavefunctions from the first stage.

Distorted-wave solutions satisfy the equation

$$\hat{H}^{(1)}\Psi_{1,\alpha}^{(+)}(\vec{R},\boldsymbol{q};\vec{k}_{\alpha}) = E\Psi_{1,\alpha}^{(+)}(\vec{R},\boldsymbol{q};\vec{k}_{\alpha})$$

where $\hat{H}^{(1)} = \hat{H}_0 + V^{(1)}$, with scattering boundary conditions.

To proceed, we consider cases when the distortion potential function is of form $V^{(1)}(\vec{R})$ and independent of other variables $\boldsymbol{q}$, which may be described by the internal states $\phi_n(\boldsymbol{q})$ of energy E_n, taken to be noninteracting during this first interaction. The relative motion distorted wave $\chi_{1,\alpha}^{(+)}(\vec{R})$ can be constructed from the Green function $G_{0,\alpha}^{(+)}(\vec{R},\vec{R}')$ for the free motion Hamiltonian $\hat{H}_0 = \hat{K}_R + \hat{H}_{int}$ with outgoing-waves boundary conditions, as

$$\chi_{1,\alpha}^{(+)}(\vec{R};\vec{k}_{\alpha}) = \chi_{1,\alpha}^{(0)}(\vec{R};\vec{k}_{\alpha}) + \int d^3R\, G_{0,\alpha}^{(+)}(\vec{R},\vec{R}')V^{(1)}(\vec{R}')\chi_{1,\alpha}^{(+)}(\vec{R}';\vec{k}_{\alpha})$$

with $\chi_{1,\alpha}^{(0)}(\vec{R};\vec{k}_{\alpha}) = (2\pi)^{-\frac{3}{2}}\exp(i\vec{k}_{\alpha}\cdot\vec{R}))$. The full distorted wave is $\Psi_{1,\alpha}^{(+)}(\vec{R},\boldsymbol{q}) = \chi_{1,\alpha}^{(+)}(\vec{R})\phi_{\alpha}(\boldsymbol{q})$ and the total wavefunction

$$\Psi_{\alpha}^{(+)}(\vec{R},\boldsymbol{q}) = \chi_{1,\alpha}^{(+)}(\vec{R})\phi_{\alpha}(\boldsymbol{q}) + \Psi_{2,\alpha}^{(+)}(\vec{R},\boldsymbol{q})$$

satisfies $(\hat{H}^{(1)} + V^{(2)})\left(\Psi_{1,\alpha}^{(+)} + \Psi_{2,\alpha}^{(+)}\right) = E\left(\Psi_{1,\alpha}^{(+)} + \Psi_{2,\alpha}^{(+)}\right)$ or

$$(E - \hat{H}^{(1)})\Psi_{2,\alpha}^{(+)} = V^{(2)}\left(\Psi_{1,\alpha}^{(+)} + \Psi_{2,\alpha}^{(+)}\right).$$

This can be formally solved in terms of the distorted-wave resolvent (or Green function) operator satisfying $(E - \hat{H}^{(1)})\hat{G}_1^{(+)} = \hat{I}$ where $\hat{I} = \sum_n|\phi_n\rangle\langle\phi_n|$, in the bracket notation and assuming that the internal states form a complete expansion set. Expanding $\hat{G}_1^{(+)}(\vec{R},\vec{R}') = \sum_n G_{1,n}^{(+)}(\vec{R},\vec{R}')|\phi_n\rangle\langle\phi_n|$ gives

$$\left[\frac{\hbar^2}{2M}\left(k_{\alpha}^2 + \nabla_R^2\right) + V^{(1)}(\vec{R})\right]G_{1,n}^{(+)}(\vec{R},\vec{R}') = \delta(\vec{R} - \vec{R}')$$

and from this it follows that

$$\left|\Psi_{2,\alpha}^{(+)}(\vec{R})\right\rangle = \int d^3R'\,\hat{G}_1^{(+)}(\vec{R},\vec{R}')\hat{V}^{(2)}(\vec{R}')\left[\left|\Psi_{1,\alpha}^{(+)}(\vec{R}')\right\rangle + \left|\Psi_{2,\alpha}^{(+)}(\vec{R}')\right\rangle\right].$$

The term containing $V^{(2)}\Psi_{2,\alpha}^{(+)}$ to the right can be dropped to first order when $V^{(2)}$ is small, and the total wavefunction $\langle\boldsymbol{q}|\Psi_{\alpha}^{(+)}(\vec{R})\rangle = \Psi_{\alpha}^{(+)}(\vec{R},\boldsymbol{q}) = \sum_{all\,\alpha'}\psi_{\alpha'\alpha}^{(+)}(\vec{R})\,\phi_{\alpha'}(\boldsymbol{q})$, can then be obtained from the distorted waves above, as a distorted-wave approximation to first order, or *DW1*, with components

$$\psi_{\alpha'\alpha}^{(+)}(\vec{R}) = \chi_{1,\alpha}^{(+)}(\vec{R})\delta_{\alpha',\alpha} + \int d^3R' G_{1,\alpha'}^{(+)}(\vec{R},\vec{R}')\langle\phi_{\alpha'}|V^{(2)}(\vec{R}')|\phi_{\alpha}\rangle\chi_{1,\alpha}^{(+)}(\vec{R}')$$

and scattering limit

$$\psi_{\alpha'\alpha}^{(+)}(\vec{R}) \approx (2\pi)^{-\frac{3}{2}}\exp(i\vec{k}_{\alpha}\cdot\vec{R}) + (2\pi)^{-\frac{3}{2}}f_{\alpha'\alpha}^{(DW1)}(\Omega)\exp\left(ik_{\alpha'}R\right)R^{-1}$$

$$f_{\alpha'\alpha}^{(DW1)}(\Omega) = (2\pi)^2 \frac{M}{\hbar^2} \int d^3R' \chi_{1,\alpha'}^{(-)}(\vec{R}')^* \langle \phi_{\alpha'} | V^{(2)}(\vec{R}') | \phi_\alpha \rangle \chi_{1,\alpha}^{(+)}(\vec{R}')$$

where the asymptotic form of $G_{1,\alpha'}^{(+)}(\vec{R}, \vec{R}')$ was used from Section 3.2.2. This provides the differential cross sections

$$\left(\frac{d\sigma_{\alpha'\alpha}}{d\Omega} \right)^{(DW1)} = \frac{k_{\alpha'}}{k_\alpha} \left| f_{\alpha'\alpha}^{(DW1)}(\Omega) \right|^2$$

A higher order solution DWn, $n > 1$, can also be generated by iteration of the above integral equation for $\Psi_{2,\alpha}^{(+)}$.

A more specific choice of the distortion potential can be done separating diagonal and nondiagonal parts of the potential energy function represented by a matrix $\boldsymbol{V}(\vec{R})$ with elements $V_{r'n}(\vec{R})$, which can be organized into diagonal and nondiagonal components as $\boldsymbol{V} = \boldsymbol{V}_D + \boldsymbol{V}_N$. The diagonal elements describe events where an internal state ϕ_α does not change during collisions in a potential $V_{\alpha\alpha}(\vec{R})$, while nondiagonal potential matrix elements account for state-to-state transitions. The above developments can be directly carried out in terms of state and operator matrices in the space of internal states, with $\Psi_\alpha^{(+)}(\vec{R}, \boldsymbol{q}) = \sum_{all\ \alpha} \psi_{\alpha'\alpha}^{(+)}(\vec{R})\, \phi_{\alpha'}(\boldsymbol{q})$, with $\boldsymbol{V}_D$ and $\boldsymbol{V}_N$ in place of $V^{(1)}$ and $V^{(2)}$, to obtain the column matrices $\boldsymbol{\psi}_\alpha^{(+)}(\vec{R}; \vec{k}_\alpha) = [\psi_{\alpha'\alpha}^{(+)}(\vec{R}; \vec{k}_\alpha)]$ by first generating distorted-wave matrices $\chi_\alpha^{(+)}(\vec{R}; \vec{k}_\alpha) = [\chi_{\alpha'\alpha}^{(+)}(\vec{R}; \vec{k}_\alpha)]$ using the potential energy matrix $\boldsymbol{V}_D(\vec{R})$.

3.5.2 Multichannel Distorted Partial-Wave Treatment

Here we develop a more detailed treatment for the scattering functions $\psi_{\lambda'\lambda}(R; k_\alpha)$, which follow from an expansion of the total wavefunction $\Psi_\alpha^{(+)}(\vec{R}, \boldsymbol{q}; \vec{k}_\alpha)$ in the set of partial wave states $\Phi_\lambda(\Omega, \boldsymbol{q}) = Y_{lm}(\Omega)\phi_n(\boldsymbol{q})$ and momentum orientation functions $Y_{lm}(\Omega_\alpha)^*$, as done in Section 3.2. Here we consider only open channels λ, for simplicity. The potential energy elements $U_{\lambda'\lambda}(R) = (2M/\hbar^2)\langle \Phi_{\lambda'} | V | \Phi_\lambda \rangle_R$ describe couplings between partial-wave states for each set of internal and relative angular momentum quantum numbers. Decomposing the matrix $\boldsymbol{U}(R)$ into diagonal and non diagonal components, $\boldsymbol{U}(R) = \boldsymbol{U}_D(R) + \boldsymbol{U}_N(R)$ with $\boldsymbol{U}_D = \left[U_{\lambda\lambda} \delta_{\lambda'\lambda} \right]$ and $\boldsymbol{U}_N = \left[U_{\lambda'\lambda} \left(1 - \delta_{\lambda'\lambda} \right) \right]$, we have from $\widehat{\boldsymbol{D}}_R \boldsymbol{\psi}^{(+)}(R) = \boldsymbol{U}(R) \boldsymbol{\psi}^{(+)}(R)$ in Section 3.3 that

$$\left[\widehat{\boldsymbol{D}}_R - \boldsymbol{U}_D(R) \right] \boldsymbol{\psi}^{(+)}(R) = \boldsymbol{U}_N(R) \boldsymbol{\psi}^{(+)}(R).$$

with outgoing-wave scattering conditions in $\boldsymbol{\psi}^{(+)}(R) = \left[\psi_{\lambda'\lambda}(R) \right]$. Considering $\boldsymbol{U}_D$ and $\boldsymbol{U}_N$ as the $V^{(1)}$ and $V^{(2)}$ in the previous subsection, we can develop a two-potential treatment with $\boldsymbol{U}_D$ as the distortion potential, to obtain $\boldsymbol{\psi}^{(+)}(R)$ from a distorted-wave matrix $\boldsymbol{\chi}^{(+)}(R) = \left[\chi_{\lambda'\lambda}(R) \right]$.

However, we choose next to work with standing-wave real-valued boundary conditions for computational convenience, to obtain $\boldsymbol{\psi}^{(K)}(R)$ and a scattering

$\boldsymbol{K}$-matrix to begin with. We start with solutions of $[\hat{\boldsymbol{D}}_R - \boldsymbol{U}_D(R)]\boldsymbol{\chi}^{(K)}(R) = \boldsymbol{0}$, where $\boldsymbol{\chi}^{(K)}$ is a diagonal matrix of distorted waves, with standing-wave scattering functions $\chi_\lambda^{(K)}(R)$ as elements. These functions are needed with regular and irregular asymptotic conditions, called here $\chi_\lambda^{(K)}$ and $\overline{\chi}_\lambda^{(K)}$, given by

$$\chi_\lambda^{(K)}(R) \approx u_l(k_\alpha R) + v_l(k_\alpha R)K_{\lambda\lambda}^{(D)}$$
$$\overline{\chi}_\lambda^{(K)}(R) \approx v_l(k_\alpha R)$$

where $\chi_\lambda^{(K)}(R)$ is generated integrating the differential equation with the potential function $U_{\lambda\lambda}(R)$ from zero at a small value of R to a large value when $U_{\lambda\lambda}(R) \approx 0$, while $\overline{\chi}_\lambda^{(K)}(R)$ is generated integrating back from the large value of R. Here $u_l(x) \approx \sin(x - l\pi/2)$ and $v_l(x) \approx \cos(x - l\pi/2)$, and $K_{\lambda\lambda}^{(D)}$ can be extracted by matching as done for elastic scattering calculations, from which $\boldsymbol{\chi}^{(K)}(R) \approx [\boldsymbol{I}^{(s)}(R) + \boldsymbol{I}^{(c)}(R)\boldsymbol{K}^{(D)}]\boldsymbol{k}^{1/2}$ in the notation of Section 3.3.

These solutions can be used to construct the distorted standing-wave Green functions $G_{D,\lambda}^{(K)}(R, R')$ forming a diagonal matrix $\boldsymbol{G}_D^{(K)}(R, R')$ satisfying

$$[\hat{\boldsymbol{D}}_R - \boldsymbol{U}_D(R)]\boldsymbol{G}_D^{(K)}(R, R') = \mathbf{I}\,\delta(R - R')$$
$$G_{D,\lambda}^{(K)}(R, R') = k_\lambda^{-1}\left[\chi_\lambda^{(K)}(R)\overline{\chi}_\lambda^{(K)}(R') - \overline{\chi}_\lambda^{(K)}(R)\chi_\lambda^{(K)}(R')\right], \quad R' \le R; = 0, \quad R' > R.$$

This provides a solution to $[\hat{\boldsymbol{D}}_R - \boldsymbol{U}_D(R)]\boldsymbol{\psi}^{(K)}(R) = \boldsymbol{U}_N(R)\boldsymbol{\psi}^{(K)}(R)$ as

$$\boldsymbol{\psi}^{(K)}(R) = \boldsymbol{\chi}^{(K)}(R) + \int_0^R dR'\boldsymbol{G}_D^{(K)}(R, R')\boldsymbol{U}_N(R')\boldsymbol{\psi}^{(K)}(R')$$

where $\boldsymbol{k}$ is a diagonal matrix with elements k_λ. The $\boldsymbol{K} = \left[K_{\lambda'\lambda}\right]$ matrix can be obtained comparing this with $\boldsymbol{\psi}^{(K)}(R) \approx [\boldsymbol{I}^{(s)}(R) + \boldsymbol{I}^{(c)}(R)\boldsymbol{K}]\boldsymbol{k}^{1/2}$.

Its form to first order in $\boldsymbol{U}_N$ is obtained approximating $\boldsymbol{\psi}^{(K)}(R')$ to the right with $\boldsymbol{\chi}^{(K)}(R')$. The first-order solution $\boldsymbol{\psi}^{(K1)}$ is

$$\boldsymbol{\psi}^{(K1)}(R) = \left[\boldsymbol{\chi}^{(K)}(R) + \boldsymbol{\chi}^{(K)}(R)\boldsymbol{k}^{-1}\int_0^\infty dR'\overline{\boldsymbol{\chi}}^{(K)}(R')\boldsymbol{U}_N(R')\boldsymbol{\chi}^{(K)}(R')\right.$$
$$\left.-\overline{\boldsymbol{\chi}}^{(K)}(R)\boldsymbol{k}^{-1}\int_0^\infty dR'\,\boldsymbol{\chi}^{(K)}(R')\boldsymbol{U}_N(R')\boldsymbol{\chi}^{(K)}(R')\right]\boldsymbol{A}$$

with the matrix of coefficients $\boldsymbol{A}$ obtained using that asymptotically $\boldsymbol{\chi}^{(K)}(R) \approx [\boldsymbol{I}^{(s)}(R) + \boldsymbol{I}^{(c)}(R)\boldsymbol{K}^{(D)}]\boldsymbol{k}^{1/2}$ and $\overline{\boldsymbol{\chi}}^{(K)}(R) \approx \boldsymbol{I}^{(c)}(R)\boldsymbol{k}^{1/2}$. Comparing this to $\boldsymbol{\psi}^{(K1)}(R) \approx [\boldsymbol{I}^{(s)}(R) + \boldsymbol{I}^{(c)}(R)\boldsymbol{K}^{(DW1)}]\boldsymbol{k}^{1/2}$, we find that

$$\boldsymbol{K}^{(DW1)} = \boldsymbol{K}^{(D)} - \boldsymbol{k}^{-\frac{1}{2}}\int_0^\infty dR'\,\boldsymbol{\chi}^{(K)}(R')\,\boldsymbol{U}_N(R')\boldsymbol{\chi}^{(K)}(R')\boldsymbol{k}^{-\frac{1}{2}}/\boldsymbol{D}$$
$$\boldsymbol{D} = \left[\boldsymbol{I} + \boldsymbol{k}^{-\frac{1}{2}}\int_0^\infty dR'\overline{\boldsymbol{\chi}}^{(K)}(R')\,\boldsymbol{U}_N(R')\boldsymbol{\chi}^{(K)}(R')\boldsymbol{k}^{-\frac{1}{2}}\right] \cong \boldsymbol{I}$$

to lowest order in $\boldsymbol{U}_N$. This approximate matrix $\boldsymbol{K}^{(DW1)}$ can then be used to construct an approximate $\tilde{\boldsymbol{T}}$ matrix or an approximate $\tilde{\boldsymbol{S}}$ matrix satisfying unitarity properties as in

$$\tilde{\boldsymbol{T}} = \boldsymbol{K}^{(DW1)}(\mathbf{1} - i\boldsymbol{K}^{(DW1)})^{-1}$$

$$\tilde{\boldsymbol{S}} = (\mathbf{1} + i\boldsymbol{K}^{(DW1)})(\mathbf{1} - i\boldsymbol{K}^{(DW1)})^{-1}$$

An alternative treatment using distorted waves $\chi^{(+)}(R)$ with outgoing boundary conditions provides an approximate $\boldsymbol{S}^{(DW1)}$ matrix. When the elements in the matrix $i\boldsymbol{A}^{(DW1)} = \boldsymbol{S}^{(DW1)} - \mathbf{1}$ are small, it is convenient to construct an approximate exponential form $\overline{\boldsymbol{S}} = \exp[i\boldsymbol{A}^{(DW1)}]$, unitary insofar $\boldsymbol{A}^{(DW1)}$ is Hermitian, useful in applications to inelastic collisions to extend the validity of distorted-wave calculations [5].

3.6 Optical Potential Models

3.6.1 Physical Models of Optical Potentials

Optical potential models of molecular inelastic and reactive collisions have been developed to maintain the simplicity of two-body collision treatments in the presence of many coupled intermolecular states [2, 28]. Here we follow the treatments in [10, 29]. At a physical level, a simple complex-valued optical potential

$$\mathcal{V}_{opt}(R) = V(R) - iW(R)$$

can be introduced to account for scattering where a wavefunction $\Psi_\alpha^{(-)}(\vec{R}, \boldsymbol{q})$ undergoes flux loss from a chosen scattering channel $\alpha = (\vec{k}, n)$ which undergoes transitions into channels $\alpha' \neq \alpha$. The internal variables $\boldsymbol{q}$ change during a collision with a distribution of values $w(\boldsymbol{q}; \vec{R}) = a(\boldsymbol{q}; \vec{R})^2$ at each intermolecular position $\vec{R}$. Taking an average over this distribution, we introduce an optical wavefunction $\psi_\alpha^{(+)}(\vec{R}) = \int d\boldsymbol{q}\, a(\boldsymbol{q}; \vec{R})\Psi_\alpha^{(+)}(\vec{R}, \boldsymbol{q})$ to describe scattering with relative kinetic energy E, satisfying the equation

$$\left[-(\hbar^2/2M)\nabla_{\vec{R}}^2 + \mathcal{V}_{opt}(R) - E\right]\psi_\alpha^{(+)}(\vec{R}) = 0.$$

The flux at location $\vec{R}$ is

$$\vec{J}_\alpha(\vec{R}) = (\hbar/2Mi)\left[\left(\psi_\alpha^{(+)}\right)^* \nabla_{\vec{R}}\psi_\alpha^{(+)} - \psi_\alpha^{(+)}\nabla_{\vec{R}}\left(\psi_\alpha^{(+)}\right)^*\right]$$

with a local divergence

$$\nabla_{\vec{R}} \cdot \vec{J}_\alpha(\vec{R}) = (\hbar/2Mi)\left[\left(\psi_\alpha^{(+)}\right)^* \nabla_{\vec{R}}^2\psi_\alpha^{(+)} - \psi_\alpha^{(+)}\nabla_{\vec{R}}^2\left(\psi_\alpha^{(+)}\right)^*\right].$$

Replacing $\nabla^2_{\vec{R}}\psi_\alpha^{(+)}$ from the differential equation above, this gives

$$\nabla_{\vec{R}} \cdot \vec{J}_\alpha(\vec{R}) = -(2/\hbar)W(R)\left|\psi_\alpha^{(+)}(\vec{R})\right|^2$$

hence one obtains a negative divergence, or flux loss, if $W(R) > 0$ and a flux gain if $W(R) < 0$. Local flux loss depends then on both the magnitude of the imaginary potential term in a shell around distance R and on the anisotropy of $|\psi_\alpha^{(+)}(\vec{R})|^2$ for varying solid angles Ω of $\vec{R}$. On physical grounds, the relationship between flux loss and the imaginary potential function can be used to construct $W(R)$ and reproduce the flux loss measured in an experiment [30], integrating the two sides of the above equation over solid angles for each fixed distance R.

A partial wave analysis can be developed along the same lines as used for elastic scattering, with $\psi_\alpha^{(+)}(\vec{R}) = \psi_{\vec{k}}^{(+)}(\vec{R})$ for a fixed internal state n. In the same notation, the expansion

$$\psi_{\vec{k}}^{(+)}(\vec{R}) = \sum_{l=0}^{\infty}(2l+1)(kR)^{-1}\psi_l^{(+)}(R;k)P_l(\cos\Theta)$$

gives a radial ordinary differential equation for $\psi_l^{(+)}(R;k)$ with the optical potential $\mathcal{V}_{opt}(R)$, solved with boundary conditions

$$\psi_l^{(+)}(R_{in};k) = 0$$

$$\psi_l^{(+)}(R;k) \approx u_l(kR) + t_l^{(+)}(k)w_l^{(+)}(kR)$$

using the Riccati functions u_l and $w_l^{(+)}$. The scattering amplitude $f_k^{(+)}(\Omega)$ is obtained from the transition coefficients $t_l^{(+)}(k)$, as for elastic scattering, or by the scattering amplitudes $S_l^{(+)}(k) = 1 + 2it_l^{(+)}(k)$ but with a complex-valued phase shift $\delta_l = \xi_l + i\zeta_l$ so that $S_l^{(+)} = \exp[2i(\xi_l + i\zeta_l)]$, as

$$f_k^{(+)}(\Omega) = k^{-1}\sum_{l=0}^{\infty}(2l+1)t_l^{(+)}(k)P_l(\cos\Theta)$$

from which $d\sigma_{el}/d\Omega = \left|f_k^{(+)}(\Omega)\right|^2$ is a new elastic differential cross section containing now purely elastic, diffraction, and interference terms.

Integral elastic cross sections are given by

$$\sigma_{el}(v) = (\pi/k^2)\sum_{l=0}^{\infty}(2l+1)\left[S_l^{(+)}(k) - 1\right]^2 = \sigma_{pot}(v) + \sigma_{dif}(v) + \sigma_{int}(v)$$

where purely potential and diffraction terms can be obtained from phase shifts ξ_l and ζ_l, respectively. Integral absorption cross sections are calculated by integrating the flux loss within a sphere of radius $R = a$. Writing the flux divergence from its definition in terms of $\psi_\alpha^{(+)}$ and its complex conjugate, and using its partial-wave decomposition in the integral, as $a \to \infty$,

$$\sigma_{abs}(v) = -v^{-1} \int_0^a dR\, R^2 \int_{4\pi} d\Omega\, \nabla_{\vec{R}} \cdot \vec{J}_\alpha(\vec{R}) \approx (\pi/k^2) \sum_{l=0}^{\infty} (2l+1) \left[1 - \left|S_l^{(+)}(k)\right|^2\right]$$

which can also be given in terms of the opacity function $p_l(k) = 1 - |S_l^{(+)}(k)|^2 = 1 - \exp(-4\zeta_l)$.

An application example can be found in [31], where results from an optical potential are shown for the real- and imaginary-valued phase shifts and the opacity vs partial wave numbers, and a comparison is given for elastic differential cross sections from the real part of the potential and from the complex optical potential, for Li + HBr at a collision velocity $v = 1.5 \times 10^2$ m/sec. The differential cross sections are significantly different for the physically chosen optical potential with a 12-6 real part $V(R)$ dependence on R with well-depth energy parameters $\epsilon = 4.85 \times 10^{-14}$ erg and minimum potential at $R_m = 3.97 \times 10^{-8}$ cm and an imaginary part $W(R) = c\,\epsilon\,(R_m/R)^{12}$, with $c = 0.44$.

3.6.2 Basis-Set Partitioning Method

A theoretical derivation and treatment of optical potentials can be done in terms of projection operators in the functional space of the internal states $\phi_n(\boldsymbol{q})$ introduced in Section 3.1.3. Such formal theory can provide insights on the functional form and energy dependence of the real and imaginary parts of the optical potentials and help in their semiempirical construction as needed to interpret experimental measurements of inelastic cross sections.

Given the unperturbed Hamiltonian $\hat{H}_0 = \hat{K}_R + \hat{H}_{int}$, and

$$\hat{H}_0 \Psi_\alpha^0(\vec{R}, \boldsymbol{q}; \vec{k}_\alpha) = E \Psi_\alpha^0(\vec{R}, \boldsymbol{q}; \vec{k}_\alpha)$$

satisfied by the free-motion eigenfunction $\Psi_\alpha^0(\vec{R}, \boldsymbol{q}; \vec{k}_\alpha) = \psi^0(\vec{R}; \vec{k}_\alpha)\phi_\alpha(\boldsymbol{q})$, for a collision starting in internal state ϕ_a, with total energy $E = E_a + \hbar^2 k_\alpha^2/(2M)$, we introduce a projection operator $\hat{P}_n = |\phi_n\rangle\langle\phi_n|$ in a bracket notation indicating states dependent on the internal coordinates $\boldsymbol{q}$ so that $\phi_n(\boldsymbol{q}) = \langle \boldsymbol{q}|\phi_n\rangle$ and $\langle\phi_n|\phi_{n'}\rangle = \delta_{nn'}$, and therefore $\hat{P}_n|\Psi_\alpha^0\rangle = \delta_{na}|\Psi_\alpha^0\rangle$. This operator therefore selects an internal state and commutes with $\hat{H}_0$.

In a collision involving transitions induced by an interaction potential energy $V(\vec{R}, \boldsymbol{q})$, with $\hat{H} = \hat{H}_0 + \hat{V}$, the scattering wavefunction satisfies $\hat{H}\Psi_\alpha^{(+)}(\vec{R}, \boldsymbol{q}; \vec{k}_\alpha) = E\Psi_\alpha^{(+)}(\vec{R}, \boldsymbol{q}; \vec{k}_\alpha)$ with scattering boundary conditions and is given in general by an expansion in all (energetically open and closed) internal states as

$$\Psi_\alpha^{(+)}(\vec{R}, \boldsymbol{q}) = \sum_{\text{all } n'} \psi_{\alpha'\alpha}^{(+)}(\vec{R})\, \phi_{n'}(\boldsymbol{q})$$

with $\alpha' = (\vec{k}_{\alpha'}, n')$. However, in a specific collision, this wavefunction will usually involve direct transitions only within a limited set of internal states $\{n_1 \le n \le n_2\}$, including the initial state, comprising all the states active in the collision. This set can be selected with the projection operator $\hat{P} = \sum_{n_1}^{n_2} |\phi_n\rangle\langle\phi_n|$, which satisfies $\hat{P} = (\hat{P})^2$ and $\hat{P} = \hat{P}^\dagger$ and has a complementary projector $\hat{Q}$ so that $\hat{P} + \hat{Q} = \hat{I}$, the identity operator in the space of internal states. This projector $\hat{Q}$ selects intermediate states that provide indirect couplings among the active ones and also commutes with $\hat{H}_0$.

The wavefunction and Hamiltonian equation $(E - \hat{H})\Psi_\alpha^{(+)} = 0$ can be partitioned applying the two projectors, with components in $\Psi_\alpha^{(+)} = \hat{P}\Psi_\alpha^{(+)} + \hat{Q}\Psi_\alpha^{(+)}$ coupled as

$$E\hat{P}\Psi_\alpha^{(+)} - \hat{P}\,\hat{H}\hat{P}\Psi_\alpha^{(+)} - \hat{P}\hat{H}\hat{Q}\Psi_\alpha^{(+)} = 0$$

$$E\hat{Q}\Psi_\alpha^{(+)} - \hat{Q}\,\hat{H}\hat{P}\Psi_\alpha^{(+)} - \hat{Q}\hat{H}\hat{Q}\Psi_\alpha^{(+)} = 0.$$

The second line can be solved for $\hat{Q}\Psi_\alpha^{(+)}$ and substituted in the first line, using that $\hat{P}\hat{H}\hat{Q} = \hat{P}\hat{V}\hat{Q}$ and introducing the resolvent $\hat{G}_Q^{(+)}(E)$ for outgoing waves in

$$(E\hat{Q} - \hat{Q}\hat{H}\hat{Q}\,)\hat{G}_Q^{(+)}(E) = \hat{Q}\,.$$

It can be formally expressed as an inverse operator with the energy extended to the complex plane and $E^{(+)} = E + i\eta,\ \eta \to 0+$, to avoid singularities for real energies [3]. The resolvent $\hat{G}_Q^{(+)}(E) = (E^{(+)}\hat{Q} - \hat{Q}\hat{H}\hat{Q}\,)^{-1}$ must first be evaluated in a state basis set before letting $\eta \to 0+$. The result for $\hat{P}\Psi_\alpha^{(+)}$ is

$$(E - \hat{\mathcal{H}}_P)\hat{P}\Psi_\alpha^{(+)} = 0$$

$$\hat{\mathcal{H}}_P(E) = \hat{P}\,\hat{H}\hat{P} + \hat{P}\hat{V}\hat{Q}\hat{G}_Q^{(+)}(E)\hat{Q}\,\hat{V}\hat{P}$$

where $\hat{\mathcal{H}}_P(E)$ is an effective Hamiltonian operator in the space of active states, containing a first term with direct couplings of the active states, and a second term displaying indirect couplings of active states through the $\hat{Q}$ complementary states. This second term contains real and imaginary-valued terms and is energy dependent. The derivation gives an effective potential energy operator

$$\hat{\mathcal{V}}_P(E) = \hat{P}\,\hat{V}\hat{P} + \hat{P}\hat{V}\hat{Q}\hat{G}_Q^{(+)}(E)\hat{Q}\,\hat{V}\hat{P}$$

that can be calculated and compared with semiempirical optical potentials $\mathcal{V}_{opt}(R) = V(R) - iW(R)$. Details on calculation procedures can be found in [10, 29]. This same treatment can be extended to electronically diabatic and to reaction collisions. A related general formulation is given in Chapter 7.

References

1 H. Goldstein, C. P. Poole Jr., and J. L. Safko, *Classical Mechanics* (3rd edition), San Francisco, CA: Addison-Wesley, 2002.

2 N. F. Mott and H. S. W. Massey, *The Theory of Atomic Collisions, Oxford*, England: Clarendon Press, 1965.

3 A. Messiah, *Quantum Mechanics Volumes 1 and 2*, Amsterdam: North-Holland, 1962.

4 D. A. Micha, *Molecular Interactions: Concepts and Methods*, Hoboken, NJ, USA: John Wiley & Sons, Inc., 2020.

5 M. Child, *Molecular Collision Theory*, New York: Academic Press, 1974.

6 R. N. Porter and L. M. Raff, "Classical trajectory methods in molecular collisions," in *Dynamics of Molecular Collisions Part B*, (Editor: W. H. Miller), New York: Plenum Press, 1976, pp. 1–52.

7 B. H. Bransden, *Atomic Collision Theory*, Reading MA, USA: Benjamin-Cummings, 1983.

8 E. Vilallonga and D. A. Micha, "The calculation of time-correlation functions for molecular collisions," *Phys. Rep.*, vol. 212, p. 329, 1992.

9 R. G. Newton, *Scattering Theory of Waves and Particles* (2nd edition), New York: Springer-Verlag, 1982.

10 D. A. Micha, "Effective hamiltonian methods for molecular collisions," in *Adv. Quantum Chem.*, Volume 8, New York: Academic Press Inc , 1974, p. 211.

11 M. D. Pattengill, "Rotational excitation III. Clasical trajectory methods," in *Atom-Molecule Collision Theory*, (Editor: R. B. Bernstein), New York: Plenum Press, 1979, pp. 359–376.

12 W. R. Gentry, "Vibrational excitation II: Classical and semiclassical methods (Chapter 12)," in *Atom-Molecule Collision Theory*, (Editor: R. B. Bernstein), New York: Plenum Press, 1979, pp. 391–426.

13 D. G. Truhlar and J. T. Muckerman, "Reactive scattering cross sections III. Quasiclassical and semiclassical methods," in *Atom-Molecule Collision Theory*, (Editor: R. B. Bernstein), New York: Plenum Press, 1979, pp. 505–565.

14 W. H. Miller, "Classical-limit quantum mechanics and the theory of molecular collisions," in *Adv. Chem. Phys.*, vol. 25, New York: John Wiley & Sons, 1974, pp. 69–177.

15 A. R. Edmonds, *Angular Momentum in Quantum Mechanics*, Princeton, NJ, USA: Princeton University Press, 1960.

16 P. Atkins and J. dePaula, *Physical Chemistry* (9th edition), New York: WH Freeman and Co., 2010.

17 G. Balint-Kurti and A. Palov, *Theory of Molecular Collisions*, London, England: Royal Society of Chemistry, 2015.

18 D. A. Micha, "A self-consistent eikonal treatment of electronic transitions in molecular collisions," *J. Chem. Phys.*, vol. 78, p. 7138, 1983.

19 G. D. Billing and K. V. Mikkelsen, *Advanced Molecular Dynamics and Chemical Kinetics*, New York: Wiley-Interscience Inc., 1997.

20 B. H. Bransden and M. R. C. McDowell, *Charge Exchange and the Theory of Ion-Atom Collisions*, Oxford, England: Oxford University Press, 1992.

21 J. C. Tully, "Mixed quantum-classical dynamics," *Faraday Discuss.*, vol. 110, p. 407, 1998.

22 A. Messiah, *Quantum Mechanics*, Volume 1, Amsterdam: North-Holland, 1962.

23 S. J. Farlow, *Partial Differential Equations for Scientists and Engineers*, New York: Dover Publications, 1993.

24 D. A. Micha and I. Burghardt (Editors), *Quantum Dynamics of Complex Molecular Systems*, Heidelberg, Germany: Springer-Verlag, 2007.

25 W. N. Sams and D. J. Kouri, "Noniterative solutions of integral equations for scattering. II. Coupled channels," *J. Chem. Phys.*, vol. 51, p. 4815, 1969.

26 H. Feshbach, "A unified theory of nuclear reactions. II," *Ann. Phys.*, vol. 19, p. 287, 1962.

27 D. A. Micha, "Compound state resonances in atom-diatomic molecule collisions," *Physiol. Rev.*, vol. 162, pp. 88–97, 1967.

28 L. S. Rodberg and R. M. Thaler, *Introduction to the Quantum Theory of Scattering*, New York: Academic Press, 1967.

29 D. A. Micha, "Optical models in molecular collision theory," in *Dynamics of Molecular Collisions, Part A*, (Editor: W. H. Miller), New York: Plenum Press, 1976, pp. 81–129.

30 C. Schlier, *Molecular Beams and Reaction Kinetics (Proc. E. Fermi School, Course XLIV)*, New York: Academic Press, 1970.

31 R. Marriott and D. A. Micha, "Potential for Li+HBr collisions," *Phys. Rev.*, vol. 180, p. 120, 1969.

4

Inelastic Collisions: Adiabatic Energy Transfer

CONTENTS

Quantum Molecular Dynamics: Energy Transfer and Reactivity, First Edition. David A. Micha.

4.1 Adiabatic Energy Transfer Cross Sections

4.1.1 Kinematics of Adiabatic Energy Transfer

Collisional energy transfer of atoms and molecules in their ground electronic states involves their rotational and vibrational motions and their exchange of energy with the kinetic energy of relative motion. Typically, given the energy relations $\Delta E = hc\tilde{\nu}$, with wavenumbers $\tilde{\nu}$ in cm^{-1}, rotational energy levels and their changes are of the order of 1.0–100 cm^{-1} and are much smaller than vibrational energy levels and their changes, which are around 500–5000 cm^{-1}. These correspond to rotational and vibrational temperatures $\langle E \rangle_{eq} = k_B T$ around 1–100 K and 500–5000 K, respectively [1]. Therefore, at low gas temperatures or for low collisional energies, exchange of relative translational energy into internal motions of colliding partners is below the threshold for vibrational excitation, and vibrational motion remains in its ground state while collision-induced rotational transitions occur. It is helpful to first consider translational–rotational (or T–R) transitions at low collision energies, while assuming that vibrational degrees of freedom (or bond distances and angles) are constants at their average values, over collision times.

We consider here inelastic collision processes between molecules A and B in internal rovibrational states I and J, changing as

$$\mathrm{A}(I) + \mathrm{B}(J) \rightarrow \mathrm{A}(I') + \mathrm{B}(J')$$

at a fixed total energy E in a laboratory (LAB) reference frame, where $E_I^{(A)}$ and $E_J^{(B)}$ are their state internal energies. Their center-of-mass and relative positions are $\vec{R}_{CM} = (M_A\vec{R}_A + M_B\vec{R}_B)/M_{tot}$ and $\vec{R} = \vec{R}_A - \vec{R}_B$, while center of mass (CM) positions are $\vec{R}_A = \vec{R}_{CM} + M_B\vec{R}/M_{tot}$ and $\vec{R}_B = \vec{R}_{CM} - M_A\vec{R}/M_{tot}$, with $M_{tot} = M_A + M_B$. In the CM-SF (or CM-Space Fixed) frame, with its origin at $\vec{R}_{CM} = 0$, one finds that $\vec{P}_{CM} = \vec{L}_{CM} = E_{CM} = 0$. It follows that $E_{kin}^A + E_{kin}^B = E_{rel} = P^2/(2M)$ for the initial collisional event, with E'_{rel} the final value. The amount of energy transfer is $\varepsilon = E'_{rel} - E_{rel} = E_n - E_{n'}$, for initial and final internal states $n = (I, J)$ and $n' = (I', J')$.

Choosing A to indicate initial and final directions in an experimental setup, collisions are forward if $\vec{v} \cdot \vec{v}' > 0$ and backward if $\vec{v} \cdot \vec{v}' < 0$, and a usual description involves only the angular and energy transfer distribution given by the double-differential cross section $d^3\sigma_{\alpha'\alpha}/(d^2\Omega d\varepsilon)$, with Ω the solid angle direction of $\vec{v}'$ with respect to $\vec{v}$. A further simplification arises when one is interested only in the initial and final internal energies $E_{int}^{(AB)} = E_I^{(A)} + E_J^{(B)}$ and $E_{int}^{(AB)'} = E_{int}^{(AB)} - \varepsilon$, for given initial relative kinetic energy, with a cross section $d^3\sigma\left(\Omega, \varepsilon; E_{int}^{(AB)}, E_{rel}\right)/\left(d^2\Omega d\varepsilon\right)$, which can be obtained in classical [2] or quantal descriptions [3].

The total wavefunction of a molecular pair can be expanded in a basis set of N_{el} molecular electronic states $\Phi_\Gamma(\mathbf{X}; \mathbf{Q}) = \langle \mathbf{X} \mid \Phi_\Gamma(\mathbf{Q}) \rangle$ in the Dirac notation, with states $|\Phi_\Gamma(\mathbf{Q})\rangle$ in the electronic variables $\mathbf{X}$ dependent parametrically on the set $\mathbf{Q}$ of all atomic variables. Here we consider molecular collisions at relative velocities low compared with electron velocities, involving atoms and molecules with closed electronic shells. In these cases, a system evolving in its ground electronic state $\Phi_g(\mathbf{X};\, \mathbf{Q})$ undergoes an electronically adiabatic process on the potential energy surface $E(\mathbf{Q}) = \langle \Phi_g | \hat{H}_\mathbf{Q} | \Phi_g \rangle$, with brackets indicating integration over electronic variables, for fixed $\mathbf{Q} = (\vec{R}, \boldsymbol{q})$, with $\boldsymbol{q} = (\boldsymbol{q}_A, \boldsymbol{q}_B)$ the set of internal coordinates of species A and B and $\vec{R}$ providing the CM positions $\vec{R}_A = M_B\vec{R}/M_{tot}$ and $\vec{R}_B = -M_A\vec{R}/M_{tot}$. The Hamiltonian $\hat{H}_\mathbf{Q}$ contains the Coulomb interaction energy of all ion-core and electronic charges for fixed core positions so that atomic kinetic energies are excluded [4].

An electronic state Γ, evolving adiabatically in a scattering event without electronic transitions, must be chosen to go asymptotically to products of molecular states, so that $\Phi_\Gamma(\mathbf{X};\mathbf{Q}) \approx \Phi_\gamma^{A,B}(\mathbf{X};\boldsymbol{q}_A,\boldsymbol{q}_B) = \Phi_{\Gamma_A}^A(\mathbf{X}_A;\boldsymbol{q}_A)\Phi_{\Gamma_B}^B(\mathbf{X}_B;\boldsymbol{q}_B)$ as the CM intermolecular distance $R \to \infty$. The potential energies of the separate molecular states are $E_{\Gamma_A}^A(\boldsymbol{q}_A)$ and $E_{\Gamma_B}^B(\boldsymbol{q}_B)$. The interaction potential energy for the given electronic state of the molecular pair is then $V(\mathbf{Q}) = E_\Gamma(\mathbf{Q}) - E_{\Gamma_A}^A(\boldsymbol{q}_A) - E_{\Gamma_B}^B(\boldsymbol{q}_B) \approx 0$ for $R \to \infty$.

For example, in the interaction of atom A with diatomic BC with ion cores at $\vec{r}_b$ and $\vec{r}_c$ and diatomic relative distance $\vec{r} = \vec{r}_b - \vec{r}_c$, the potential energy is $V(\mathbf{Q}) = V(\vec{R},\vec{r}) = V(R,r,\theta)$ with θ the angle between $\vec{R}$ and $\vec{r}$. In the case of $He(^1S) + H_2(^1\Sigma)$, the triatomic state is a singlet, the rotational quantum numbers of H_2 are (j, m_j), and its vibrational quantum number is v, so that $J = (v, j, m_j)$ is the diatomic rovibrational state.

4.1.2 Quantal and Semiclassical Equations for Rovibrational Energy Transfer

4.1.2.1 Differential Equations and Boundary Conditions

Chapter 3 on the dynamics of inelastic collisions described the equations and scattering boundary conditions that can be used to calculate cross sections of adiabatic rovibrational energy transfer in state-to-state transitions. Quantal and semiclassical approaches are available for calculations and comparisons with experimental measurements. This chapter deals with related numerical methods and results for several molecular systems, involving collisions of atoms with molecules and molecule–molecule collisions. The following text collects the equations needed later on while discussing numerical methods.

Calculations for electronically adiabatic energy transfer between species A and B require knowledge of the interaction potential energy surface $V(\boldsymbol{Q})$ as a function of all the atomic positions in $\boldsymbol{Q} = (\vec{R}, \boldsymbol{q}_A, \boldsymbol{q}_B)$. For $V \neq 0$, the total Hamiltonian describing the collision dynamics is $\hat{H} = \hat{K}_R + \hat{H}_{int} + V$, with the internal Hamiltonian $\hat{H}_{int}$ containing molecular Coulomb energies of electrons and ion cores plus the atomic kinetic energies of ion cores in the CM frames of each molecule A and B, and with $\widehat{K}_R$ the kinetic energy for the relative motion of the molecular CMs. The scattering wavefunction $\Psi_\alpha^{(+)}(\vec{R}, \boldsymbol{q})$ starting in channel $\alpha = (\vec{k}, n) = 1$ to N_o, the total number of open channels, with $n = (I, J)$, and relative momentum $\vec{P} = \hbar\vec{k}_\alpha$, is in general a superposition of internal states $\phi_{n'}(\boldsymbol{q})$, as in

$$\Psi_\alpha^{(+)}(\vec{R}, \boldsymbol{q}) = \sum_{all\ n'} \psi_{\alpha'\alpha}^{(+)}(\vec{R})\, \phi_{n'}(\boldsymbol{q})$$

which extends over both energetically open and closed rovibrational states $\alpha' = (\vec{k}', n'), n' = (I', J') = 1$ to $N, N = N_o + N_c$, for open plus closed channels, satisfying the equation

$$\widehat{H}\Psi_\alpha^{(+)}(\vec{R}, \boldsymbol{q}; \vec{k}_\alpha) = E\Psi_\alpha^{(+)}(\vec{R}, \boldsymbol{q}; \vec{k}_\alpha)$$

with asymptotic conditions

$$\psi_{\alpha'\alpha}^{(+)} \approx \psi^0(\vec{R}; \vec{k}_\alpha)\delta_{\alpha'\alpha} + (2\pi)^{-\frac{3}{2}} f_{\alpha'\alpha}^{(+)}(\Omega) R^{-1} e^{ik_{\alpha'}R}, \quad n' \text{ open}$$

$$\psi_{\alpha'\alpha}^{(+)} \approx C_{\alpha'\alpha}(\Omega) R^{-1} e^{-\kappa_{\alpha'}R}, \quad n' \text{ closed}$$

where

$$f_{\alpha'\alpha}^{(+)}(\Omega) = -\frac{(2\pi)^2 M}{\hbar^2} \left\langle \Psi_\alpha^0(\vec{k}_{\alpha'}) | V | \Psi_\alpha^{(+)}(\vec{k}_\alpha) \right\rangle$$

is the scattering amplitude for the $\alpha \to \alpha'$ rovibrational transition in a collision with final relative momentum $\hbar\vec{k}_{\alpha'}$ in the Ω direction, for a chosen CM–SF reference frame. The set of coupled differential equations for the amplitudes $\psi_{\alpha'\alpha}^{(+)}(\vec{R})$, with $E = \left(\hbar^2 k_{\alpha'}^2\right)/2M + E_{n'}$, is

$$\frac{\hbar^2}{2M}\left(k_{\alpha'}^2 + \nabla_R^2\right) \psi_{\alpha'\alpha}^{(+)}(\vec{R}) = \sum_{\alpha''} V_{n'n''}(\vec{R}) \psi_{\alpha''\alpha}^{(+)}(\vec{R})$$

where $V_{n'n''}(\vec{R}) = \langle \phi_{n'} | V | \phi_{n''} \rangle$.

As detailed in Chapter 3, a partial-wave expansion with a basis set of states $\Phi_\lambda(\Omega, \boldsymbol{q}) = Y_{lm}(\Omega)\phi_n(\boldsymbol{q})$, where (lm) are the orbital angular momentum quantum numbers and $\lambda = (lmn)$, gives

$$\Psi_\alpha^{(+)}(\vec{R}, \boldsymbol{q}; \vec{k}_\alpha) = \sum_{\lambda'} \psi_{\lambda'\alpha}^{(+)}(R; \vec{k}_\alpha) \Phi_{\lambda'}(\Omega, \boldsymbol{q})$$

$$\psi_{\lambda'\alpha}^{(+)}(R; \vec{k}_\alpha) = (2\pi)^{-\frac{3}{2}} \sum_{l,m} 4\pi i^l (k_\alpha R)^{-1} \psi_{\lambda'\lambda}(R; k_\alpha) Y_{lm}(\Omega_\alpha)^*$$

containing the radial scattering functions $\psi_{\lambda'\lambda}(R; k_\alpha)$. In the absence of coupling, $\psi_{\lambda'\lambda}(R; k_\alpha) = u_l(k_\alpha R)\delta_{\lambda'\lambda} \approx \sin(k_\alpha R - l\pi/2)\delta_{\lambda'\lambda}$, where $u_l(x)$ is the regular Riccati–Bessel function. Replacing this expansion in the equation for $\Psi_\alpha^{(+)}(\vec{R}, \boldsymbol{q}; \vec{k}_\alpha)$, one arrives at the set of coupled ordinary differential equations

$$\left[k_{\alpha'}^2 + \frac{d^2}{dR^2} - \frac{l'(l'+1)}{R^2}\right]\psi_{\lambda'\lambda}(R;k_\alpha) = \sum_{\lambda''} U_{\lambda'\lambda''}(R)\psi_{\lambda''\lambda'}(R;k_\alpha)$$

$$U_{\lambda'\lambda''}(R) = (2M/\hbar^2)\langle\Phi_{\lambda'}|V|\Phi_{\lambda''}\rangle_R$$

to be solved with scattering asymptotic conditions

$$k_\alpha^{-1/2}\psi_{\lambda'\lambda}(R;k_\alpha) \approx k_\alpha^{-1/2}u_l(k_\alpha R)\delta_{\lambda'\lambda} + k_{\alpha'}^{-1/2}w_l^{(+)}\left(k_{\alpha'}R\right)f_{\lambda'\lambda}(k_\alpha), \quad \lambda' \text{ open}$$

$$k_\alpha^{-1/2}\psi_{\lambda'\lambda}(R;k_\alpha) \approx k_{\alpha'}^{-1/2}e^{\frac{il'\pi}{2}}w_l^{(+)}\left(i\kappa_{\alpha'}R\right)f_{\lambda'\lambda}(k_\alpha), \quad \lambda' \text{ closed}$$

where $w_l^{(+)}(x) \approx \exp\left[i(x - l\pi/2)\right]$ is the Riccati–Hankel function.

It is convenient for computational work to introduce matrix equations, as described in Chapter 3, and repeated here for treatments of adiabatic energy transfer. Letting $N = N_o + N_c$, for the channel index λ, running over N_o values for open channels and N_c values for closed channels, $N \times N$ matrices with operator-valued or function elements are defined in

$$\widehat{\boldsymbol{D}}_R\boldsymbol{\psi}^{(+)}(R) = \boldsymbol{U}(R)\boldsymbol{\psi}^{(+)}(R)$$

$$\widehat{\boldsymbol{D}}_R = \left[\delta_{\lambda'\lambda}\left(k_\alpha^2 + \frac{d^2}{dR^2} - \frac{l(l+1)}{R^2}\right)\right],$$

$$\boldsymbol{U}(R) = \left[U_{\lambda'\lambda}(R)\right], \boldsymbol{\psi}^{(+)}(R) = \left[\psi_{\lambda'\lambda}(R)\right].$$

Boundary conditions for scattering are given for the block matrix

$$\boldsymbol{\psi}^{(+)} = \begin{pmatrix}\boldsymbol{\psi}_{oo}\\ \boldsymbol{\psi}_{co}\end{pmatrix}$$

with $N_o \times N_o$ and $N_c \times N_o$ blocks for open and closed components.

Choosing to work with real-valued expressions, the asymptotic amplitudes for the open channels are

$$\boldsymbol{\psi}_{oo}^{(K)}(R) \approx [\boldsymbol{I}^{(s)}(R) + \boldsymbol{I}^{(c)}(R)\boldsymbol{K}_{oo}]\boldsymbol{k}_o^{1/2}$$

$$\boldsymbol{I}^{(s)}(R) = \left[\delta_{\lambda'\lambda}k_\alpha^{-\frac{1}{2}}u_l(k_\alpha R)\right]$$

$$\boldsymbol{I}^{(c)}(R) = \left[\delta_{\lambda'\lambda}k_\alpha^{-\frac{1}{2}}v_l(k_\alpha R)\right]$$

with $\boldsymbol{K}_{oo}$ the K-matrix of dimension $N_o \times N_o$ from which T-matrix and S-matrix forms can be obtained, as explained in Chapter 3. The matrix $\boldsymbol{K}_{oo}$ is real-valued and symmetric when the potential matrix $\boldsymbol{U}(R)$ is Hermitian and symmetric.

The transition and scattering matrices follow by comparing asymptotic forms in $\boldsymbol{\psi}_{oo}(R) \approx \boldsymbol{\psi}_{oo}^{(K)}(R)\boldsymbol{A}^{(K)} \approx \boldsymbol{\psi}_{oo}^{(S)}(R)\boldsymbol{A}^{(S)}$ as

$$\boldsymbol{S}_{oo} = (\mathbf{1}_{oo} + i\boldsymbol{K}_{oo})(\mathbf{1}_{oo} - i\boldsymbol{K}_{oo})^{-1}$$

$$\boldsymbol{T}_{oo} = \boldsymbol{K}_{oo}(\mathbf{1}_{oo} - i\boldsymbol{K}_{oo})^{-1} = (2i)^{-1}(\boldsymbol{S}_{oo} - \mathbf{1}_{oo}).$$

The closed-channel amplitudes may be expressed in terms of the exponentially growing and decaying $N_c \times N_c$ matrices

$$\boldsymbol{I}^{(g)}(R) = \left[\delta_{\lambda'\lambda} k_\alpha^{-\frac{1}{2}} e^{-i(l+1)\pi/2}\, u_l(i|k_\alpha|R)\right]$$

$$\approx \left[\delta_{\lambda'\lambda} k_\alpha^{-\frac{1}{2}} exp(\kappa_\alpha R)/2\right]$$

$$\boldsymbol{I}^{(d)}(R) = \left[\delta_{\lambda'\lambda} k_\alpha^{-\frac{1}{2}} e^{il\pi/2} w_l^{(+)}(i|k_\alpha|R)\right]$$

$$\approx \left[\delta_{\lambda'\lambda} k_\alpha^{-\frac{1}{2}} exp(-\kappa_\alpha R)\right]$$

by means of which the block of decaying relative motion wavefunctions take the asymptotic form

$$\boldsymbol{\psi}_{co}^{(K)}(R) \approx \boldsymbol{I}^{(d)}(R)\boldsymbol{K}_{co}\boldsymbol{k}_o^{1/2}.$$

The set of $N = N_o + N_c$-coupled second-order differential equations, together with regular boundary conditions at the origin $R = 0$ and the above asymptotic conditions for open and closed channels as $R \to \infty$, define a unique mathematical solution of the scattering problem.

4.1.2.2 Integral Equations

The $\boldsymbol{\psi}^{(+)}(R) = \left[\psi_{\lambda'\lambda}(R)\right]$ physical solutions for an initial channel λ and a final channel λ' can be expressed for each initial channel λ as linear combinations of $N_o + N_c$ numerical linear independent regular solutions $\boldsymbol{y}_\lambda(R)$ of $\widehat{\boldsymbol{D}}_R \boldsymbol{y}_\lambda(R) = \boldsymbol{U}(R)\boldsymbol{y}_\lambda(R)$ organized as a column matrix $\boldsymbol{y}_\lambda(R) = \left[y_{\lambda'\lambda}(R)\right]$. The physical solutions are of block form

$$\boldsymbol{\psi}^{(+)} = \begin{pmatrix} \boldsymbol{\psi}_{oo} \\ \boldsymbol{\psi}_{co} \end{pmatrix} = \begin{pmatrix} \boldsymbol{y}_{oo} & \boldsymbol{y}_{oc} \\ \boldsymbol{y}_{co} & \boldsymbol{y}_{cc} \end{pmatrix} \begin{pmatrix} \boldsymbol{C}_{oo} \\ \boldsymbol{C}_{co} \end{pmatrix}$$

with blocks $\boldsymbol{C}_{oo}$ and $\boldsymbol{C}_{co}$ providing scattering matrices $\boldsymbol{T}_{oo}$ and $\boldsymbol{T}_{co}$, respectively, from asymptotic conditions at large distances. They are given by a Fredholm-type integral equation

$$\boldsymbol{\psi}^{(+)}(R) = \boldsymbol{\psi}^{(0)}(R) + \int_0^\infty dR'\, \boldsymbol{G}^{(0)}(R, R')\boldsymbol{U}(R')\boldsymbol{\psi}^{(+)}(R')$$

$$\boldsymbol{\psi}^{(0)}(R) = \begin{pmatrix} \boldsymbol{I}^{(s)}(R) & \mathbf{0} \\ \mathbf{0} & \boldsymbol{I}^{(d)}(R) \end{pmatrix} \begin{pmatrix} \mathbf{I}_{oo} \\ \mathbf{0}_{co} \end{pmatrix}.$$

where the free motion $\boldsymbol{G}^{(0)}(R, R')$ propagator is a diagonal matrix containing oscillating and exponentially increasing and decreasing elements. However, the physical solutions $\boldsymbol{\psi}^{(+)}$ contain asymptotically only exponentially decreasing amplitudes in the closed channels and must be constructed choosing the coefficient matrices $\boldsymbol{C}_{oo}$ and $\boldsymbol{C}_{co}$ so that the exponentially increasing components are cancelled.

For computational purposes, it is more convenient to work with real-valued amplitudes and to first construct solutions $\boldsymbol{y}_{\lambda}(R)$ of Volterra-type integral equations [5]. Free Green functions $\mathbf{g}_{oo}$ and $\mathbf{g}_{cc}$, satisfying $\widehat{\boldsymbol{D}}_R(R,R')\mathbf{g}_{aa} = \delta(R-R')\mathbf{I}_{aa}$, given by

$$\mathbf{g}_{oo}(R,R') = \boldsymbol{I}^{(s)}(R)\boldsymbol{I}^{(c)}(R') - \boldsymbol{I}^{(c)}(R)\boldsymbol{I}^{(s)}(R'), R > R', \quad \mathbf{g}_{oo}(R,R') = 0, R < R'$$

$$\mathbf{g}_{cc}(R,R') = \boldsymbol{I}^{(g)}(R)\boldsymbol{I}^{(d)}(R') - \boldsymbol{I}^{(d)}(R)\boldsymbol{I}^{(g)}(R'), R > R', \quad \mathbf{g}_{cc}(R,R') = 0, R < R'$$

provide the total free radial Green function matrix

$$\mathbf{g}_0(R,R') = \begin{pmatrix} \mathbf{g}_{oo}(R,R') & \mathbf{0} \\ \mathbf{0} & \mathbf{g}_{cc}(R,R') \end{pmatrix}.$$

Numerical solutions are given by

$$\boldsymbol{y}(R) = \boldsymbol{y}^{(0)}(R) + \int_0^R dR'\, \mathbf{g}_0(R,R')\boldsymbol{U}(R')\boldsymbol{y}(R')$$

$$\boldsymbol{y}^{(0)}(R) = \begin{pmatrix} \boldsymbol{I}^{(s)}(R) & \mathbf{0} \\ \mathbf{0} & \boldsymbol{I}^{(d)}(R) \end{pmatrix}$$

with the free motion radial Green function $\mathbf{g}_0$ satisfying $\widehat{\boldsymbol{D}}_R\mathbf{g}_0(R,R') = \delta(R-R')\mathbf{I}$.

At large distances,

$$\begin{pmatrix} \boldsymbol{y}_{oo} & \boldsymbol{y}_{oc} \\ \boldsymbol{y}_{co} & \boldsymbol{y}_{cc} \end{pmatrix} \approx \begin{pmatrix} \boldsymbol{I}^{(s)} & \mathbf{0} \\ \mathbf{0} & \boldsymbol{I}^{(g)} \end{pmatrix} \boldsymbol{a} + \begin{pmatrix} \boldsymbol{I}^{(c)} & \mathbf{0} \\ \mathbf{0} & \boldsymbol{I}^{(d)} \end{pmatrix} \boldsymbol{b}, \quad R \to \infty$$

where $\boldsymbol{a}$ and $\boldsymbol{b}$ are constant $N \times N$ matrices

$$\boldsymbol{a} = \mathbf{I} + \int_0^\infty dR' \begin{pmatrix} \boldsymbol{I}^{(c)} & \mathbf{0} \\ \mathbf{0} & \boldsymbol{I}^{(d)} \end{pmatrix}_{R'} \boldsymbol{U}(R')\boldsymbol{y}(R'),$$

$$\boldsymbol{b} = -\int_0^\infty dR' \begin{pmatrix} \boldsymbol{I}^{(s)} & \mathbf{0} \\ \mathbf{0} & \boldsymbol{I}^{(g)} \end{pmatrix}_{R'} \boldsymbol{U}(R')\boldsymbol{y}(R').$$

Matching to the asymptotic conditions for real-valued outgoing waves and no exponentially increasing terms in

$$\boldsymbol{\psi}^{(K)} \approx \begin{pmatrix} \boldsymbol{y}_{oo} & \boldsymbol{y}_{oc} \\ \boldsymbol{y}_{co} & \boldsymbol{y}_{cc} \end{pmatrix} \begin{pmatrix} \boldsymbol{C}_{oo} \\ \boldsymbol{C}_{co} \end{pmatrix} \approx \begin{pmatrix} \boldsymbol{I}^{(s)}(R)\boldsymbol{k}_o^{1/2} \\ \mathbf{0} \end{pmatrix} + \begin{pmatrix} \boldsymbol{I}^{(c)}(R)\boldsymbol{K}_{oo}\boldsymbol{k}_o^{1/2} \\ \boldsymbol{I}^{(d)}(R)\boldsymbol{K}_{co}\boldsymbol{k}_o^{1/2} \end{pmatrix}$$

provides the relations

$$\boldsymbol{a}\begin{pmatrix}\boldsymbol{C}_{oo}\\ \boldsymbol{C}_{co}\end{pmatrix}=\begin{pmatrix}\mathbf{I}_{oo}\boldsymbol{k}_o^{-1/2}\\ \mathbf{0}_{co}\end{pmatrix},\quad \boldsymbol{b}\begin{pmatrix}\boldsymbol{C}_{oo}\\ \boldsymbol{C}_{co}\end{pmatrix}=\begin{pmatrix}\boldsymbol{K}_{oo}\boldsymbol{k}_o^{-1/2}\\ \boldsymbol{K}_{co}\boldsymbol{k}_o^{-1/2}\end{pmatrix}$$

and a set of $2\times N\times N_o$ algebraic equations for the same number of unknowns in $\boldsymbol{C}_{oo}$, $\boldsymbol{C}_{co}$, $\boldsymbol{K}_{oo}$, and $\boldsymbol{K}_{co}$. They give

$$\begin{pmatrix}\boldsymbol{K}_{oo}\\ \boldsymbol{K}_{co}\end{pmatrix}=\boldsymbol{b}\boldsymbol{a}^{-1}\begin{pmatrix}\mathbf{I}_{oo}\\ \mathbf{0}_{co}\end{pmatrix}$$

and also $\boldsymbol{\psi}^{(K)}(R)$ from the known $\boldsymbol{C}_{oo}$, $\boldsymbol{C}_{co}$ coefficients, and $\boldsymbol{y}(R)$. The scattering function $\boldsymbol{\psi}^{(+)}(R)$ follows from the relations between the $N\times N_o$ matrices $\boldsymbol{K}$ and $\boldsymbol{T}$.

4.1.2.3 Distorted-Wave Approximation

It is convenient in calculations to break up the treatment into two stages, with the first one involving a portion $V^{(1)}$ of the interaction potential energy $V(\vec{R},\boldsymbol{q})$ suitable for accurate calculations of distorted waves (DWs), to be used in a second stage that accounts for the effects of the remaining potential energy function $V^{(2)}=V-V^{(1)}$ [3, 6, 7], in a perturbation expansion of the wavefunction $\Psi_\alpha^{(+)}(\mathbf{Q})$ in powers of $V^{(2)}$ with terms constructed from the known DWs from the first stage.

As detailed in Chapter 3, a relative motion for potential energy $V^{(1)}$ with DW $\chi_{1,\alpha}^{(+)}(\vec{R})$ can be constructed from the Green function $G_{0,\alpha}^{(+)}(\vec{R},\vec{R}')$ for the free motion Hamiltonian $\widehat{H}_0=\widehat{K}_R+\widehat{H}_{int}$, with outgoing-waves boundary conditions, as

$$\chi_{1,\alpha}^{(+)}(\vec{R};\vec{k}_\alpha)=\chi_{1,\alpha}^{(0)}(\vec{R};\vec{k}_\alpha)+\int d^3R\,G_{0,\alpha}^{(+)}(\vec{R},\vec{R}')V^{(1)}(\vec{R}')\chi_{1,\alpha}^{(+)}(\vec{R}';\vec{k}_\alpha)$$

with $\chi_{1,\alpha}^{(0)}(\vec{R};\vec{k}_\alpha)=(2\pi)^{-\frac{3}{2}}\exp(i\vec{k}_\alpha\cdot\vec{R}))$. The total wavefunction is $\Psi_\alpha^{(+)}(\vec{R},\boldsymbol{q})=\Psi_{1,\alpha}^{(+)}(\vec{R},\boldsymbol{q})+\Psi_{2,\alpha}^{(+)}(\vec{R},\boldsymbol{q})$, with $\Psi_{1,\alpha}^{(+)}(\vec{R},\boldsymbol{q})=\chi_{1,\alpha}^{(+)}(\vec{R})\phi_1(\boldsymbol{q})$, and $\Psi_{2,\alpha}^{(+)}(\vec{R},\boldsymbol{q})$ obtained from the integral equation for $\Psi_\alpha^{(+)}$ containing $V^{(2)}\Psi_{2,\alpha}^{(+)}$. This can be dropped to first order when $V^{(2)}$ is small, and the total wavefunction $\Psi_\alpha^{(+)}(\vec{R},\boldsymbol{q})=\sum_{all\ n'}\psi_{\alpha'\alpha}^{(+)}(\vec{R})\,\phi_{n'}(\boldsymbol{q})$ can be obtained from the DWs above, as a DW approximation to first order, or DW1, giving a scattering amplitude

$$f_{\alpha'\alpha}^{(DW1)}(\Omega)=(2\pi)^2\frac{M}{\hbar^2}\int d^3R'\chi_{1,\alpha'}^{(-)}(\vec{\mathrm{R}}')^*\langle\phi_{n'}|V^{(2)}(\vec{R}')|\phi_n\rangle\chi_{1,\alpha}^{(+)}(\vec{R}')$$

and the differential cross sections

$$\left(\frac{d\sigma_{\alpha'\alpha}}{d\Omega}\right)^{(DW1)}=\frac{k_{\alpha'}}{k_\alpha}\left|f_{\alpha'\alpha}^{(DW1)}(\Omega)\right|^2.$$

A specific choice of distortion potentials separates diagonal and non-diagonal parts of the potential energy function represented by a matrix $\boldsymbol{V}(\vec{R})$ with elements

$V_{\alpha'\alpha}(\vec{R})$, as $\boldsymbol{V} = \boldsymbol{V}_D + \boldsymbol{V}_N$, with $\boldsymbol{V}_D$ and $\boldsymbol{V}_N$ in place of $V^{(1)}$ and $V^{(2)}$. A calculation proceeds by first generating DW matrices $\chi_\alpha^{(+)}(\vec{R};\vec{k}_\alpha) = \left[\chi_{\alpha\,\alpha}^{(+)}(\vec{R};\vec{k}_\alpha)\right]$ using the potential energy matrix $\boldsymbol{V}_D(\vec{R})$.

Numerical work is conveniently done after a wavefunction expansion in partial-wave states $\Phi_\lambda(\Omega,\ \boldsymbol{q}) = Y_{lm}(\Omega)\phi_\alpha(\boldsymbol{q})$, to derive coupled radial differential equations, and is best done for real-valued scattering solutions. The potential energy elements $U_{\lambda'\lambda}(R) = (2M/\hbar^2)\langle\Phi_{\lambda'}|V|\Phi_\lambda\rangle_R$ describe couplings between partial-wave states for each set of internal and relative angular momentum quantum numbers. Decomposing the matrix $\boldsymbol{U}(R)$ into diagonal and non-diagonal components, $\boldsymbol{U}(R) = \boldsymbol{U}_D(R) + \boldsymbol{U}_N(R)$ with $\boldsymbol{U}_D = \left[U_{\lambda\lambda}\delta_{\lambda'\lambda}\right]$ and $\boldsymbol{U}_N = \left[U_{\lambda'\lambda}\left(1-\delta_{\lambda'\lambda}\right)\right]$, we choose to work with standing-wave real-valued boundary conditions for computational convenience to obtain $\boldsymbol{\psi}^{(K)}(R)$ and a scattering $\boldsymbol{K}$-matrix to begin with. We start with solutions of $[\widehat{\boldsymbol{D}}_R - \boldsymbol{U}_D(R)]\boldsymbol{\chi}^{(K)}(R) = \boldsymbol{0}$, where $\boldsymbol{\chi}^{(K)}$ is a diagonal matrix of DWs, with standing-wave scattering functions $\chi_\lambda^{(K)}(R)$ as elements. These functions are needed with regular and irregular asymptotic conditions, called here $\chi_\lambda^{(K)}$ and $\overline{\chi}_\lambda^{(K)}$, given by

$$\chi_\lambda^{(K)}(R) \approx u_l(k_\alpha R) + v_l(k_\alpha R)K_{\lambda\lambda}^{(D)}$$

$$\overline{\chi}_\lambda^{(K)}(R) \approx v_l(k_\alpha R)$$

where $\chi_\lambda^{(K)}(R)$ is generated integrating the differential equation with the potential function $U_{\lambda\lambda}(R)$ from zero at a small value of R to a large value when $U_{\lambda\lambda}(R) \approx 0$, while $\overline{\chi}_\lambda^{(K)}(R)$ is generated integrating back from the large value of R. Here $u_l(x) \approx \sin(x - l\pi/2)$ and $v_l(x) \approx \cos(x - l\pi/2)$, and $K_{\lambda\lambda}^{(D)}$ can be extracted by matching as done for elastic scattering calculations. In matrix notation, $\boldsymbol{\chi}^{(K)}(R) \approx [\boldsymbol{I}^{(s)}(R) + \boldsymbol{I}^{(c)}(R)\boldsymbol{K}^{(D)}]\boldsymbol{k}^{1/2}$.

These solutions can be used to construct the distorted standing-wave Green functions $G_{D,\lambda}^{(K)}(R,R')$ forming a diagonal matrix $\boldsymbol{G}_D^{(K)}(R,R')$ satisfying

$$[\widehat{\boldsymbol{D}}_R - \boldsymbol{U}_D(R)]\boldsymbol{G}_D^{(K)}(R,R') = \mathbf{I}\,\delta(R-R')$$

$$G_{D,\lambda}^{(K)}(R,R') = k_\lambda^{-1}\left[\chi_\lambda^{(K)}(R)\overline{\chi}_\lambda^{(K)}(R') - \overline{\chi}_\lambda^{(K)}(R)\chi_\lambda^{(K)}(R')\right],\quad R' \le R; = 0, R' > R.$$

This provides a solution to $[\widehat{\boldsymbol{D}}_R - \boldsymbol{U}_D(R)]\boldsymbol{\psi}^{(K)}(R) = \boldsymbol{U}_N(R)\boldsymbol{\psi}^{(K)}(R)$ as

$$\boldsymbol{\psi}^{(K)}(R) = \boldsymbol{\chi}^{(K)}(R) + \int_0^R dR'\boldsymbol{G}_D^{(K)}(R,R')\boldsymbol{U}_N(R')\boldsymbol{\psi}^{(K)}(R')$$

The $\boldsymbol{K} = \left[K_{\lambda'\lambda}\right]$ matrix can be obtained comparing this with $\boldsymbol{\psi}^{(K)}(R) \approx [\boldsymbol{I}^{(s)}(R) + \boldsymbol{I}^{(c)}(R)\boldsymbol{K}]\boldsymbol{k}^{1/2}$, where $\boldsymbol{k}$ is a diagonal matrix with elements k_λ. Its form to first order in $\boldsymbol{U}_N$ is obtained approximating $\boldsymbol{\psi}^{(K)}(R')$ to the right with $\boldsymbol{\chi}^{(K)}(R')$. The first-order solution $\boldsymbol{\psi}^{(K1)}$ is

$$\boldsymbol{\psi}^{(K1)}(R) = \left[\boldsymbol{\chi}^{(K)}(R) + \boldsymbol{\chi}^{(K)}(R)\boldsymbol{k}^{-1}\int_0^\infty dR'\overline{\boldsymbol{\chi}}^{(K)}(R')\boldsymbol{U}_N(R')\boldsymbol{\chi}^{(K)}(R') \right.$$
$$\left. -\overline{\boldsymbol{\chi}}^{(K)}(R)\boldsymbol{k}^{-1}\int_0^\infty dR'\,\boldsymbol{\chi}^{(K)}(R')\boldsymbol{U}_N(R')\boldsymbol{\chi}^{(K)}(R')\right]\boldsymbol{A}$$

with the matrix of coefficients $\boldsymbol{A}$ obtained using that asymptotically $\boldsymbol{\chi}^{(K)}(R) \approx [\boldsymbol{I}^{(s)}(R) + \boldsymbol{I}^{(c)}(R)\boldsymbol{K}^{(D)}]\boldsymbol{k}^{1/2}$ and $\overline{\boldsymbol{\chi}}^{(K)}(R) \approx \boldsymbol{I}^{(c)}(R)\boldsymbol{k}^{1/2}$. Comparing this to $\boldsymbol{\psi}^{(K1)}(R) \approx [\boldsymbol{I}^{(s)}(R) + \boldsymbol{I}^{(c)}(R)\boldsymbol{K}^{(DW1)}]\boldsymbol{k}^{1/2}$, we find that

$$\boldsymbol{K}^{(DW1)} = \boldsymbol{K}^{(D)} - \boldsymbol{k}^{-\frac{1}{2}}\int_0^\infty dR'\,\boldsymbol{\chi}^{(K)}(R')\,\boldsymbol{U}_N(R')\boldsymbol{\chi}^{(K)}(R')\boldsymbol{k}^{-\frac{1}{2}}/\boldsymbol{D}$$

$$\boldsymbol{D} = \left[\boldsymbol{I} + \boldsymbol{k}^{-\frac{1}{2}}\int_0^\infty dR'\overline{\boldsymbol{\chi}}^{(K)}(R')\,\boldsymbol{U}_N(R')\boldsymbol{\chi}^{(K)}(R')\boldsymbol{k}^{-\frac{1}{2}}\right] \cong \boldsymbol{I}$$

to lowest order in $\boldsymbol{U}_N$. This approximate $\boldsymbol{K}^{(DW1)}$ can be used to construct an approximate T matrix or an approximate S matrix satisfying unitarity properties as in

$$\tilde{\boldsymbol{T}} = \boldsymbol{K}^{(DW1)}(\boldsymbol{1} - i\boldsymbol{K}^{(DW1)})^{-1}$$
$$\tilde{\boldsymbol{S}} = (\boldsymbol{1} + i\boldsymbol{K}^{(DW1)})(\boldsymbol{1} - i\boldsymbol{K}^{(DW1)})^{-1}.$$

A treatment using DWs $\boldsymbol{\chi}^{(+)}(R)$ with outgoing boundary conditions provides an approximate $\boldsymbol{S}^{(DW1)}$ matrix to first order in the couplings $\boldsymbol{U}_N$. When the elements in the matrix $\boldsymbol{S}^{(DW1)} - \boldsymbol{1} = i\boldsymbol{A}^{(DW1)}$ are small, it is possible to construct an approximate exponential form $\tilde{\boldsymbol{S}} = \exp[i\boldsymbol{A}^{(DW1)}]$, unitary insofar $\boldsymbol{A}^{(DW1)}$ is Hermitian, useful in applications to inelastic collisions [6].

4.1.2.4 Semiclassical Equations

Semiclassical treatments involve trajectories for positions and momenta that satisfy the Hamiltonian equations

$$\frac{d\vec{R}}{dt} = \frac{\partial\mathcal{H}}{\partial\vec{P}}, \quad \frac{d\vec{P}}{dt} = -\frac{\partial\mathcal{H}}{\partial\vec{R}}$$

with

$$\mathcal{H}(\vec{P},\vec{R}) = \vec{P}\cdot\vec{P}/(2M) + \mathcal{V}(\vec{P},\vec{R})$$

for an effective potential energy $\mathcal{V}(\vec{P},\vec{R})$, to be solved for each initial set of conditions $(\vec{P}_s,\vec{R}_s)$, $s = 1$ to N_{in}, starting at a time t_{in} to generate the bundle of trajectories $\{\vec{R}(t;\vec{P}_s,\vec{R}_s), \vec{P}(t;\vec{P}_s,\vec{R}_s)\}$ in phase space, for constant total energy $\mathcal{H}(\vec{P},\vec{R}) = E$. This is illustrated in Figure 4.1 for an atom–atom collision in a CM–SF frame using relative position and momenta variables $(\vec{r},\vec{p})$, showing a trajectory starting at a time t_1 from an initial position $\vec{r}_1 = (z_1,\rho_1,\varphi_1)$ in cylindrical coordinates and with momentum $\vec{p}_1$. An initial sector of area $\rho_1 d\rho_1 d\varphi_1$, shown as shaded, is the origin of a bundle of trajectories evolving in time as depicted.

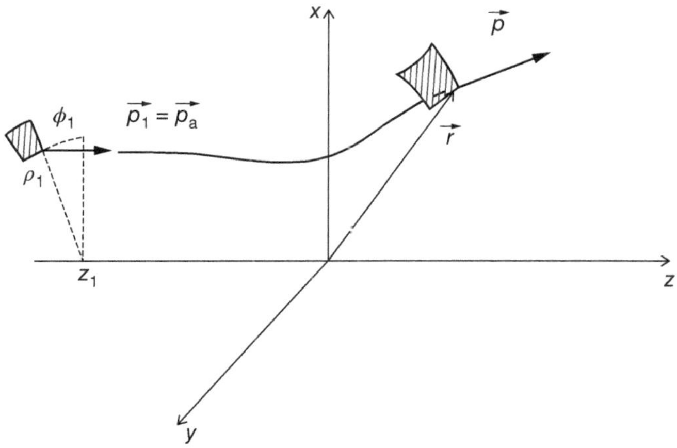

Figure 4.1 An atom–atom collision trajectory starting at a time t_1 from an initial position $\vec{r}_1 = (z_1, \rho_1, \phi_1)$, in cylindrical coordinates, and with momentum $\vec{p}_1$. An initial sector of area $\rho_1 d\rho_1 d\phi_1$, shown as shaded, is the origin of a bundle of trajectories evolving in time as depicted. Source: [8] / AIP Publishing.

As described in the previous chapter, the collisional dynamics is obtained from the wavefunction of the system as a superposition of eikonal forms $\chi(\vec{R}, \boldsymbol{q}) \exp[iS(\vec{R})/\hbar]$ [8, 9], where $S(\vec{R})$ is a real-valued phase function, identified with a mechanical action [2] in an effective potential energy $\mathcal{V}(\vec{R})$ constructed from physical considerations. Action $S(\vec{R})$ and amplitude $\chi(\vec{R}, \boldsymbol{q})$ functions in the eikonal wavefunction [8, 10] can be written as functions of time and initial variables, given by $S(\vec{R}) = S(t; \vec{P}_s, \vec{R}_s)$ and $\chi(\vec{R}, \boldsymbol{q}) = \chi(t, \boldsymbol{q}; \vec{P}_s, \vec{R}_s)$.

The trajectories give $S(t; \vec{P}_s, \vec{R}_s)$ from the line integral in

$$S(t) = S(\vec{R}_s) + \int_{\vec{R}_s}^{\vec{R}(t)} d\vec{R} \cdot \vec{P} = \hbar\vec{k} \cdot \vec{R}(t) + \int_{t_{in}}^{t} dt \frac{d\vec{R}}{dt} \cdot (\vec{P} - \hbar\vec{k})$$

and give amplitudes, in the approximation of short wavelengths, as solutions of

$$\left[-i\hbar\frac{d}{dt} + \widehat{H}_{\boldsymbol{q}}(\vec{R}) - \mathcal{V}(\vec{R}) - iU_{\vec{k}}(\vec{R})\right] |\chi(t)\rangle = 0$$

with $U_{\vec{k}}(\vec{R}) = (\hbar/M)\nabla \cdot \vec{P}$ and a chosen effective potential energy $\mathcal{V}(\vec{R})$. Further letting

$$|\chi(t, \boldsymbol{q})\rangle = \sum_{n'} A_{\alpha'\alpha}(t)\, \phi_{n'}(\boldsymbol{q}) \exp\left\{-\frac{i}{\hbar}\int_{t_{in}}^{t} dt'[\mathcal{V}(\vec{R}') - iU_{\vec{k}}(\vec{R}')]\right\}$$

with $\vec{R}' = \vec{R}(t'; \vec{P}_s, \vec{R}_s)$ gives a set of coupled ordinary differential equations for the channel amplitudes $A_{\alpha'\alpha}(t)$,

$$-i\hbar\frac{dA_{\alpha'\alpha}}{dt} + \sum_{\alpha''} H_{n'n''}(t) A_{\alpha''\alpha}(t) = 0$$

with $H_{n'n''}(t) = \langle \phi_{n'} | \hat{H}_{\mathbf{q}}[\vec{R}(t)] | \phi_{n''} \rangle = E_{n'} \delta_{n'n''} + V_{n'n''}[\vec{R}(t)]$. The initial conditions at $t = t_{in}$ follow from $\vec{R}(t_{in}) = R_{in} \to \infty$ and $\vec{P}_{in} = \hbar \vec{k}_\alpha$ for each bundle component. In a compact matrix notation,

$$-i\hbar \frac{d\mathbf{A}_\alpha}{dt} + \mathbf{H}(t)\mathbf{A}_\alpha = \mathbf{0}$$

with $\mathbf{A}_\alpha = [A_{\alpha'\alpha}]$ an $N \times 1$ column matrix and $\mathbf{H}(t) = [H_{n'n''}(t)]$ an $N \times N$ square matrix.

4.2 Numerical Methods

4.2.1 Step-by-Step Propagation of Wavefunctions

Step-by-step propagation along the variable R is a convenient procedure when the wavelengths of the oscillating $\psi_{\lambda'\lambda}(R)$ functions are of magnitude comparable to molecular bond distances. This can happen when collisions between heavy particles (atoms and molecules) occur at very low (milli-Kelvin) energies or when the projectile is an electron scattering at room temperature energies. Large wavelengths allow large step sizes, and this helps to decrease the amount of numerical calculations needed to advance the solutions from initial to asymptotic values of R.

The physical solution may be expressed as a combination of $N = N_o + N_c$ sets of linearly independent regular and real-valued solutions $y_{\lambda'\lambda}(R)$ of N-coupled differential equations, with N_o for open channels and N_c for closed channels and each set collected in the column matrix $\boldsymbol{y}_n(R)$, $n = 1$ to N. A row of these solutions forms the $N \times N$ matrix

$$\boldsymbol{y} = (\boldsymbol{y}_1, \boldsymbol{y}_2, \ldots, \boldsymbol{y}_N) = \begin{pmatrix} \boldsymbol{y}_{oo} & \boldsymbol{y}_{co} \\ \boldsymbol{y}_{oc} & \boldsymbol{y}_{cc} \end{pmatrix}$$

given here also in block form. Linear independence requires that the determinant of the matrix $[\boldsymbol{y}_n^T \boldsymbol{y}_{n'}]$ from transposed matrices times given columns satisfies $det\,[\boldsymbol{y}_n^T \boldsymbol{y}_{n'}] \neq 0$. All the channel-to-channel transition amplitudes may be obtained taking linear combinations of the mathematical solutions $\boldsymbol{y}_n(R)$ with coefficients $C_{n\lambda}$, $\lambda = 1$ to N_o, to construct the physical solutions $\boldsymbol{\psi}_\lambda(R) = \boldsymbol{y}(R)\boldsymbol{C}_\lambda$ for each initial channel. In block form, $\boldsymbol{\psi}^{(+)}(R) = \boldsymbol{y}(R)\boldsymbol{C}$, or in more detail

$$\boldsymbol{\psi}^{(+)} = \begin{pmatrix} \boldsymbol{\psi}_{oo} \\ \boldsymbol{\psi}_{co} \end{pmatrix} = \begin{pmatrix} \boldsymbol{y}_{oo} & \boldsymbol{y}_{co} \\ \boldsymbol{y}_{oc} & \boldsymbol{y}_{cc} \end{pmatrix} \begin{pmatrix} \boldsymbol{C}_{oo} \\ \boldsymbol{C}_{co} \end{pmatrix}$$

with blocks $\boldsymbol{C}_{oo}$ and $\boldsymbol{C}_{co}$, which can be related to scattering matrices $\boldsymbol{T}_{oo}$ and $\boldsymbol{T}_{co}$ at large distances, respectively.

This can be conveniently done for the real-valued physical solutions $\boldsymbol{\psi}_{oo}^{(K)}(R) = [\boldsymbol{I}^{(s)}(R) + \boldsymbol{I}^{(c)}(R)\boldsymbol{K}_{oo}]\boldsymbol{k}_o^{1/2}$ and $\boldsymbol{\psi}_{co}^{(K)}(R) = \boldsymbol{I}^{(d)}(R)\boldsymbol{K}_{co}\boldsymbol{k}_o^{1/2}$ at large R, equating combined functions and their derivatives as in the blocks

$$\psi^{(K)}(R) = \mathbf{y}(R)\mathbf{C}^{(K)}, \quad d\psi^{(K)}/dR = (d\mathbf{y}/dR)\mathbf{C}^{(K)}.$$

At large distances, this gives $2N \times N_o$ linear algebraic equations to be solved for the $N \times N_o$ unknowns in the real-valued matrix $\mathbf{C}^{(K)}$ and the $N \times N_o$ unknowns in a matrix $\mathbf{K} = [\mathbf{K}_{oo}, \mathbf{K}_{co}]$. This leads to the $\mathbf{T}_{oo}$ and $\mathbf{S}_{oo}$ scattering matrices, whose symmetry properties can be verified as a check on the numerical procedures.

Procedures for starting the integration at short distances and for propagation to large distances are similar to the ones described for elastic scattering, but are now done in terms of matrices. The matrix differential equation for the numerical solutions is

$$\mathbf{y}''(R) + \mathbf{Q}(R)\mathbf{y}(R) = \mathbf{0}$$

with $\mathbf{Q}(R) = \left[Q_{\lambda\lambda'}(R)\right]$ and $Q_{\lambda\lambda'} = \left[k_\alpha^2 - l(l+1)/R^2\right]\delta_{\lambda\lambda'} - U_{\lambda\lambda'}(R)$. The potential energy functions $U_{\lambda\lambda}(R)$ display large positive repulsion cores at short distances, with classical turning points $R_{\lambda,t}$ for each collisional kinetic energy. A suitable distance $R_{in} < \min\,(R_{\lambda,t})$ and starting constants in $y_{\lambda,n}(R_{in} - \Delta R) = \delta_{\lambda,n}a_\lambda$ and $y'_{\lambda,n}(R_{in} - \Delta R) = \delta_{\lambda,n}a'_\lambda$ can be chosen by trial so that the numerical matrix $\mathbf{y}_{in}$ starts with very small values (for computational purposes) and its propagation to larger distances remains of small magnitude inside the potential energy cores.

Step-wise propagation is done by analogy with the treatment for elastic scattering with the de Vogaleare or Numerov algorithms, presented in the chapter on elastic collisions, replacing functions $y(x)$ there with matrices $\mathbf{y}(R)$ here, and advancing solutions as $\mathbf{y}_s = \mathbf{y}(R - s\,h)$, at step s with length $h = \delta R$, starting at $R_{in} - \Delta R$ [11, 12].

At large distances, where $U_{\lambda\lambda'}(R) \ll [l(l+1)/R^2]\delta_{\lambda\lambda'}$, the physical solutions $\psi_\lambda(R) = \mathbf{y}(R)\mathbf{C}_\lambda$ are constructed as mentioned above to extract the matrix $\mathbf{K} = [\mathbf{K}_{oo}, \mathbf{K}_{co}]$ from the asymptotic conditions.

4.2.2 Piecewise Propagation

Piecewise propagation of the matrix of physical solutions $\psi_\lambda(R)$ can be done by extension of the treatments described in the chapter on elastic collisions, done here within a matrix formulation. The above matrix function $\mathbf{y}(R)$, including the centrifugal potentials for each l, is divided into pieces, or sectors, within left and right boundaries $[R_l < R_r]$ of a sector and is approximated locally by a reference $\mathbf{Q}_0(R)$ function, which can be constant, linear, or quadratic in R for convenience and accuracy. Introducing two linearly independent dimensionless solutions $\mathbf{F}(R) = \mathbf{A}(R), \mathbf{B}(R)$ of the matrix differential equation

$$\mathbf{F}''(R) + \mathbf{Q}_0(R)\mathbf{F}(R) = \mathbf{0}, \qquad \mathbf{F}(R) = \mathbf{A}(R), \mathbf{B}(R)$$

they are used to define coefficient functions $\boldsymbol{a}(R)$ and $\boldsymbol{b}(R)$ for the dimensionless numerical solutions $\boldsymbol{y}(R)$ so that

$$\boldsymbol{y}(R) = \boldsymbol{A}(R)\boldsymbol{a}(R) + \boldsymbol{B}(R)\boldsymbol{b}(R)$$
$$\boldsymbol{y}'(R) = \boldsymbol{A}'(R)\boldsymbol{a}(R) + \boldsymbol{B}'(R)\boldsymbol{b}(R)$$

with primes for derivatives. Differentiating the first line and using the second line, one finds (i) $\boldsymbol{A}(R)\boldsymbol{a}'(R)+\boldsymbol{B}(R)\boldsymbol{b}'(R)=\boldsymbol{0}$, while differentiating the second line and using the differential equations for $\boldsymbol{y}(R)$, $\boldsymbol{A}(R)$ and $\boldsymbol{B}(R)$, one finds (ii) $\boldsymbol{A}'(R)\boldsymbol{a}'(R)+\boldsymbol{B}'(R)\boldsymbol{b}'(R)=[\boldsymbol{Q}_0(R)-\boldsymbol{Q}(R)][\boldsymbol{A}(R)\boldsymbol{a}(R)+\boldsymbol{B}(R)\boldsymbol{b}(R)]$. Equations (i) and (ii) provide expressions for $\boldsymbol{a}'(R)$ and $\boldsymbol{b}'(R)$ in terms of the difference $\Delta\boldsymbol{Q}(R) = \boldsymbol{Q}(R)-\boldsymbol{Q}_0(R)$ and of the Wronskian $\boldsymbol{W} = \boldsymbol{A}(R)\boldsymbol{B}'(R)-\boldsymbol{A}'(R)\boldsymbol{B}(R)$ satisfying $d\boldsymbol{W}/dR = 0$, as

$$\boldsymbol{a}' = \boldsymbol{W}^{-1}\boldsymbol{B}\Delta\boldsymbol{Q}\,(\boldsymbol{A}\boldsymbol{a}+\boldsymbol{B}\boldsymbol{b})$$
$$\boldsymbol{b}' = -\boldsymbol{W}^{-1}\boldsymbol{A}\Delta\boldsymbol{Q}\,(\boldsymbol{A}\boldsymbol{a}+\boldsymbol{B}\boldsymbol{b})$$

with all the matrices as functions of R. Solving these differential equations for $\boldsymbol{a}(R)$ and $\boldsymbol{b}(R)$ within the sector $[R_l, R_r]$ allows piecewise propagation from $\{\boldsymbol{a}_l=\boldsymbol{a}(R_l), \boldsymbol{b}_l=\boldsymbol{b}(R_l)\}$ to $\{\boldsymbol{a}_r=\boldsymbol{a}(R_r), \boldsymbol{b}_r=\boldsymbol{b}(R_r)\}$ [13]. The sector interval can be chosen small enough so that $\Delta\boldsymbol{Q}(R)$ is small, and the solutions can be accurately constructed to first order in $\Delta\boldsymbol{Q}$. Then by direct integration,

$$\boldsymbol{a}_r - \boldsymbol{a}_l = \boldsymbol{I}_{BA}\boldsymbol{a}_l + \boldsymbol{I}_{BB}\boldsymbol{b}_l$$
$$\boldsymbol{b}_r - \boldsymbol{b}_l = \boldsymbol{I}_{AA}\boldsymbol{a}_l + \boldsymbol{I}_{AB}\boldsymbol{b}_l$$

where

$$\boldsymbol{I}_{BA} = \int_{R_l}^{R_r} dR\,\boldsymbol{B}(R)\Delta\boldsymbol{Q}(R)\boldsymbol{A}(R)$$

and so on for $\boldsymbol{I}_{BB}$, $\boldsymbol{I}_{AA}$, and $\boldsymbol{I}_{AB}$.

In a simple case where the matrix $\boldsymbol{Q}_0$ is diagonal and constant over R, the solution matrices $\boldsymbol{A}(R)$ and $\boldsymbol{B}(R)$ are also diagonal, with trigonometric functions for open channels or exponential functions for closed channels, as matrix elements, and the integrals can be readily done. More generally, given $\boldsymbol{Q}(R)$, a new $\tilde{\boldsymbol{Q}}_0$ can be constructed as follows. Define the sector midpoint $\overline{R} = (R_l + R_r)/2$ and construct a transformation unitary matrix $\boldsymbol{M}$ so that $\boldsymbol{M}\boldsymbol{Q}(\overline{R})\boldsymbol{M}^{-1} = \tilde{\boldsymbol{Q}}_0$ is diagonal and $\Delta\tilde{\boldsymbol{Q}}(R) = \tilde{\boldsymbol{Q}}(R) - \tilde{\boldsymbol{Q}}_0$ after the transformation. This matrix transforms the previous reference solutions to $\tilde{\boldsymbol{A}}(R)$, $\tilde{\boldsymbol{B}}(R)$, and $\tilde{\boldsymbol{y}}(R)$, so that $\tilde{\boldsymbol{y}}(R) = \tilde{\boldsymbol{A}}(R)\tilde{\boldsymbol{a}}(R) + \tilde{\boldsymbol{B}}(R)\tilde{\boldsymbol{b}}(R)$ with $\tilde{\boldsymbol{a}} = \boldsymbol{M}\boldsymbol{a}$ and $\tilde{\boldsymbol{b}} = \boldsymbol{M}\boldsymbol{b}$. The new solution matrices $\tilde{\boldsymbol{A}}(R)$, $\tilde{\boldsymbol{B}}(R)$ are again diagonal with trigonometric or exponential functions along diagonals. Piecewise propagation from sector n to $n+1$ follows from $\tilde{\boldsymbol{a}}_l(n+1) = \boldsymbol{M}(n+1)\boldsymbol{M}(n)^{-1}\tilde{\boldsymbol{a}}_r(n)$ and $\tilde{\boldsymbol{b}}_l(n+1) = \boldsymbol{M}(n+1)\boldsymbol{M}(n)^{-1}\tilde{\boldsymbol{b}}_r(n)$. More details, such as instead choosing a reference $\boldsymbol{Q}_0(R)$ linear in R, and how to choose the size of the sectors, can be found in the literature [13].

A very convenient alternative procedure introduces piecewise potentials to solve the coupled integral equations for scattering [14, 15]. This again is a generalization of the procedure described for elastic scattering, which is done here in terms of matrices. It is notably stable for propagation of wavefunctions with coupled open and closed channels. Using again the reference local solutions $\boldsymbol{F}(R) = \boldsymbol{A}(R), \boldsymbol{B}(R)$, a local Green function (or propagator) is

$$\boldsymbol{G}_{0,loc}(R, R') = \boldsymbol{k}^{-1}[\boldsymbol{A}(R)\boldsymbol{B}(R') - \boldsymbol{A}(R')\boldsymbol{B}(R)]$$

with $\boldsymbol{k} = \left[\delta_{\lambda'\lambda}k_\alpha\right]$ for $R > R'$ and $\boldsymbol{G}_{0,loc}(R, R') = \boldsymbol{0}$ for $R' > R$. Replacing this in the local integral equation,

$$\boldsymbol{y}(R) = \boldsymbol{y}(R_l) + \int_{R_l}^{R} dR' \boldsymbol{G}_{0,loc}(R, R')\boldsymbol{U}_0(R')\boldsymbol{y}(R')$$

where as before $\boldsymbol{y}(R') = \boldsymbol{A}(R')\boldsymbol{a}(R') + \boldsymbol{B}(R')\boldsymbol{b}(R')$, and letting $R = R_r$ to obtain $\boldsymbol{y}(R_r)$, each sector is found to contain integrals of type

$$\boldsymbol{I}_{FG} = \int_{R_l}^{R_r} dR' \boldsymbol{F}(R')\boldsymbol{U}_0(R')\boldsymbol{G}(R')$$

with $\boldsymbol{F}(R')$ and $\boldsymbol{G}(R')$ either $\boldsymbol{A}(R)$ or $\boldsymbol{B}(R)$ [14]. The solution for $\boldsymbol{y}(R)$ is a sum over sectors from an initial $R_l = R_{in}$ up to a large $R_r = R$. In particular, when $\boldsymbol{U}_0$ can be chosen to be constant inside a sector, the matrices $\boldsymbol{A}(R)$ and $\boldsymbol{B}(R)$ can be obtained from $\sin(k_\alpha R)$ and $\cos(k_\alpha R)$ functions for energetically open channels and from $\exp(\pm\kappa_\alpha R)$ functions for closed channels, as treated in Section 4.3. Hence the integrals are simple analytical expressions for constant piecewise sectors. Compared to step-by-step propagation, this treatment decreases the number of terms appearing in integrals to the right and also avoids having to generate Riccati functions at quadrature points. It also shows stability when exponential functions appear locally due to the presence of classically forbidden regions of the potential energy vs R at some total energies E [15].

Another procedure also found to be stable in the presence of classically forbidden (or closed) channels is the **R**-matrix sector propagation method [16]. The $\boldsymbol{R}^{(i)}$ matrix in sector i of length $\Delta R = R_r - R_l$ is defined by $\boldsymbol{y}^{(i)}(R_r) = \boldsymbol{R}^{(i)}(d\boldsymbol{y}^{(i)}/dR)_{R_r}$. This can be related to $\boldsymbol{R}^{(i-1)}$ in the previous sector in a recurrent process leading from R_{in} to a final large. The relationship can be derived in terms of a local Green function $\boldsymbol{G}^{(i)}(R', R)$ [17] and gives, with $\epsilon \to 0+$,

$$\begin{aligned}\boldsymbol{R}^{(i)} &= -\boldsymbol{G}^{(i)}(R_r - \epsilon, R_r) - \boldsymbol{G}^{(i)}(R_r, R_l)[\boldsymbol{R}^{(i-1)} - \boldsymbol{G}^{(i)}(R_l + \epsilon, R_l)]^{-1} \\ &\quad \times \boldsymbol{G}^{(i)}(R_l, R_r)\end{aligned}$$

At the end of propagation, one obtains the final $\boldsymbol{y}^{(i)}(R)$, with a stable procedure insofar any closed-channel intermediate exponential functions need not be generated.

4.2.3 Propagation of Amplitude Densities

Again as an extension of the treatment for elastic scattering, an alternative to the piecewise expansion in the integral equation for the wavefunction column matrix $\boldsymbol{y}_\lambda(R)$ for initial channel λ is to instead introduce an integral equation for the $1 \times N$ column matrix of amplitude densities $\boldsymbol{\xi}_\lambda(R) = \boldsymbol{U}(R)\boldsymbol{y}_\lambda(R)$ and solve for it first [5]. Matrix elements of this new matrix function contain products like $U_{\lambda''\lambda'}y_{\lambda'\lambda}(R)$, which go asymptotically to zero as $\boldsymbol{U}(R)$ does and also go to zero at some initial R_{in}, where $\boldsymbol{y}_\lambda(R_{in}) = \boldsymbol{0}$. They are usually square-integrable and can be expanded in an orthonormal basis set $\{\eta_n(R),\ n = 1 \text{ to } N_B\}$ of L^2 functions [5, 18]. This gives $\boldsymbol{\xi}_\lambda(R) = \sum_n \eta_n(R)\boldsymbol{A}_n^{(\lambda)}$ with expansion coefficients in a $1 \times N$ matrix $\boldsymbol{A}_n^{(\lambda)}$, to be obtained solving coupled algebraic equations.

From the equation for $\boldsymbol{y}(R)$ in Section 4.3, multiplied to the left times $\boldsymbol{U}(R)$, one finds the Volterra-type integral equation

$$\boldsymbol{\xi}_\lambda(R) = \boldsymbol{U}(R)\boldsymbol{y}_\lambda^{(0)}(R) + \boldsymbol{U}(R)\int_0^R dR'\, \mathbf{g}_0(R, R')\boldsymbol{\xi}_\lambda(R')$$

and expanding the two sides in the basis set, the matrix algebraic equation

$$\sum_{n'}(\boldsymbol{\delta}_{nn'} + \boldsymbol{B}_{nn'})\boldsymbol{A}_{n'}^{(\lambda)} = \boldsymbol{A}_n^{(\lambda,0)}$$

$$\boldsymbol{A}_n^{(\lambda,0)} = \int_0^\infty dR\eta_n(R)^*\boldsymbol{U}(R)\boldsymbol{y}_\lambda^{(0)}(R)$$

$$\boldsymbol{B}_{nn'} = \int_0^\infty dR\eta_n(R)^*\boldsymbol{U}(R)\int_0^R dR'\, \mathbf{g}_0(R, R')\eta_{n'}(R')$$

to be solved for the set $\left\{\boldsymbol{A}_{n'}^{(\lambda)}\right\}$. Once the $\boldsymbol{A}_{n'}^{(\lambda)}$ coefficients are known, one finds

$$\boldsymbol{y}_\lambda(R) = \boldsymbol{y}_\lambda^{(0)}(R) + \int_0^R dR'\, \mathbf{g}_0(R, R')\sum_n \eta_n(R')\boldsymbol{A}_n^{(\lambda)}$$

giving the scattering wavefunction $\boldsymbol{\psi}^{(K)}(R) = \boldsymbol{y}(R)\boldsymbol{C}^{(K)}$, and the $\boldsymbol{K}$-matrix is readily obtained.

4.2.4 Variational Calculation of Scattering Amplitudes

State-to-state inelastic transition amplitudes $T_{\alpha'\alpha}\left(\vec{k}_{\alpha'}, \vec{k}_\alpha\right) = \left\langle \Psi_\alpha^0\left(\vec{k}_{\alpha'}\right) |V| \Psi_\alpha^{(+)}\left(\vec{k}_\alpha\right)\right\rangle$ can be obtained from a variational treatment as a generalization of the procedures described for elastic scattering [19–21]. A Kohn-type variational functional is given by

$$\mathcal{T}_{\alpha'\alpha}\left(\vec{k}_{\alpha'}, \vec{k}_\alpha\right) = \left\langle \tilde{\Psi}_{\alpha'}^{(-)}\left(\vec{k}_{\alpha'}\right) |(E - \hat{H})| \tilde{\Psi}_\alpha^{(+)}\left(\vec{k}_\alpha\right)\right\rangle$$

where $\tilde{\Psi}_\alpha^{(\pm)}\left(\vec{R}, \boldsymbol{q}; \vec{k}_\alpha\right) = \left\langle \vec{R}, \boldsymbol{q} \middle| \tilde{\Psi}_\alpha^{(\pm)}\left(\vec{k}_\alpha\right)\right\rangle$ is a trial wavefunction dependent on internal and relative position variables. Its variations $\delta\Psi_\alpha^{(\pm)}\left(\vec{k}_\alpha\right) = \tilde{\Psi}_\alpha^{(\pm)}\left(\vec{k}_\alpha\right) - \Psi_\alpha^{(\pm)}\left(\vec{k}_\alpha\right)$

from the exact wavefunction lead to changes in the functional satisfying [22]

$$\delta\left(\mathcal{T}_{\alpha'\alpha} - T_{\alpha'\alpha}\right) = \mathcal{O}[\delta\tilde{\Psi}]^2$$

so that the functional approximates the exact transition amplitude $T_{\alpha'\alpha}$ to second order in the trial function variation. This provides ways to improve the transition amplitude given a known trial function. Numerical applications of this approach can be done in more detail after partial wave expansions of trial functions and can be expressed in terms of column matrices $\tilde{\boldsymbol{\psi}}_{\lambda}^{(\pm)}(R) = \left[\tilde{\psi}_{\lambda'\lambda}^{(\pm)}(R)\right]$ for each initial scattering channel λ, introduced in Section 4.2. The functional is now

$$\mathcal{T}_{\lambda'\lambda}\left(\vec{k}_{\alpha'}, \vec{k}_{\alpha}\right) = \int dR\, \tilde{\boldsymbol{\psi}}_{\lambda'}^{(-)}\left(R; \vec{k}_{\alpha'}\right)^{\dagger} [\hat{\boldsymbol{D}}_R - \boldsymbol{U}(R)]\tilde{\boldsymbol{\psi}}_{\lambda}^{(+)}(R; \vec{k}_{\alpha})$$

where λ' and λ refer to open channels, with variations giving $\left(\mathcal{T}_{\lambda'\lambda} - T_{\lambda'\lambda}\right) = \mathcal{O}[\delta\tilde{\boldsymbol{\psi}}]^2$.

As done for elastic scattering, the above variational result can be used to calculate an improved scattering amplitude $T_{\lambda'\lambda}$ starting with a good trial function, so that the exact $\boldsymbol{\psi} = \tilde{\boldsymbol{\psi}} + \delta\tilde{\boldsymbol{\psi}}$ and $T_{\lambda'\lambda} = \tilde{T}_{\lambda'\lambda} + \delta T_{\lambda'\lambda}$ to first order in the trial function accuracy. Then insofar $\delta\mathcal{T}_{\lambda'\lambda} = \mathcal{T}_{\lambda'\lambda}[\boldsymbol{\psi}] - \mathcal{T}_{\lambda'\lambda}[\tilde{\boldsymbol{\psi}}] = -\mathcal{T}_{\lambda'\lambda}[\tilde{\boldsymbol{\psi}}]$ for the exact solution, we find that $\delta T_{\lambda'\lambda} = T_{\lambda'\lambda} - \tilde{T}_{\lambda'\lambda} = -\mathcal{T}_{\lambda'\lambda}[\tilde{\boldsymbol{\psi}}]$ and

$$T_{\lambda'\lambda} = \tilde{T}_{\lambda'\lambda} - \mathcal{T}_{\lambda'\lambda}[\tilde{\boldsymbol{\psi}}] + \mathcal{O}[\boldsymbol{\psi} - \tilde{\boldsymbol{\psi}}]^2$$

provides a desired improvement.

Again as before, a practical way to construct a good trial function is to parametrize it with a set of variables $\{v_j, j = 1 \text{ to } N\}$ plus a variable amplitude $\tilde{T}$ so that $\mathcal{T}_{\lambda'\lambda}[\tilde{\boldsymbol{\psi}}] = F_{\lambda'\lambda}(\{v_j\}, \tilde{T})$. One procedure then requires that $F_{\lambda'\lambda}(\{v_j\}, \tilde{T}) = 0$ is solved to give a transition amplitude $\tilde{T}_{\lambda'\lambda}(\{v_j\})$ to order $[\delta\tilde{\boldsymbol{\psi}}(\{v_j\})]^2$. A second procedure, somewhat simpler numerically, imposes that the first derivatives of the function $\mathcal{T}_{\lambda'\lambda}[\tilde{\boldsymbol{\psi}}] = F_{\lambda'\lambda}(\{v_j\}, \tilde{T})$ vanish with respect to variation of its parameters. The $N+1$ equations $\partial F_{\lambda'\lambda}/\partial v_j = 0$ and $\partial F_{\lambda'\lambda}/\partial \tilde{T} = 0$ give solutions for the $N+1$ parameters, and an improved $T_{\lambda'\lambda} = \tilde{T}_{\lambda'\lambda} - F_{\lambda'\lambda}(\{v_j\}, \tilde{T})$ accurate to first order in parameter variations.

These developments can also be adapted to calculations of improved transition amplitudes from the $N_o + N_c$ column matrix of trial radial scattering functions $\tilde{\boldsymbol{\psi}}_{\lambda}^{(\pm)}(R)$ including open- and closed-channel functions for each initial channel state λ [21]. Following the notation in Section 4.3, a possible choice contains an expansion in a basis set $\{\eta_n(R)\}$ of the inner component of the trial function excluding the incoming and outgoing waves

$$\tilde{\boldsymbol{\psi}}_{\lambda}^{(\pm)}(R) = \begin{pmatrix} \tilde{\boldsymbol{\psi}}_{o\lambda}(R) \\ \tilde{\boldsymbol{\psi}}_{c\lambda}(R) \end{pmatrix} = \begin{pmatrix} \boldsymbol{I}^{(s)}(R)\boldsymbol{k}_{\lambda}^{1/2} \\ \boldsymbol{0} \end{pmatrix} + \begin{pmatrix} \boldsymbol{I}^{(+)}(R)\tilde{\boldsymbol{T}}_{o\lambda}(R)\boldsymbol{k}_{\lambda}^{1/2} \\ \boldsymbol{I}^{(d)}(R)\tilde{\boldsymbol{T}}_{c\lambda}(R)\boldsymbol{k}_{\lambda}^{1/2} \end{pmatrix} - \sum_{n=1}^{N} \tilde{\boldsymbol{c}}_n \eta_n(R)$$

where the basis set and trial matrices $\tilde{\boldsymbol{T}}_{o\lambda}(R)$ and $\tilde{\boldsymbol{T}}_{c\lambda}(R)$ are chosen so that $\tilde{\boldsymbol{\psi}}_{\lambda}^{(\pm)}(0) = \boldsymbol{0}$ and $\tilde{\boldsymbol{c}}_n$ is a $N_o + N_c$ column matrix of expansion coefficients to be found.

Replacing these trial expressions in the transition functional $\mathcal{T}_{\lambda'\lambda}\left(\vec{k}_{\alpha'}, \vec{k}_{\alpha}\right)$ for each pair $(\lambda'\lambda)$ of initial and final channels, and integrating over R give an expression $\mathcal{T}_{\lambda'\lambda}[\tilde{\boldsymbol{\psi}}_{\lambda}] = \boldsymbol{F}_{\lambda'\lambda}(\{\tilde{\mathbf{c}}_n\}, \tilde{\boldsymbol{T}}_{o\lambda}, \tilde{\boldsymbol{T}}_{c\lambda})$ bilinear in the set of matrices $\{\tilde{\mathbf{c}}_n\}$. Solving for this set, with one of the two procedures described above, gives an improved transition amplitude $T_{\lambda'\lambda}$.

4.2.5 Integration of Semiclassical Coupled Equations

Semiclassical dynamics requires the simultaneous integration of the differential Hamiltonian equations for trajectories $\{\vec{R}(t; \vec{P}_s, \vec{R}_s), \vec{P}(t; \vec{P}_s, \vec{R}_s)\}$ and coupled matrix equations for channel amplitudes $\mathbf{A}_{\alpha}(t)$. Trajectory equations are

$$\frac{d\vec{R}}{dt} = \frac{\partial \mathcal{H}}{\partial \vec{P}}, \quad \frac{d\vec{P}}{dt} = -\frac{\partial \mathcal{H}}{\partial \vec{R}}$$

for $\mathcal{H}(\vec{P}, \vec{R}) = \vec{P} \cdot \vec{P}/(2M) + \mathcal{V}(\vec{P}, \vec{R})$ with an effective potential energy $\mathcal{V}(\vec{P}, \vec{R})$, to be solved for each initial set of conditions $(\vec{P}_s, \vec{R}_s)$, $s = 1$ to N_{in}, starting at a time t_{in} to generate the bundle of trajectories $\{\vec{R}(t; \vec{P}_s, \vec{R}_s), \vec{P}(t; \vec{P}_s, \vec{R}_s)\}$ in phase space, for constant total energy $\mathcal{H}(\vec{P}, \vec{R}) = E$. Numerical methods for solving these equations have been described in Chapter 2 for elastic collisions and are also applicable here. The Hamiltonian equations for all six variables $(\vec{P}, \vec{R})$ in a set $\mathbf{Y} = (Y_1, \dots, Y_6)$ are formally given by

$$d\mathbf{Y}/dt = \mathbf{F}[\mathbf{Y}(t)]$$

with $\mathbf{F} = (F_1, \dots, F_6)$ the set of Hamiltonian gradients, which can be solved by a numerical algorithm starting with initial conditions $\mathbf{Y}_{in} = \mathbf{Y}(t_{in})$. The algorithm can be a single-step procedure of the Runge–Kutta type, with $\mathbf{Y}(t + \Delta t)$ at a time increment Δt, obtained from $\mathbf{Y}(t)$ and its time derivatives. An alternative is a multistep procedure, where $\mathbf{Y}(t_n + \Delta t)$ is obtained from $\mathbf{Y}(t_{n-p})$ for $p = 0$ to m. Such a frequently used procedure is the Verlet algorithm, which employs relative position, velocity, and acceleration $\vec{a} = -M^{-1}\partial V/\partial \vec{R}$, of functions $\vec{R}(t), \vec{v}(t), \vec{a}(t)$ in a sequence which allows for testing of energy conservation at $t + \Delta t$, as already described in Chapter 2.

It was mentioned there that an alternative general treatment of multistep procedures has been developed for Hamiltonian equations using a Liouvillian propagator and the Trotter factorization of its exponential, which leads to reversible multiple time-scale dynamics [23]. These procedures can be chosen to optimize a trade-off between high accuracy and computational efficiency. Semiclassical scattering amplitudes can be obtained by integration of the set of coupled ordinary differential equations for the channel amplitudes collected in a column $1 \times N$ matrix $\mathbf{A}_{\alpha}(t)$ in

$$-i\hbar \frac{d\mathbf{A}_{\alpha}}{dt} + \mathbf{H}(t)\mathbf{A}_{\alpha}(t) = \mathbf{0}$$

with initial conditions $\mathbf{A}_\alpha(t_{in}) = (2\pi)^{-\frac{3}{2}}[\delta_{\alpha'\alpha}]$, where $\alpha = 1$ to N_o and $\alpha' = 1$ to N, $N = N_o + N_c$. This set can be solved using single-step or multiple-step time integration methods, extending to matrices the algorithms described in the chapter on elastic collisions or it can be obtained introducing a time-evolution operator.

In a time-step procedure, the coupled equations can be solved by a numerical algorithm starting with initial conditions $\mathbf{A}_\alpha(t_{in}) = \mathbf{A}_{in}$ for an initial state. The algorithm can be a single-step procedure of the Runge–Kutta type, with $\mathbf{A}(t+\Delta t)$ at a time increment Δt, obtained from $\mathbf{A}(t)$ and its time derivatives. This is a self-starting procedure, which however requires generation of derivatives of $\mathbf{HA} = \mathbf{F}$ with respect to t, a computationally inefficient step. It can be avoided using instead a finite quadrature formula such as

$$\mathbf{A}(t_n + \Delta t) = \mathbf{A}(t_n) + \Delta t \sum_{i=0}^{r-1} w_i\, \mathbf{F}(t_n + \zeta_i \Delta t)$$

where w_i and ζ_i are weights and location parameters for a quadrature of order r in Δt, to advance a time step after t_n. Parameters have been derived, in particular, for $r = 4$, a usual choice for high accuracy [24]. An alternative is a multistep procedure, where $\mathbf{A}(t_n + \Delta t)$ is obtained from $\mathbf{A}(t_{n-p})$ for $p = 0$ to m frequently combined with predictor–corrector methods [24]. Alternatively using a time-evolution operator procedure has the advantage of preserving the norm of the amplitude $|\chi(t)\rangle$ and conserving flux from it. Writing the amplitude in a sequence of intervals from time t_1 to a $t_2 = t_1 + \tau$, as $\mathbf{A}(t_2) = \mathbf{U}(t_2, t_1)\mathbf{A}(t_1)$, with the time-evolution $N \times N$ matrix $\mathbf{U}(t_2, t_1) = \exp_T\left[i\int_{t_1}^{t_2} dt'\mathbf{H}(t')/\hbar\right]$, an exponential given formally by a time-ordered expansion [20]. This can be constructed from an exponential cumulant series in powers of τ as a Magnus expansion [25]. Starting with the decomposition of the matrix $\mathbf{H}(t)$ into diagonal and nondiagonal terms as $\mathbf{H} = \mathbf{H}_D + \mathbf{H}_N$, introduce the easily calculated diagonal time-evolution operator $\mathbf{T}_D(t, t_1)$ satisfying $-i\hbar d\mathbf{T}_D/dt = \boldsymbol{H}_D(t)\mathbf{T}_D(t, t_1)$ and given by $\mathbf{T}_D(t, t_1) = \exp\left[i\int_{t_1}^{t} dt'\boldsymbol{H}_D(t')/\hbar\right]$. New amplitudes in $\mathbf{A}(t) = \mathbf{T}_D(t, t_1)\mathbf{B}(t)$ satisfying

$$\frac{d\mathbf{B}}{dt} = i\,\boldsymbol{\omega}(t, t_1)\mathbf{B}(t), \quad \boldsymbol{\omega}(t, t_1) = \mathbf{T}_D(t, t_1)^\dagger \mathbf{H}_N(t)\,\mathbf{T}_D(t, t_1)/\hbar$$

can be obtained at time t_2 from a Magnus expansion [25] in

$$\mathbf{B}(t_2) = \exp[\mathbf{L}(t_2, t_1)]\mathbf{B}(t_1)$$

$$\mathbf{L}(t_2, t_1) = \int_{t_1}^{t_2} dt\, i\boldsymbol{\omega}(t, t_1) + \frac{1}{2}\int_{t_1}^{t_2} dt \int_{t_1}^{t} dt'\, [i\boldsymbol{\omega}(t, t_1), i\boldsymbol{\omega}(t', t_1)] + \mathcal{O}(\tau\|\overline{\boldsymbol{\omega}}\|)^3$$

with $\overline{\boldsymbol{\omega}}$ an average over the interval τ. This is simplified by expanding the integrands in the terms above around an average time t_a with $t_1 = t_a - \tau/2$ and

$t_2 = t_a + \tau/2$, and integrating to obtain

$$\mathbf{L}(t_2, t_1) = i\boldsymbol{\omega}_a\tau + \frac{\tau^3}{24}i\ddot{\boldsymbol{\omega}}_a + \frac{\tau^3}{12}[i\dot{\boldsymbol{\omega}}_a, i\boldsymbol{\omega}_a] + \mathcal{O}(\tau\|\overline{\boldsymbol{\omega}}\|)^4$$

where $\boldsymbol{\omega}_a = \boldsymbol{\omega}(t_a, t_1)$ and dots indicate derivatives with respect to time.

For a small interval $\tau = (t_2 - t_1)$, it is accurate to keep only the first term with $\boldsymbol{\omega}(t, t_1) = \boldsymbol{\omega}_a + \dot{\boldsymbol{\omega}}_a \cdot (t - t_a) + h.o.t.$ in the integral, simply giving

$$\mathbf{B}(t_2) = \exp[i\,\boldsymbol{\omega}_a \cdot \tau]\mathbf{B}(t_1) = \mathbf{T}_D(t_a, t_1)^\dagger \exp[i\mathbf{H}_N(t_a)\tau/\hbar]\mathbf{T}_D(t_a, t_1)$$

which can readily be calculated. This can be repeated in a sequence of intervals of duration τ between initial t_{in} and final t_{fn} times to propagate the amplitudes $\mathbf{A}(t)$ during a collision. It provides a fast-collision, or sudden, approximation.

4.3 Electronically Adiabatic Rotational Transitions

4.3.1 Expansion in a Basis Set of Rotor States

To start with a simple case, we consider an inelastic collision between an atom A and a rotating diatomic BC with ion cores at $\vec{r}_b$ and $\vec{r}_c$, where energy is transferred between the relative kinetic motion of the two colliding species and the rotational motion of the diatomic while the diatomic bond distance is fixed at its equilibrium value $r_e = |\vec{r}_b - \vec{r}_c|$. The internal motion Hamiltonian is simply a rotational Hamiltonian $\hat{H}_R = \hbar^2\hat{j}^2/(2I_{BC})$, where $\hat{\vec{j}}$ is the rotational operator with $I_{BC} = M_{BC}r_e^{\,2}$ the moment of inertia, and the internal rotational states of BC are eigenfunctions $\phi_{j,m_j}(\vec{r}) = Y_{j,m_j}(\omega) = \langle\omega|j, m_j\rangle$ of the rotational operators $\hat{j}^2$ and $\hat{j}_z$ with respective rotational quantum numbers j, m_j. Here $Y_{j,m_j}(\omega)$ is a spherical harmonic function of the solid angles $\omega = (\vartheta, \varphi)$, with eigenenergy $E_j = \hbar^2 j(j+1)/(2I_{BC})$.

The interaction potential energy of the atom A and diatomic BC is in general $V(\mathbf{Q}) = V(\vec{R}, \vec{r}) = V(R, r, \theta)$, where θ is the angle between $\vec{R}$ and $\vec{r}$ and $\cos(\theta) = \vec{R}\cdot\vec{r}/Rr$. This can be expanded in Legendre polynomials $P_\mu(\cos\theta)$ so that

$$V(R, r, \theta) = \sum_{\mu=0}^{\mu_{max}} v_\mu(R, r)P_\mu(\cos\theta)$$

where the term with $v_0(R, r)$ describes an isotropic energy component, and following terms provide the anisotropies. The expansion factors can be decomposed into short-range and long-range parts and parametrized as

$$v_\mu(R, r) = v_\mu^{(S)}(R, r) + v_\mu^{(L)}(R, r)$$

$$v_\mu^{(S)}(R, r) = A_\mu(r)\exp[-b_\mu(r)R]$$

$$v_\mu^{(L)}(R, r) = -\sum_{n\geq 1} C_\mu^{(n)}(r)f_{dmp}^{(n)}(R)R^{-n}$$

where the short range is exponentially repulsive and the long range contains attractive dispersion energy terms damped at short distances. Alternatively, the potential function can be expressed as a sum over pair and triplet atomic terms from ion cores and localized electronic charges [4].

Relative motion with a momentum vector $\vec{P} = M\vec{v} = \hbar\vec{k}$ can be described in a space-fixed CM–SF reference frame with its Z-axis along $\vec{P}$, and quantal scattering follows the treatment in Section 4.1 for scattering states $\Psi_{\alpha}^{(+)}(\vec{R}, \boldsymbol{q}; \vec{k})$, with $\boldsymbol{q} = \vec{r}$, expanded in a basis set of rotational states $\phi_{\alpha'}(\boldsymbol{q})$ where $\alpha = (j, m_j)$. Expanding $\Psi_{j,m_j}^{(+)}(\vec{R}, \vec{r}; \vec{k})$, with the BC bond distance assumed to be constant at r_e, in a basis set $\{Y_{j,m_j}(\omega)\}$ of the rotational states of BC gives

$$\Psi_{j,m_j}^{(+)}(\vec{R}, \vec{r}; \vec{k}) = \sum_{all\, j',m_{j'}} \psi_{j',m_{j'},j,m_j}^{(+)}(\vec{R}; \vec{k}) Y_{j',m_{j'}}(\omega)$$

with state-to-state scattering amplitudes $\psi_{j',m_{j'},j,m_j}^{(+)}$ describing rotational energy transfer and reorientation, and the expansion in (j', m_j') extending over all the rotational states needed in calculations, for both open and closed channels. Its dependence on the orientation angles Ω and ω_k of $\vec{R}$ and $\vec{k}$ can be expanded in spherical harmonics $Y_{l,m_l}(\Omega) = \langle \Omega | l, m_l \rangle$ and $Y_{p,q}(\omega_k)$, respectively, with (l, m_l) orbital quantum numbers for the angular momentum operators $\hat{L}^2$ and $\hat{L}_z$ of the relative motion angular momentum $\vec{L}$. The state-to-state amplitudes are then

$$\psi_{j',m_j',j,m_j}^{(+)}(\vec{R}; \vec{k}) = \sum_{l',m_{l'}} \sum_{p,q} \psi_{\lambda',\kappa}^{(+)}(R; k) Y_{l',m_{l'}}(\Omega) Y_{p,q}(\omega_k)^*$$

with $\lambda' = (l', m_l', j', m_j')$ and initial conditions given by quantum numbers $(p, q, j, m_j) = \kappa$. The angular dependence of $\Psi_{j,m_j}^{(+)}$ is therefore expanded in a basis set $\{\Phi_{\lambda}(\Omega, \omega) = Y_{l,m_l}(\Omega) Y_{j,m_j}(\omega)\}$ with $\lambda = (l, m_l, j, m_j)$, so that

$$\Psi_{j,m_j}^{(+)}(\vec{R}, \vec{r}; \vec{k}) = \sum_{all\, \lambda'} \sum_{p,q} \psi_{\lambda',\kappa}^{(+)}(R; k) \Phi_{\lambda'}(\Omega, \omega) Y_{p,q}(\omega_k)^*.$$

Substituting this into the differential equation for $\Psi_{j,m_j}^{(+)}(\vec{R}, \vec{r}; \vec{k})$ and expanding both equation sides in the basis sets $\{\Phi_{\lambda}(\Omega, \omega)\}$ and $\{Y_{p,q}(\omega_k)\}$, we arrive at the set of coupled ordinary differential equations for the radial amplitudes $\psi_{\lambda',\kappa}^{(+)}(R; k)$, in matrix notation with a column $\boldsymbol{\psi}_{\kappa}^{(+)} = \left[\psi_{\lambda',\kappa}^{(+)}\right]$, as

$$\mathcal{D}_R \boldsymbol{\psi}_{\kappa}^{(+)}(R) = \boldsymbol{U}(R) \boldsymbol{\psi}_{\kappa}^{(+)}(R)$$

$$\mathcal{D}_R = \left[\delta_{\lambda\lambda'} \left(k_j^2 + \frac{d^2}{dR^2} - \frac{l(l+1)}{R^2}\right)\right], \quad \boldsymbol{U}(R) = [U_{\lambda\lambda'}(R)]$$

where the relative motion wavevector has its magnitude given by $k_j^2 = (2M/\hbar^2)(E - E_j)$. The radial potential energy matrix elements in

$$U_{\lambda\lambda'}(R) = \frac{2M}{\hbar^2} \int d\Omega\, d\omega\, \Phi_{\lambda'}(\Omega, \omega) V(R, r, \theta) \Phi_\lambda(\Omega, \omega)$$

can be obtained from the spherical harmonic addition theorem giving $P_\mu(\cos\theta) = 4\pi(2\mu+1)^{-1} \sum_\nu Y_{\mu,\nu}(\omega)^* Y_{\mu,\nu}(\Omega)$ and $3-j$ Wigner coefficients for integrals involving a product of three spherical harmonics [20, 26] as in

$$\int_0^\pi \int_0^{2\pi} \sin(\vartheta) d\vartheta\, d\varphi Y_{l_1 m_1}(\vartheta, \varphi) Y_{l_2 m_2}(\vartheta, \varphi) Y_{l_3 m_3}(\vartheta, \varphi)$$

$$= \left[\frac{(2l_1+1)(2l_2+1)(2l_3+1)}{4\pi}\right]^{1/2} \begin{pmatrix} l_1 & l_2 & l_3 \\ 0 & 0 & 0 \end{pmatrix} \begin{pmatrix} l_1 & l_2 & l_3 \\ m_1 & m_2 & m_3 \end{pmatrix}$$

in terms of $3-l$ coefficients, with the left one null if $l_1 + l_2 + l_3$ is odd valued.

That set of coupled ordinary differential equations must be solved with asymptotic conditions for both open ($E \geq E_j$) and closed ($E < E_j$) channels, as described in Section 4.2. The total wavefunction in matrix form is

$$\Psi_{j,m_j}^{(+)}(\vec{R}, \vec{r}; \vec{k}) = \sum_{p,q} \mathbf{\Phi}(\Omega, \omega) \boldsymbol{\psi}_\kappa^{(+)}(R) Y_{p,q}(\omega_k)^*$$

where $\mathbf{\Phi}(\Omega, \omega) = [\Phi_{\lambda'}(\Omega, \omega)]$ is a row matrix, and its asymptotic form provides scattering amplitudes $f_{j',m_{j'},j,m_j}^{(+)}(\Omega)$. Scattering amplitudes are given following Section 4.3, with $a = (j, m_j)$ and initial conditions with $(p, q) = (l, m)$ here, by

$$f_{j',m_{j'},j,m_j}^{(+)}(\Omega; \vec{k}) = \frac{1}{\sqrt{k_j k_{j'}}} \sum_{l,m} \sum_{l',m'} 4\pi^{l-l'} f_{\lambda'\lambda}(k_a) Y_{l'm'}(\Omega) Y_{lm}(\omega_k)^*.$$

However, in the present cases of rotational transition phenomena, it is convenient to develop an alternative treatment that incorporates conservation of the total, or orbital plus rotational, angular momentum during collisions, as follows.

The treatment of translational–rotational energy transfer can be done in a wavefunction representation that accounts for conservation of the total angular momentum vector $\vec{J} = \vec{L} + \vec{j}$, with operators $\hat{J}^2$ and $\hat{J}_z$ and eigenstates $|l, j, J, M_J\rangle$ of total quantum numbers (J, M_J), with $|l-j| \leq J \leq (l+j)$ and $-J \leq M_J \leq J$ [27]. Introducing Clebsch–Gordan coefficients $\langle l, m_l, j, m_j \,|\, l, j, J, M_J\rangle = C_{\lambda,\Lambda}$ in a short notation, where $\Lambda = (l, j, J, M_J)$ [26] to transform the original orbital–rotational states $\Phi_\lambda(\Omega, \omega)$ into new total angular momentum states $\Phi_\Lambda(\Omega, \omega) = \Phi_{l,j}^{(J,M_J)}(\Omega, \omega) = \sum_\lambda \Phi_\lambda(\Omega, \omega) C_{\lambda,\Lambda}$, previous matrices in the radial equations can be transformed using the unitary transformation matrix $\boldsymbol{C} = [C_{\lambda,\Lambda}]$, which satisfies $(\boldsymbol{C})^{-1} = \boldsymbol{C}^\dagger$, a Hermitian conjugate.

Introducing a new row matrix $[\Phi_\Lambda(\Omega, \omega)] = \mathbf{\Phi}^{(C)}(\Omega, \omega) = \mathbf{\Phi}(\Omega, \omega)\boldsymbol{C}$ and a new column matrix $\boldsymbol{\psi}_\kappa^{(C)}(R) = (\boldsymbol{C})^{-1} \boldsymbol{\psi}_\kappa^{(+)}(R)$ so that $\Psi_{j,m_j}^{(+)} = \sum_{p,q} \mathbf{\Phi}^{(C)} \boldsymbol{\psi}_\kappa^{(C)} Y_{p,q}(\omega_k)^*$, a transformed set of coupled differential equations is now $\mathcal{D}_R^{(C)}\boldsymbol{\psi}_\kappa^{(C)}(R) = \boldsymbol{U}^{(C)}(R)\boldsymbol{\psi}_\kappa^{(C)}(R)$, where

$$\mathcal{D}_R^{(C)} = (\boldsymbol{C})^{-1}\mathcal{D}_R\boldsymbol{C} = \left[\delta_{\Lambda\Lambda'}\left(k_j^2 + \frac{d^2}{dR^2} - \frac{l(l+1)}{R^2}\right)\right],$$

$$\boldsymbol{U}^{(C)}(R) = (\boldsymbol{C})^{-1}\boldsymbol{U}(R)\boldsymbol{C} = [U_{\Lambda\Lambda'}(R)],$$

It is now possible to use the rotational invariance property of the interaction potential energy $V(R, r, \theta)$, with $\cos(\theta) = \vec{R}\cdot\vec{r}/(Rr)$ a function of Ω and ω. This potential energy function is invariant under rotations generated by the operators $\hat{J}^2$ and $\hat{J}_z$ and commutes with them, which leads to diagonal matrix elements in the quantum numbers (J, M_J) [3, 26] giving

$$U_{\Lambda\Lambda'}(R) = \frac{2M}{\hbar^2}\int d\Omega\, d\omega\, \Phi_\Lambda^* V \Phi_{\Lambda'} = \frac{2M}{\hbar^2}\delta_{JJ'}\delta_{M_J,M_{J'}}V_{lj,l'j'}^{(J)}(R).$$

This displays the advantages of using the total angular momentum representation and shows that the coupled differential equations can be solved independently for each set (J, M_J).

The new coupled differential equations must be solved with scattering asymptotic conditions for open (o) and closed (c) channels, matching the long-range numerical solutions to combinations of the asymptotic oscillating functions $\boldsymbol{I}^{(s)}(R)$ and $\boldsymbol{I}^{(c)}(R)$ and to the decay functions in the diagonal matrices $\boldsymbol{I}^{(d)}(R)$, as described in Section 4.3, but here transformed into $\boldsymbol{I}_C^{(s)}(R) = (\boldsymbol{C})^{-1}\boldsymbol{I}^{(s)}(R)$, and similar $\boldsymbol{I}_C^{(c)}(R)$, and $\boldsymbol{I}_C^{(d)}(R)$. By similarity with Section 4.3.1,

$$\boldsymbol{\psi}_{oo}^{(C)}(R) \approx \left[\boldsymbol{I}_C^{(s)}(R) + \boldsymbol{I}_C^{(+)}(R)\boldsymbol{T}_{oo}^{(C)}\right]\boldsymbol{k}_o^{1/2}$$

in the present basis set, where $\boldsymbol{T}_{co}^{(C)} = \left[t_{\Lambda'\Lambda}\right]$ and $\boldsymbol{k}_o = \left[\delta_{\Lambda'\Lambda}k_j\right]$, and in more detail $t_{\Lambda'\Lambda} = \delta_{JJ'}\delta_{M_J,M_{J'}}t_{lj,l'j'}^{(J)}$. Scattering matrix elements are given by $S_{\Lambda'\Lambda} = \delta_{\Lambda'\Lambda} + 2it_{\Lambda'\Lambda}$ and the state-to-state scattering amplitude at solid angle Ω between wavevectors $\vec{k}$ and $\vec{k}'$ is, with initial $(p, q) = (l, m_l)$,

$$f_{j',m_{j'},j,m_j}^{(+)}(\vec{k}',\vec{k}) = \frac{4\pi}{(k_jk_{j'})^{1/2}}\sum_{l',m_l',l,m_l} Y_{l',m_l'}(\omega_k')\left[\boldsymbol{C}\boldsymbol{T}_{oo}^{(C)}\boldsymbol{C}^\dagger\right]_{\lambda',\lambda} Y_{l,m_l}(\omega_k)^*.$$

This symmetric expression simplifies when $\vec{k}$ is along the CM Z-axis so that $Y_{l,m_l}(0,0) = [(2l+1)/(4\pi)]^{1/2}\delta_{m_l,0}$ for $\omega_k = (0,0)$, and $\omega_k' = \Omega$ can be used.

Also, the angular momentum coupling coefficients in $\boldsymbol{C}$ are null except if $M_J = m_l + m_j = m_j$ and $m_l' = M_J - m_{j'} = m_j - m_{j'}$, which eliminates the summations over M_J and m_l'. Making use of the Clebsch–Gordan coefficients $C_{\lambda,\Lambda} = \langle l, m_l, j, m_j \mid l, j, J, M_J\rangle$, an explicit form for the scattering amplitudes is [27]

$$f_{j',m_{j'},j,m_j}^{(+)}(\Omega) = \frac{(4\pi)^{1/2}}{(k_jk_{j'})^{1/2}}\sum_{l\geq 0}\sum_{l'\geq 0} a_{j',m_{j'},j,m_j}(l',l)\, Y_{l',m_j-m_{j'}}(\Omega)$$

$$a_{j',m_j',j,m_j}(l',l) = \sum_J (2l+1)^{1/2} \left\langle l', m_j - m_{j'}, j', m_j' | l'j'Jm_j \right\rangle t^{(J)}_{l'j',lj} \langle ljJm_j | l, 0, j, m_j \rangle^*$$

with $Y_{l',m_j-m_{j'}}(\Omega)$ giving the angular dependence due to the rotor reorientation, for a space-fixed CM reference frame. The summations for fixed J are restricted by $J+j \leq l \leq |J-j|$ and $J+j' \leq l' \leq |J-j'|$. The state-to-state differential cross section is

$$\frac{d\sigma_{j',m_{j'},j,m_j}}{d\Omega} = \frac{k_{j'}}{k_j} \left| f^{(+)}_{j',m_{j'},j,m_j}(\Omega) \right|^2$$

from which the integral cross section $\sigma_{j',m_{j'},j,m_j}(k_j) = \left(k_{j'}/k_j\right) \int d\Omega \left| f^{(+)}_{j',m_{j'},j,m_j}(\Omega) \right|^2$ can be obtained. For an unpolarized beam of diatomics, averaging over m_j and adding over $m_{j'}$, and using product properties of the Clebsch–Gordan coefficients [26] one finds

$$\sigma_{j',j}(k_j) = \frac{4\pi}{k_j^2} \frac{1}{2j+1} \sum_{J=0}^{\infty} (2J+1) \sum_{l=|J-j|}^{l=J+j} \sum_{l'=|J-j'|}^{l'=|J+j'|} \left| t^{(J)}_{lj,l'j'}(k_j) \right|^2$$

showing that the cross section is given by a sum over total angular momentum components or, in a limit of nearly continuous J values, as an integral over an impact parameter $b = \left(J + \frac{1}{2}\right)/k_j$.

4.3.2 Helicity Representation and the Body-Fixed Frame

An alternative expression for the angle dependence in $f^{(+)}_{j',m_{j'},j,m_j}(\Omega)$ can be obtained by introducing the helicity representation of the internal angular momentum states of the moving rotor BC [28, 29]. It involves the projection j_z of the vector $\vec{j}$ along the relative momentum $\vec{P} = \hbar\vec{k}$, with quantum values $\hbar\mu_j, -j \leq \mu_j \leq j$. This can be described in a CM($\vec{k}$) frame, with its Z-axis along $\vec{k}$. The corresponding final frame CM($\vec{k}'$) is used for the outgoing relative motions with wavevector $\vec{k}'$ and helicities along a rotated Z'-axis. The transformation of states from the original space-fixed CM–SF to the new body-fixed CM–BF frame can be accomplished using the rotation operator $\hat{D}(\alpha, \beta, \gamma)$ on angular momentum states [26], dependent on the Euler rotation angles (α, β, γ) that rotate CM($\vec{k}$) into CM($\vec{k}'$). Rotor states $Y_{j,m_j}(\omega) = \langle \omega | j, m_j \rangle$ in the helicity representation are given after rotation by $\langle \overline{\omega} | j, \mu_j \rangle$, with $\overline{\omega}$ the solid angle in the new, rotated, frame. The helicity representation is a natural one to describe experimental measurements where initial and final rotor orientations are detected along the directions of relative motion. It also gives a simple angle relation between incoming and outgoing states. However, scattering calculations are done in a reference frame common to the initial and final states and are preferred in numerical work. This is accomplished with helicity states introducing a body-fixed or CM–BF reference frame with the same origin as the space-fixed CM–SF frame, but with a new Z'-axis along the relative position $\vec{R}$ and rotating with it from initial to final $\vec{k}$.

The angular dependence of $\Psi^{(+)}_{j,\mu_j}$ is expanded in a BF basis set $\{\overline{\Phi}_{\overline{\lambda}}(\overline{\Omega},\overline{\omega}) = Y_{l,\mu_l}(\overline{\Omega})Y_{j,\mu_j}(\overline{\omega})\}$ of helicity states with $\overline{\lambda} = (l, \mu_l, j, \mu_j)$, so that

$$\Psi^{(+)}_{j,\mu_j}(\vec{R}, \vec{r}; \vec{k}\,) = \sum_{\text{all } \overline{\lambda}'} \sum_{p,q} \overline{\psi}^{(-)}_{\overline{\lambda}',\overline{\kappa}}(R; k)\overline{\Phi}_{\overline{\lambda}'}(\overline{\Omega},\overline{\omega})Y_{p,q}(\overline{\omega}_k)^*$$

where bar-accent quantities refer to the BF frame. The coefficients are radial functions $\overline{\psi}^{(+)}_{\overline{\lambda}',\overline{\kappa}}(R)$ to be obtained from coupled differential equations with scattering boundary conditions. In the CM–BF frame, the eigenvalues of $\hat{J}_{Z'}$ are labelled as $-J \leq K \leq J$. As done before, orbital and rotational angular momenta can be coupled into a total angular momentum, but now in the CM-BF frame and using helicity labels. After angular momentum coupling, new states are labelled by $\overline{\Lambda} = (l, j, J, K)$ and new basis set states are $\overline{\Phi}_{\overline{\Lambda}}(\overline{\Omega},\overline{\omega}) = \langle\overline{\Omega},\overline{\omega}|l, j, J, K\rangle$. The previous treatment in the CM–SF frame can be repeated in the CM–BF frame, with new matrix representations for the interaction potential energy and for the centrifugal kinetic energy.

The potential energy matrix is simpler here because axial symmetry around the new Z'-axis implies that $K = K'$. However, the orbital kinetic energy contains the angular momentum operator $\hat{L}^2 = (\vec{J} - \vec{j})^2 = \hat{J}^2 + \hat{j}^2 - \left(2\hat{J}_{Z'}\hat{j}_{Z'} + \hat{J}_+\hat{j}_- + \hat{J}_-\hat{j}_+\right)$, written in terms of step-up and -down operators like $\hat{J}_\pm = \hat{J}_{X'} \pm i\hat{J}_{Y'}$ with $\hat{J}_\pm|\,J, K\rangle = [J(J+1) - K(K+1)]^{\frac{1}{2}}|\ J,\ K \pm 1\rangle$ [20]. The orbital kinetic energy matrix of elements $[\hbar^2/(2MR^2)]\langle l, j, J, K|\hat{L}^2|l', j', J', K'\rangle$ is therefore non-diagonal. It involves

$$\langle l, j, J, K|\hat{L}^2|l, j, J, K\rangle = J(J+1) - K^2 + j(j+1)$$

along its diagonal, as well as nondiagonal Coriolis long-range couplings of K and $K' = K \pm 1$ [30], given by

$$\langle l, j, J, K|\hat{L}^2|l', j', J', K \pm 1\rangle = \lambda_\pm(J, K)\lambda_\mp(j, K \pm 1)$$

with

$$\lambda_\pm(J, K) = [(J \pm K + 1)(J \mp K)]^{1/2}.$$

These matrix elements together with the potential energy matrix elements

$$U_{\overline{\Lambda}\,\overline{\Lambda}'}(R) = \frac{2M}{\hbar^2}\delta_{JJ'}\delta_{KK'}V^{(J)}_{lj,l'j'}(R)$$

provide the Hamiltonian matrix in the helicity representation, to be used in the differential equations for coupled-channel calculations.

The connection between the original CM–SF state-to-state scattering amplitude and a new one in the CM–BF and a helicity representation, $f^{(+)}_{j',\mu'_j,j,\mu_j}(\overline{\Omega})$, can be derived comparing the total transition amplitudes from initial rotor state $(j, m_j) = (j, \mu_j)$, as given in the original and in the helicity representations by

$$\sum_{m_j'} f^{(+)}_{j',m'_j,j,m_j}(\Omega;\vec{k})Y_{j',m'_j}(\omega) = \sum_{\mu'_j} f^{(+)}_{j',\mu'_j,j,\mu_j}(\overline{\Omega};\vec{k})Y_{j',\mu'_j}(\overline{\omega})$$

$$f^{(+)}_{j',\mu_{j'},j,\mu_j}(\overline{\Omega}) = \frac{(4\pi)^{1/2}}{(k_j k_{j'})^{1/2}} \sum_l \sum_{l'} a_{j',\mu'_j,j,\mu_j}(l',l)\, Y_{l',\mu_j-\mu_{j'}}(\overline{\Omega})$$

with new coefficients $a_{j',\mu'_j,j,\mu_j}(l',l)$ to be related to the $a_{j',m'_j,j,m_j}(l',l)$. To do this, the function $Y_{l',n_j}(\Omega)$, with $n_j = m_j - m_{j'}$, can be transformed from the initial to the final CM frame by a rotation of angle ϑ around an axis perpendicular to the plane of $\vec{k}$ and $\vec{k}'$. The rotational change of $Y_{l',n_j}(\Omega) = \langle \Omega | l', n_j \rangle$ from the original CM–SF to the new CM–BF frame can be accomplished using the rotation operator $\widehat{D}(\alpha, \beta, \gamma)$ on angular momentum states [26], dependent on the Euler rotation angles (α, β, γ) equal here to $(0, \vartheta, 0)$, with inverse $\widehat{D}^{-1} = \widehat{D}^{\dagger}$. Writing $\langle \Omega | l', n_j \rangle = \langle \Omega | \widehat{D}^{\dagger} \widehat{D} | l', n_j \rangle = \langle \overline{\Omega} | l', \nu_j \rangle$, with $\left|\overline{\Omega}\right\rangle = \widehat{D} | \,\Omega \rangle$ and a new ket $\left| l', \nu_j \right\rangle = \widehat{D}(0, \vartheta, 0) \left| l', n_j \right\rangle$ of helicity transfer quantum number $\nu_j = \mu_j - \mu_{j'}$ along the Z'-axis, this can be expanded in a basis set of helicity states $\left\langle \overline{\Omega} | l', \nu'_j \right\rangle$ as

$$Y_{l',\nu_j}(\overline{\Omega}) = \langle \overline{\Omega} | l', \nu_j \rangle = \sum_{\nu_j'} \left\langle \overline{\Omega} \middle| l', \nu'_j \right\rangle d^{(l')}_{\nu'_j,\nu_j}(\vartheta)$$

where $d^{(l')}_{\nu'_j,\nu_j}(\vartheta) = \left\langle l', \nu'_j \middle| \widehat{D}(0,\vartheta,0) \middle| l', \nu_j \right\rangle$ is a reduced rotation matrix element, known from the treatment of quantal rotations [26]. Similarly, $Y_{j',m'_j}(\omega) = \left\langle \omega | j', m'_j \right\rangle = \left\langle \overline{\omega} | j', \mu'_j \right\rangle$ provides the relation between scattering amplitudes $f^{(+)}_{j',m_j',j,m_j}$ and $f^{(+)}_{j',\mu'_j,j,\mu_j}$. The transformed $f^{(+)}_{j',\mu'_j,j,\mu_j}(\overline{\Omega}) = f^{(+)}_{j',\mu'_j,j,\mu_j}(\vartheta)$ amplitude between helicity rotor states $|j, \mu_j\rangle$ is, as a result, simply given in the CM–BF frame in terms of $d^{(l')}_{\nu'_j,\mu_j-\mu'_j}(\vartheta)$ and the coefficients $a_{j',\mu'_j,j,\mu_j}(l',l)$.

4.3.3 Atom–Polyatomic and Molecule–Molecule Collisions

The treatment of T–R collisional transfer can be extended to collisions of an atom with a rotating polyatomic molecule and to collisions between two rotating molecules. They involve introduction of related interaction potential energy functions and expansions of both scattering wavefunctions and operators in their differential equations, in related basis sets of rotational internal states.

For an atom–polyatomic A–M collision, the polyatomic molecule (with bond distances fixed at equilibrium values) can be described as a rigid body or top with three moments of inertial I_A, I_B, and I_C for rotations around three body-fixed perpendicular axes a, b, and c, chosen so that $I_A \leq I_B \leq I_C$. A linear top such as in HCN has $I_A = 0$ and $I_B = I_C$, a spherical top has $I_A = I_B = I_C$ (such as CH_4 and SF_6), a symmetric top has two equal moments of inertia (such as prolate CH_3Cl and oblate BF_3), and asymmetric tops have all unequal moments such as H_2O.

Their quantized energies and wavefunctions have been thoroughly studied [31]. The rotational Hamiltonian is now

$$\hat{H}_{rot} = \frac{\hbar^2}{2}\left(\frac{\hat{j}_a^2}{I_A} + \frac{\hat{j}_b^2}{I_B} + \frac{\hat{j}_c^2}{I_C}\right)$$

in terms of the rotational operator components along the body-fixed axes, in units of $\hbar$. Rotational motions, described in a reference frame with its origin at the CM of the rigid rotor and with axes parallel to the (X, Y, Z) axes of the CM of the whole (atom–molecule) system, involve three angles of rotation of the frame with (a, b, c) axes attached to the rotor. These can be chosen to be the Euler angles $\Gamma = (\alpha, \beta, \gamma)$ of (a, b, c) with respect to the (X, Y, Z) axes, so that the interaction potential energy is $V(\mathbf{Q}) = V(\vec{R}, \Gamma)$, or alternatively can be chosen as the angles (ϑ, φ) giving the orientation of the top inertial axis a, plus the angle χ for the rotations of the top around this axis.

A well-studied example is the rotational motion of a prolate top, with $I_B = I_C$, in which case using $\hat{j}_b^2 + \hat{j}_c^2 = \hat{j}^2 - \hat{j}_a^2$ one finds that

$$\hat{H}_{rot} = \frac{\hbar^2}{2}\left[\frac{\hat{j}^2}{I_B} + \left(\frac{1}{I_A} - \frac{1}{I_B}\right)\hat{j}_a^2\right]$$

which is an operator commuting with $\hat{j}^2, \hat{j}_a$, and $\hat{j}_Z$. They all have a common eigenfunction $\langle \vartheta, \varphi, \chi \mid j, k, m_j\rangle$ with respective eigenvalues $j(j+1)$, $-j \leq k \leq j$, and m_j. This gives $\hat{H}_{rot}\left|j, k, m_j\right\rangle = E_{j,k}\left|j, k, m_j\right\rangle$, with $E_{j,k} = Bj(j+1) + (A - B)k^2$. Furthermore, the eigenfunctions are found to be the well-known finite rotation matrices $D^{(j)}_{k,m_j}(\vartheta, \varphi, \chi)$, also called here symmetric-top functions [26, 31]. These matrix elements form an orthogonalized basis set for expansions of the potential energy function and of the scattering function, by analogy with the procedure using expansions in Legendre polynomials for a diatomic collider. They satisfy the normalization

$$\int_0^{2\pi}\int_0^{\pi}\int_0^{2\pi} d\alpha \sin(\beta) d\beta d\gamma D^{(j_1)}_{k_1,m_{j1}}(\alpha, \beta, \gamma)^* D^{(j_2)}_{k_2,m_{j2}}(\alpha, \beta, \gamma)$$
$$= \delta_{k_1k_2}\delta_{m_{j1}m_{j2}}\delta_{j_1j_2}\frac{8\pi^2}{2j_2 + 1}.$$

The basis set of states $\phi_{\alpha'}(\boldsymbol{q}) = D^{(j)}_{k,m_j}(\alpha, \beta, \gamma)$ can be used to expand the atom–polyatomic potential energy function $V(\vec{R}, \alpha, \beta, \gamma)$ and to expand the scattering wavefunction $\Psi^{(+)}_{j,m_j,k}(\vec{R}, \alpha, \beta, \gamma; \vec{k})$, with the polyatomic bond distances and angles assumed to be constants. This gives

$$\Psi^{(+)}_{j,m_j,k}(\vec{R},\alpha,\beta,\gamma;\vec{k}\,) = \sum_{all\ j',m_j',k'} \psi^{(+)}_{j',m_{j'},k'\,j,m_j,k}(\vec{R};\vec{k}\,)D^{(j)}_{k',m_{j'}}(\alpha,\beta,\gamma)$$

with state-to-state scattering amplitudes $\psi^{(+)}_{j',m_{j'},k'\,j,m_j,k}$ describing rotational energy transfer, reorientation, and spinning. Calculations follow the same steps as in Section 4.3, generating coupled ordinary differential equations for radial expansion coefficients to be solved with scattering boundary conditions for open and closed channels. Asymptotic solutions can be expanded into final top states to extract state-to-state amplitudes $f^{(+)}_{j',m_{j'},k'\,j,m_j,k}(\Omega)$. Following the same treatment as before for a diatomic collider, cross sections are now

$$\frac{d\sigma_{j',m_{j'},k'\,j,m_j,k}}{d\Omega} = \frac{k_{j'k'}}{k_{jk}}\left|f^{(+)}_{j',m_{j'},k'\,j,m_j,k}(\Omega)\right|^2$$

where $(\hbar k_{jk})^2 = 2M_{\mathrm{A,M}}(E - E_{j,k})$. The treatment for an asymmetric top such as H_2O is somewhat more involved insofar its rotational wavefunctions are linear combinations of the above symmetric top functions.

The collision between diatomic molecules M (or AB) and N (or CD) can be described with a rotational-state basis set composed of the rotational states of each collider, $Y_{j_M,m_{jM}}(\omega_M)$ and $Y_{j_N,m_{jN}}(\omega_N)$. Depending on the strength of their interaction potential energy and on collisional impact parameters and relative velocities, each diatomic rotational state may separately undergo T–R$_\mathrm{M}$ and T–R$_\mathrm{N}$ translation–rotation transitions, or the two diatomics may form a compound MN and undergo T–R$_\mathrm{MN}$ transitions. In the first case, the basis set is $\left\{\phi^{(M,N)}_n(\boldsymbol{q}) = Y_{j_M,m_{jM}}(\omega_M)Y_{j_N,m_{jN}}(\omega_N)\right\}$, while for compound systems, it is given by angular momentum coupled states $\phi^{(MN)}_n(\boldsymbol{q}) = \mathcal{Y}^{(j_{MN},m_{jMN})}_{j_M j_N}(\omega_M,\omega_N) = \sum_{m_{jM},m_{jN}}\langle j_M,m_{jM},j_N,m_{jN}|j_{MN},m_{jMN}\rangle Y_{j_M,m_{jM}}(\omega_M)Y_{j_N,m_{jN}}(\omega_N)$, obtained using Clebsch–Gordan coefficients to couple two diatomic states into diatom–diatom states of total rotational quantum numbers (j_{MN},m_{jMN}).

The collision dynamics can be described in a CM reference frame with a relative position vector $\vec{R}$ between the mass centers of the diatomics and with vectors $\vec{r}_M$ and $\vec{r}_N$ giving the diatomic orientations and bond distances in their local reference frame with axes parallel to the collisional CM frame. The potential energy function $V(\vec{R},\vec{r}_M,\vec{r}_N)$ is given, for equilibrium bond distances, as an expansion in spherical harmonic functions of solid angles ω_M and ω_N. The scattering function for a given initial state, $\Psi^{(+)}_{j_M,m_{jM}j_N,m_{jN}}(\vec{R},\vec{r}_M,\vec{r}_N;\vec{k}\,)$, is expanded in a suitable basis set of $\phi_\alpha(\boldsymbol{q})$ rotational states for T–R transfer, either for two independent diatomics or for transfer into the diatomic–diatomic compound. Its asymptotic form, obtained from matching to an expansion in final states $Y_{j_{M'},m_{jM'}}(\omega_M)Y_{j_{N'},m_{jN'}}(\omega_N)$, provides the scattering amplitudes that are needed for state-to-state differential cross sections. Further expansion of the scattering function in the basis set $\{Y_{l,m_l}(\Omega)\}$ gives its coefficients as functions of the relative distance R, which are solutions of a set of coupled ordinary differential equations. The formalism in terms of matrices is quite similar to what has been presented above for the atom–diatomic T–R transfer. They give cross sections

$$\frac{d\sigma_{\alpha'_{M,N},\alpha_{M,N}}}{d\Omega} = \frac{k'_{MN}}{k_{MN}} \left| f^{(+)}_{\alpha'_{M,N},\alpha_{M,N}}(\Omega) \right|^2$$

with $\alpha_{M,N} = (j_M, m_{jM}, j_N, m_{jN})$ for the rotational and orientation quantum numbers of the diatomics.

4.4 T–R Transfer Calculations and Comparisons with Experimental Values

4.4.1 Approximate Treatments

4.4.1.1 Distorted-Wave Approximation

Calculations of state-to-state cross sections for collisional T–R energy transfer involve expansions of scattering wavefunctions into combinations of many orbital–rotational states $\Phi_\lambda(\Omega, \omega)$. These are coupled through the interaction potential energy $V(R, \boldsymbol{q})$, with R the relative distance and $\boldsymbol{q}$ for all the other (orbital and rotational) variables, which leads to large sets of coupled differential equations for scattering amplitudes. The strength of couplings depends on the long- and short-range rotational torques from the potential function, and the number of coupled states depends on the magnitude of collisional energies relative to rotational transition energies. A calculation of numerically converged cross sections involves expanding the scattering wavefunction in a basis set of rotational states with energies E_j up to some maximum $E_{j,M}$. This number increases rapidly due to the orientational degeneracy $2j + 1$ associated with each level. For a maximum rotational quantum number j_M, it leads to as many as $N_{rot} = \sum_{j=0}^{j_M}(2j + 1) = (j_M + 1)^2$ coupled states and as many coupled differential equations to be integrated. Even for as few as 20 rotational levels, this gives 21^2 coupled equations to be numerically integrated to desired accuracy, a very challenging computational undertaking. Practical computational approaches therefore restrict the number of coupled rotational states, in approximations suitable for different potentials and collisional conditions.

When the potential energy coupling is small between relative motion and rotational variables, a suitable procedure simply involves direct state-to-state coupling to first order in the coupling, as treated in general in Section 3.5 on DWs. The DWs must be obtained from a distortion-potential function $V^{(D)}(R)$

containing the short-range energy repulsion between colliders, with the reminder $V'(R, \boldsymbol{q}) = V(R, \boldsymbol{q}) - V^{(D)}(R)$ leading to rotational transitions. Several choices can be made for $V^{(D)}(R)$ to assure that the reminder is small during scattering [32]. Generally, given a physical statistical distribution $w(\boldsymbol{q})$ of orbital-internal variables, such as follows from a weighted sum of probabilities of active rotational states, a possible choice is $V^{(D)}(R) = \int d\boldsymbol{q}\ w(\boldsymbol{q})V(R, \boldsymbol{q})$. Scattering amplitudes $f^{(DW)}_{\alpha'\alpha}(\Omega)$ and cross sections follow from the treatment in Section 3.5, with $V^{(D)}(R)$ and $V'(R, \boldsymbol{q})$ in place of $V^{(1)}$ and $V^{(2)}$ there. They are given by

$$f^{(DW)}_{\alpha'\alpha}(\Omega; \vec{k}_\alpha) = \frac{1}{\sqrt{k_\alpha k_{\alpha'}}}\sum_{l,m}\sum_{l',m'} 4\pi^{l-l'} f^{(DW)}_{\lambda'\lambda}(k_\alpha) Y_{l'm'}(\Omega) Y_{lm}(\Omega_\alpha)^*$$

$$f^{(DW)}_{\lambda'\lambda}(k_\alpha) = (2\pi)^2 \frac{M}{\hbar^2}\int dR'\, d\boldsymbol{q}\, \chi^{(-)}_{DW,l'\alpha'}(R')^* \Phi_{\lambda'}(\boldsymbol{q})^* V'(R', \boldsymbol{q}) \Phi_\lambda(\boldsymbol{q}) \chi^{(+)}_{DW,l\alpha}(R')$$

where $\chi^{(+)}_{DW,l\alpha}(R')$ is the radial component of the wave distorted by the $V^{(D)}(R) = \hbar^2 U^{(D)}(R)/(2M)$ potential function. It is obtained from

$$\left[k_\alpha^2 + \frac{d^2}{dR^2} - \frac{l(l+1)}{R^2} - U^{(D)}(R)\right] \chi^{(+)}_{DW,l\alpha}(R; k_\alpha) = 0$$

and outgoing scattering boundary conditions for $\chi^{(+)}_{DW,l\alpha}(R)$. This equation can be numerically solved with the methods described in the chapter on elastic collisions. The scattering amplitude then simply involves radial integrals with DWs and the radial matrix elements $V'_{\lambda'\lambda}(R) = \langle \Phi_{\lambda'}|V'|\Phi_\lambda\rangle_R$. It is possible to go beyond the first-order coupling implied by the above DW approximation, by using an expansion of the scattering wavefunction in powers of V' or by exponentiation of the matrix of $(\lambda'\lambda)$ integrals involving V' [6].

4.4.1.2 Coupled-States and Infinite-Order Sudden Approximations

Several other approximations have been developed and applied to many molecular pairs undergoing T–R collisions, to reduce the number of coupled radial differential equations to be numerically solved [33, 34]. Among these, notable ones are the so-called coupled-states (or CS) approximation [35] and the infinite-order sudden (or IOS) approximation [36].

The CS approximation for an atom–rotor collision is introduced in a CM–BF frame and by using helicity states $\langle \overline{\omega}|j, \mu_j\rangle$, as previously described. It relies on two approximations to greatly simplify the set of coupled radial equations to be solved. The first approximation is applicable when the torques exerted by the atomic projectile on the rotor are small so that the reorientation of the rotor and of the orbital plane can both be neglected. Then, the coupling is limited to transitions with $\mu_j = \mu_j'$, in a $j_{Z'}$-conserving coupled-states approximation, and also limited to $\mu_l = \mu_l'$ corresponding to a conserved orientation of the orbital angular momentum vector $\vec{L}$. This means that the number of coupled rotational states is now

$N_{rot} = j_M + 1 \ll (j_M + 1)^2$, and the solutions of the set of coupled radial differential equations to be solved are more manageable. The projection quantum number $K = \mu_l + \mu_j$ remains unchanged in the CM–BF frame, which simplifies the orbital kinetic energy term in the differential equations. Here the only matrix elements of $\hat{L}^2 = (\vec{J} - \vec{j})^2 = \hat{J}^2 + \hat{j}^2 - \left(2\hat{J}_{Z'}\hat{j}_{Z'} + \hat{J}_+\hat{j}_- + \hat{J}_-\hat{j}_+\right)$ taken to be different from zero are of form $\langle l,j,J,K|\hat{L}^2|l,j,J,K\rangle = J(J+1) + j(j+1) - 2K\langle l,j,J,K|\hat{j}_{Z'}|l,j,J,K\rangle$, available from Clebsch–Gordan coefficients. A second assumption in the CS approximation is that $\langle l,j,J,K|\hat{L}^2|l,j,J,K\rangle \approx l(l+1)$, acceptable insofar $J(J+1) \approx l(l+1) \gg j(j-1)$ in most collisional states. The two approximations lead to a much smaller and simplified set of coupled differential equations. The second assumption amounts to neglecting the Coriolis coupling of states $|l, j, J, K\rangle$ and $|l, j, J, K'\rangle$ for $K' = K \pm 1$ and can be improved by adding Coriolis coupling terms to the coupled differential equations.

The IOS approximation for an atom–rotor collision assumes that, at high collision energies relative to rotational transition energies, the collision is sudden and the rotor remains fixed in its orientation $\omega_0 = (\vartheta_0, \varphi_0)$ given by the vector $\vec{r}$, while being hit by the projectile atom. In a CM–BF reference frame, this means that the angle $\theta = \vartheta$ between $\vec{R}$ and $\vec{r}$ is fixed during the encounter and can be treated as a potential energy parameter. The interaction potential energy can then be expressed in terms of variable parameters like $A^{(l)}(\theta)$, $b^{(l)}(\theta)$, $R^{(l)}(\theta)$. Scattering wavefunction calculations are done for variable R and fixed θ as in elastic scattering, to obtain a scattering amplitude $f_{IOS}^{(+)}(\overline{\Omega}; \omega_0)$ from which one gets helicity transition amplitudes

$$f_{j',\mu_{j'},j,\mu_j}^{(+)}(\overline{\Omega}) = \left\langle j', \mu_j' | f_{IOS}^{(+)}(\overline{\Omega}; \omega_0) | j, \mu_j \right\rangle$$

where ω_0 is a function of $\overline{\Omega}$ and $\overline{\omega}$ and with the bracket indicating integration over $\overline{\omega}$, the orientation angles present in the rotational wavefunctions $\langle \overline{\omega} | j, \mu_j \rangle$. The treatment can be developed in the context of differential equations [36] or of integral equations [34]. The equations to be solved follow from our previous partial differential equations, with the orbital and rotational kinetic energy operators fixed with averages $\langle \hat{L}^2 \rangle = \bar{l}(\bar{l}+1)$ and $\langle \hat{j}^2 \rangle = \bar{j}(\bar{j}+1)$ to be chosen for convenience.

The formalism of T–R collisions has also been extended to include collisions of atoms with open-shell molecules in spin-multiplet states, which requires introduction of electronic spin operators of the molecules and their eigenfunctions [37, 38]. These involve electronic transitions usually treated in an electronic diabatic representation of potentials and couplings and are considered in the following chapter on non-adiabatic collisions.

4.4.1.3 Semiclassical Dynamics Treatments

An early treatment of T–R transfer within a semiclassical sudden approximation was developed and used to calculate action integrals, as they appear in our eikonal treatment, to obtain rotational excitation cross sections vs collision energies for Ar +Tl F, giving results in accordance with experimental measurements [39]. Extensions of this time-dependent sudden approximation to time-dependent coupled states and orientation-conserving treatments have been reviewed [40], and connections of theoretical treatments with experimental measurements, including inversion of experimental data and sensitivity analysis, have also been surveyed in the literature [41].

4.4.2 Numerical Results

4.4.2.1 Numerical Results for Atom–Diatom Collisions

Expansions of the interaction potential energy in a basis set of orbital–rotational functions $\Phi_\Lambda(\Omega, \omega)$, coupled to total angular momenta with quantum numbers (J, M_J) and where $\Lambda = (l, j, J, M_J)$, provide matrix representations $\boldsymbol{U}^{(C)}(R)$ of the energy couplings. Expansion of the scattering wavefunction in the same basis set of orbital-rotational states displays coefficients $\psi_{\Lambda\kappa}^{(C)}(R)$ dependent on the radial distance R and allows construction of the matrix differential equation

$$\hat{\boldsymbol{D}}_R^{(C)} \boldsymbol{\psi}_\kappa^{(C)}(R) = \boldsymbol{U}^{(C)}(R) \boldsymbol{\psi}_\kappa^{(C)}(R)$$

where $\hat{\boldsymbol{D}}_R^{(C)}$ contains second-order derivatives in R, to be solved with scattering boundary conditions. As described in Section 4.2, a variety of numerical procedures are available to integrate this set of coupled radial differential equations starting with finite values of the radial functions near the origin $R = 0$ and propagating the solutions to large distances, where they can be matched to outgoing waves [33]. Procedures involve step-by-step propagation, useful for waves with large wavelengths, or propagation with piecewise potentials and piecewise propagation of boundary conditions (or $\boldsymbol{R}$ matrix). Alternatively, the coupled differential equations can be converted into Volterra-type integral equations for radial amplitude densities, with kernels suitable for discretization and propagation. Some of these procedures were covered in Section 4.2.

Calculations of cross sections have been done for many atom–diatomic systems, usually with approximations that decrease the number of coupled differential equations to be solved. As an example, the $He + H_2$ T–R transfer has been calculated in detail within the CS approximation. Figure 4.2 (a) shows the $j = 0 \rightarrow j' = 2$ inelastic integral cross section vs the orbital quantum number l for $He + H_2$ at three collision energies $E = 0.2$ eV, 0.4 eV, and 0.9 eV. Figure 4.2 (b) shows the $j = 0 \rightarrow j' = 2$ inelastic differential cross section vs scattering angle Θ at collision energies of 0.1 and 0.2 eV. Comparison of results of the approximate CS calculations in full lines with accurate coupled-channel numerical integration of the coupled equations (dashed lines) displays very good agreement.

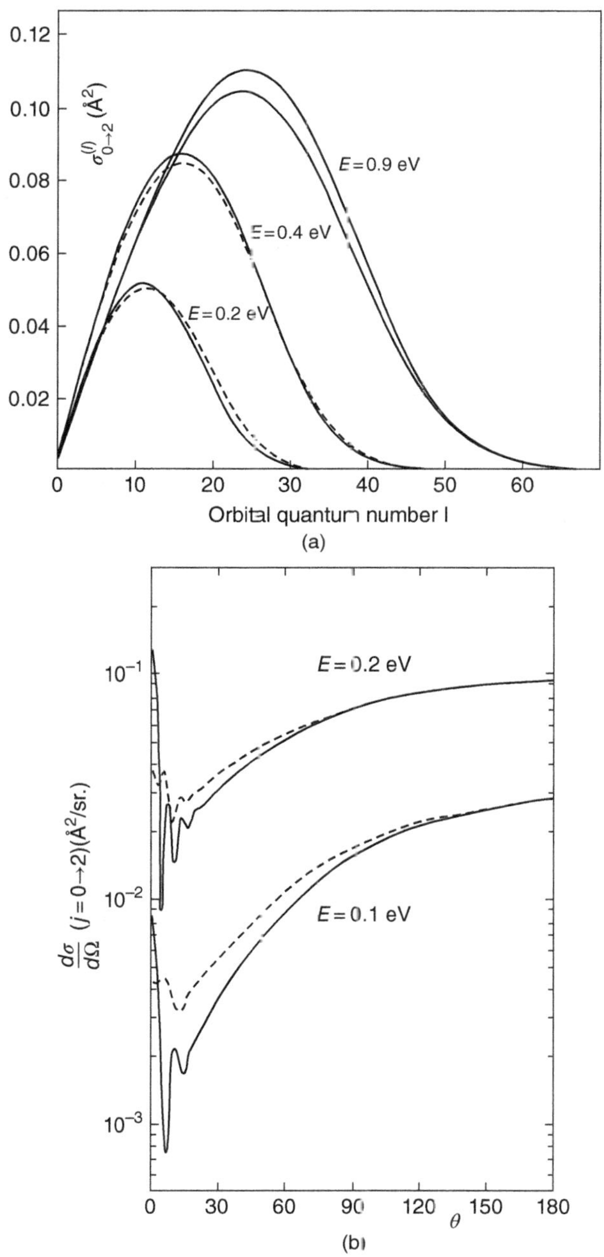

Figure 4.2 (a) The $j = 0 \to j' = 2$ inelastic integral cross section vs the orbital quantum number l for He + H_2 at three collision energies $E = 0.2$ eV, 0.4 eV, and 0.9 eV. (b) The $j = 0 \to j' = 2$ inelastic differential cross section vs scattering angle Θ at collision energies of 0.1 and 0.2 eV. Source: [35] / AIP Publishing.

Here, the atom–rotor interaction potential energy was chosen of the form $V(R, \theta) = V^{(0)}(R) + V^{(2)}(R)P_2(\cos\theta)$ with the $V^{(l)}(R) = A^{(l)}\exp[-b^{(l)}(R - R^{(l)})]$ exponential functions parametrized as in the Kraus–Mies potential function [35]. Integral cross sections are found to contain contributions from many partial waves, with their number increased with collision energies. The shown differential cross sections are mostly backward-peaked at the low energies but become more forward at larger collision energies.

A treatment for He–N_2 has compared detailed theoretical and experimental cross sections [42], with the calculated values obtained from a potential energy function including attractive long-range Van der Waals dispersion terms with a short-range cut-off, plus a short-range exponential repulsion of the Born–Mayer type. The angle dependence of the potential for this homonuclear diatomic molecule contains only even Legendre polynomials, which leads to transitions only between states of the same parity.

Cross sections were obtained with an expansion of the scattering wavefunction into five rotational states of N_2 and by solving the coupled-channel (state-to-state) radial differential equations, with a numerical procedure involving step-by-step propagation. The comparison of measured and calculated rotationally inelastic differential cross section vs scattering angle for He – N_2 collisions in the CM frame, for rotational transitions $j = 0 \rightarrow 2$ and $j = 1 \rightarrow 3$ in N_2, is shown in Figure 4.3. This indicates excellent agreement when coupled-channel converged results can be done, as presented there, at relatively low collision energies.

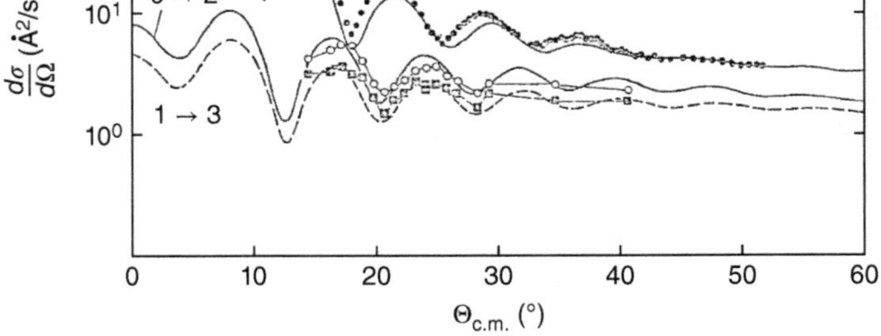

Figure 4.3 Measured (dots) and calculated (lines) rotationally inelastic differential cross section vs scattering angle for He–N_2 collisions in the CM frame, for the transitions $j = 0 \rightarrow 2$ and $j = 1 \rightarrow 3$ in N_2. Source: Adapted from [42].

4.4.2.2 Numerical Results for Diatom–Diatom Collisions

The interaction potential energy $V(\vec{R}, \vec{r}_M, \vec{r}_N)$ for two interacting diatomics M and N with fixed bond distances and orientation angles ω_M and ω_N can be expanded as

$$\begin{aligned} V(\vec{R}, \vec{r}_M, \vec{r}_N) = & \sum_{\lambda,\lambda_M,\lambda_N} A_{\lambda,\lambda_M,\lambda_N}(R, r_M, r_N) \\ & \times \sum_{\mu_M,\mu_N,\mu} \langle \lambda_M, \mu_M, \lambda_N, \mu_N | \lambda_M, \lambda_N, \lambda, \mu \rangle \\ & \times Y_{\lambda_M,\mu_M}(\omega_M) Y_{\lambda_N,\mu_N}(\omega_N) Y_{\lambda,\mu}(\Omega)^* \end{aligned}$$

in terms of spherical harmonics and Clebsch–Gordan coefficients. This expansion has been implemented in detailed ab initio calculations for $H_2 + H_2$ and employed in coupled-channel calculations of differential cross sections to be compared with experimental measurements for $D_2 + H_2$ [43]. Figure 4.4 gives the leading expansion coefficients $A_{\lambda,\lambda_M,\lambda_N}$ vs R for fixed bond distances, with a H—H bond distance which accounts for vibrational–rotational distortion. The coefficient $A_{0,0,0}$ gives the isotropic potential energy component, with a full line from calculations. The coefficient $A_{0,2,2}$ gives the leading anisotropy and mediates the $(j_M, j_N) = (0, 0) \rightarrow (j'_M, j'_N) = (0, 2)$ transitions, while $A_{2,2,4}$ is dominant for reorientations at large distances. The same calculated potential energies can be used for isotopic variants $D_2 + H_2$ and $HD + D_2$, however, allowing for small changes in average bond distances. The dashed line for $A_{0,0,0}$ gives the experimentally extracted term from $HD + D_2$ cross-sectional measurements.

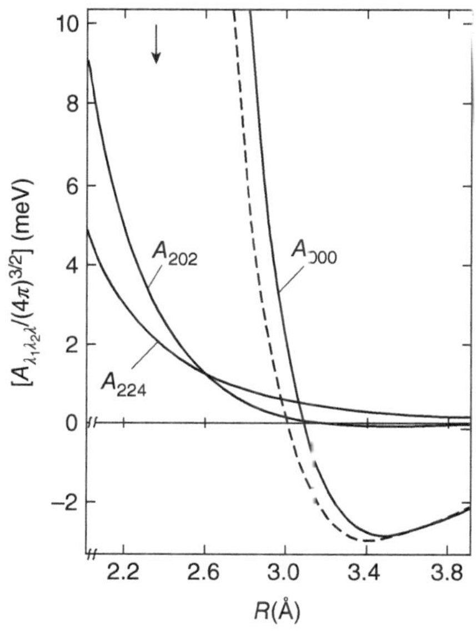

Figure 4.4 Leading potential energy expansion coefficients $A_{\lambda,\lambda_M,\lambda_N}$ vs R for a H–H bond distance which accounts for vibrational–rotational distortion. The coefficient $A_{0,0,0}$ gives the isotropic component, with a full line from calculations and a dashed line for the experimentally extracted term. Source: [43] / Royal Society of Chemistry.

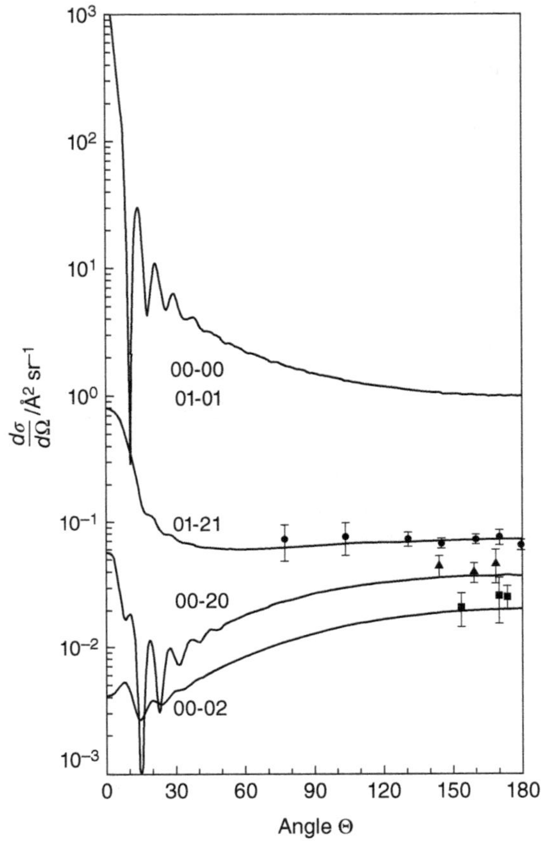

Figure 4.5 Calculated and measured state-to-state differential cross sections vs scattering angles for transitions between pairs of rotational states $(j_M,j_N) \rightarrow (j'_M,j'_N)$ in $D_2 + H_2$ at a 89.1 meV collisions energy. The experimental values are shown as circles, triangles, and squares for (0,1) → (2,1), (0,0) → (2,0), and (0,0) → (0,2), respectively. Source: [43] / Royal Society of Chemistry.

Figure 4.5 shows calculated and measured state-to-state differential cross sections vs scattering angles for transitions between pairs of rotational states $(j_M,j_N) \rightarrow (j'_M,j'_N)$ in $D_2 + H_2$ at a 89.1 meV collisions energy. The experimental values are shown as circles, triangles, and squares for (0,1) → (2,1), (0,0) → (2,0), and (0,0) → (0,2), respectively. The (0,1) → (2,1) transition is larger at small angles due to the importance of the long-range $A_{2,2,4}$ coefficient to promote forward inelastic scattering.

4.5 Translational–Rotational–Vibrational (T–R–V) Transfer

4.5.1 Landau–Teller Model of T–V Transfer

A simple model for an atom–diatomic A + BC collision leading to vibrational transitions $n \rightarrow n'$ is the so-called Landau–Teller (or L–T) model for the interaction potential energy and dynamics [6]. It assumes that the three atoms are moving colinearly on an axis (here a z-axis) which remains fixed in orientation, with coordinates (z_A, z_B, z_C) and atom A interacting only with B. The diatomic BC undergoes harmonic vibrations in a potential energy $v_{BC}(r) = -f \cdot (r - r_e)^2/2$ with $r = |z_B - z_C|$ and a force constant f for vibrations around a bond equilibrium distance r_e. The interaction potential energy function is a repulsive $V(R, r) = A\exp(-a\, r_{AB}) = A\exp[-a(R - \lambda r)]$, where R is the distance between A and the CM of BC, $r = r_{BC}$, and $\lambda = m_C/(m_B + m_C)$.

Scattering solutions of the Hamiltonian equation $\hat{H}\,\Psi^{(+)}(R, r) = E\Psi^{(+)}(R, r)$ with this L–T model potential can be obtained at different levels of accuracy, with a physically meaningful treatment based on the DW approximation or with accurate results from a coupled-channels treatment. Their cross sections provide insights on trends of the magnitude of vibrational energy transfer for varying repulsion potential magnitude and slope and for varying collision kinetic energies $Mv^2/2$ compared to the vibrational excitation quantum $\hbar\omega_V$. However, accurate values of rates of collisional energy transfer for these strongly repulsive potential energies require accurate calculations [44].

In outline, with details found in the literature [6], a DW treatment introduces the unit of vibrational frequency $\omega_V = (f/m)^{1/2}$ with $m = m_B m_C/(m_B + m_C)$ and unit of length $d = [\hbar/(m\omega_V)]^{1/2}$, and reduced space variables $X = R/(\lambda d)$ and $x = (r - r_e)/d$. Harmonic oscillator wavefunctions $u_n(r) = d^{1/2}\ \varphi_n(x)$ are given by $\varphi_n(x) = \pi^{-1/4}(2^n n!)^{-\frac{1}{2}} H_n(x)\exp(-x^2/2)$, $n = 0, 1, \ldots$, with $H_n(x)$ the Hermite polynomial [45] and oscillator energies $E_n^{(V)} = \hbar\omega_V(n + 1/2)$. Matrix elements of the interaction potential are

$$V_{n'n}(R) = A\exp(\alpha x_e)\exp(-\alpha X) I_{n'n}$$

$$I_{n'n} = \int dx\, \varphi_{n'}(x)\exp(-\alpha x)\varphi_n(x)$$

with $\alpha = \lambda a d$. Typically, magnitudes of constants are $a = 10\ \text{nm}^{-1}$, $d = 0.01$ nm, $\lambda = 1.0$, giving $\alpha = 0.1$ and $I_{nn} \approx 1.0$. DWs can be obtained for $V_{nn}(R)$, and DW transition probabilities are given by $P_{n\to n'} = 4|\tilde{T}_{n'n}|^2$ with $\tilde{\boldsymbol{T}} = \boldsymbol{K}^{(DW1)}(1 - i\boldsymbol{K}^{(DW1)})^{-1}$ from Section 4.1. In the special case of high collision energies and small momentum transfer for $(k_0 - k_1) \ll k_n$, where $(\hbar k_n)^2 = 2M\left(E - E_n^{(V)}\right) = (Mv_n)^2$, one finds that

$$P_{0\to 1} = (8\pi\mu^2/\alpha^2)\exp[-2\pi\omega_V/(a\bar{v})]$$

with $\mu = m_A m_C/(M m_B)$ and $\bar{v} = (v_0 + v_1)/2$. This shows how the probability changes with the slope α of the interaction potential, vibrational quantum $\hbar\omega_V$, relative mass, and relative collision velocities. In particular, $\log(P_{0\to 1})$ increases rapidly as the total energy E increases. Trends are qualitatively correct, but magnitudes of the probabilities are overestimated by the DW approximation, as shown in comparisons with accurate coupled-channel calculations for $He + H_2$ and similar systems [44].

4.5.2 Atom–Diatomic T–R–V Transfer

We again consider the simple case of an inelastic collision between an atom A and a rotating and vibrating diatomic BC with ion cores at $\vec{r}_b$ and $\vec{r}_c$ and bond vector position $\vec{r} = \vec{r}_b - \vec{r}_c$, with orientation angles (ϑ, φ). In a collision event, energy is transferred between the relative kinetic motion of the two colliding species and the rovibrational motion of the diatomic while it rotates and vibrates around its equilibrium value $r_e = |\vec{r}_b - \vec{r}_c|$. The internal Hamiltonian of BC is $\hat{H}_{int} = \hat{H}_R + \hat{H}_V + \hat{H}_{RV}$, a sum of rotational and vibrational Hamiltonians plus their coupling. The rotational Hamiltonian is $\hat{H}_R = \hbar^2\hat{j}^2/(2I_{BC})$, where $\hat{j}(\vartheta, \varphi)$ is the rotational operator with $I_{BC} = M_{BC}{r_e}^2$ the moment of inertia, and $\hat{H}_V = k_V(r - r_e)^2/2$ is the Hamiltonian for harmonic vibrations, while $\hat{H}_{RV} = W_{RV}(r, \vartheta, \varphi)$ describes coupling energies of rotational and vibrational variables, including moment-of-inertia centrifugal distortions and anharmonic effects. This coupling can be obtained from calculations of the BC electronic energy or by fitting its measured diatomic spectra [31, 46].

The interaction of an atom A with a rotating and vibrating diatomic BC involves a potential energy function $V(R, r, \theta)$ with a variable bond distance changing around an equilibrium value r_e so that

$$V(R, r, \theta) = \sum_{\mu=0}^{\mu_{max}} v_\mu(R, r)P_\mu(\cos\theta)$$

$$v_\mu(R, r) = v_\mu(R, r_e) + v_\mu^{(1)}(R, r_e)(r - r_e) + v_\mu^{(2)}(R, r_e)(r - r_e)^2/2 + \cdots$$

for each μ-anisotropy. The term with $v_\mu^{(2)}(R, r_e) = k_\mu(R)$ leads to a change of the vibrational constant to a local value $k_V + k_\mu(R)$ in the BC harmonic potential energy, while $v_\mu^{(1)}(R, r_e) = -f_\mu(R)$ describes a vibrational driving force, as R change during a collision, leading to anharmonicity.

A straightforward treatment of translational–rotational–vibrational (or T–R–V) inelastic transitions could in principle be based on the full quantal coupled-channel equations of Section 4.2, with the scattering wavefunction $\Psi_\alpha^{(+)}(\vec{R}, \boldsymbol{q})$ expanded in a basis set $\{\phi_\alpha(\boldsymbol{q})\}$ of rotational–vibrational states. Internal motions include rotational and vibrational variables $\boldsymbol{q} = (\boldsymbol{q}_R, \boldsymbol{q}_V)$ of the internal motions of colliding molecules, and internal states $\alpha = (\alpha_R, \alpha_V)$ involve both rotational and vibrational quantum numbers. However, a coupled-channels approach leads to very large sets of coupled differential equations for the scattering wavefunction because many rotational energy levels $E_\alpha^{(R)}$ are accessible at the collision energies involved in vibrational transitions, and this involves expansion in very many rotational states with varying orientations.

A qualitative discussion of energy transfer involving both vibrational and rotational molecular transitions can be based on time scales of relative, rotational, and vibrational motions. As an example, in the collision of a diatomic leading to its excitation between energy levels with rotational quantum numbers $j = 0 \rightarrow 1$ and vibrational quantum numbers $n = 0 \rightarrow 1$, the corresponding excitation energies and related frequencies of oscillation between levels are $\Delta E_{rot} = E_1^{(R)} - E_0^{(R)} = \hbar\omega_R$ and $\Delta E_{vib} = E_1^{(V)} - E_0^{(V)} = \hbar\omega_V$, with associated transition times $\tau_R = 2\pi/\omega_R$ and $\tau_V = 2\pi/\omega_V$. Typically, with rotational energy changes much smaller than vibrational ones, one finds that $\tau_R \gg \tau_V$, so that vibrational transitions may be pictured as happening over short times while rotational orientations are almost unchanged.

A related model of vibrational-rotational excitations $(n, j, m_j) \rightarrow (n', j', m_j')$ can be derived for the collision between atom A and diatomic BC in states $\phi_\alpha(\vec{r}) = u_n(r) Y_{jm_j}(\omega)$, for rotational states $Y_{jm_j}(\omega)$, and harmonic vibrational states $u_n(r)$ satisfying $\widehat{H}_V u_n = \hbar\omega_n\, u_n, n = 0, 1, 2, \ldots, \omega_n = \omega_V(n + 1/2)$, by considering that rotational motions are slow compared to vibrational ones, at the high sudden collision energies that induce vibrational transitions. Introducing again the unit of length $= [\hbar/(m\omega_V)]^{1/2}$, the reduced displacement variable $x = (r - r_e)/d$ and harmonic oscillator wavefunctions $u_n(r) = d^{1/2}\ \varphi_n(x)$ given by $\varphi_n(x) = \pi^{-1/4}(2^n n!)^{-\frac{1}{2}} H_n(x) \exp(-x^2/2)$, $n = 0, 1, \ldots$, with $H_n(x)$ the Hermite polynomial satisfying the recurrence relation $2xH_n(x) = H_{n+1}(x) + 2nH_{n-1}(x)$, it is found from the expansion of the orientation coefficients $v_\mu(R, r)$ in powers of x that the linear term in the interaction potential energy leads to vibrational couplings $n \leftrightarrow n \pm 1$, and the quadratic term gives $n \leftrightarrow n \pm 2$ couplings. Coupled-channel differential equations contain, as a result, couplings of the same nature.

In the IOS approximation previously described, the atom–diatomic collision is assumed to be sudden and the diatomic remains fixed in its orientation while being hit by the projectile atom. Assuming that the diatomic orientation is fixed in a CM–BF reference frame at BC orientation angles $\omega_0 = (\theta_0, \phi_0)$, an inelastic vibration transition amplitude $f_{n \rightarrow n'}(\Omega; \omega_0)$, parametrized for the fixed orientation of BC, can be calculated to approximate the vibrational–rotational transition amplitude as

$$f_{n,j,m_j \rightarrow n',j',m_{j'}}(\Omega) = \int d\omega_0\ Y_{j'm_{j'}}(\omega_0)^* f_{n \rightarrow n'}(\Omega; \omega_0) Y_{jm_j}(\omega_0)$$

This greatly simplifies the T–R–V calculations since only wavefunction expansions in a basis set $\{u_n(r),\ n = 0 - n_V\}$ of vibrational states are needed in a coupled-channel treatment.

This amplitude can be derived from the full coupled channels treatment of collisions by generalizing, from T–R to T–R–V energy transfer, the coupled-states (or CS) approximation given before, within a CM–BF description using vibrational-helicity states $|n, l, j, J, K\rangle$. More generally, T–R–V transition amplitudes can be derived from a CS treatment including now vibrational states $u_n(r)$

in the expansion basis set for the total wavefunction and introducing the approximations previously described for T–R transfer, to eliminate rotational reorientation and orbital couplings in the resulting coupled-states equations: (a) coupling is limited to transitions with helicities $\mu_j = \mu_j'$, in a $j_{Z'}$-conserving coupled-states approximation, and also limited to $\mu_l = \mu_l'$ corresponding to a conserved orientation of the orbital angular momentum vector $\vec{L}$, and (b) that $\langle n, l, j, J, K|\hat{L}^2|n, l, j, J, K'\rangle \approx l(l+1)\delta_{KK'}$, therefore dropping Coriolis couplings.

A practical procedure, applicable to a variety of processes like diatom–diatom and atom–polyatomic T–R–V transitions, relies on the coupled-states approximation previously described for T–R transfer. It greatly decreases the number of coupled vibrational–rotational states needed in expansions of the wavefunction and can be improved by addition of basis-set states or by inclusion of angular momentum couplings in the partial differential equations. A comparison of vibrational relaxation cross sections in $He + H_2$ between the accurate coupled-channels and the coupled-states treatments shows that they are in very good agreement at low relative kinetic energies [47]. This has been done for an anisotropic interaction potential energy

$$V(\vec{R}, \vec{\rho}) = C\exp(-\alpha_0 R + \alpha_1 R\rho)[(1 + \gamma_0\rho) + \beta(1 + \gamma_2\rho)P_2(\hat{R}\cdot\hat{\rho})]$$

with $\vec{\rho} = \vec{r} - \vec{r}_e$, parametrized for He + H_2, and integral cross sections where vibrationally excited $(n,j) = (1,0)$ relaxes to ground vibrational states (0,0), (0,2), (0,4), with results shown in Figure 4.6, where close-coupling is shown as full lines and close-states as dashed lines.

T–R–V transition amplitudes obtained from a CS treatment introducing the two approximations previously described for T–R transfer, that eliminate rotational reorientation and orbital couplings in the resulting coupled state equations, can be improved adding to the differential equations the Coriolis couplings $\langle n, l, j, J, K|\hat{L}^2|n, l, j, J, K'\rangle$ with $K' = K \pm 1$. This has been done in calculations with an extended CS (or ECS) treatment applied to rovibrationally inelastic scattering between an atom and a triatomic molecules, with results obtained for H_2O collisions with rare gas atoms, where the triatomic rovibrational states are labelled by quantum numbers $(n_1, n_2, n_3) = (j, k, \lambda)$ for the three vibrational modes and rotating top quantum numbers. The ECS has been found to converge toward accurate coupled-channels results for a model of $He + H_2O$ as the number of helicity matrix blocks of Coriolis coupling is increased in the Hamiltonian matrix differential equation for the wavefunction amplitudes [48].

4.5.3 T–R–V Transfer Involving Polyatomic Molecules

Collisions involving polyatomic molecules at high relative kinetic energies lead to both single and multiple vibrational transitions. Internal vibrational motions can

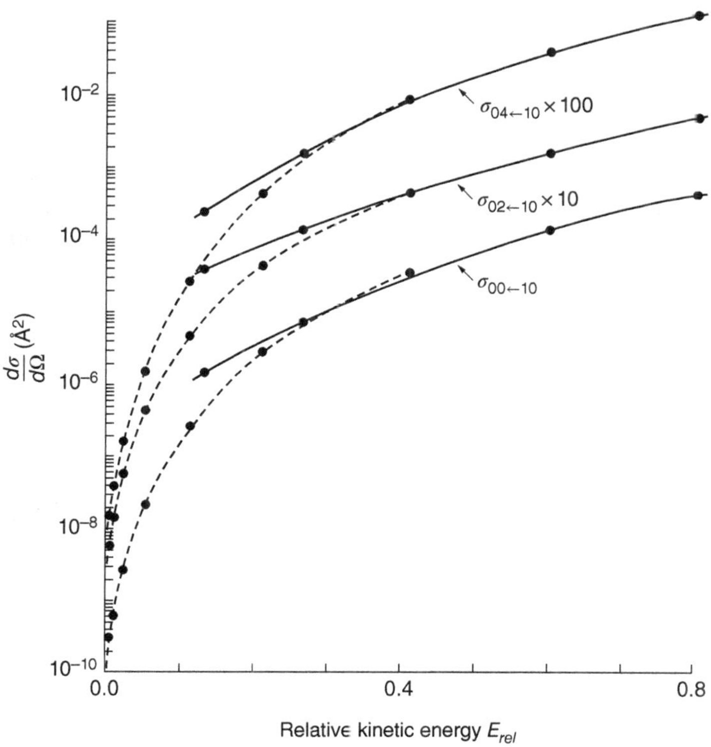

Figure 4.6 Vibrational relaxation integral cross sections for He + H_2 $(n,j) = (1,0) \rightarrow (0,j)$, with close-coupling results shown as full lines and close-states results as dashed lines. Source: [47] / AIP Publishing.

be expressed in terms of symmetry molecular variables for bond distances, bond angles, and dihedral angles for each molecule $I = A, B$ forming a set $\{s_i, i = 1$ to $f_S\}$, where the number $f_S = 3N_I - 6$ or $3N_I - 5$ depending on whether the species $I = A, B$ is nonlinear or linear [49]. What follows is a summary of the procedures leading to vibrational coupling matrix elements.

Starting with the vibrational Hamiltonian of classical mechanics for polyatomic molecule I, with a quadratic potential energy $V_V = \mathbf{s}^T\ \mathbf{K}_S\mathbf{s}/2$, where $\mathbf{s}$ is a column matrix of the f_S symmetry variables and $\mathbf{K}_S$ a square matrix of force constants, a transformation can be done to re-express the molecular potential energy in terms of the N_I atomic displacements $\vec{r}_m - \vec{r}_{m,eq} = \vec{x}_m$ labelled by components $\{x_j, j = 1$ to $f_S\}$, as linear combinations of the $\{s_i\}$. Introducing mass-weighted variables $\left\{y_j = m_j^{\frac{1}{2}} x_j, j = 1 \text{ to} f_S\right\}$, followed by a transformation to normal vibrational coordinates $\mathbf{q}_I = \{q_\nu, \nu = 1 \text{ to} f_S\}$, this gives for molecule I the vibrational Hamiltonian as $H_V^{(I)} = \dot{\mathbf{q}}_I{}^T \dot{\mathbf{q}}_I/2 + \mathbf{q}_I{}^T \mathbf{K}_I \mathbf{q}_I/2$, with $\mathbf{q}_I$ a column matrix, $\dot{\mathbf{q}} = d\mathbf{q}/dt$, and with a normal-motion diagonal force matrix $\mathbf{K}_I$, so that all normal mode variables q_v are decoupled. This Hamiltonian generates harmonic oscillator eigenfunctions $u_{n_I}^{(I)}\left(q_{v_I}^{(I)}\right)$ with eigenenergies $\hbar\omega_V^{(I)}(n_I + 1/2)$ for each normal mode. The product of all these normal mode eigenfunctions form a vibrational expansion state $\boldsymbol{u}_{\boldsymbol{n}_I}^{(I)}(\boldsymbol{q}_I)$.

The interaction potential energy can be expressed in terms of those normal vibrational coordinates and of Euler orientation angles $\Gamma_I = (\alpha,\beta,\gamma)_I$, with expansions using rotational matrices in

$$V(R, \Gamma_A, \mathbf{q}_A, \Gamma_B, \mathbf{q}_B) = \sum_{j_A,k_A,m_{jA}} \sum_{j_B,k_B,m_{jB}} D_{k_A,m_{jA}}^{(j_A)}(\Gamma_A) V_{k_A,m_{jA},k_B,m_{jB}}^{(j_A j_B)}(R, \mathbf{q}_A, \mathbf{q}_B) D_{k_B,m_{jB}}^{(j_B)}(\Gamma_B).$$

Here, the factor $V_{k_A,m_{jA},k_B,m_{jB}}^{(j_A j_B)}$ can be further expanded in powers of the vibrational displacements $\mathbf{q}_I$, with linear and quadratic terms leading to potential energy couplings $n_I \leftrightarrow n_I \pm 1$, and $n_I \leftrightarrow n_I \pm 2$ couplings, and intramolecular excitation-de-excitation. In addition, bilinear expansion terms containing $q_{v_A}^{(A)} q_{v_B}^{(B)}$ generate couplings $(n_{v_A}, n_{v_B}) \leftrightarrow (n_{v_A} \pm 1, n_{v_B} \mp 1)$ between normal modes of the two species, and intermolecular energy transfer.

A rovibrational basis set $\{\Phi_\lambda(\Omega, \Gamma_A, \mathbf{q}_A, \Gamma_B, \mathbf{q}_B), \lambda = (l, m_l, j_A, k_A, m_{jA}, \boldsymbol{n}_A, j_B, k_B, m_{jB}, \boldsymbol{n}_B)\}$ for an expansion of the scattering wavefunction $\Psi_\lambda^{(+)}(\vec{R}, \Gamma_A, \mathbf{q}_A, \Gamma_B, \mathbf{q}_B; \vec{k})$ contains products $\boldsymbol{u}_{\boldsymbol{n}_I}^{(I)}(\boldsymbol{q}_I)$ of normal harmonic vibration functions and molecular rotational functions $D_{k_I,m_{jI}}^{(j_I)}(\Gamma_I)$ times the spherical harmonic $Y_{l,m_l}(\Omega) = \langle \Omega | l, m_l \rangle$ for the orientation of the relative position vector $\vec{R}$. This expansion can lead to a very large number of coupled-channel equations at the large collision energies for vibrational excitation. Practical ways to carry out calculations of measurable cross sections involve separation of rotational and vibrational motions as in the IOS approximation or simplifications of the rotational motions as in the extended coupled-states approximation.

As an example, slow collisions of Ar with vibrationally excited H_2O lead to rovibrational reorientation and de-excitation. The target H_2O has three vibrational modes with quantum numbers (v_1, v_2, v_3) and rotates in states labelled as $j_{k_a k_c}$ for the triatomic with a total angular momentum vector of rotational quantum number j and projections of quantum numbers (k_a, k_c) along SF and BF axes.

Figure 4.7 shows results for the integral cross section (ICS) of rovibrationally inelastic collisions of $Ar + H_2O$ vs collision energy E_c comparing the rigorous time-independent coupled-channels (TICC) and the extended coupled-states (ECS) approximations [48], showing excellent agreement over most collision energies. The symbols $\Delta = 0, 1$ refer to two different ways of accounting for Coriolis couplings extending the coupled-states treatment. Those results were

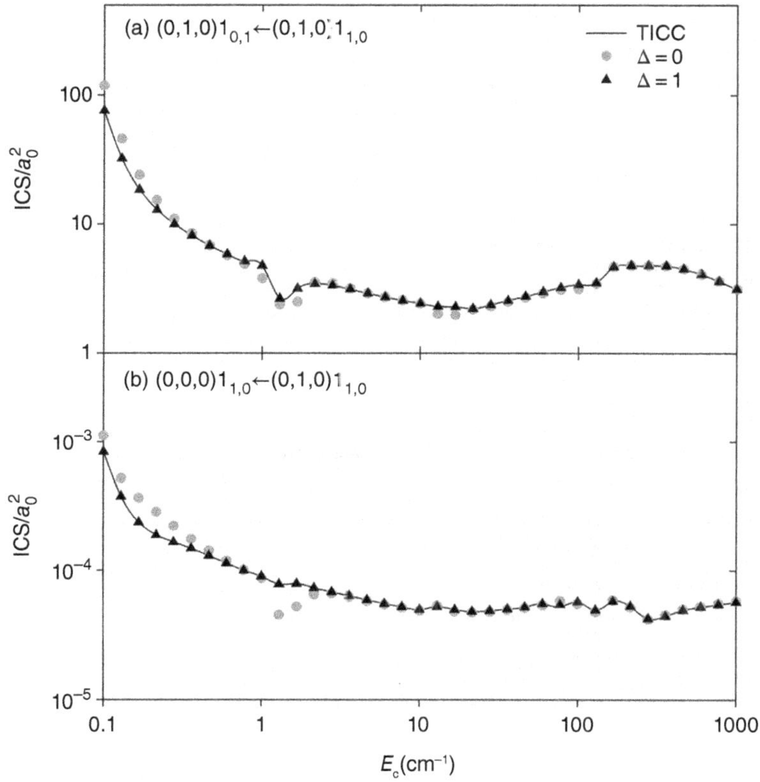

Figure 4.7 Comparison of state-to-state ICSs for Ar + H_2O from the initial state $(0,1,0)1_{1,0}$ to the final one $(0,1,0)1_{0,1}$ in rotational reorientation [panel (a)] and to $(0,0,0)1_{1,0}$ in vibrational de-excitation [panel (b)] using the rigorous time-independent coupled-channels (TICC) and the extended coupled-states (ECS) approximations. Source: [48] / AIP Publishing.

obtained using a potential energy surface generated by electronic coupled-cluster calculations.

4.5.4 Rates of Energy Transfer

4.5.4.1 Reduced Dimensionality and Optical Potential Treatments

It is sometimes possible to simplify the treatment of inelastic collisions by decreasing the number of relevant degrees of freedom in the dynamics. This can be done by selecting certain active degrees of freedom, such as active molecular rotations, bendings, or vibrations to be treated in detail while ignoring other degrees of freedom in the dynamics. It then involves potential energy surfaces of reduced dimensionality [41, 50].

An example is given in atom–diatom collisions A + BC at low relative kinetic energies, below the threshold for excitations of vibrational levels of BC. In this case, the dynamics involves only relative and diatomic rotational motions and one needs PESs dependent only on the relative atom–diatomic position and on the rotational angles of the diatomic.

In scattering involving polyatomic molecules, it is expected that collision-induced vibrational excitation is localized and involves only a small number of vibrational modes. Theories introducing local vibrational modes [51] are relevant to the calculation of localized energy transfer in these cases.

4.5.4.2 Thermal Rates from Cross Sections

Collisions involving many coupled rovibrational states α of energy E_α initially present in an experiment with a statistical distribution w_α lead to measurements of averaged cross sections usually added over a final distribution of states n'. In a typical experiment measuring by time-of-flight methods the initial and final relative momenta $\vec{P} = \hbar\vec{k}$ and $\vec{P}' = \hbar\vec{k}'$, the state-to-state rates of energy transfer $R(\vec{k}, \alpha \to \vec{k}', \alpha')$ add up to total inclusive rates

$$R(\vec{k} \to \vec{k}') = \sum_\alpha w_\alpha \sum_{\alpha'} R(\vec{k}, \alpha \to \vec{k}', \alpha')$$

Rates are frequently measured in systems at thermal equilibrium with a temperature T. In these cases, the statistical distribution is a Boltzmann function $w_\alpha(T) = \exp[-E_\alpha/k_BT]/\sum_\alpha \exp[-E_\alpha/k_BT]$.

These rates can be related to double differential cross sections per unit solid angle and per unit of collisional energy transfer $\varepsilon = (P'^2 - P^2)/(2M)$, with M the collisional reduced mass. The relation

$$\frac{d^2\sigma}{d\varepsilon d\Omega} = \left(\frac{2\pi}{\hbar}\right)^3 M^2 \frac{k'}{k} R\left(\vec{k} \to \vec{k}'\right)$$

has been derived in [52]. There the rate $R(\vec{k} \to \vec{k}')$ has been related to the Fourier transform of the time correlation function of transition operators $\hat{T}_{\vec{k}\vec{k}'}$, providing an alternative way to calculate rates from algebras of operators instead of solving coupled differential equations.

References

1 P. Atkins and J. dePaula, *Physical Chemistry* (9th edition), New York: WH. Freeman and Co., 2010.

2 H. Goldstein, C. P. Poole Jr., and J. L. Safko, *Classical Mechanics* (3rd edition), San Francisco, CA: Addison-Wesley, 2002.

3 A. Messiah, *Quantum Mechanics*, Volumes 1 and 2, Amsterdam: North-Holland, 1962.

4 D. A. Micha, *Molecular Interactions: Concepts and Methods*, Hoboken, NJ, USA: John Wiley & Sons, Inc., 2020.

5 W. N. Sams and D. J. Kouri, "Noniterative solutions of integral equations for scattering. II. Coupled channels," *J. Chem. Phys.*, vol. 51, p. 4815, 1969.

6 M. Child, *Molecular Collision Theory*, New York: Academic Press, 1974.

7 D. A. Micha, "Effective hamiltonian methods for molecular collisions," in *Advances in Quantum Chemistry*, Volume 8, New York: Academic Press Inc., 1974, p. 211.

8 D. A. Micha, "A self-consistent eikonal treatment of electronic transitions in molecular collisions," *J. Chem. Phys.*, vol. 78, p. 7138, 1983.

9 A. Messiah, *Quantum Mechanics*, Volume 1, Amsterdam: North-Holland, 1962.

10 W. R. Gentry, "Vibrational excitation II: Classical and semiclassical methods (Chapter 12)," in *Atom-Molecule Collision Theory*, (Editor: E. R. B. Bernstein), New York: Plenum Press, 1979, pp. 391–426.

11 W. A. Lester and R. B. Bernstein, "Computational procedure for the close-coupled rotational excitation problem: Scattering of diatomic molecules by atoms," *J. Chem. Phys.*, vol. 48, p. 4896, 1968.

12 K. Smith, *Calculation of Atomic Collision Processes*, New York: Wiley Publ. Co., 1971.

13 R. G. Gordon, "New method for constructing wavefunctions for bound states and scattering," *J. Chem. Phys.*, vol. 51, p. 14, 1969.

14 M. J. Redmon and D. A. Micha, "A computational method fro multichannel scattering: Applications to rotational excitation and long-lived states of He-N_2," *Chem. Phys. Lett.*, vol. 28, p. 341, 1974.

15 M. J. Redmon and D. A. Micha, "Interaction potential and dynamics for Li-F collisions," *Int. J. Quantum Chem. Symp.*, vol. 8, no. Symp. 8, p. 253, 1974.

16 J. C. Light and R. B. Walker, "An R-matrix approach to the solution of coupled equations for atom-molecule reactive scattering," *J. Chem. Phys.*, vol. 65, p. 4272, 1976.

17 B. I. Schneider and R. B. Walker, "The coupled-channel R-matrix propagation method," *J. Chem. Phys.*, vol. 70, p. 2466, 1979.

18 J. Z. H. Zhang, D. J. Kouri, K. Haug, D. W. Schwenke, Y. Shima and D. G. Truhlar, "L2 amplitude density method for multichannel inelastic and rearrangement collisions," *J. Chem. Phys.*, vol. 88, p. 2492, 1988.

19 B. Lipman and J. Schwinger, "Variational principles for scattering processes. I," *Phys. Rev.*, vol. 79, p. 469, 1949.

20 A. Messiah, *Quantum Mechanics*, Volume 2, Amsterdam: North-Holland, 1962.

21 B. H. Bransden, *Atomic Collision Theory*, Reading, MA, USA: Benjamin-Cummings, 1983.

22 W. Kohn, "Variational methods in nuclear collision problems," *Phys. Rev.*, vol. 74, p. 1763, 1948.

23 M. Tuckerman and B. J. Berne, "Reversible multiple time scale molecular dynamics," *J. Chem. Phys.*, vol. 97, p. 1990, 1992.

24 C.-E. Froeberg, *Introduction to Numerical Analysis* (2nd edition), Reading, MA, USA: Addison-Wesley, 1969.

25 W. Magnus, "On the exponential solution of differential equations for a linear operator," *Commun. Pure Appl. Math.*, vol. 7, p. 649, 1954.

26 A. R. Edmonds, *Angular Momentum in Quantum Mechanics*, Princeton, NJ, USA: Princeton University Press, 1960.

27 A. M. Arthurs and A. Dalgarno, "The theory of scattering by a rigid rotator," *Proc. Roy. Soc. London Ser. A*, vol. 256, p. 540, 1960.

28 M. Jacob and G. C. Wick, "On the general theory of collisions for particles with spin," *Ann. Phys.*, vol. 7, p. 404, 1959.

29 H. Klar, "Theory of scattering between two rigid rotors," *Z. Phys. A: Hadrons Nucl.*, vol. 228, pp. 59–67, 1969.

30 R. T. Pack, "Space-fixed vs body-fixed axes in atom-diatomic molecule scattering. Sudden approximations," *J. Chem. Phys.*, vol. 60, p. 633, 1974.

31 P. F. Bernath, *Spectra of Atoms and Molecules*, Oxford, England: Oxford University Press, 1995.

32 J. N. L. Connor, "The distorted wave theory of chemical reactions," in *The Theory of Chemical Reaction Dynamics*, (Editor: D. C. Clary), Dordrecht, Holland: Reidel Publ. Co., 1985, p. 247.

33 D. Secrest, "Rotational excitations I: The quantal treatment," in *Atom-Molecule Collision Theory*, (Editor: R. B. Bernstein), New York: Plenum Press, 1979, pp. 265–299.

34 D. J. Kouri, "Rotational excitation II: Approximation methods," in *Atom-Molecule Collision Theory*, (Editor: R. B. Bernstein), New York: Plenum Press, 1979, pp. 301–358.

35 P. McGuire and D. J. Kouri, "Quantum mechanical close coupling approach to molecular collisions: jz conserving coupled states approximation," *J. Chem. Phys.*, vol. 60, p. 2488, 1974.

36 G. A. Parker and R. T. Pack, "Rotationally and vibrationally inelastic scattering in the rotational IOS approximation. Ultrasimple calculation of total (differential, integral, and transport) cross sections for non-spherical molecules," *J. Chem. Phys.*, vol. 68, p. 1585, 1978.

37 M. H. Alexander, "Rotationally inelastic collisions between a diatomic molecule in a doublet-sigma electronic state and a structureless target," *J. Chem. Phys.*, vol. 76, p. 3637, 1982.

38 M. H. Alexander, "Rotationally inelastic collisions between a diatomic molecule in a doublet-pi electronic state and a structureless target," *J. Chem. Phys.*, vol. 76, p. 5974, 1982.

39 R. B. Bernstein and K. H. Kramer, "Sudden approximation applied to rotational excitationof molecules by atoms. II. Scattering of polar diatomics," *J. Phys. Chem.*, vol. 44, p. 4473, 1966.

40 A. S. Dickinson, "Non-reactive heavy-particle collision calculations," *Comput. Phys. Commun.*, vol. 17, p. 51, 1979.

41 A. E. DePristo and H. Rabitz, "Vibrational and rotational collision processes," *Adv. Chem. Phys.*, vol. 42, p. 271, 1980.

42 M. Faubel, K.-H. Kohl, J. P. Toennies, K. T. Tang and Y. Y. Yung, "The He–N_2 anisotropic Van der Waals potential. Test of a simple model using state-to-state differential scattering cross-sections," *Faraday Discuss. Chem. Soc.*, vol. 73, pp. 205–220, 1982.

43 U. Buck, "Rotationally inelastic scattering of hydrogen molecules and the non-spherical Interaction," *Faraday Discuss. Chem. Soc.*, vol. 73, p. 187, 1982.

44 D. Secrest and B. R. Johnson, "Exact quantum mechanical calculation of a collinear collision of a particle with a harmonic oscillator," *J. Chem. Phys.*, vol. 45, p. 4556, 1966.

45 M. Abramowitz and I. A. Stegun, *Handbook of Mathematical Functions*, New York: Dover Publ. Inc., 1965.

46 R. B. Bernstein (Editor), *Atom-Molecule Collision Theory: A Guide to the Experimentalist*, New York: Plenum Press, 1979.

47 C. S. Lin and D. Sercrest, "Vibrationally inelastic integral cross sections for the scattering of He from H_2," *J. Chem. Phys.*, p. 199, 1979.

48 D. Yang, D. Xie and H. Guo, "Extended coupled states approximation for full-dimensional quantum treatments of rovibrationally inelastic scattering between atoms and triatomic molecules," *J. Chem. Phys.*, vol. 157, 164111, 2022.

49 E. Bright Wilson, J. C. Decius and P. C. Cross, *Molecular Vibrations*, New York: Dover, 1955.

50 J. Bowman, "Reduced dimensionality theory of quantum reactive scattering," in *Advances in Chemical Physics*, Volume 61, New York: Wiley, 1985, p. 115.

51 E. Kraka, M. Quintano, H. W. La Force, J. J. Antonio and M. Freindorf, "The local vibrational mode theory and its place in the vibrational spectroscopy arena," *J. Phys. Chem A*, vol. 126, p. 8781, 2022.

52 E. F. Vilallonga and D. A. Micha, "The calculation of time-correlation functions for molecular collisions," *Phys. Rep.*, vol. 212, pp. 329–390, 1992.

5

Electronically Diabatic Collisions

CONTENTS

Quantum Molecular Dynamics: Energy Transfer and Reactivity, First Edition. David A. Micha.

5.1 Expansion in an Electronic Basis Set

We consider here inelastic collision processes between molecules A and B in internal states p and q, specified by electronic quantum numbers I and vibrational–rotational quantum numbers in μ_I, with $p = (I, \mu_I)$ and also $q = (J, \mu_J)$, leading to electronic transitions

$$A(p) + B(q) \rightarrow A(p') + B(q')$$

with relative momentum changes $\vec{P} = \hbar\vec{k} \rightarrow \vec{P'}$. The collection of internal quantum numbers and relative wavevector will be called $\alpha = (\vec{k}, p, q)$. Further consider to begin with the simple case of collisions between two atoms, A and B, with atomic states in the L–S coupling scheme labeled by electronic orbital, spin, and parity quantum numbers $\Gamma_A = (LM_L SM_S \pi)_A$ and Γ_B in a transition

$$A(\Gamma_A) + B(\Gamma_B) \rightarrow A\left(\Gamma'_A\right) + B\left(\Gamma'_B\right).$$

Diatomic electronic states are labeled by the total electronic angular momentum L, its projections ML on a space-fixed (SF) Z-axis and M'_L on a body-fixed (BF) Z'-axis going through the nuclei, with values $|LA - LB| \leq L \leq |LA + LB|$, $-L \leq ML$, $M'_L \leq L$. The diatomic states can be labelled with indices $\Gamma \approx (\Gamma_A, \Gamma_B)$, which go asymptotically for large distances into the free atomic states. A diabatic (or electronically nonadiabatic) collision involves transitions $\Gamma \rightarrow \Gamma'$.

Figures 5.1–5.3 show respectively a (separated atoms)–(united atom) correlation diagram of energy levels for a one-electron homonuclear diatomic, the calculated electronic energies for H_2^+ vs internuclear distance, and the total (electronic plus nuclear repulsion) energies for H_2^+.

Figures 5.4 and 5.5 similarly show a correlation diagram for a heteronuclear diatomic and calculated electronic energies vs internuclear distance for the one-electron heteronuclear diatomic HeH^{2+} [3]. Regions of crossings and avoided crossings of electronic energies vs interatomic distance are particularly important in the description of collision-induced electronic rearrangement [4].

An atom pair can be described as a collection of two atomic ions, with tightly bound electrons in closed shells, and of delocalized electrons making up molecular orbitals. Indicating with **X** the collection of (position and spin) electronic variables and with $\vec{R}$ the relative position of the atomic nuclei in a center-of-mass frame, the Schrodinger equation is

$$(\hat{H} - E)\Psi_\alpha^{(+)}(\vec{R}, \boldsymbol{X}) = 0$$

for the wavefunction $\Psi_\alpha^{(+)}$, $\alpha = (\vec{k}, a)$, starting in internal state $a = (\Gamma_A, \Gamma_B)_{in}$. The state $\Psi_\alpha^{(+)}$ must satisfy asymptotic boundary conditions for a colliding system at large R.

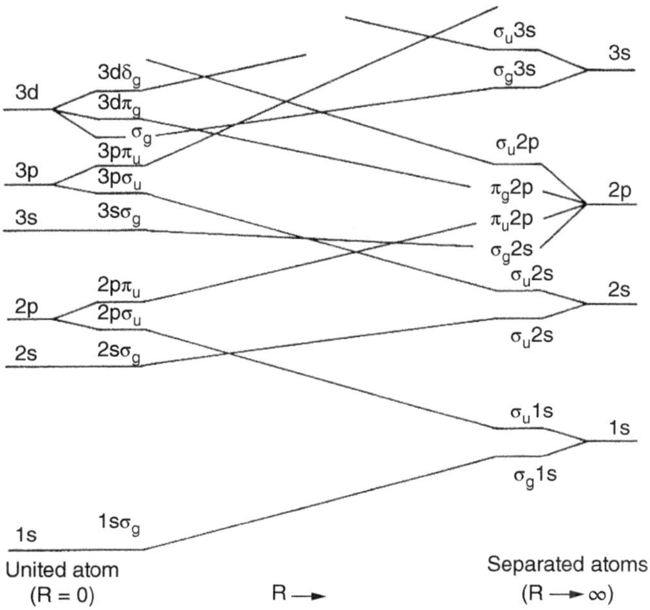

Figure 5.1 Orbital energy correlation diagram from united atoms to separated atoms for a homonuclear diatomic. Source: [1] / Oxford University Press.

Figure 5.2 Calculated electronic energies for H_2^+ vs internuclear distance. Source: [1] / Oxford University Press.

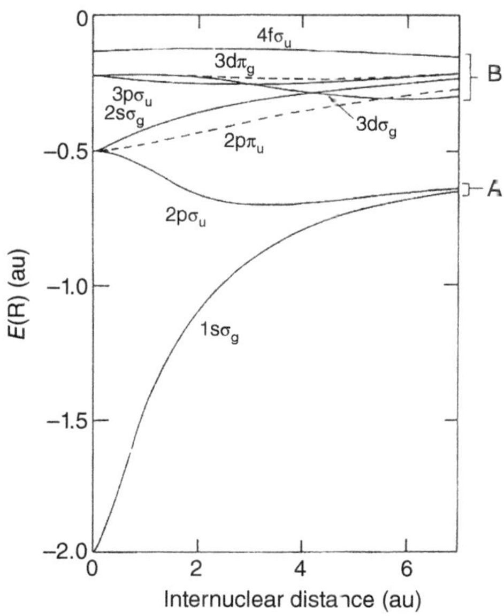

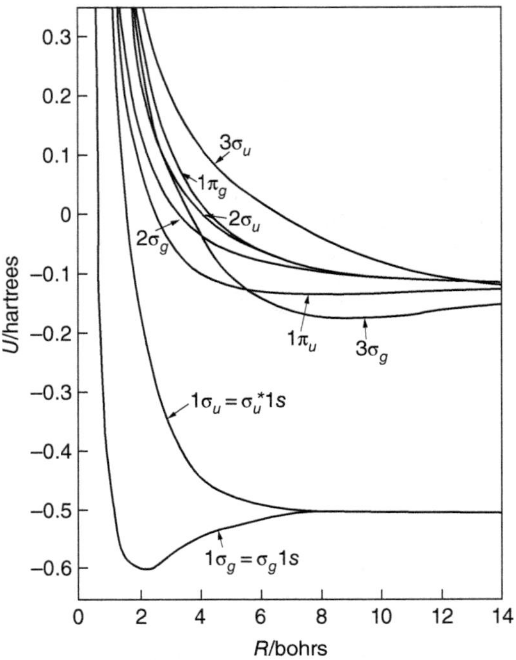

Figure 5.3 Total (electronic plus nuclear repulsion) energies for H_2^+ vs internuclear distance. Source: From [2].

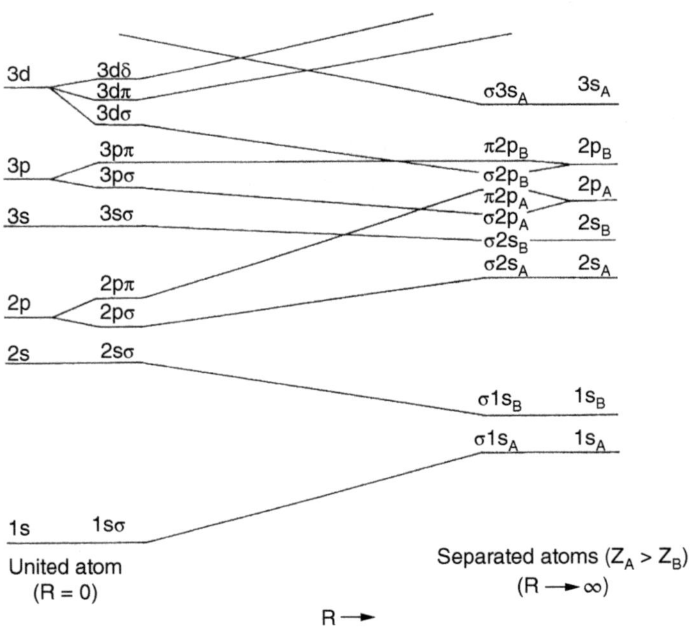

Figure 5.4 Energy level correlation diagram for a heteronuclear diatomic. Source: [1] / Oxford University Press.

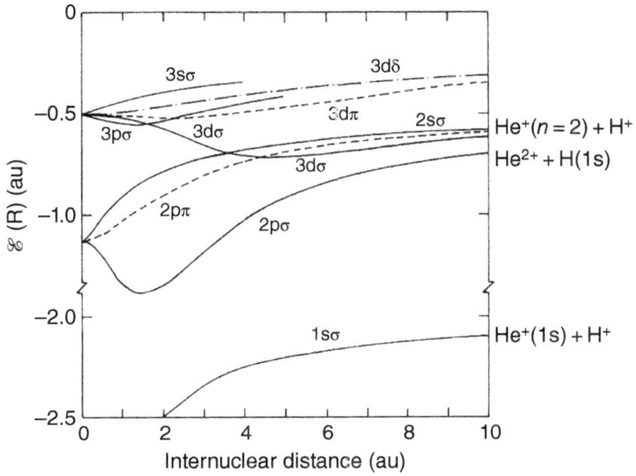

Figure 5.5 Calculated electronic energies vs internuclear distance for the one-electron heteronuclear diatomic $(HeH)^{2+}$. Source: Adapted from [3].

The Hamiltonian $\widehat{H} = \widehat{K}_{at} + \widehat{H}_R$ contains the relative kinetic energy operator $\widehat{K}_{at}$ of atomic ions and the remaining total energy for fixed nuclei in $\widehat{H}_R$. In more detail,

$$\widehat{H}_R = \widehat{K}_{el} + \widehat{V}_{el,el} + \widehat{V}_{at,el} + \widehat{V}_{at,at}$$

where the potential energies involve interactions among electrons and ions and possibly spin–orbit couplings. The atomic kinetic energy of relative motion is given by the square of the atomic momentum operator as $\widehat{K}_{at} = -\hbar^2(\vec{\nabla}_R)^2/(2\,M)$ with M the reduced mass and $\vec{\nabla}_R$ the vector gradient operator. The potential energies $\widehat{V}_{at,at}$ and $\widehat{V}_{at,el}$ are functions of $\vec{R}$ and contain long-range Coulomb interactions and short-range repulsive interactions from the closed electronic shells, and possibly a sum of atomic spin–orbit coupling energies $\widehat{H}_{SO} = \sum_{at}\widehat{h}_{SO}^{(at)}$ [5]. Furthermore at large distances when A and B are not interacting, $\widehat{H}_R \approx \widehat{H}_R^{(0)} = \widehat{H}_A + \widehat{H}_B$, the sum of the two internal Hamiltonians.

For atomic collisions at relative velocities low compared with electron velocities, it is convenient to expand the scattering state in a basis set of *Nel* orthonormalized molecular electronic states $\Phi_\Gamma(\boldsymbol{X};\vec{R}), \Gamma = 1$ to *Nel*, each a function of electronic variables for fixed nuclei, in the SF reference frame with axes (X, Y, Z) defined in a laboratory setup. The choice of a basis set of eigenfunctions of $\widehat{H}_R$ provides the so-called adiabatic electronic representation, but other representations are also introduced for collisional phenomena [6]. The electronic states in a scattering event must be chosen to go asymptotically to products of atomic states, so that $\Phi_\Gamma(\boldsymbol{X};\vec{R}) \approx \Phi_\Gamma^{A,B}(\boldsymbol{X};\vec{R}_A,\vec{R}_B) = \Phi_{\Gamma_A}^A(\boldsymbol{X}_A;\vec{R}_A)\Phi_{\Gamma_B}^B(\boldsymbol{X}_B;\vec{R}_B)$ as $R \to \infty$. The expansion is then, with $\gamma = (\vec{k}', \Gamma)$ for a final collisional wavevector $\vec{k}'$,

$$\Psi_{\alpha}^{(+)}(\vec{R}, \boldsymbol{X}) = \sum_{\Gamma} \psi_{\gamma,\alpha}^{(+)}(\vec{R})\Phi_{\Gamma}(\boldsymbol{X}; \vec{R})$$

$$\approx \Psi_{\alpha}^{(0)}(\vec{R}, \boldsymbol{X}) + \sum_{\Gamma} \psi_{\gamma,\alpha}^{(sc)}(\vec{R})\Phi_{\Gamma}^{A,B}(\boldsymbol{X}), \text{ as } R \to \infty$$

with expansion coefficients $\psi_{\gamma\alpha}^{(+)}$ dependent on the internuclear position vector. The asymptotic boundary conditions contain the free motion state $\Psi_{\alpha}^{(0)}(\vec{R}, \boldsymbol{X}) = \psi_{\alpha}^{(0)}(\vec{R})\Phi_{\alpha}^{A,B}(\boldsymbol{X})$, with $\psi_{\alpha}^{(0)}(\vec{R}) = (2\pi)^{-\frac{3}{2}} e^{i\vec{k}_{\alpha}\cdot\vec{R}}$, and the outgoing scattering functions

$$\psi_{\gamma,\alpha}^{(sc)}(\vec{R}) \approx (2\pi)^{-\frac{3}{2}} f_{\gamma\alpha}^{(+)}(\Omega) R^{-1} e^{ik'R}.$$

During a collision, one must allow for the interaction of electronic and nuclear degrees of freedom due to moving nuclei [6–13]. We describe it in a reference frame with SF axes of coordinates (X, Y, Z) and its origin at the center of mass (CM) of the nuclei (CMN), assuming that electron masses are negligible, in a CMN-SF reference frame. To simplify the relative motion, we consider here only two moving nuclei a and b with masses m_a and m_b. The reduced diatomic mass for the relative atom–atom motion is $M = m_a m_b/(m_a + m_b)$ and their relative position is $\vec{R}$. The set of electron variables is given by $\boldsymbol{X} = \{\vec{r}_i, \zeta_i\}$ in terms of the position vector and spin variables of electrons i.

Many-electron states of molecules can be constructed from many-electron atomic states, in an approach known as the valence-bond (VB) method, or from atomic orbitals combined to form molecular orbitals, in the so-called molecular orbital (MO) method [5]. This however is done in a CMN-BF reference frame with a Z-axis oriented along the interatomic (or in general intermolecular) position vector $\vec{R}$ and with axis coordinates (X', Y', Z'). The calculation of scattering cross sections involves integrals over electronic variables, and this requires transformations between the CMN-BF and CMN-SF frames, as shown below. We follow here the treatment in Chapter 5 of [5].

Using the Dirac notation for general electronic states, not necessarily molecular stationary states, labelled by a $\Gamma = k$ index, we write $\Phi_k(\boldsymbol{X}; \vec{R}) = \langle \boldsymbol{X} | \Phi_k(\vec{R}) \rangle$ and work in a linear space of K states, introducing an orthonormal and complete basis set of such states for fixed nuclei; the expansion is

$$\Psi^{(+)}(\vec{R}, \boldsymbol{X}) = \sum_{k=1}^{K} \Phi_k(\boldsymbol{X}; \vec{R})\psi_k(\vec{R})$$

$$\langle \Phi_k(\vec{R}) | \Phi_l(\vec{R}) \rangle = \delta_{kl}$$

$$\sum_{k=1}^{K} |\Phi_k(\vec{R})\rangle\langle\Phi_k(\vec{R})| = \hat{I}.$$

The bracket notation here and in what follows signifies integration and sum over all the electron variables, for fixed nuclei as shown. The last line gives an assumed completeness relation, with $\widehat{I}$ the identity operator. Several choices are possible for the basis set of electronic states, which lead to different electronic representations of potential energies and nuclear amplitudes. These representations are related by matrix transformations which can be conveniently introduced to simplify treatments of molecular spectroscopy and dynamics [14–18].

Introducing a matrix notation with rows and columns shown as boldface symbols, we have that

$$\left|\Psi^{(+)}(\vec{R})\right\rangle = \left|\mathbf{\Phi}(\vec{R})\right\rangle \boldsymbol{\psi}(\vec{R})$$

where the first factor is a $1 \times K$ row of states Φ_k and the second factor is a $K \times 1$ column of amplitudes ψ_k. The Schrodinger equation can now be put in matrix form. Operating on Ψ with the momentum operator $\widehat{\vec{P}_R} = (\hbar/i)\nabla_R$, we find a sum of two terms, and using the completeness relation of the electronic basis in front of the second term, we have

$$\begin{aligned}\frac{\hbar}{i}\nabla_R\left|\Psi(\vec{R})\right\rangle &= \left|\mathbf{\Phi}(\vec{R})\right\rangle \frac{\hbar}{i}\nabla_R\boldsymbol{\psi}(\vec{R}) + \left[\frac{\hbar}{i}\nabla_R\left|\mathbf{\Phi}(\vec{R})\right\rangle\right]\boldsymbol{\psi}(\vec{R}) \\ &= \left|\mathbf{\Phi}(\vec{R})\right\rangle\left[\mathbf{I}\frac{\hbar}{i}\nabla_R + \mathbf{G}(\vec{R})\right]\boldsymbol{\psi}(\vec{R})\end{aligned}$$

$$\mathbf{G}(\vec{R}) = (\hbar/i)\langle\mathbf{\Phi}(\vec{R})|\nabla_R\mathbf{\Phi}(\vec{R})\rangle$$

where $\mathbf{I}$ is the identity matrix and $\mathbf{G}$ is a vector-like quantity with three components each equal to a square $K \times K$ matrix $(\hbar/i)\langle\mathbf{\Phi}(\vec{R})|\partial\mathbf{\Phi}(\vec{R})/\partial\mu\rangle$, $\mu = X, Y, Z$. The symbol $\langle\mathbf{\Phi}(\vec{R})|$ represents a $K \times 1$ column matrix containing the complex conjugate elements Φ_k^*. The elements $\vec{G}_{kl}$ of the momentum-coupling matrix $\mathbf{G}$ are vectors that have the meaning of momentum fluctuations, resulting from the electronic transitions induced by nuclear displacements. This matrix is Hermitian, as can be seen by taking the gradient of the normalization condition for the electronic basis, giving

$$\vec{G}_{kl}(\vec{R}) = (\hbar/i)\langle\Phi_k(\vec{R})|\nabla_R\Phi_l(\vec{R})\rangle = -(\hbar/i)\langle\Phi_l(\vec{R})|\nabla_R\Phi_k(\vec{R})\rangle^* = \vec{G}_{lk}(\vec{R})^*.$$

Its diagonal elements are null when the electronic functions are real-valued, as follows from $\nabla_R\langle\Phi_k(\vec{R})|\Phi_k(\vec{R})\rangle = 0$, but they are not necessarily zero when they are complex-valued.

Since the effect of operating with the gradient is to introduce a square matrix with vector-valued elements between the two factors making Ψ, operating twice with the gradient simply repeats the effect and one ends up with the vector product of two square matrices; the elements of this product matrix are scalar products of two vectors. The original Schrodinger equations takes the matrix form

$$\left\{\frac{1}{2M}\left[\mathbf{I}\frac{\hbar}{i}\nabla_R+\mathbf{G}(\vec{R})\right]^2+\mathbf{H}_R-\mathbf{I}E\right\}\boldsymbol{\psi}(\vec{R})=0$$

$$\mathbf{H}_R=\langle\boldsymbol{\Phi}(\vec{R})|\hat{H}_R|\boldsymbol{\Phi}(\vec{R})\rangle$$

where $\mathbf{H}_R$ is a square $K\times K$ matrix, and the second power of the square bracket signifies a scalar product of vector-valued μ-components, each with a $K\times K$ matrix. This is a set of coupled differential equations that must be solved with the appropriate boundary conditions for the $\boldsymbol{\psi}$ amplitudes, for bound or unbound (decaying or scattering) states.

It is of interest to analyze the vector product of the total momentum appearing in this equation. Expanding it

$$\left[\mathbf{I}\frac{\hbar}{i}\nabla_R+\mathbf{G}(\vec{R})\right]^2\boldsymbol{\psi}(\vec{R})=\left\{-\hbar^2{\nabla_R}^2+\frac{\hbar}{i}\left[\mathbf{G}\cdot\nabla_R+\left(\nabla_R\cdot\mathbf{G}\right)\right]+\mathbf{G}\cdot\mathbf{G}\right\}\boldsymbol{\psi}(\vec{R})$$

we find that the two terms in the square bracket to the right add up to a Hermitian operator but each of them is, separately, non-Hermitian. Hence it is safer to keep them together as a sum so that the matrix elements satisfy Hermiticity symmetries.

A generalization of this treatment, to deal with the interaction of two molecules A and B with internal (vibrational–rotational) degrees of freedom $\boldsymbol{q}_A$ and $\boldsymbol{q}_B$, introduces scattering states $\Psi_\alpha^{(+)}(\boldsymbol{Q},\boldsymbol{X})$ of the system with N atoms, dependent on electronic variables $\boldsymbol{X}$ and nuclear position variables $\boldsymbol{Q}=\{Q_n, n=1 \text{ to } 3N-3\}$ in a CM frame. The set of nuclear variables $\boldsymbol{Q}=(\vec{R},\boldsymbol{q})$ can be further separated into the relative position vector $\vec{R}$ with a reduced mass M for relative motion and other nuclear internal variables $\boldsymbol{q}=(\boldsymbol{q}_A,\boldsymbol{q}_B)$ describing rotational and vibrational motions in internal states $\varphi^A_{\gamma_A}(\boldsymbol{q}_A)$ and $\varphi^B_{\gamma_B}(\boldsymbol{q}_B)$ of A and B.

The Hamiltonian of the system contains the kinetic energy operator $\hat{K}_Q$ containing the nuclear (or atomic ion) position variables ρ_n, each one with a mass m_n, $n=1$ to $3N-3$, in a CM frame. These variables can be mass-scaled to simplify the equations, using instead variables $Q_n=(m_n/M)^{1/2}\rho_n$, so that $\hat{K}_Q=-\hbar^2(\vec{\nabla}_Q)^2/(2M)$, with $\vec{\nabla}_Q$ a multidimensional gradient operator. The time-independent Schrodinger equation for the total wavefunction $\Psi(\boldsymbol{Q},\boldsymbol{X})=\langle\boldsymbol{X}\,|\,\Psi(\boldsymbol{Q})\rangle$ is, in the bracket notation, given by

$$(\hat{K}_Q+\hat{H}_Q-E)\,|\Psi(\boldsymbol{Q})\rangle=0$$

with the electronic Hamiltonian operator $\hat{H}_Q$ for fixed nuclei, and a nuclear kinetic energy operator giving $\hat{K}_Q\,|\Psi\rangle=-(\hbar^2/2M)\partial^2\,|\Psi\rangle\,/\partial\boldsymbol{Q}^2=-(\hbar^2/2M)\partial^2\,|\Psi\rangle\,/\partial\vec{R}^2+\hat{K}_{\boldsymbol{q}}\,|\Psi\rangle$.

Then a scattering state can be expressed as

$$\Psi_{\alpha}^{(+)}(\boldsymbol{Q},\boldsymbol{X}) = \sum_{\Gamma} \psi_{\gamma,\alpha}^{(+)}(\boldsymbol{Q})\Phi_{\Gamma}(\boldsymbol{X};\boldsymbol{Q})$$
$$\approx \Psi_{\alpha}^{(0)}(\boldsymbol{Q},\boldsymbol{X}) + \sum_{\Gamma} \psi_{\gamma,\alpha}^{(sc)}(\vec{R},\boldsymbol{q})\Phi_{\Gamma}^{A,B}(\boldsymbol{X};\boldsymbol{q}),$$

where

$$\psi_{\gamma,\alpha}^{(sc)}(\vec{R},\boldsymbol{q}) = (2\pi)^{-\frac{3}{2}} f_{\gamma\alpha}^{(+)}(\Omega)R^{-1}e^{ik'R}\varphi_{\gamma_A}^{A}(\boldsymbol{q}_A)\varphi_{\gamma_B}^{B}(\boldsymbol{q}_B)$$

and $\Phi_{\Gamma}(\boldsymbol{X};\boldsymbol{Q}) \approx \Phi_{\Gamma_A}^{A}(\boldsymbol{X}_A;\boldsymbol{q}_A)\Phi_{\Gamma_B}^{B}(\boldsymbol{X}_B;\boldsymbol{q}_B)$ as $R \to \infty$. The above developments similarly follow by the substitution of ∇_R with the multidimensional gradient operator $\nabla_{\mathbf{Q}} = \nabla_R + \nabla_{\mathbf{q}}$.

5.2 Electronic Representations

5.2.1 Adiabatic Representation and Momentum Couplings

This and following subsections are based on Chapter 5 in [5] and are again presented for two interacting atoms, for simplicity.

In this representation, the Hamiltonian matrix is diagonal, and we choose electronic states $\Phi_k = \Phi_k^{(a)}$ so that for fixed nuclei,

$$\left\langle \Phi_k^{(a)} \middle| \hat{H}_R \middle| \Phi_l^{(a)} \right\rangle = E_k^{(a)}(\vec{R})\,\delta_{kl}$$

where we have specified the representation with a superscript a. The total scattering wavefunction is now given by

$$\Psi^{(+)}(\vec{R},\boldsymbol{X}) = \sum_{k=1}^{K} \Phi_k^{(a)}(\boldsymbol{X};\vec{R})\,\psi_k^{(a)}(\vec{R})$$

and the nuclear amplitudes satisfy corresponding matrix equations

$$\left\{ \frac{1}{2M}\left[\mathbf{I}\frac{\hbar}{i}\nabla_R + \mathbf{G}^{(a)}(\vec{R})\right]^2 + \mathbf{H}_R^{(a)} - \mathbf{I}E \right\} \psi^{(a)}(\vec{R}) = 0$$

with the momentum and Hamiltonian matrices $\mathbf{G}^{(a)}, \mathbf{H}_R^{(a)}$ calculated in the adiabatic representation. The vector-like matrix $\mathbf{G}^{(a)} = (\hbar/i)\left[\left\langle \Phi_k^{(a)}(\vec{R}) \middle| \nabla_R \Phi_l^{(a)}(\vec{R}) \right\rangle\right]$ contains momentum couplings of adiabatic states and $\mathbf{H}_R^{(a)}$ is diagonal by construction of the basis set. The gradient coupling $\left\langle \Phi_k^{(a)}(\vec{R}) \middle| \nabla_R \Phi_l^{(a)}(\vec{R}) \right\rangle$ can be obtained by taking first derivatives in $\hat{H}_R \left| \Phi_l^{(a)}(\vec{R}) \right\rangle = E_l^{(a)}(\vec{R}) \left| \Phi_l^{(a)}(\vec{R}) \right\rangle$ and projecting on $\left\langle \Phi_k^{(a)}(\vec{R}) \right|$. This gives a well-known relationship [5].

$$\left\langle \Phi_k^{(a)}(\vec{R}) \middle| \nabla_R \Phi_l^{(a)}(\vec{R}) \right\rangle = \left(E_k^{(a)} - E_l^{(a)}\right)^{-1} \left\langle \Phi_k^{(a)}(\vec{R}) \middle| \nabla_R \left(\hat{H}_R - E_l^{(a)}\right)\right] \left| \Phi_l^{(a)}(\vec{R}) \right\rangle$$

which involves only adiabatic potential energies and their gradients.

The adiabatic wavefunctions can be approximated by expanding them in a given basis set of N_D many-electron determinantal functions $D_J(\boldsymbol{X};\vec{R}) = \langle \boldsymbol{X}|D_J(\vec{R})\rangle$, $J = 1$ to N_D, generally dependent on the intermolecular position variables, with the expansion coefficients determined by a variational procedure. We have that

$$\Phi_k^{(a)}(\vec{R}) = \sum_J D_J(\vec{R})c_{Jk}(\vec{R}) = \boldsymbol{D}(\vec{R})\mathbf{c}_k(\vec{R})$$

$$\mathbf{c}_k^\dagger \Delta \mathbf{c}_l = \delta_{kl}$$

where $\boldsymbol{D}$ is a $1 \times N_D$ row matrix and $\mathbf{c}_k(\vec{R})$ is a $N_D \times 1$ column matrix representing the adiabatic functions, with elements dependent on the intermolecular position variables, while $\Delta(\vec{R}) = \langle \boldsymbol{D}|\boldsymbol{D}\rangle$ indicates the overlap matrix of the given basis set functions. The coefficients follow from the diagonalization of the electronic Hamiltonian in the original basis,

$$\mathbf{H}_R \mathbf{c}_k(\vec{R}) = \Delta(\vec{R})\, \mathbf{c}_k(\vec{R}) E_k^{(a)}(\vec{R})$$

and the matrix of the electronic Hamiltonian is diagonal because $H_{kl}^{(a)} = \mathbf{c}_k^\dagger \mathbf{H}_R\, \mathbf{c}_l = E_k^{(a)}\delta_{kl}$. The momentum coupling matrix has vector components

$$\vec{G}_{kl}^{(a)} = \frac{\hbar}{i}\left[\mathbf{c}_k^\dagger \langle D_k(\vec{R})|\nabla_R D_l(\vec{R})\rangle\, \mathbf{c}_l + \mathbf{c}_k^\dagger \Delta(\vec{R})\, (\nabla_R \mathbf{c}_l)\right]$$

where the first term describes the momentum coupling contributed by the displacement of the basis functions, and the second term comes from the changes of coefficients with positions. The gradients of coefficients in the second term can be obtained by differentiation of the secular equation for the coefficients, as

$$\nabla_R \mathbf{c}_l = \left(\mathbf{H}_R - E_l^{(a)}\mathbf{I}\right)^{-1}\left[\nabla_R\left(\mathbf{H}_R - E_l^{(a)}\mathbf{I}\right)\right]\mathbf{c}_l$$

which only involves gradients of adiabatic potential energies.

The equation of motion for the nuclear amplitudes in the adiabatic representation, with a diagonal matrix of adiabatic energy potentials $\mathbf{E}^{(a)}(\vec{R}) = \left[E_k^{(a)}(\vec{R})\delta_{kl}\right]$, is

$$\left\{\frac{1}{2M}\left[\mathbf{I}\frac{\hbar}{i}\nabla_R + \mathbf{G}^{(a)}(\vec{R})\right]^2 + \mathbf{E}^{(a)}(\vec{R}) - \mathbf{I}E\right\}\boldsymbol{\psi}^{(a)}(\vec{R}) = 0$$

which shows that the amplitudes are interacting only through the momentum couplings. Ways to generate nonadiabatic representations have been given in the literature [10, 11, 14, 15].

5.2.2 Nonadiabatic Representations

Nonadiabatic representations can be defined in a variety of ways. A specific choice is the strictly diabatic or P-representation where new electronic states $\left|\Phi_k^{(d)}(\vec{R})\right\rangle$ are constructed from given N_D basis functions such that

$$\left\langle \Phi_k^{(d)} \middle| \nabla_R \Phi_l^{(d)} \right\rangle = \delta_{kl}$$

so that there is no coupling of different electronic states due to displacements of nuclear positions. These new states can be related to the previous adiabatic ones by means of a $N_D \times N_D$ matrix transformation $\mathbf{A}(\vec{R})$. At a large intermolecular distance R_0, where neither electronic nor momentum interactions are present, both representations must coincide. We write in matrix notation, the row equation

$$\left|\boldsymbol{\Phi}^{(d)}(\vec{R})\right\rangle = \left|\boldsymbol{\Phi}^{(a)}(\vec{R})\right\rangle \mathbf{A}(\vec{R};\vec{R}_0)$$

with $\mathbf{A}(\vec{R}_0;\vec{R}_0) = \mathbf{I}$ and the unitarity condition $\mathbf{A}^\dagger = \mathbf{A}^{-1}$. From the condition defining the diabatic basis set, we have the differential equation

$$(\hbar/i)\nabla_R \mathbf{A}(\vec{R};\vec{R}_0) + \mathbf{G}^{(a)}(\vec{R})\mathbf{A}(\vec{R};\vec{R}_0) = 0$$

which can be solved to obtain the transformation matrix from given adiabatic momentum couplings. The total molecular wavefunction in the new basis set is

$$\Psi(\vec{R},\boldsymbol{X}) = \sum_{k=1}^{M} \Phi_k^{(d)}(\boldsymbol{X};\vec{R})\psi_k^{(d)}(\vec{R}) = \langle \boldsymbol{X}|\boldsymbol{\Phi}^{(d)}(\vec{R})\rangle \boldsymbol{\psi}^{(d)}(\vec{R})$$

with nuclear motion amplitudes $\boldsymbol{\psi}^{(d)}(\vec{R}) = \mathbf{A}(\vec{R};\vec{R}_0)^{-1}\, \boldsymbol{\psi}^{(a)}(\vec{R})$. Replacing this in the equation for Ψ,

$$\begin{aligned}
\frac{\hbar}{i}\nabla_R \left|\Psi(\vec{R})\right\rangle &= \left|\boldsymbol{\Phi}^{(a)}(\vec{R})\right\rangle \left[\mathbf{I}\frac{\hbar}{i}\nabla_R + \mathbf{G}^{(a)}(\vec{R})\right] \boldsymbol{\psi}^{(a)}(\vec{R}) \\
&= \left|\boldsymbol{\Phi}^{(a)}(\vec{R})\right\rangle \left[\mathbf{I}\frac{\hbar}{i}\nabla_R \mathbf{A} + \mathbf{G}^{(a)}(\vec{R})\mathbf{A} + \mathbf{A}\cdot \mathbf{I}\frac{\hbar}{i}\nabla_R\right] \boldsymbol{\psi}^{(d)}(\vec{R}) \\
&= \left|\boldsymbol{\Phi}^{(d)}(\vec{R})\right\rangle \frac{\hbar}{i}\nabla_R \boldsymbol{\psi}^{(d)}(\vec{R})
\end{aligned}$$

that gives

$$\left\{ -\frac{\hbar^2}{2M}\nabla_R^2 + \mathbf{H}_R^{(d)} - \mathbf{I}E \right\} \boldsymbol{\psi}^{(d)}(\vec{R}) = 0$$

which shows them coupled through the non-diagonal electronic matrix $\mathbf{H}_R^{(d)} = \mathbf{A}^\dagger \mathbf{H}_R^{(a)} \mathbf{A} = \mathbf{A}^\dagger \mathbf{E}^{(a)} \mathbf{A}$. These couplings are typically smoother that the ones arising from momentum couplings so that the diabatic representation is frequently used to treat electronic rearrangements in molecular spectroscopy and collisions, usually introducing semiclassical approximations for the nuclear amplitudes [18, 19]. The transformation between adiabatic and diabatic representations in a space of N_D dimensions can be generally treated introducing an algebra of skew symmetric matrices [20].

An alternative to the P-diabatic representation is given by the Mulliken–Hush treatment of pseudo-crossing of potential energy surfaces (PESs) that deals with electronic state couplings through transition dipoles in analogy but as alternatives to the above momentum couplings. This treatment is suitable for molecular spectroscopy studies and for localized electron transfer phenomena, including electronic transitions in solvated species [21, 22].

5.2.3 Two-State Case

A solution to the differential equation for **A** at all distances can be easily obtained for a simple case with two real-valued coupled electronic states, $\Phi_j^{(a)}$, $j = 1, 2$, and only one radial variable. Introducing the gradient coupling on the adiabatic representation, $\tau^{(a)}(R) = \left\langle \Phi_1^{(a)} \middle| \partial \Phi_2^{(a)} / \partial R \right\rangle$, the momentum coupling matrix has diagonal elements $G_{jj}^{(a)} = 0$ and off-diagonal ones $G_{12}^{(a)} = G_{21}^{(a)*} = (\hbar/i)\tau^{(a)}$ and by construction $H_{12}^{(a)} = H_{21}^{(a)} = 0, H_{jj}^{(a)}(R) = E_j^{(a)}(R)$.

We introduce a 2×2 transformation $\mathbf{A}(R; R_0)$ for a large R_0, where there are no couplings between states 1 and 2, such that $\mathbf{A}(R_0; R_0) = \boldsymbol{I}$, the identity matrix. When the two states and therefore $\tau^{(a)}$ are real valued, one can choose

$$\boldsymbol{A}(R; R_0) = \begin{bmatrix} \cos(\gamma) & \sin(\gamma) \\ -\sin(\gamma) & \cos(\gamma) \end{bmatrix}.$$

It follows from the differential equation for $\boldsymbol{A}(R; R_0)$ that the angle $\gamma(R; R_0)$ is a function satisfying the differential equation $d\gamma/dR = \tau^{(a)}(R)$ with the boundary condition $\gamma(R_0; R_0) = 0$. Its solution

$$\gamma(R; R_0) = \int_{R_0}^{R} dR\, \tau^{(a)}(R)$$

can be used to construct the Hamiltonian matrix elements in the strictly diabatic representation where $G_{jk}^{(d)} = 0$ for all elements, as

$$H_{11}^{(d)}(R) = E_1^{(a)}(R)\cos^2(\gamma) + E_2^{(a)}(R)\sin^2(\gamma)$$

$$H_{22}^{(d)}(R) = E_1^{(a)}(R)\sin^2(\gamma) + E_2^{(a)}(R)\cos^2(\gamma)$$

$$H_{12}^{(d)}(R) = \left[E_1^{(a)}(R) - E_2^{(a)}(R)\right] \sin(\gamma)\cos(\gamma).$$

These expressions allow us to consider what happens when the adiabatic potentials cross at a point $R_\times < R_0$, with R_0 large and $\tau^{(a)}(R)$ negligible there. Then for large R, one finds $H_{11}^{(d)}(R) = E_1^{(a)}(R)$ and $H_{22}^{(d)}(R) = E_2^{(a)}(R)$. Around the crossing point, $H_{12}^{(d)}(R)$ is small and for even shorter distances, the diabatic elements differ significantly from the adiabatic energies.

We also find that the diagonal diabatic potential energies $H_{jj}^{(d)}(R)$ will cross whenever $\gamma = \pm(n/2 + 1/4)\pi$ and that this can happen more than once on the way to small R if the adiabatic momentum couplings are large. This is not physically meaningful and is a consequence of the definition of the strictly diabatic representation. An alternative more physically meaningful diabatic representation can be introduced by parametrizing the transformation angle to a form $\gamma(R;\ \alpha)$ with the parameter α chosen to minimize the momentum coupling in a new "m" representation [17]. This generates new smooth diabatic potentials $H_{jk}^{(m)}(R)$ and a new smaller momentum coupling $\tau^{(m)}(R;\ \alpha)$, which can be used for molecular spectroscopy and collision treatments in the new representation. Figure 5.6 shows an example constructed from Morse-type potentials.

The new states are related to adiabatic ones through the transformation matrix as

$$\left|\Phi_1^{(d)}(\vec{R})\right\rangle = \left|\Phi_1^{(a)}(\vec{R})\right\rangle \cos(\gamma) - \left|\Phi_2^{(a)}(\vec{R})\right\rangle \sin(\gamma)$$

$$\left|\Phi_2^{(d)}(\vec{R})\right\rangle = \left|\Phi_1^{(a)}(\vec{R})\right\rangle \sin(\gamma) + \left|\Phi_2^{(a)}(\vec{R})\right\rangle \cos(\gamma)$$

and similarly for the $\Phi_j^{(m)}$. This can be generalized when the coupling $\tau^{(a)}(R) = |\,\tau^{(a)}(R)\,| \exp[i\chi(R)]$ is a complex function, writing instead the diabatic states as combinations of the adiabatic basis functions $\left|\Phi_1^{(a)}(\vec{R})\right\rangle \exp[-i\chi(R)/2]$ and $\left|\Phi_2^{(a)}(\vec{R})\right\rangle \exp[i\chi(R)/2]$.

Extensions of the treatment to several variables and more than two electronic states have been presented in the literature [12]. For example, the NaH molecule in its ground and excited $^1\Sigma^+$ states shows multiple pairs of avoided crossings, one for each excited potential energy that breaks into Na(nl) – H(1s) after it pseudo-crosses the $Na^+ + H^-$ attractive potential energy, as seen in Figure 5.7 [23].

5.2.4 Fixed-Nuclei, Adiabatic, and Condon Approximations

The previous description of alternative representations can give accurate results for properties such as molecular spectra and scattering cross sections by introducing a large number N_D of functions in the basis set of many-electron functions $\Phi_k(\boldsymbol{X}; \vec{R})$ and by calculating nuclear amplitudes $\psi_k(\vec{R})$ as solutions of coupled differential equations with scattering boundary conditions, including the electronic state couplings. However, in many cases, it is sufficient to introduce approximations where some of the momentum couplings are neglected.

The fixed-nuclei approximation (also called the Born–Oppenheimer approximation) starts with the adiabatic representation and sets all the momentum couplings $\vec{G}_{kl}^{(a)} = 0$, so that the equation for the nuclear motion in each electronic state k is

$$\left\{ -\frac{\hbar^2}{2M}\nabla_R^2 + E_k^{(a)}(R) - E \right\} \psi_k^{(BO)}(\vec{R}) = 0$$

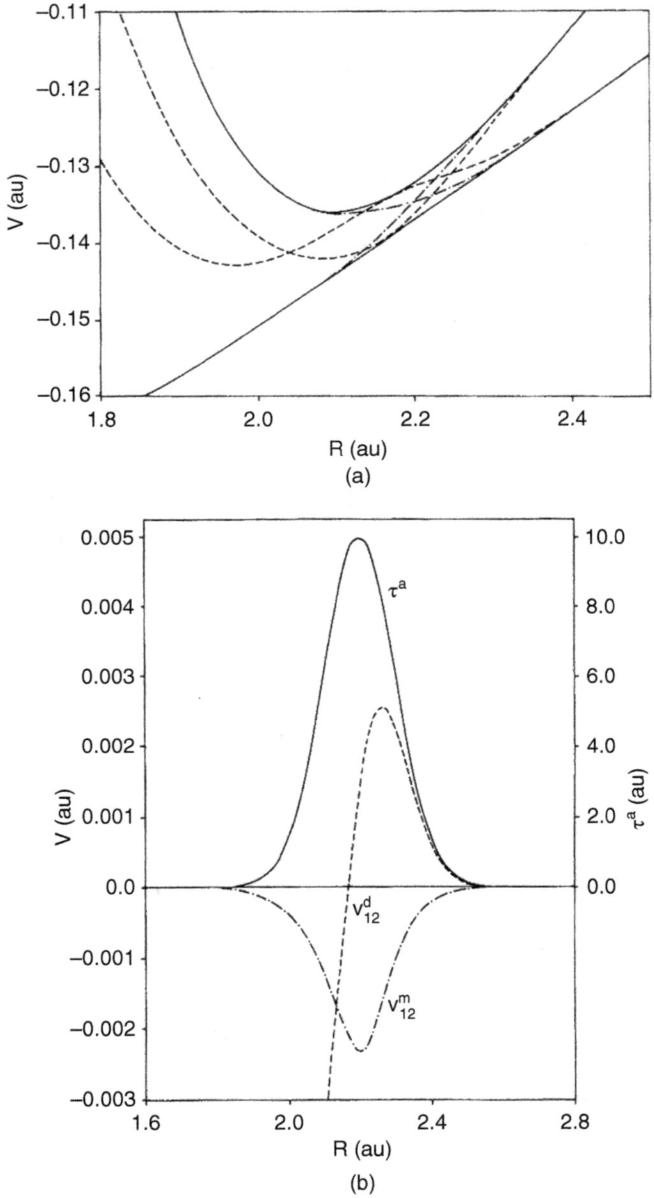

Figure 5.6 (a) Adiabatic (full line), diabatic (dashed line), and minimum coupling (dash-dotted line) potential energies for two states and (b) their couplings in the three electronic representations. Source: [18] / AIP Publishing.

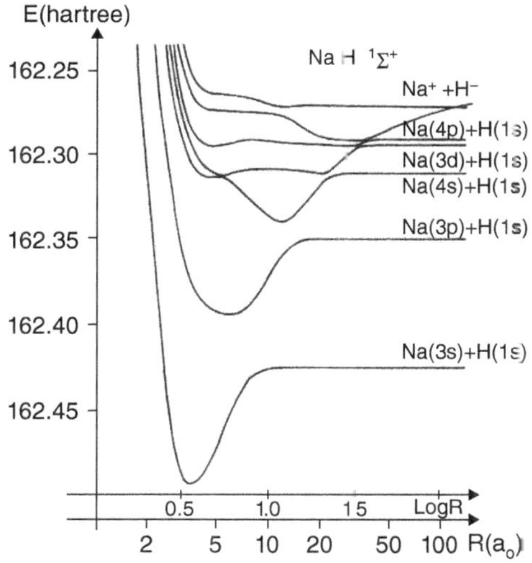

Figure 5.7 NaH potential energies showing multiple pairs of avoided crossings. Source: [23] / John Wiley & Sons.

to be solved with scattering boundary conditions. Transitions among BO scattering states follow by including perturbation couplings $\vec{G}_{kl}(\vec{R}) = (\hbar/i)\langle\Phi_k(\vec{R})|\nabla_R\Phi_l(\vec{R})\rangle$ with a perturbed total state $\Psi^{(BO)}(\vec{R},\boldsymbol{X}) = \sum_k \Phi_k^{(a)}(\boldsymbol{X};\vec{R})\psi_k^{(BO)}(\vec{R})$.

In the adiabatic approximation, one again starts with the adiabatic representation but with $\Phi_k^{(a)}$ a real-valued normalized function for which $\left\langle \Phi_k^{(a)} \middle| \nabla_R \Phi_k^{(a)} \right\rangle = 0$. When this is replaced in the Schrodinger equation, one finds

$$\left\{ -\frac{\hbar^2}{2M}\nabla_R^2 + E_k^{(a)}(R) + W_k^{(a)}(R) - E \right\} \psi_k^{(a)}(\vec{R}) = 0$$

where

$$W_k^{(a)}(R) = -(\hbar^2/2M)\left\langle \Phi_k^{(a)}(\vec{R}) \middle| \nabla_R^2 \Phi_k^{(a)}(\vec{R}) \right\rangle = (\hbar^2/2M)\left\langle \nabla_R \Phi_k^{(a)}(\vec{R}) \middle| \nabla_R \Phi_k^{(a)}(\vec{R}) \right\rangle$$

is a repulsive potential energy addition, with the last equality obtained by partial integration. Collision-induced electronic transitions are again mediated by momentum couplings $\vec{G}_{kl}(\vec{R})$. The total molecular scattering wavefunction is now $\Psi^{(ad)}(\vec{R},\boldsymbol{X}) = \sum_k \Phi_k^{(a)}(\boldsymbol{X};\vec{R})\psi_k^{(a)}(\vec{R})$ and contains nonadiabatic effects.

The differences in scattering cross sections between these two approximations are small but can be detected in high-precision measurements of populations of final electronic states. The differences are not large in most calculations of cross sections, but nonadiabatic couplings are essential in some phenomena due to photo-induced or collision-induced electronic rearrangement.

Another nonadiabatic approximation useful in molecular scattering (as well as in spectroscopy) for molecular colliders involves internal variables $q_n^{(eq)}$, which remain near their equilibrium values in the initial electronic state, is the so-called Condon approximation. Here a basis set of electronic functions $\Phi_k^{(C)}(\boldsymbol{X};\vec{R},\boldsymbol{q}_{eq})$, with $\boldsymbol{q}_{eq} = \left\{ q_n^{(eq)}, n = 1 \text{ to } 3N-6 \right\}$ in a CM frame, is used in an expansion of the molecular states, where all the expansion functions have distances kept constant at the equilibrium values in $\boldsymbol{q}_{eq}$, and the variation of energies with displacements is described expanding the electronic Hamiltonian as a function of displacements around the equilibrium conformations. This generates a vibronic Hamiltonian displaying couplings of electronic and nuclear degrees of freedom in molecular dynamics.

Given the set of internal coordinates $\boldsymbol{q} = \{q_n, n = 1 \text{ to } 3N-6\}$, the total molecular wavefunction is expanded in the basis set of Condon states as

$$\Psi(\vec{R}, \boldsymbol{q}, \boldsymbol{X}) = \sum_{k=1}^{K} \Phi_k^{(C)}(\boldsymbol{X}; \vec{R}, \boldsymbol{q}_{eq})\, \psi_k^{(C)}(\vec{R}, \boldsymbol{q})$$

while the electronic Hamiltonian is expanded around equilibrium values of internal variables in powers of displacements $\boldsymbol{s} = \boldsymbol{q} - \boldsymbol{q}_{eq} = \{s_n\}$, giving

$$\hat{H}_Q = \hat{H}_{eq} - \hat{\boldsymbol{F}}_{eq} \cdot \mathbf{s} + \mathbf{s}^{\dagger} \cdot \hat{\boldsymbol{K}}_{eq} \cdot \mathbf{s}/2$$

in a matrix notation where $\mathbf{s}$ is a column matrix, $\hat{\boldsymbol{F}}_{eq}$ is a row matrix of force operators, and $\hat{\boldsymbol{K}}_{eq}$ is a square matrix of potential energy curvatures, also operators, both dependent on $\vec{R}$. A description of vibronic dynamics can be based on a perturbation calculation of the coefficient functions $\psi_k^{(C)}$, starting with an expansion of the equation for $\Psi(\vec{R}, \boldsymbol{q}, \boldsymbol{X})$ in the basis set $\left\{ \Phi_k^{(C)}(\boldsymbol{X}; \vec{R}, \boldsymbol{q}_{eq}) \right\}$. This leads to equations containing force matrix elements $F_{kl}^{(n)}$ multiplying the q_n displacements and curvature matrix elements $K_{kl}^{(nn')}$ multiplying the products $q_n q_{n'}$.

For a simple case of an atom colliding with a diatomic with bond distance r, this amounts to the expansion

$$\Psi(\vec{R}, \vec{r}, \boldsymbol{X}) = \sum_{k=1}^{K} \Phi_k^{(C)}(\boldsymbol{X}; \vec{R}, \vec{r}_{eq})\, \psi_k^{(C)}(\vec{R}, \vec{r})$$

and calculations of molecular energies as functions of the displacement $s = r - r_{eq}$ doing perturbation theory with the Hamiltonian perturbation operator $(\partial \hat{H}_R/\partial r)_{eq} s$. This leads to vibronic coupling among electronic–nuclear states, with electronic coupling matrix elements $\left\langle \Phi_k^{(C)} \middle| (\partial \hat{H}_R/\partial r)_{eq} s \middle| \Phi_l^{(C)} \right\rangle = -F_{kl}^{(r)} s$ and corresponding coupling of nuclear motion states through $\left\langle \psi_k^{(C)} \middle| s \middle| \psi_l^{(C)} \right\rangle_{\vec{R}}$. The transition force $F_{kl}^{(r)}$ is different from zero only when the product of point group symmetry representations of states k and l and of s contains the totally symmetric representation of the molecular point group [24].

The interaction of two species during scattering also creates internal electric dipoles which change with intermolecular distances and leads to transitions calculated in terms of Franck–Condon overlaps $\left\langle \Phi_k^{(C)} \psi_k^{(C)} \middle| \vec{D} \middle| \Phi_l^{(C)} \psi_l^{(C)} \right\rangle$, where $\vec{D}$ is the molecular electric dipole vector operator. This can be expanded in powers of s, which gives electronic couplings $\left\langle \Phi_k^{(C)} \middle| \vec{D} \middle| \Phi_l^{(C)} \right\rangle_{\vec{R}} \approx \vec{D}_{kl}(\vec{R}, \vec{r}_{eq}) + \vec{P}_{kl}(\vec{R}, \vec{r}_{eq}) s$ with $\vec{D}_{kl}(\vec{R}, \vec{r}_{eq}) = \left\langle \Phi_k^{(C)} \middle| \vec{D} \middle| \Phi_l^{(C)} \right\rangle_{eq}$ and $\vec{P}_{kl}(\vec{R}, \vec{r}_{eq}) = \left\langle \Phi_k^{(C)} \middle| (\partial \vec{D}/\partial r)_{eq} \middle| \Phi_l^{(C)} \right\rangle$.

When the collision-induced excitation occurs at large impact parameters, the vibronic displacement factor $\left\langle \psi_k^{(C)} \middle| s \middle| \psi_l^{(C)} \right\rangle_{\vec{R}}$ is small for $k \neq l$ and one can use amplitudes $\psi_k^{(C)}(\vec{R}, \vec{r}_{eq})$ obtained from a Schrodinger equation with a potential energy function $H_{kk}^{(C)}(R, \vec{r}_{eq})$ to obtain scattering rates. A similar situation arises for collisions of two molecules, with collision-induced intramolecular, and also intermolecular, electric dipoles both of which lead to electronic excitations during scattering.

5.2.5 Approximate Nonadiabatic Coupling for Same-Symmetry States

The calculation of nonadiabatic couplings from electronic structures can be very demanding, especially for many-atom systems. Approximations have been developed to account for the couplings, and in particular, a recent treatment provides an alternative to calculation of the momentum couplings for the same symmetry states [25], that can be implemented in computational work [26]. These papers contain many references to related previous work.

In short, a momentum coupling matrix element $\vec{G}_{kl}(\vec{R}) = (\hbar/i)\langle \Phi_k(\vec{R}) | \nabla_R \Phi_l(\vec{R}) \rangle$ between two states of the same symmetry, along a relative coordinate Q so that $\nabla_R \Phi_l = d\Phi_l/dQ$ can be approximated assuming that $G_{kl}(Q)$ around an avoided crossing at location Q_c can be accurately described in the crossing region with a Lorentzian function of form $F_{kl}^{(Lo)}(q) = \alpha_c / \left[2\left(1 + \alpha_c^2 q^2\right)\right]$, where $q = Q - Q_c$. The displacement coupling $\kappa = \langle \Phi_k(Q) \,|\, d\Phi_l/dQ \rangle$ can be related to the second derivative $d^2\Delta/dQ^2$ of the energy gap function $\Delta(Q) = E_j^{(a)}(Q) - E_k^{(a)}(Q)$ by taking the second derivative of the eigenvalue equation for $\Phi_l(Q)$ [25]. Choosing the avoided crossing location Q_c to be at $(d\Delta/dQ)_c = 0$, it has been found in computational work that values of the Lorentzian $F_{kl}^{(Lo)}(q)$ are well reproduced for several interatomic avoided crossings in selected species (LiF, C_2, NH_3Cl, C_6H_5SH), using the relations [25]:

$$G_{kl}(Q) = (\hbar/i)\langle \Phi_k(Q) | d\Phi_l/dQ \rangle = \alpha_c / \left[2\left(1 + \alpha_c^2 q^2\right)\right]$$

$$\alpha_c = \frac{2\kappa_c}{\Delta_c}, \qquad \kappa_c = \left(\Delta_c \cdot \frac{d^2\Delta_c}{dq^2}\right)^{\frac{1}{2}} /2.$$

These coupling and parametrization, which bypass the need to calculate nonadiabatic couplings, have been used in extensive model calculations with a quantum-classical treatment. In particular, the calculated population decay of photoexcited ethylene [26], with the above parametrization, shows very good agreement compared to calculations with explicit momentum couplings.

5.3 Collisional Coupling of Molecular Electronic States

5.3.1 Quantal Transition Amplitudes and Cross Sections

A scattering state involving several coupled electronic states $\Phi_\Gamma(\mathbf{X};\vec{R})$ can be expanded as

$$\Psi^{(+)}(\vec{R},\mathbf{X}) = \sum_\Gamma \psi_\Gamma^{(+)}(\vec{R})\Phi_\Gamma(\mathbf{X};\vec{R})$$

with expansion coefficients $\psi_\Gamma^{(+)}$ dependent on the internuclear position vector, in a SF reference frame with axes for coordinates (X, Y, Z).

Collecting the expansion coefficient functions in a column $N_{el}\times 1$ matrix $\psi^{(+)}\left(\vec{R}\right)$ and introducing the $N_{el}\times N_{el}$ matrices $\mathbf{H}_R = \left[\left\langle \Phi_\Gamma \middle| \widehat{H}_R \middle| \Phi_{\Gamma'} \right\rangle\right]$ for fixed internuclear positions and the vector-like momentum coupling matrix $\vec{\mathbf{G}}_R = -i\hbar\left[\left\langle \Phi_\Gamma \middle| \vec{\nabla}_R \Phi_{\Gamma'} \right\rangle\right]$ (with brackets indicating integrals over electronic variables), it follows that

$$\left[(2M)^{-1}\left(-i\hbar\nabla_R + \vec{\mathbf{G}}_R\right)^2 + \mathbf{H}_R - \mathrm{E}\right]\psi^{(+)}(\vec{R}) = \mathbf{0}.$$

To proceed further, we must transform to derivatives in a BF reference frame with coordinates (X', Y', Z') because the electronic states are given in this frame. We use here primed quantities when they refer to the *BF* frame. For Z' along the internuclear axis, with $\boldsymbol{X}'$ indicating all the electronic variables (coordinates and spins) in the BF frame, the electronic states are $\Phi'_{\boldsymbol{n}M'_{el}}(\boldsymbol{X}';R)$, with the projection of the electronic angular momentum on the internuclear axis equal to $\hbar M'_{el}$, and $\boldsymbol{n}$ indicating a collection of other quantum numbers. In the adiabatic electronic representation, those states are eigenfunctions of the Hamiltonian $\widehat{H}'_R$ and of the electronic angular momentum operator $\vec{L}'_{el}$ with axial components $\left(\widehat{L}'_{el,Z'}, \widehat{L}'_{el,\pm}\right)$, of quantum numbers (L_{el}, M'_{el}), with corresponding eigenenergy $E_{\boldsymbol{n}\Lambda'}(R)$ for $\Lambda' = |M'_{el}|$, a function of the internuclear distance. The relation between electronic states in SF and BF frames is given by a rotation with Euler angles $(\alpha, \beta, \gamma) = (\Phi, \Theta, 0)$, where (Φ, Θ) are the orientation angles of vector $\vec{R}$, and by a rotation matrix $D^{(L_{el})}_{\Lambda M_{el}}(\Phi,\Theta,0)$ [27] as

$$\Phi_{\boldsymbol{n}LM_{el}}(\mathbf{X};\vec{R}) = \sum_{M'_{el}} D^{(L_{el})}_{M'_{el},M_{el}}(\Phi,\Theta,0)\Phi'_{\boldsymbol{n}M'_{el}}(\boldsymbol{X}';R).$$

The squared term $\left(-i\hbar\nabla_R + \vec{\mathbf{G}}_R\right)^2$ in the previous equation is a nuclear kinetic energy including the effect of electronic momentum couplings. These couplings can be large in the adiabatic electronic representation when the potential energies cross or are close to each other.

The basis set of electronic wavefunctions can be constructed so that the momentum couplings are null, in a diabatic electronic representation (or DER). This is particularly useful to obtain electronic energies, which vary smoothly with internuclear variables.

Here we further choose to work in the DER where $\vec{\mathbf{G}}_R = \mathbf{0}$ so that the kinetic energy operator simplifies to

$$\widehat{K}_R = -\frac{\hbar^2}{2M}\nabla_R^2 = -\frac{\hbar^2}{2M}\frac{1}{R}\frac{\partial^2}{\partial R^2}R + \frac{\widehat{N}^2}{2MR^2}$$

where $\widehat{\vec{N}}$ is the present operator notation for the nuclear angular momentum vector containing derivates $(\partial/\partial\Theta, \partial/\partial\Phi)$. The orientation angle differentiations in the angular momentum operator can be converted from partial differentiation at fixed $\boldsymbol{X}$ to fixed $\boldsymbol{X}'$ in the BF space.

To this end, we use the relations $\theta_i = \theta_i' + \Theta$ and $\phi_i = \phi_i' + \Phi$ for each electron, and the partial derivatives

$$\left(\frac{\partial}{\partial\Theta}\right)_{\boldsymbol{X}} = \left(\frac{\partial}{\partial\Theta}\right)_{\boldsymbol{X}'} + \sum_{el\cdot i}\frac{\partial\theta_i'}{\partial\Theta}\frac{\partial}{\partial\theta_i'}$$

$$\left(\frac{\partial}{\partial\Phi}\right)_{\boldsymbol{X}} = \left(\frac{\partial}{\partial\Phi}\right)_{\boldsymbol{X}'} + \sum_{el\cdot i}\frac{\partial\phi_i'}{\partial\Phi}\frac{\partial}{\partial\phi_i'}$$

to write $\vec{N} = \vec{N}' - \vec{L}_{el}'$, where primed quantities signify differentiation in the BF frame. With the second term to the right identified as the electronic angular momentum, it is found that the first term to the right is the total angular momentum (electronic plus nuclear) expressed in the BF frame. Here, using harmonic oscillator functions $Y_{lm'}(\Omega')$,

$$\widehat{N}'^2 Y_{lm'}(\Omega') = l(l+1)Y_{lm'}(\Omega')$$

and electronic–rotational states in the BF frame can be given as $\Phi_{\lambda'}'(\Omega', \boldsymbol{X}'; R) = Y_{lm'}\left((\Omega')\,\Phi_{nM_{el}}'(\boldsymbol{X}'; R)\right)$, which are eigenstates of the Hamiltonian operator $\hbar^2\widehat{N}'^2/(2M) + \widehat{H}_R'$ with eigenenergies $E_{\lambda'}(R) = \hbar^2 l(l+1)/(2MR^2) + E_{\boldsymbol{n}\Lambda'}(R)$. They form an expansion set coupled by the interatomic angular momentum operator, with matrix elements in axial components $(\pm, Z')$,

$$\left\langle \Phi_{\boldsymbol{n}M_{el}}' \middle| \widehat{L}_{el,Z}' \middle| \Phi_{\boldsymbol{n}'M_{el}'}' \right\rangle = \delta_{\boldsymbol{n},\boldsymbol{n}'}\delta_{M_{el},M_{el}'}\hbar M_{el}$$

$$\left\langle \Phi_{\boldsymbol{n}M_{el}}' \middle| \widehat{L}_{el,\pm}' \middle| \Phi_{\boldsymbol{n}'M_{el}'}' \right\rangle = \delta_{M_{el},M_{el}'\pm 1}\hbar C_{\boldsymbol{nn}'}^{(\pm)}(M_{el})$$

and with the $C_{nn'}^{(\pm)}$ centrifugal (or Coriolis) coefficients obtained from the overlap of the two internal states. Furthermore, the SF frame operator $\hat{N}^2$ can be expressed in terms of spherical components by means of

$$\hat{N}^2 = \hat{N}'^2 - \left(\vec{L}'_{el} \cdot \vec{N}' + \vec{N}' \cdot \vec{L}'_{el}\right) + \vec{L}'^{2}_{el}$$

and evaluated with the basis set in the BF frame.

We next introduce a basis set in the SF-reference frame, $\{\Phi_\lambda(\Omega, \mathbf{X}; R) = Y_{lm}(\Omega)\Phi_{\boldsymbol{n}M_{el}}(\boldsymbol{X}; R)\}$, which describes rotational–electronic states for fixed radial distance R, with the index $\lambda = (lm\boldsymbol{n}L_{el}M_{el})$. It can be re-expressed in terms of BF electronic states by means of the rotational matrices, to calculate matrix elements of the electronic angular momentum operator and then return to the SF-frame. A scattering state $\Psi_\alpha^{(+)}$ starting in electronic state $k = a$ with relative momentum $\hbar\vec{k}$ and outgoing waves is given by the double expansion

$$\Psi_\alpha^{(+)}(\vec{R}, \boldsymbol{X}) = \sum_\lambda \Phi_\lambda(\Omega, \boldsymbol{X}; R)\psi_{\lambda,a}^{(+)}(\vec{R}; \vec{k})/R$$

$$\psi_{\lambda,a}^{(+)}(\vec{R}; \vec{k}) = \sum_{l_0 m_0} Y_{l_0 m_0}(\Omega_k)\psi_{\lambda\lambda_0}^{(+)}(R; k)$$

with $\alpha = (\vec{k}, a)$, $\lambda_0 = (l_0 m_0 a)$, and Ω_k giving the orientation of $\vec{k}$. The radial coefficient functions $\psi_{\lambda\lambda_0}^{(+)}$ satisfy a set of coupled radial differential equations including electronic, radial, and angular momentum couplings.

In what follows, we assume that a transformation has been done from the original adiabatic electronic representation to a diabatic representation where momentum couplings between basis functions are eliminated by construction. The diabatic basis functions are instead coupled by the Coulomb interaction potentials between nuclei and electrons and by the nuclear rotation operators. This gives the set of coupled radial differential equations

$$\sum_{\lambda'} \left\{ \left[E + \frac{\hbar^2}{2M}\left(\frac{d^2}{dR^2} - \frac{l(l+1) - \Lambda^2}{R^2}\right)\right]\delta_{\lambda\lambda'} - \left[H_{\lambda\lambda'}(R) + \frac{\hbar^2 C_{\lambda\lambda'}}{2MR^2}\right]\right\} \times \psi_{\lambda',\lambda_0}^{(+)}(R) = 0$$

where $H_{\lambda\lambda'} = \left\langle \Phi_\lambda \middle| \hat{H}_R \middle| \Phi_{\lambda'} \right\rangle$, with $\hat{H}_R = \hat{K}_{el} + \hat{V}_{el,el} + \hat{V}_{at,el} + \hat{V}_{at,at}$, and the centrifugal (Coriolis) coupling $C_{\lambda\lambda'} = \left\langle \Phi_\lambda \middle| \hat{L}'_{el,\pm} \middle| \Phi_{\lambda'} \right\rangle$ is present when $M_{el} = M'_{el} \pm 1$. The $C_{\lambda\lambda'}$ can be obtained for pairs of electronic states in terms of the matrix elements of the electronic angular momentum given above. The Coriolis coupling acts out to long range and is important at low collision energies but can sometimes be neglected at high collision energies when compared to strong repulsive potential energies [28].

In a matrix notation, we introduce the constant $K^2 = 2ME/\hbar^2$ and the matrix elements

$$W_{\lambda\lambda'}(R) = \frac{2M}{\hbar^2} H_{\lambda\lambda'}(R) + \frac{C_{\lambda\lambda'}}{R^2} + \frac{l(l+1) - \Lambda^2}{R^2} \delta_{\lambda\lambda'}$$

within boldface matrix symbols and write

$$\left[\left(K^2 + \frac{d^2}{dR^2} \right) \mathbf{I} - \mathbf{W}(R) \right] \boldsymbol{\psi}^{(+)}(R) = \mathbf{0}$$

where now the matrices have dimension $N \times N$, with $N = N_{el} N_{nu}$ the dimension of the electronic–nuclear basis set for $\lambda = 1$ to N, and $\mathbf{I}$ is the unit matrix. At large distances, we have that $W_{\lambda\lambda'}(R) \approx (2M/\hbar^2) \left(E_\alpha^{(A)} + E_\beta^{(B)} \right) \delta_{\lambda\lambda'} = W_\lambda^{(0)}, R \to \infty$, from a sum of atomic energies $E_\alpha^{(A)}$ and $E_\beta^{(B)}$.

Cross sections are obtained from the asymptotic form of the radial wavefunctions in the square matrix $\boldsymbol{\psi}^{(+)}$, which gives the $\mathbf{K}$ or $\mathbf{S}$ scattering matrices depending on the chosen asymptotic form. Using the notation in Chapter 4, the open channel to open channel amplitudes $\boldsymbol{\psi}_{oo}$ go asymptotically as

$$\boldsymbol{\psi}_{oo}(R) \approx [\mathbf{I}^{(s)}(R) + \mathbf{I}^{(+)}(R)\mathbf{T}_{oo}]\boldsymbol{k}_o^{1/2}$$

where $\mathbf{T}_{oo} = [t_{\lambda'\lambda}]$ and $\boldsymbol{k}_o = [\delta_{\lambda'\lambda} k_\alpha]$, with indices running over open channels.

Other useful equivalent forms are

$$\boldsymbol{\psi}_{oo}^{(K)}(R) \approx [\mathbf{I}^{(s)}(R) + \mathbf{I}^{(c)}(R)\mathbf{K}_{oo}]\boldsymbol{k}_o^{1/2}$$

and

$$\boldsymbol{\psi}_{oo}^{(S)}(R) \approx [\mathbf{I}^{(-)}(R) - \mathbf{I}^{(+)}(R)\mathbf{S}_{oo}]\boldsymbol{k}_o^{1/2}$$

from which

$$\mathbf{T}_{oo} = \mathbf{K}_{oo}(\mathbf{1}_{oo} - \mathbf{K}_{oo})^{-1} = (2i)^{-1}(\mathbf{S}_{oo} - \mathbf{1}_{oo}).$$

Alternatively, the transition amplitudes follow from the transition integral

$$\mathbf{T}_{\mathbf{oo}} = \int_0^\infty dR\, \boldsymbol{\psi}^{(0)}(R)^* [\mathbf{W}(R) - \mathbf{W}^{(0)}] \boldsymbol{\psi}^{(+)}(R)$$

where $\boldsymbol{\psi}^{(0)}$ is the diagonal matrix of all the final free-motion states and asymptotically $\mathbf{W}(R) \approx \mathbf{W}^{(0)}$.

Cross sections are obtained transforming back from $\psi_{\lambda\lambda_0}^{(+)}(R;k)$ to $\psi_{\lambda,a}^{(+)}(\vec{R};\vec{k})$, and they follow from the scattering flux calculated for $\psi_{\lambda,a}^{(+)}$ through an element of surface $dS = R^2 d\Omega$, as shown before for inelastic scattering, or from the transition matrix elements $T_{\alpha'\alpha}$ obtained from the $T_{\lambda'\lambda}$ giving

$$\frac{d\sigma_{\alpha'\alpha}}{d\Omega} = \frac{k_{\alpha'}}{k_\alpha} \left| f_{\alpha'\alpha}^{(+)}(\Omega) \right|^2$$

with $f_{\alpha'\alpha}^{(+)}(\Omega) = -(2\pi)^2 (M/\hbar^2) T_{\alpha'\alpha} \left(\vec{k}_{\alpha'}, \vec{k}_\alpha \right)$.

5.3.2 Non-Crossing Rule

Matrix elements H_{kl} of the Hamiltonian operator for fixed nuclei, in an original electronic basis set $\{\Phi_k\}$ which may describe adiabatic or nonadiabatic electronic states and is taken to be orthonormalized, also depend on the nuclear variables, and as they vary, the diagonal elements H_{kk} and H_{ll} may cross or show avoided crossings at specific points or lines. These locations are important for the description of molecular spectra and interaction dynamics, showing the structure conformations where light absorption–emission may happen or where atomic rearrangement occurs for electronic transitions between k and l.

General statements can be made about the location and nature of interaction energies at these special points. To simplify, we consider first the case with only one internal degree of freedom for two atoms at the internuclear distance R and where only two states $k = 1, 2$ are involved. Examples are the lowest two $^1\Sigma$ states of LiF or of NaH, where ionic electronic distributions give states of lower energy around their ground equilibrium distance, but covalent distributions give lower energy states at larger distances.

A special point R_c of a crossing or avoided crossing is located by finding where the difference $|H_{11}(R) - H_{22}(R)|$ shows a minimum as R is varied. Writing $R = R_c + x$ and $H_{kl}(R_c) = H_{kl}^{(c)}$, matrix elements can be expanded around $x = 0$ as

$$H_{kl}(x) = H_{kl}^{(c)} + x\,\frac{\partial H_{kl}}{\partial R_c} + \frac{1}{2}x^2\frac{\partial^2 H_{kl}}{\partial R_c{}^2} + \cdots$$

with $H_{11}^{(c)} = H_{22}^{(c)}$ if there is a crossing when working with the original basis set. A non-diagonal Hamiltonian matrix element $H_{12}^{(c)}$ appears if the representation is nonadiabatic. Near the special point, an electronically adiabatic representation is generated by the linear combinations $\Phi_k^{(a)} = \Phi_1 C_{1k} + \Phi_2 C_{2k}$ with the coefficients obtained by diagonalizing the Hamiltonian matrix at each R so that the new elements are $H_{kl}^{(a)}(R) = \delta_{kl}E_k^{(a)}(x)$. These two energies are given by the well-known roots of the second-order equation $det[\boldsymbol{H} - E\boldsymbol{I}] = 0$, where 2×2 matrices are shown in boldface, as

$$E_{1,2}^{(a)}(x) = \frac{1}{2}[H_{11}(x) + H_{22}(x)] \pm \frac{1}{2}\left\{[H_{11}(x) - H_{22}(x)]^2 + 4|H_{12}(x)|^2\right\}^{1/2}$$

and become equal only when the radical $L(x) = [H_{11}(x) - H_{22}(x)]^2 + 4|H_{12}(x)|^2 = 0$ or when both $H_{11}(x) - H_{22}(x) = 0$ and $H_{12}(x) = 0$. These are two equations in x to be satisfied by a single solution root x_0, and there is no real-valued answer, except if in a particular system, one of the two equations is already obeyed for all x, such as when the states 1 and 2 are degenerate and $H_{11}(x) = H_{22}(x)$. For two different diagonal element functions of x and in the absence of accidental vanishing of their coupling H_{12}, the two adiabatic energies $E_{1,2}^{(a)}(x)$ are different for all $R = R_c + x$ around $x = 0$. This is the non-crossing rule for adiabatic potential energies.

In this case, the potentials show an avoided crossing, and the difference between the new potential energies, $\Delta E(x) = E_1^{(a)}(x) - E_2^{(a)}(x)$, has a minimum at the gap located at x_m, where $\partial(\Delta E)/\partial x = 0$, equal to

$$\Delta E_m = \left[(\Delta H_m)^2 + 4|H_{Cm}|^2\right]^{1/2}$$

with $\Delta H_m = H_{11}(x_m) - H_{22}(x_m)$ and $H_{Cm} = H_{12}(x_m)$. Introducing derivatives $f_{kl}^{(c)} = \partial H_{kl}/\partial R_c$ and $g_{kl}^{(c)} = \partial^2 H_{kl}/\partial R_c{}^2$, the energies $E_{1,2}^{(a)}(x)$ can be expanded up to second order in x and are found to be opposing parabolas glancing at x_m.

The momentum coupling between the two adiabatic states is a special case of the previous relation described for approximate adiabatic states, where $\left|\Phi_k^{(a)}(R)\right\rangle = |\boldsymbol{\Phi}(R)\rangle\, \boldsymbol{C}_k(R)$ in matrix notation with $\boldsymbol{C}_k$ a 2×1 column matrix, giving

$$G_{12}^{(a)} = \frac{\hbar}{i}\left[\boldsymbol{C}_1^{\dagger}\langle\boldsymbol{\Phi}(R)|\nabla_R\boldsymbol{\Phi}(R)\rangle\, \boldsymbol{C}_2 + \boldsymbol{C}_1^{\dagger}\langle\boldsymbol{\Phi}(R)|\boldsymbol{\Phi}(R)\rangle\, (\nabla_R\boldsymbol{C}_2)\right]$$

$$\nabla_R\boldsymbol{C}_2 = \left(E_1^{(a)} - E_2^{(a)}\right)^{-1}\left[\nabla_R\left(\boldsymbol{H}_R - E_2^{(a)}\boldsymbol{I}\right)\right]\boldsymbol{C}_2.$$

The denominator in the second line has a minimum value ΔE_c at the avoided crossing, showing that this is where the momentum coupling will be large.

Conditions under which $H_{12}(x) \equiv 0$ can give state potential energy crossings and can be found from considerations of state symmetry. This happens if (i) the two original states have different electronic spin quantum numbers or (ii) the two original states belong to different irreducible representations of their point group symmetry. The groups are $\mathcal{D}_{\infty h}$ for homonuclear diatomics and $C_{\infty v}$ for heteronuclear diatomics [24]. When the symmetry species (or labels) of the diatomic states are different, then $H_{12}(x) \equiv 0$, and if the $E_k(R) = H_{kk}(x)$ are equal at a real-valued x_0 in the original representation, it follows that a crossing of adiabatic potential energies $E_{1,2}^{(a)}$ may occur at a distance $R = R_c + x_0$.

If instead the two states have the same spin quantum numbers and point group symmetry, then in general $H_{12}(x) \neq 0$. In this case, to find the root where adiabatic energies are equal, one must require that the radical $L(x) = L_0 + xL_1 + x^2L_2 = 0$, giving in general a complex-valued root $x_0' + ix_0''$ and a crossing in the complex plane of position variables.

For a complex valued coupling $H_{12}(x) = |H_{12}(x)| \exp[i\alpha(x)]$, the two new adiabatic states around an avoided crossing are [29]

$$\left|\Phi_1^{(a)}(R)\right\rangle = |\Phi_1(R)\rangle \exp\left(-\frac{i\alpha}{2}\right)\cos(\beta) + |\Phi_2(R)\rangle \exp\left(\frac{i\alpha}{2}\right)\sin(\beta)$$

$$\left|\Phi_2^{(a)}(R)\right\rangle = -\,|\Phi_1(R)\rangle \exp\left(-\frac{i\alpha}{2}\right)\sin(\beta) + |\Phi_2(R)\rangle \exp\left(\frac{i\alpha}{2}\right)\cos(\beta)$$

with

$$\tan[2\beta(x)] = 2|H_{12}(x)|/[H_{11}(x) - H_{22}(x)]$$

taking a large value at the avoided crossing, leading there to a rapid change of $\beta(x)$ and rapid switch between states.

This analysis can be extended to cover interactions involving spin–orbit couplings in a Hamiltonian, in which case the relevant point symmetry groups are extended groups with representations generated by the direct product of spin-dependent and space-dependent basis functions. In particular double groups arise when orbital angular momentum states are extended to total angular momentum states to include half-integer spin states, and their representation labels can be used to derive conclusions from molecular symmetry arguments [24].

5.3.3 Quantal Cross Sections of Atom–Atom Collisions

To simplify what follows, we assume here that the electronic angular momenta are small and their contributions to the centrifugal potentials are negligible compared to the nuclear centrifugal term, so that the nuclear angular momentum is conserved, implying that $(l', m') = (l, m)$, and that energy coupling matrix elements are independent of m.

Therefore states are also independent of m and we label them with a quantum number l and the channel index $j = 1$ to N. The radial potential can be written as $W_{\lambda'\lambda} = W^{(l)}_{j'j}$, with matrix elements $W_{jj'}(R) = (2M/\hbar^2)H_{jj'}(R)$ and

$$W^{(l)}_{j'j}(R) = W_{jj'}(R) + \frac{C_{jj'}}{R^2} + \frac{l(l+1) - \Lambda^2}{R^2}\delta_{jj'},$$

with wavefunction matrix elements as $\psi^{(+)}_{\lambda',\lambda}(R) = \psi^{(l)}_{j',j}(R)$ for scattering starting in channel j, and transition matrix elements are given by $t_{\lambda',\lambda} = \delta_{l',l}\delta_{m',m}t^{(l)}_{j'j}(k_j)$, with $t^{(l)}_{j'j} = \left(S^{(l)}_{j'j} - \delta_{j'j}\right)/(2i)$ in terms of the S-matrix elements. The coupled differential equations are now

$$\left[\left(K^2 + \frac{\partial^2}{\partial R^2}\right)\mathbf{I} - \mathbf{W}^{(l)}(R)\right]\boldsymbol{\psi}^{(l)}(R) = \mathbf{0}$$

and their asymptotic solutions provide the scattering amplitudes. For a wave incoming along the z-direction in channel j, where the internal energy of the atom pair is E_j, the scattering amplitude $f^{(+)}_{j'j}(\Omega)$ for waves emerging within solid angle Ω, follows from the asymptotic form

$$k_j^{-1/2}\psi^{(l)}_{j'j}(R) \approx k_j^{-\frac{1}{2}}\sin\left(k_jR - \frac{l\pi}{2}\right)\delta_{j'j} + k_{j'}^{-\frac{1}{2}}\exp\left(k_{j'}R - \frac{l\pi}{2}\right)t^{(l)}_{j'j}$$

$$f^{(+)}_{j'j}(\Omega) = \left(k_jk_{j'}\right)^{-\frac{1}{2}}\sum_l (2l+1)t^{(l)}_{j'j}P_l(\cos\Theta)$$

with $k_j = [2M(E - E_j)/\hbar^2]^{1/2}$, which gives the differential cross sections $d\sigma_{j'j}/d\Omega = \left(k_{j'}/k_j\right)\left|f^{(+)}_{j'j}(\Omega)\right|^2$ and the integral cross sections

$$\sigma_{j'j}(k_j) = \left(4\pi/k_j^2\right)\sum_j (2l+1)\left|t^{(l)}_{j'j}\right|^2.$$

Solutions can be obtained in approximate schemes such as the distorted-wave approximation or accurately by numerical integration. The equations can also be solved in a semiclassical limit valid for short deBroglie wavelengths. An approximate solution to the coupled equations is obtained in the distorted wave approximation (DWA), where equations are first uncoupled to obtain the distorted waves $\psi_{jj}^{(l)}(R)_{DW} \cong \chi_j^{(l)}(R)$, which are solutions to the single-channel equations, neglecting usually small electronic angular momenta,

$$\{K^2 + d^2/dR^2 - [\,W_{jj}(R) + l(l+1)/R^2]\}\chi_j^{(l)}(R) = 0$$

$$\chi_j^{(l)}(R) \approx \sin(k_j R - l\pi/2 + \delta_{l,j})$$

with $\delta_{l,j}$ a phase shift for the channel j and angular momentum quantum number l.

They are used to construct the DWA transition integrals

$$t_{j'j,DW}^{(l)} = \exp\left[i\left(\delta_{l,j'} + \delta_{l,j}\right)\right] \frac{1}{2\left(k_j k_{j'}\right)^{\frac{1}{2}}} \int_0^\infty dR\, \chi_{j'}^{(l)}(R)\left[W_{j'j}(R) - W_{j'j}^{(0)}\right]\chi_j^{(l)}(R)$$

and from these the differential cross sections $\left(d\sigma_{j'j}/d\Omega\right)_{DW} = \left(k_{j'}/k_j\right)\left|f_{j'j,DW}^{(+)}(\Omega)\right|^2$ and the integral cross sections $\sigma_{j'j}(k_j)_{DW} = \left(4\pi/k_j^2\right)\sum_j(2l+1)\left|t_{j'j,DW}^{(l)}\right|^2$.

An accurate solution can alternatively be obtained numerically, integrating the coupled radial equations for the $\psi_{j'j}^{(l)}(R)$ wavefunctions. To simplify, consider a set of two coupled equations for channels 1 and 2 and scattering starting in channel 1,

$$\{K^2 + d^2/dR^2 - [W_{11}(R) + l(l+1)/R^2]\}\psi_{11}^{(l)}(R) = W_{12}(R)\psi_{21}^{(l)}(R)$$

$$\{K^2 + d^2/dR^2 - [W_{22}(R) + l(l+1)/R^2]\}\psi_{21}^{(l)}(R) = W_{21}(R)\psi_{11}^{(l)}(R)$$

which can be solved with asymptotic conditions

$$k_1^{-1/2}\psi_{j1}^{(l)}(R) \approx k_1^{-\frac{1}{2}} \sin\left(k_1 R - \frac{l\pi}{2}\right)\delta_{j1} + k_j^{-\frac{1}{2}} \cos\left(k_j R - \frac{l\pi}{2}\right) K_{j1}^{(l)}$$

where $K_{j1}^{(l)}$ is a real-valued K-matrix element. This can be repeated starting in channel 2 to obtain the 2×2 matrix $\boldsymbol{K}^{(l)} = \left[K_{j'j}^{(l)}\right]$ from which one finds the scattering matrix $\boldsymbol{S}^{(l)} = \left[S_{jk}^{(l)}\right] = (\boldsymbol{I} + \boldsymbol{K}^{(l)})(\boldsymbol{I} - \boldsymbol{K}^{(l)})^{-1}$. This matrix contains a total of eight real numbers which are related by its properties of symmetry and unitarity, reduced however by two real conditions from $S_{12}^{(l)} = S_{21}^{(l)*}$ and three real conditions from $\sum_n S_{jn}^{(l)} S_{j'n}^{(l)*} = \delta_{jj'}$, so that $\boldsymbol{S}^{(l)}$ can be parametrized with only three real parameters, $A_l, \eta_{l,1}, \eta_{l,2}$, for each l in the form

$$S_{jj}^{(l)} = (1 - A_l)^{\frac{1}{2}} \exp(2i\eta_{l,j}), \quad j = 1, 2$$

$$S_{12}^{(l)} = S_{21}^{(l)*} = iA_l \exp[i(\eta_{l,1} + \eta_{l,2})].$$

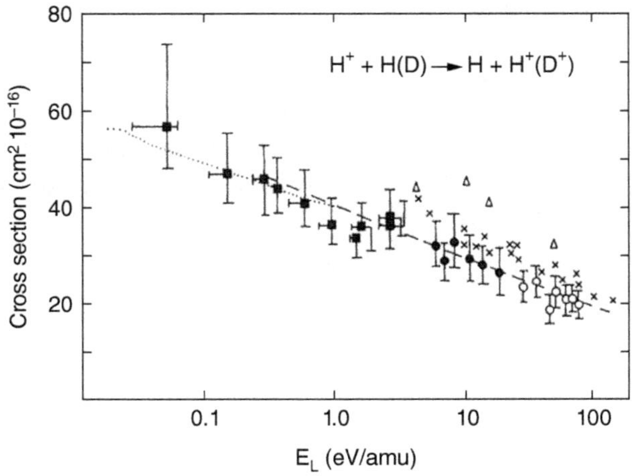

Figure 5.8 Total cross section for electron capture $\sigma_{cap,1s} = \sum_{nl} \sigma_{1s \to nl}$, for the low energy $H^+ + H(1s) \to H(nl) + H^+$. Source: [1] / Oxford University Press.

An example is $H^+ + H(1s)$ leading to electronic excitation to $H^+ + H(nl)$ or to electron capture in $H(nl) + H^+$. The potential energies obtained from electronic states of this one-electron system are shown in Figures 5.2 and 5.3.

Results of several calculations are shown in Figure 5.8 on the total cross section for electron capture $\sigma_{cap,1s} = \sum_{nl} \sigma_{1s \to nl}$ and in Figure 5.9 for the low-energy exchange $H^+ + H(1s) \to H(1s) + H^+$, with the later displaying glory and nuclear exchange symmetry oscillations at very low energies, obtained from coupled channel calculations.

Another example is the collisional electronic excitation $He^+(1s) + Ne(2p^6) \to He^+(1s) + Ne^*(2p^5 3s)$. From the potential energies for the diatomic $HeNe^+$, it is

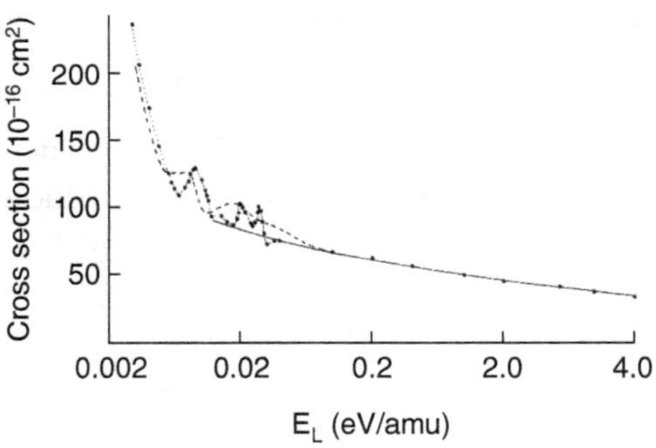

Figure 5.9 Low-energy electron exchange in $H^+ + H(1s) \to H(1s) + H^+$ displaying glory and nuclear exchange symmetry oscillations at very low energies. Source: [1] / Oxford University Press.

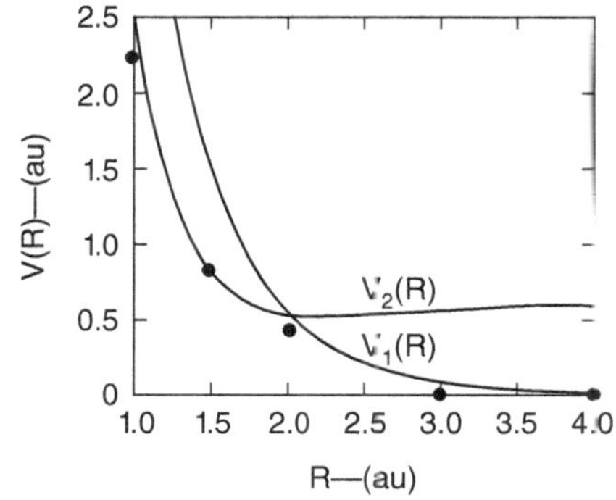

Figure 5.10 Potential energy functions B $^2\Sigma$ and C $^2\Sigma$ and their coupling, for the collisional electronic excitation $He^+(1s) + Ne(2p^6) \rightarrow He^+(1s) + Ne^*(2p^5 3s)$. Source: [30] / American Physical Society.

found that those collision channels involve the B $^2\Sigma$ and C $^2\Sigma$ potentials and their coupling. They have been parametrized as [30]

$$H_{11} = a_1 \exp(-R/a_2)R^{-1}$$

$$H_{22} = (a_1 R^{-1} - a_3)\exp(-R/a_2) + \Delta V$$

$$H_{12} = H_{21} = a_4 \exp(-R/a_5)$$

with the potential energy functions H_{jj} crossing at a distance R_c, as are shown in Figure 5.10. Calculations at a collision energy $E = 70.9$ eV give very similar results for the DWA and for accurate calculations and show that $\left|t_{j'j}^{(l)}(E)\right|^2$ oscillates with l in the range 0–350, in Figure 5.11, and that the differential cross section oscillates with scattering angle, in a plot of $\Theta \sin(\Theta) d\sigma_{21}/d\Omega$ vs $E\Theta$, as shown in Figure 5.12 from calculations and experiments [30].

5.3.4 Short-Wavelength Approximation

When the local deBroglie wavelengths $\lambda_j(R)$ of wavefunctions in all channels j are small compared with relative changes of the potentials, so that $\lambda_j(R) \ll |H_{jj}(R)/(dH_{jj}/dR)|$, it is possible to transform the coupled wave equations which involve very rapidly changing solutions in R, into coupled differential equations for wavefunction amplitudes $A_{ja}^{(\pm)}(R)$ and phases $\alpha_j(R)$, which are slowly varying with R. Let

$$q_j(R) = 2\pi/\lambda_j(R) = (2M/\hbar^2)^{1/2}[E - H_{jj}(R) - \hbar^2[l(l+1) - \Lambda^2]/(2MR^2)]^{1/2}$$

be the local wavenumber in channel j. For a given initial relative kinetic energy $E_{kin} = \hbar^2 k_1{}^2/(2M)$, internal energy E_1 in channel 1, and orbital quantum number $l \gg 1$, the centrifugal potential energy $\hbar^2[l(l+1) - \Lambda^2]/(2MR^2) = L^2/(2MR^2)$, defining the squared angular momentum L^2, can be assumed to be nearly continuous.

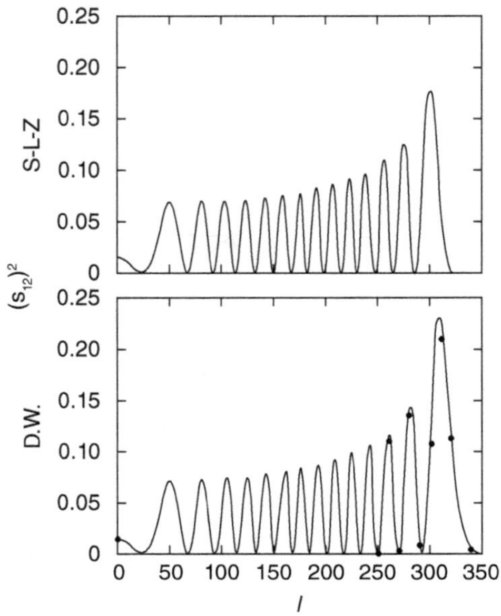

Figure 5.11 Oscillation of cross sections with orbital quantum number for $He^+(1s) + Ne(2p^6) \rightarrow He^+(1s) + Ne^*(2p^5 3s)$. Source: [30] / American Physical Society.

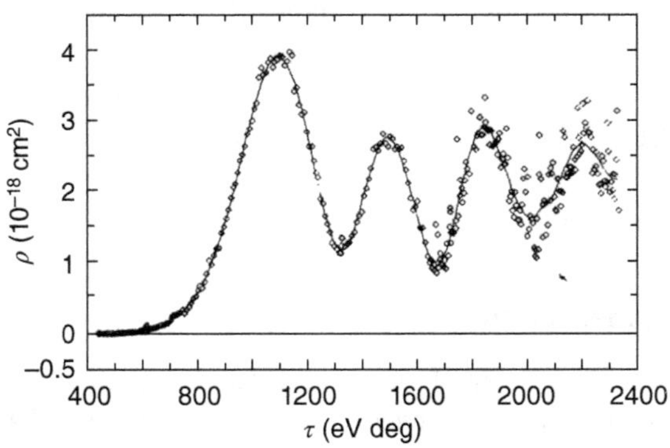

Figure 5.12 Plot of $\Theta \sin(\Theta)\, d\sigma_{21}/d\Omega$ vs $E\Theta$ for $He^+(1s) + Ne(2p^6) \rightarrow He^+(1s) + Ne^*(2p^5 3s)$. Source: [30] / American Physical Society.

Furthermore, the electronic component $\hbar^2\Lambda^2$ is usually small compared to $\hbar^2 l(l+1) \cong \hbar^2(l+1/2)^2$ and can be discarded. Here we consider a case where two states are coupled.

In the absence of a coupling H_{12}, there are no transitions and we can obtain atom-atom scattering trajectories starting at large distances going inward for each potential H_{11} and H_{22}. For given momentum $P_1 = \hbar k_1$ and impact parameter $b = L/P_1$, so that the orbital angular momentum is $L = b\hbar k_1$, we find the trajectories' turning-point radii $R_{j,L}$ solving for R in $E = H_{jj}(R) + L^2/(2MR^2)$. When the coupling is considered, transitions occur between states $j = 1, 2$; we assume here that this occurs with high probability in a region of strong coupling around a distance R_C, of width $2\Delta R_C$, at distances larger than the two turning points for $j = 1, 2$. In this case, transitions may happen in the way in and also out, as scattering progresses. In a quantal description in terms of wavefunctions, this corresponds to a superposition of waves, which can be constructed from a short-wavelength approximation.

The case where the transition region falls between turning points must be treated separately, insofar it corresponds to a coupling between a scattering and a bound state [4]. Another special case is that where the transition region is spread over a wide range of internuclear distances, and it can be treated semiclassically with analytical trajectory solutions for specific potential models [1]. Here we deal for now with the usual situation of a localized transition between two scattering states.

We next consider N coupled radial equations for scattering starting in a channel a, given total energy E and impact parameter b. We define $K^2 = (2M/\hbar^2)E$ and

$$W_{jj'}^{(l)}(R) = \frac{2M}{\hbar^2}H_{jj'}(R) + \frac{C_{jj'}}{R^2} + \frac{l(l+1) - \Lambda^2}{R^2}\delta_{jj'}$$

so that diagonal matrix elements contain the centrifugal potential and find the coupled equations for wavefunctions $\psi_{ja}(R)$,

$$\sum_{j'} \left[(K^2 + d^2/dR^2)\delta_{jj'} - W_{jj'}^{(l)}(R)\right] \psi_{j'a}(R) = 0.$$

When the collision energy is large compared to the excitation energies around the transition region, it is frequently possible to simplify the relative motion introducing a single effective potential energy function $\overline{H}(R)$. This effective potential can be defined for convenience as a weighted function $\overline{H}(R) = w_1(R)H_{11}(R) + w_2(R)H_{22}(R) \approx w_1(\infty)E_1 + w_2(\infty)E_2 = \overline{E}_{int}$ for $R \to \infty$, with the weights $w_j(R)$ chosen after physical considerations, for example simply as $w_1 = w_2 = 1/2$. The corresponding wavenumber $\overline{k}$ for relative motion is given by $E - \overline{E}_{int} = \hbar^2\overline{k}^2/(2M)$. Keeping only atomic angular momenta, with the impact parameter $b \cong (l + 1/2)/k_1$, the diagonal term $W_{jj}^{(l)}(R)$ is replaced by an effective interaction function $\overline{W}(R;b) = (2M/\hbar^2)\overline{H}(R) + (k_1 b)^2/R^2$. A corresponding local wavenumber is $\overline{q}(R) = [K^2 - \overline{W}(R;b)]^{1/2} \approx \overline{k}$ for $R \to \infty$. We introduce scaled local wavenumbers $\overline{q}_j(R) = k_j\overline{q}(R)/\overline{k} \approx k_j$ and write a short-wavelength radial function for $R \geq \overline{R}_L$, the turning point for $\overline{W}(R;b)$, as a sum of terms with incoming (−) and outgoing (+) fluxes in exponential form [4],

$$\psi_{ja}(R) = \phi_{ja}^{(+)}(R) + \phi_{ja}^{(-)}(R)$$

$$k_a^{-1/2}\phi_{ja}^{(\pm)}(R) = \overline{q}_j(R)^{-1/2}A_{ja}^{(\pm)}(R)\exp[\pm i\overline{\alpha}_j(R)]$$

$$\overline{\alpha}_j(R) = \int_{\overline{R}_L}^{R} dR'\,\overline{q}_j(R') + \pi/4.$$

Here $\overline{\alpha}_j(R)$ is a mechanical action function divided by $\hbar$, which can be rewritten using that, for free motion, $q_j^0(R) = \left[k_j^2 - (l+1/2)^2/R^2\right]^{1/2}$ and

$$\int_{R_{j,L}^0}^{R} dR'\, q_j^0(R') = k_j R - \left(l + \frac{1}{2}\right)\pi/2$$

where $R_{j,L}^0 = (l+1/2)/k_j$ is the turning point for free motion. This gives

$$\overline{\alpha}_j(R) = k_j R - l\frac{\pi}{2} + \eta_j, \qquad \eta_j = \int_{\overline{R}_L}^{\infty} dR'\, \overline{q}_j(R') - \int_{R_{j,L}^0}^{\infty} dR'\, q_j^0(R')$$

in which η_j is a short-wavelength phase shift, for the impact parameter $b = (l+1/2)/k_j$. The amplitudes $A_{ja}^{(\pm)}(R)$ must satisfy $\overline{q}_j(R)^{-1/2}\left[A_{ja}^{(+)}(R) + A_{ja}^{(-)}(R)\right] \approx 0$ as $R \to \overline{R}_L$ to insure that the wavefunction is finite at the turning point $\overline{R}_L$, which implies the boundary condition $A_{ja}^{(+)}(R_L) = -A_{ja}^{(-)}(R_L)$.

The short-wavelength functions $\psi_{ja}(R)$ can be combined asymptotically at a large matching distance R_M with coefficients M_{a1} for the two states $a = 1, 2$ to obtain the S-matrix elements $S_{j1}^{(l)}$ by means of

$$\sum_a k_a^{-\frac{1}{2}} \psi_{ja}(R) M_{a1} \approx k_1^{-\frac{1}{2}} \exp\left[-i\left(k_1 R - \frac{l\pi}{2}\right)\right]\delta_{j1} - k_j^{-\frac{1}{2}} \exp\left[i\left(k_1 R - \frac{l\pi}{2}\right)\right] S_{j1}^{(l)}$$

from which $S_{j1}^{(l)}$, for the orbital quantum number l, can be obtained in terms of the asymptotic values $A_{ja}^{(\pm)}(R_M)$ at a large matching distance R_M. To this end, the two sides of the asymptotic equality can be integrated between R_M and $R_M + \Delta R_M$, with $|k_j - k_{j'}|\Delta R_M \ll 1$, and divided by ΔR_M. We use that

$$\begin{aligned}\frac{1}{\Delta R_M}&\int_{R_M}^{R_M+\Delta R_M} dR \exp\left[i\left(k_j - k_{j'}\right)R\right] \\ &= e^{i(k_j-k_{j'})R_M}\left[e^{i(k_j-k_{j'})\Delta R_M} - 1\right] \Big/ \left[i\left(k_j - k_{j'}\right)\Delta R_M\right] \\ &\approx e^{i(k_j-k_{j'})R_M}\end{aligned}$$

and, after cancelling exponentials in the equation for the M_{a1}, we obtain the two relations

$$\sum_a k_j^{-\frac{1}{2}} \exp(-i\eta_j) A_{ja}^{(-)}(R_M) M_{a1} \approx k_1^{-\frac{1}{2}} \exp[i(k_1 - k_j)R_M]\delta_{j1}$$

$$\sum_a k_j^{-\frac{1}{2}} \exp(i\eta_j) A_{ja}^{(+)}(R_M) M_{a1} \approx -k_j^{-\frac{1}{2}} S_{j1}^{(l)}.$$

Solving this for $\left[S_{jj'}^{(l)}\right] = \mathbf{S}^{(l)}$, in a matrix notation with a diagonal $\left[\delta_{jj'}\eta_j\right] = \boldsymbol{\eta}$, and square matrices $\mathbf{M}$ and $\mathbf{A}^{(\pm)}$, which give $\mathbf{M} \approx \mathbf{A}^{(-)}(R_M)^{-1}\exp(i\boldsymbol{\eta})$ from the first line, we find

$$\mathbf{S}^{(l)} \approx -\exp(i\boldsymbol{\eta})\mathbf{A}^{(+)}(R_M)\mathbf{A}^{(-)}(R_M)^{-1}\exp(i\boldsymbol{\eta}).$$

Differential equations for the amplitudes $\phi_{ja}^{(\pm)}$ are obtained replacing $\psi_{ja}(R) = \phi_{ja}^{(+)}(R) + \phi_{ja}^{(-)}(R)$ in its differential equation. Amplitudes with (+) and (−) labels are weakly coupled, and they can be separated within an approximation which neglects the average of a rapidly oscillating term containing $\exp[\mp 2i\overline{\alpha}_j(R)]$, over a small range of R, compared to unity. One finds terms in the equations with first and second derivatives of $A_{ja}^{(\pm)}(R)$. For large and slowly changing $\overline{q}_j(R)$ around R_C so that $|\overline{q}_j \cdot d\overline{q}_j(R)/dR| \ll k_j$, it follows that

$$\frac{d^2\phi_{ja}^{(\pm)}}{dR^2} \approx \left(-\overline{q}_j^{\,2}A_{ja}^{(\pm)} \pm 2i\overline{q}_j\frac{dA_{ja}^{(\pm)}}{dR} + \frac{d^2A_{ja}^{(\pm)}}{dR^2}\right)\left(\frac{k_a}{\overline{q}_j}\right)^{-\frac{1}{2}}\exp(\pm i\overline{\alpha}_j).$$

Using this in the coupled differential equations for $\psi_{ja}(R)$, simplifications can be introduced using that for short wavelengths, $\left|d^2A_{ja}^{(\pm)}/dR^2\right| \ll \left|\overline{q}_j dA_{ja}^{(\pm)}/dR\right|$ and that $|\overline{\alpha}_{j'}(R) - \overline{\alpha}_j(R)| \ll 1$ provided $|(k_j - k_{j'})/\overline{k}| \ll 1$, when the relative value of transferred internal energy is small. The result is the set of coupled differential equations

$$\pm 2i\overline{q}_j(R)\frac{dA_{ja}^{(\pm)}}{dR} = \sum_{j'}\frac{\overline{k}}{\left(k_jk_{j'}\right)^{\frac{1}{2}}}\left[W_{jj}^{(l)}(R)\left(1 - \frac{k_j}{\overline{k}}\right)\delta_{jj'} + W_{jj'}^{(l)}(R)\left(1 - \delta_{jj'}\right)\right]A_{j'a}^{(\pm)}(R)$$

which is of first order in the unknowns and has only slowly varying coefficients. This set of equations is much easier to integrate than the original one for $\psi_{ja}(R)$. In this way, the calculation of S-matrix elements has been reduced to integrals for phases and simpler differential equations for amplitudes. Differentiating the equation for amplitudes gives an expression for $d^2A_{ja}^{(\pm)}/dR^2$, which can be used to verify that the assumption of slow variation is satisfied. The short-wavelength transition probabilities can be obtained now by numerical integration of the equations in the R variable or by changing to the time variable.

Alternatively, a more detailed and accurate eikonal approximation is obtained from the integral form of the transition amplitude, expressing first the radial wavefunction $\psi_{ja}(R)$ as a combination of the eikonal radial functions, and then calculating the integral

$$I_{ja} = \int_0^\infty dR\,\psi_j^{(0)}(R)^*\sum_{j'}\left[W_{jj'}(R) - W_{jj'}^{(0)}\right]\left[\phi_{j'a}^{(+)}(R) + \phi_{j'a}^{(-)}(R)\right]$$

where $\psi_j^{(0)}(R) = \sin(k_jR - l\pi/2)$ is the free motion wavefunction for the final channel j. This has the effect of extracting from the approximate solution a component with the correct final state. These integrals can be combined with coefficients M_{a1} to obtain the transition amplitudes $t_{j1}^{(l)} = \sum_a I_{ja} M_{a1}$, which appear in the differential cross section.

5.4 Semiclassical Description

5.4.1 Eikonal Wavefunctions

A treatment valid for phenomena involving many degrees of freedom and short-wavelength quantum dynamics can be based on a wavefunction constructed from a mechanical action integral and related trajectories [31], providing a semiclassical transition between quantum and classical mechanics. We start our treatment with the time-independent Schrodinger equation for total energy E and a formally exact decomposition for the stationary total (molecular) wavefunction $\Psi(\boldsymbol{Q}, \boldsymbol{X})$ of the system with N atoms, dependent on a set of electronic variables $\boldsymbol{X}$ and a set of nuclear (or atomic ion-core) position variables $\boldsymbol{Q} = \{Q_n, n = 1$ to $3N-3\}$ in a CM frame. As before, the set of nuclear variables can be further separated into the relative position vector $\vec{R}$ with a reduced mass M and other nuclear internal variables $\boldsymbol{q}$ describing rotational and vibrational motions. Here however we prefer to keep all $\boldsymbol{Q} = (\vec{R}, \boldsymbol{q})$ variables together, insofar phenomena like electron transfer may lead to their strong coupling.

Our first step in the description of the collisional dynamics is to write the wavefunction of the system in the eikonal form $\chi \exp(iS/\hbar)$ [32], where $S(\boldsymbol{Q})$ is a real valued phase function to be identified with a mechanical action, generalized to deal with electronic–nuclear motions [1, 18, 33], and $\chi(\boldsymbol{Q}, \boldsymbol{X})$ is an eikonal amplitude. Related momentum variables $\boldsymbol{P} = (\vec{P}, \boldsymbol{p})$ can be defined by the set $\boldsymbol{P} = \partial S/\partial \boldsymbol{Q}$ at a fixed energy E. Following here the treatment in [18, 34], a general wavefunction is written as

$$\Psi(\boldsymbol{Q}, \boldsymbol{X}) = \int d\Lambda \, \mathrm{a}(\Lambda) \chi(\boldsymbol{Q}, \boldsymbol{X}; \Lambda) \exp[iS(\boldsymbol{Q}; \Lambda)/\hbar]$$

in an eikonal representation, given as a sum of eikonal functions times a set of amplitudes $\mathrm{a}(\Lambda)$ with parameters Λ chosen to satisfy boundary conditions, with $S(\boldsymbol{Q})$ a real-valued phase function of the nuclei at relative position $\boldsymbol{Q}$ and with $\chi(\boldsymbol{Q}, \boldsymbol{X}) = \langle \boldsymbol{X} \mid \chi(\boldsymbol{Q}) \rangle$ an electronic–nuclear amplitude. Here as before in this chapter, the bracket notation signifies integration and sum over all the electron variables, for fixed nuclei. The $\chi(\boldsymbol{Q}, \boldsymbol{X})$ is a function of the collection of electronic variables $\boldsymbol{X}$ and a functional of $S(\boldsymbol{Q})$. The action function $S(\boldsymbol{Q})$ satisfies a Hamilton–Jacobi partial differential equation [31], as shown in what follows and must be supplied with boundary conditions for scattering. The differential equation for $\chi(\boldsymbol{Q}, \boldsymbol{X})$ can be derived as an extension of the treatment for inelastic scattering in Chapter 3, incorporating here momentum couplings of electronic states. In what follows, relative and internal atomic variables are considered together, inside $\boldsymbol{Q} = (\vec{R}, \boldsymbol{q})$.

The action function $S(\boldsymbol{Q})$ can be constructed as physically convenient introducing an effective potential energy $\mathcal{V}(\boldsymbol{P}, \boldsymbol{Q}) = \mathrm{tr}\left[\widehat{\omega}(\boldsymbol{P}, \boldsymbol{Q})\widehat{V}(\boldsymbol{Q})\right]$ where $\widehat{V}(\boldsymbol{Q}) = \widehat{V}_{el,el} + \widehat{V}_{at,el} + \widehat{V}_{at,at}$ is an operator over electronic variables, and $\widehat{\omega}(\boldsymbol{P}, \boldsymbol{Q})$ is a chosen density distribution, also an operator over electronic variables, over which the trace is taken. The action function is obtained by solving a time-independent Hamilton–Jacobi equation [31]

$$\mathcal{V}(\boldsymbol{P}, \boldsymbol{Q}) + \frac{1}{2M}\left(\frac{\partial S}{\partial \boldsymbol{Q}}\right)^{\dagger} \cdot \frac{\partial S}{\partial \boldsymbol{Q}} - E = 0$$

with scattering boundary conditions so that asymptotically $S(\boldsymbol{Q}) \approx \vec{P}_{in} \cdot \vec{R} + \int d\boldsymbol{q} \cdot \boldsymbol{p}(\vec{R}, \boldsymbol{q})$, a path integral. Here $\partial S/\partial \boldsymbol{Q}$ is a column matrix of gradients $\partial S/\partial Q_n$, $n = 1$ to $3N - 3$, and $(\partial S/\partial \boldsymbol{Q})^{\dagger}$ is the transpose, a row matrix.

Replacing $|\Psi(\boldsymbol{Q})\rangle = |\chi(\boldsymbol{Q})\rangle \exp[iS(\boldsymbol{Q})/\hbar]$ in the time-independent Schrodinger equation

$$(\widehat{K}_{\mathrm{Q}} + \widehat{H}_{\mathrm{Q}} - E)\,|\Psi(\boldsymbol{Q})\rangle = 0$$

with the electronic–atom core Hamiltonian operator $\widehat{H}_{\mathrm{Q}}$ for fixed nuclei, and with a nuclear kinetic energy operator giving $\widehat{K}_{\mathrm{Q}}\,|\Psi\rangle = -(\hbar^2/2M)\partial^2\,|\Psi\rangle\,/\partial \boldsymbol{Q}^2 = -(\hbar^2/2M)\partial^2\,|\Psi\rangle\,/\partial \vec{R}^2 + \widehat{K}_{\boldsymbol{q}}\,|\Psi\rangle$, where all internal variables in $\{Q_n,\ n = 1$ to $3N - 3\}$ have been mass-scaled and are non-interacting so that $\widehat{K}_{\mathrm{Q}}$ is a sum of second derivatives. Using the notation $\widehat{K}_n = -(\hbar^2/2M)\partial^2/\partial {Q_n}^2$, we have

$$\widehat{K}_n\,|\Psi\rangle = -\frac{\hbar^2}{2M}\left[\frac{\partial^2\,|\chi\rangle}{\partial {Q_n}^2} + 2\frac{\partial\,|\chi\rangle}{\partial Q_n}\frac{i}{\hbar}\frac{\partial S}{\partial Q_n} + |\chi(\boldsymbol{Q})\rangle\left(\frac{i}{\hbar}\frac{\partial^2 S}{\partial {Q_n}^2} - \frac{1}{\hbar^2}\frac{\partial S}{\partial Q_n}.\frac{\partial S}{\partial Q_n}\right)\right] \times \exp[iS(\boldsymbol{Q})/\hbar].$$

We can derive from this an equation for the electronic–nuclear amplitude $|\chi(\boldsymbol{Q})\rangle$. Introducing a set of momentum-like variables $\mathbf{P} = \{P_n,\ n = 1$ to $F\}$, $F = 3N - 3$, by means of $P_n = \partial S/\partial Q_n$, and a notation with upright boldface symbols for $1 \times F$ column matrices such as $\mathbf{P} = [P_n]$, the amplitude satisfies

$$(\widehat{L}_{\mathrm{Q}} + \widehat{H}_{\mathrm{Q}} - E)\,|\chi(\boldsymbol{Q})\rangle = 0$$

$$\widehat{L}_{\mathrm{Q}}(\mathbf{P}) = -\frac{\hbar^2}{2M}\left[\left(\frac{\partial}{\partial \boldsymbol{Q}}\right)^{\dagger} \cdot \frac{\partial}{\partial \boldsymbol{Q}} + 2\frac{i}{\hbar}\mathbf{P}^{\dagger} \cdot \frac{\partial}{\partial \boldsymbol{Q}} + \frac{i}{\hbar}\left(\frac{\partial}{\partial \boldsymbol{Q}}\right)^{\dagger} \cdot \mathbf{P} - \frac{1}{\hbar^2}\mathbf{P}^{\dagger} \cdot \mathbf{P}\right]$$

$$= \frac{1}{2M}\left[\frac{\hbar}{i}\frac{\partial}{\partial \boldsymbol{Q}} + \mathbf{P}\right]^{\dagger}\left[\frac{\hbar}{i}\frac{\partial}{\partial \boldsymbol{Q}} + \mathbf{P}\right].$$

Here $\partial/\partial\boldsymbol{Q}$ and $\mathbf{P}$ are vector-like quantities in column matrices with $F = 3N - 3$ components $\partial/\partial Q_n$ and P_n, and transposed (adjoint) quantities indicate row matrices.

This operator equation can be converted into a set of coupled differential equations in the nuclear variables, introducing an expansion of $\chi(\boldsymbol{Q}, \boldsymbol{X}) = \langle \boldsymbol{X} \mid \chi(\boldsymbol{Q})\rangle$ in a set of orthonormal electronic states for fixed atomic positions $\Phi_k(\boldsymbol{X};\ \boldsymbol{Q})$, $k = 1$ to K, with $\langle \Phi_k \mid \Phi_l \rangle = \delta_{kl}$, giving in the bracket notation $|\chi(\boldsymbol{Q})\rangle = |\boldsymbol{\Phi}(\boldsymbol{Q})\rangle \boldsymbol{\xi}(\boldsymbol{Q})$, where the first factor is a row matrix of kets and the second one is a column matrix of state amplitude functions, $\boldsymbol{\xi}(\boldsymbol{Q}) = [\xi_k(\boldsymbol{Q})]$. In a matrix notation after projections on the electronic states in $\langle \boldsymbol{\Phi}(\boldsymbol{Q})|$, this gives

$$\left\{ \frac{1}{2M} \left[\mathbf{I}\frac{\hbar}{i}\frac{\partial}{\partial \boldsymbol{Q}} + \mathbf{G}(\boldsymbol{Q}) \right]^{\dagger} \left[\mathbf{I}\frac{\hbar}{i}\frac{\partial}{\partial \boldsymbol{Q}} + \mathbf{G}(\boldsymbol{Q}) \right] + \mathbf{H}_{\mathrm{Q}} - \mathbf{I}E \right\} \boldsymbol{\xi}(\boldsymbol{Q}) = \mathbf{0}$$

where $\mathbf{H}_{\mathrm{Q}} = \langle \boldsymbol{\Phi}|\widehat{H}_{\mathrm{Q}}|\boldsymbol{\Phi}\rangle$ is a $K \times K$ matrix and $\mathbf{G}(\boldsymbol{Q})$ is a column $1 \times (3N-3)$ matrix with elements $\mathbf{G}^{(n)}(\boldsymbol{Q}) = (\hbar/i)\langle \boldsymbol{\Phi} \mid \partial\boldsymbol{\Phi}/\partial Q_n\rangle$, each again a $K \times K$ matrix. These equations look similar to the ones for inelastic collisions, but here they contain momentum couplings $\mathbf{G}(\boldsymbol{Q})$ leading to electronically diabatic transitions. The treatment can be developed within the adiabatic or a diabatic electronic representation. The resulting set of coupled differential equations is further considered in the subsection on a trajectories description.

It is further convenient to introduce amplitude phases for each state as $\xi_k(\boldsymbol{Q}) = |\xi_k(\boldsymbol{Q})| \exp[i\,\theta_k(\boldsymbol{Q})]$, with $\theta_k(\boldsymbol{Q}) = \arg[\xi_k(\boldsymbol{Q})]$ obtained from the real and imaginary value parts of $\xi_k = \xi'_k + i\xi''_k$ and $\tan(\theta_k) = \xi''_k/\xi'_k$. The phase factor can be added to the original phase action to obtain a state-dependent new action $S_k(\boldsymbol{Q}) = S(\boldsymbol{Q}) + \hbar\theta_k(\boldsymbol{Q})$. This can be used to define new canonical momenta in

$$\mathbf{P}_k = \partial S_k/\partial \boldsymbol{Q} = \mathbf{P} + \boldsymbol{\pi}_k(\boldsymbol{Q})$$

with $\boldsymbol{\pi}_k(\boldsymbol{Q}) = \hbar\partial\theta_k/\partial\boldsymbol{Q}$ and new state-dependent momentum matrices $\mathbf{P}_k = \left[P_n^{(k)}\right]$. Amplitudes $\xi_k(\mathbf{Q})$ then adapt to each electronic PES $V_{kk}(\boldsymbol{Q}) = \langle \Phi_k|\widehat{H}_{\mathrm{Q}}|\Phi_k\rangle$, by adding the set of momentum components in $\boldsymbol{\pi}_k = \left\{\pi_n^{(k)}\right\}$. Depending on the choice of electronic representation, related potential energy functions V_{kk} may change smoothly or rapidly over distances, so that the effective potential $\mathcal{V}(\boldsymbol{P}, \boldsymbol{Q})$ and the amplitude and phase of $|\chi[\boldsymbol{Q}(t)]\rangle$ change differently over atomic distances. In the diabatic representations, smooth variations imply small changes of eikonal phases and amplitudes, but in the adiabatic representation, the potentials may show avoided crossings, with resulting rapidly changing phases and amplitudes of wavefunctions there.

An explicit effective potential energy and a differential equation can be found for the action function $S(\boldsymbol{Q})$ projecting the previous equation for the amplitude function, on the bra electronic state $\langle \chi(\boldsymbol{Q})|$, with brackets indicating integration over electronic variables for fixed nuclear positions, to obtain $\langle\chi|(\widehat{L}_{\mathrm{Q}}+\widehat{H}_{\mathrm{Q}})|\chi\rangle/\langle\chi|\chi\rangle - E = 0$ and its adjoint equation containing the adjoint operator $\widehat{L}_{\mathrm{Q}}^{\dagger}$. We can extract the real part from $\left\langle\chi\right|\left(\widehat{L}_{\mathrm{Q}}+\widehat{L}_{\mathrm{Q}}^{\dagger}\right)/2+\widehat{H}_{\mathrm{Q}}\left|\chi\right\rangle/\langle\chi|\chi\rangle - E = 0$, leading to a quantum potential $\mathcal{V}(\boldsymbol{P},\boldsymbol{Q}) = V_{qu}(\boldsymbol{Q})$ in

$$V_{qu}(\boldsymbol{Q}) + \frac{1}{2M}\frac{\partial S}{\partial \boldsymbol{Q}}\cdot\frac{\partial S}{\partial \boldsymbol{Q}} - E = 0$$

where

$$V_{qu}(\boldsymbol{Q}) = \overline{V}(\boldsymbol{Q}) + V'(\boldsymbol{Q}) + V''(\boldsymbol{Q})$$

$$\overline{V}(\boldsymbol{Q}) = \langle\chi|\widehat{H}_{\mathrm{Q}}|\chi\rangle/\langle\chi|\chi\rangle$$

$$V'(\boldsymbol{Q}) = \frac{i\hbar}{2M}\left[\left\langle\frac{\partial\chi}{\partial\boldsymbol{Q}}\middle|\chi\right\rangle\cdot\mathbf{P} - \mathbf{P}^{\dagger}\cdot\left\langle\chi\middle|\frac{\partial\chi}{\partial\boldsymbol{Q}}\right\rangle\right]\Big/\langle\chi|\chi\rangle$$

$$V''(\boldsymbol{Q}) = -\frac{\hbar^2}{2M}\frac{1}{2}\left[\left\langle\frac{\partial^2\chi}{\partial\boldsymbol{Q}^2}\middle|\chi\right\rangle + \left\langle\chi\middle|\frac{\partial^2\chi}{\partial\boldsymbol{Q}^2}\right\rangle\right]\Big/\langle\chi|\chi\rangle.$$

Here the first term is an averaged potential $\overline{V}(\boldsymbol{Q})$ for nuclear motions [35], while the second and third terms give quantal corrections. They can be further developed using the expansion $|\chi(\boldsymbol{Q})\rangle = |\boldsymbol{\Phi}(\boldsymbol{Q})\rangle\boldsymbol{\xi}(\boldsymbol{Q})$, giving $\langle\chi\,|\,\chi\rangle = \boldsymbol{\xi}(\boldsymbol{Q})^{\dagger}\boldsymbol{\xi}(\boldsymbol{Q})$, and expressed in terms of square matrices $\mathbf{H}(\mathbf{Q}) = \langle\boldsymbol{\Phi}|\widehat{H}_{\mathrm{Q}}|\boldsymbol{\Phi}\rangle$ and $\mathbf{G}(\boldsymbol{Q}) = (\hbar/i)\langle\boldsymbol{\Phi}(\mathbf{Q})\,|\,\nabla_{Q}\boldsymbol{\Phi}(\mathbf{Q})\rangle$, and of column matrices $\partial\boldsymbol{\xi}/\partial\boldsymbol{Q}$ and $\partial^2\boldsymbol{\xi}/\partial\boldsymbol{Q}^2$ and their transpose.

Alternatively, solutions can be obtained from Hamiltonian equations of motion for the set of momentum $\mathbf{P} = \partial S/\partial\boldsymbol{Q}$ and position $\boldsymbol{Q}$ variables at a fixed energy E, introducing a time variable t, and initial-value conditions for trajectories (or paths) labelled by an index s for each initial set of phase-space points $(\boldsymbol{P}, \boldsymbol{Q})_s$, $s = 1$ to N_{in} at an initial time *tin*. These paths are in principle entangled over time, but their interference frequently lead to cancellation of interference effects on scattering probabilities. A simple treatment can be first developed assuming that each path is independent, and this can be extended to allow for path interference when needed.

The set of momentum and position variables in $(\boldsymbol{P} = \partial S/\partial\boldsymbol{Q}, \boldsymbol{Q})$ can be obtained for a fixed energy E and the general effective potential energy $\mathcal{V}(\boldsymbol{P}, \boldsymbol{Q})$ introducing position and momentum functions of a time variable t, satisfying the Hamiltonian equations

$$\frac{d\boldsymbol{Q}}{dt} = \frac{\partial\mathcal{H}}{\partial\boldsymbol{P}},\quad \frac{d\boldsymbol{P}}{dt} = -\frac{\partial\mathcal{H}}{\partial\boldsymbol{Q}}$$

$$\mathcal{H}(\boldsymbol{P},\boldsymbol{Q}) = \mathbf{P}^{\dagger}\cdot\mathbf{P}/(2M) + \mathcal{V}(\boldsymbol{P},\boldsymbol{Q})$$

for each initial set of conditions $(\boldsymbol{P}, \boldsymbol{Q})_s$, $s = 1$ to N_{in}, at a time *tin*. The set of all these trajectories form a bundle in phase space, constructed to satisfy the initial conditions. The trajectories give the action function over time $S_s(t) = S_s[\boldsymbol{Q}(t)]$, from the line integral

$$S_s(t) = S(\boldsymbol{Q}_s) + \int_{\boldsymbol{Q}_s}^{\boldsymbol{Q}(t)} d\boldsymbol{Q}^\dagger \cdot \mathbf{P}$$

for the total mechanical action $S_s(t)$ along each trajectory.

For each trajectory, the related amplitudes satisfying the differential equation $[\widehat{L}_{\mathrm{Q}} + \widehat{H}_{\mathrm{Q}} - E]\,|\chi(\boldsymbol{Q})\rangle = 0$ can be simplified considering that two terms within $\widehat{L}_{\mathrm{Q}}$ to the left in its definition equation are scalar functions of time and can be transformed away introducing a phase factor into $|\chi(\boldsymbol{Q})\rangle$. Letting $|\chi[\boldsymbol{Q}(t)]\rangle = \left|\overline{\chi}(\boldsymbol{Q};t)\right\rangle \exp[\eta(t)]$ with

$$\begin{aligned}\eta(t) &= -\frac{i}{\hbar}\int_{t_{in}}^{t} dt' \left[-\frac{i\hbar}{2M}\left(\frac{\partial}{\partial \boldsymbol{Q}}\right)^\dagger \cdot \mathbf{P}(t') + \frac{1}{2M}\mathbf{P}^\dagger(t')\cdot\mathbf{P}(t') - E\right]\\ &= -\frac{1}{2M}\int_{t_{in}}^{t} dt' \left(\frac{\partial}{\partial \boldsymbol{Q}}\right)^\dagger \cdot \mathbf{P}(t') - \frac{i}{2\hbar}[S(t) - S(t_{in}) - E(t - t_{in})]\end{aligned}$$

gives the simpler equation

$$\left\{-\frac{\hbar^2}{2M}\left[\left(\frac{\partial}{\partial \boldsymbol{Q}}\right)^\dagger \cdot \frac{\partial}{\partial \boldsymbol{Q}} + 2\frac{i}{\hbar}\mathbf{P}^\dagger(t)\cdot\frac{\partial}{\partial \boldsymbol{Q}}\right] + \widehat{H}_{\mathrm{Q}}\right\}\left|\overline{\chi}(\boldsymbol{Q};t)\right\rangle = 0.$$

The time evolution of the amplitude also derives from the relation $\boldsymbol{P}(t) = M\, d\boldsymbol{Q}/dt$, for each trajectory, which gives a term $-(i\hbar/M)\mathbf{P}^\dagger(t)\cdot\partial\left|\overline{\chi}\right\rangle/\partial\boldsymbol{Q} = -i\hbar d\left|\overline{\chi}\right\rangle/dt$ in its differential equation. Furthermore, for short wavelengths along a trajectory, giving the large momenta in $\boldsymbol{P}(t)$, one finds that $|\partial^2\overline{\chi}/\partial\boldsymbol{Q}^2| \ll |2(i/\hbar)\mathbf{P}^\dagger(t)\cdot\partial\overline{\chi}/\partial\boldsymbol{Q}|$ and the second derivative can be neglected or treated as a perturbation in the differential equation.

To proceed, it is convenient to introduce for each trajectory s an expansion of the corresponding $\overline{\chi}^{(s)}(\boldsymbol{Q},\boldsymbol{X})$ in the orthonormalized set of electronic states for fixed atomic positions $\{\Phi_k(\boldsymbol{X};\ \boldsymbol{Q})\}$, giving in the bracket notation $\left|\overline{\chi}^{(s)}(\boldsymbol{Q})\right\rangle = |\boldsymbol{\Phi}(\boldsymbol{Q})\rangle\,\boldsymbol{\xi}^{(s)}(\boldsymbol{Q})$, where the second factor is a column matrix of state amplitude functions, $\boldsymbol{\xi}^{(s)}(\boldsymbol{Q}) = \left[\xi_k^{(s)}(\boldsymbol{Q})\right]$. Introducing a time-dependent position variable, with $\mathbf{H}_{Q(t)} = \mathbf{H}[\boldsymbol{Q}(t)]$, $\boldsymbol{\xi}^{(s)}[\boldsymbol{Q}(t)] = \mathbf{c}(t) = \left[c_k^{(s)}(t)\right]$, $\boldsymbol{P}/M = d\boldsymbol{Q}/dt$, and $-(i\hbar/M)\mathbf{P}^\dagger(t)\cdot(\partial\boldsymbol{\xi}^{(s)}/\partial\mathbf{Q})[\mathbf{Q}(t)] = -i\hbar d\mathbf{c}/dt$, the previous equation for $\boldsymbol{\xi}(\boldsymbol{Q})$ becomes

$$\left(-i\hbar\frac{d}{dt} + \frac{d\mathbf{Q}^\dagger}{dt}\cdot\mathbf{G}[\boldsymbol{Q}(t)] + \mathbf{H}[\boldsymbol{Q}(t)]\right)\mathbf{c}(t) = \mathbf{0}$$

for each initial set of conditions $(\boldsymbol{P},\ \boldsymbol{Q})_s$, $s = 1$ to N_{in}, at an initial time t_{in}. The coefficient solutions $\mathbf{c}(t)$ provide the eikonal wavefunction amplitude needed in the calculation of cross sections. The original eikonal amplitude in any electronic representation, $|\chi[Q(t)]\rangle = |\Phi[Q(t)]\rangle c(t) \exp[\eta(t)]$, contains a normalization factor within the time integral $\eta(t)$ involving the divergence of trajectories in $(\partial/\partial \mathbf{Q})^{\dagger} \cdot \mathbf{P}$. The equation is similar to the one for inelastic scattering, but here it also contains time-dependent momentum couplings $\mathbf{G}[Q(t)]$ along the trajectories. The effective potential energy $\mathcal{V}(P, Q) = V_{qu}(Q)$ for the atomic motions in the same approximation of short wavelengths gives $|V''(Q)| \ll |V'(Q)|$ and

$$\mathcal{V}(P, Q) \cong \overline{V}(Q) + V'(Q)$$

$$\overline{V}(Q) = \mathbf{c}^{\dagger}(t)\overline{\mathbf{V}}(Q)\mathbf{c}(t)/[\mathbf{c}^{\dagger}(t)\mathbf{c}(t)]$$

$$V'(Q) = (1/2M) \cdot \mathbf{c}^{\dagger}(t)\{\mathbf{P}^{\dagger} \cdot \mathbf{G}[Q(t)] + \mathbf{G}[Q(t)]^{\dagger} \cdot \mathbf{P}\}\mathbf{c}(t)/[\mathbf{c}^{\dagger}(t)\mathbf{c}(t)].$$

This shows, from the definition of G, that $V'(Q)$ is identically null when the electronic wavefunctions are real-valued.

The treatment of the collision dynamics is radically simplified in the diabatic electronic representation described in detail in Section 5.2.2, where, starting with the adiabatic basis set, one transforms it to a diabatic basis set as

$$\left|\Phi^{(d)}(Q)\right\rangle = \left|\Phi^{(a)}(Q)\right\rangle A(Q; Q_0)$$

with the basis transformation satisfying $A(Q_0; Q_0) = \mathbf{I}$ and the unitarity condition $A^{\dagger} = A^{-1}$. This has been chosen so that

$$(\hbar/i)\nabla_{\mathrm{Q}} A(Q; Q_0) + \mathbf{G}^{(a)}(Q)^{\dagger} A(Q; Q_0) = \mathbf{0}$$

with $\nabla_{\mathrm{Q}} = \partial/\partial \mathbf{Q}$ to eliminate the $\mathbf{G}^{(a)}(Q)$ terms in the differential equation for $|\chi(Q)\rangle$ and to transform adiabatic potentials into diabatic ones, with $\mathbf{G}^{(d)}(Q) = \mathbf{0}$. The total wavefunction in the diabatic basis set is

$$\Psi(Q, X) = \Phi^{(d)}(Q, X)\xi^{(d)}(Q)\exp[iS^{(d)}(Q)/\hbar]$$

with $S^{(d)}(Q)$ the action function for the new diabatic potential terms in $V_{qu}^{(d)}(t)$. This simplifies the equation for $\xi^{(d)}(Q)$ as

$$\left\{ \frac{1}{2M}\left(\mathbf{I}\frac{\hbar}{i}\nabla_{\mathrm{Q}}\right)^2 + \frac{1}{M}\mathbf{P}^{\dagger}(t) \cdot \left(\mathbf{I}\frac{\hbar}{i}\nabla_{\mathrm{Q}}\right) + \mathbf{H}_{\mathrm{Q}}^{(d)} \right\} \xi^{(d)}(Q) = \mathbf{0}.$$

Further introducing the trajectory $Q(t)$ after differentiation and $\xi^{(d)}[Q(t)] = \mathbf{c}^{(d)}(t)$, this now satisfies, again for short wavelengths along trajectories,

$$\left(-i\hbar\frac{d}{dt} + \mathbf{H}^{(d)}[Q(t)]\right)\mathbf{c}^{(d)}(t) = \mathbf{0}.$$

The effective potential energy $\mathcal{V}(P, Q) = V_{qu}^{(d)}$, in the self-consistent derivation, gives the quantum potential for trajectories as before, but for the diabatic basis set, it is, for short wavelengths,

$$\mathcal{V}(P, Q) \cong \mathbf{c}^{\dagger}(t)[\overline{\mathbf{V}}(Q) + \mathbf{V}'(Q)]\mathbf{c}(t)/[\mathbf{c}^{\dagger}(t)\mathbf{c}(t)]$$

with matrices $\overline{\boldsymbol{V}} = \langle \boldsymbol{\Phi}^{(d)} | \widehat{H}_Q | \boldsymbol{\Phi}^{(d)} \rangle$ and

$$\boldsymbol{V}'(\boldsymbol{Q}) = \frac{i\hbar}{2M}\left[\left\langle \frac{\partial \boldsymbol{\Phi}^{(d)}}{\partial \mathbf{Q}} \middle| \boldsymbol{\Phi}^{(d)} \right\rangle \cdot \mathbf{P} - \mathbf{P}^{\dagger} \cdot \left\langle \boldsymbol{\Phi}^{(d)} \middle| \frac{\partial \boldsymbol{\Phi}^{(d)}}{\partial \mathbf{Q}} \right\rangle\right] = \mathbf{0}.$$

This shows that $\boldsymbol{V}'(\boldsymbol{Q})$ is by construction identically null for diabatic states.

Reintroducing a bundle of trajectories with labels s and boundary-condition parameters Λ_s and for a general electronic representation with states $|\boldsymbol{\Phi}(\boldsymbol{Q})\rangle$, the general scattering state $|\Psi(\boldsymbol{Q})\rangle$ includes the superposition of trajectory components as

$$|\Psi(\boldsymbol{Q})\rangle = \sum_s \Lambda_s \left|\chi^{(s)}(\boldsymbol{Q})\right\rangle \exp[i\, S_s(\boldsymbol{Q})/\hbar] = |\boldsymbol{\Phi}(\boldsymbol{Q})\rangle\, \mathbf{B}(\boldsymbol{Q})$$

where $\mathbf{B}(\boldsymbol{Q}) = \sum_s \Lambda_s \boldsymbol{\xi}^{(s)}(\boldsymbol{Q}) \exp[i\; S_s(\boldsymbol{Q})/\hbar]$. The trajectory coefficients Λ_s can be obtained from the spatial boundary conditions for scattering. The differential equations in the time variable for the coefficients $\xi_k^{(s)}[\boldsymbol{Q}(t)] = c_k^{(s)}(t)$ must be supplemented with the set of initial conditions $\{\boldsymbol{P}(t_{in}),\, \boldsymbol{Q}(t_{in})\}_s$ for the trajectories, which can be chosen from a quantal distribution of initial values for positions and momenta of the atoms in the molecules, and are provided with an initial $\boldsymbol{c}_{in}^{(s)} = \boldsymbol{c}^{(s)}(t_{in})$ matrix stating the initial electronic state. The time-dependence of the $c_k^{(s)}(t)$ state amplitudes is guided by the phase space points $\{\boldsymbol{P}(t),\, \boldsymbol{Q}(t)\}_s$ starting at each initial condition, obtained from the original action $S_s(\boldsymbol{Q})$, and satisfying the previous equations of motion with the Hamiltonian $\mathcal{H}$. Amplitude functions and trajectories are self-consistently changing over time, as given by their coupled differential equations [18], which include the effect of coherence among trajectories s as they couple, now with the state-dependent momenta in matrices $\mathbf{P}_k^{(s)} = \mathbf{P}^{(s)} + \boldsymbol{\pi}_k^{(s)}(\boldsymbol{Q})$.

The scattering boundary conditions follow from a comparison of the shape of the free-motion scattering state $\Psi_\alpha^{(0)}(\boldsymbol{Q},\boldsymbol{X})$ with the eikonal wavefunction $\Psi_\alpha^{(0)}(\boldsymbol{Q}_s,\boldsymbol{X};\boldsymbol{P}_s)$ originating at time t_{in} in $(\boldsymbol{P},\, \boldsymbol{Q})_s$, for initial channel $\alpha = (\vec{k}_\alpha, \mu, a)$. This is specified by the initial relative intermolecular momentum $\vec{p}_\alpha = \hbar\vec{k}_\alpha$, by a vibrational–rotational function $\varphi_\mu(\boldsymbol{q}) = \varphi_{\mu_A}^{(A)}(\boldsymbol{q}_{\boldsymbol{A}})\varphi_{\mu_B}^{(B)}(\boldsymbol{q}_{\boldsymbol{B}})$ given by a product of internal (vibrational–rotational) states for A and for B, and by an initial electronic state $\Gamma = a$. The free motion states are $\Psi_\alpha^{(0)}(\boldsymbol{Q},\boldsymbol{X}) = \psi_\alpha^{(0)}(\boldsymbol{Q})\Phi_a^{A,B}(\boldsymbol{X};\boldsymbol{Q})$, where $\psi_\alpha^{(0)}(\boldsymbol{Q}) = (2\pi)^{-3/2} e^{i\vec{k}_\alpha.\vec{R}} \varphi_\mu(\boldsymbol{q})$ outside the reach of interaction potentials. In this free motion region, the electronic state Φ_a has energy E_a, $\mathcal{V} = E_a$ and $k_\alpha^2 = 2M(E - E_a)/\hbar^2$. Comparison with the initial eikonal form

$$\Psi_\alpha^{(0)}(\boldsymbol{Q},\boldsymbol{X}) = \boldsymbol{\Phi}_a^{(0)}(\boldsymbol{X};\boldsymbol{Q})\mathbf{B}_\alpha^{(0)}[\boldsymbol{Q}(t_{in})]$$

gives a column matrix

$$\mathbf{B}_\alpha^{(0)}[\boldsymbol{Q}(t_{in})] = [B_{ka}(t_{in})] = \left[\delta_{ka}(2\pi)^{-3/2}\varphi_\mu(\boldsymbol{q})e^{i\vec{k}_\alpha.\vec{R}}\right]$$

so that for a trajectory s starting with internal state $\varphi_\mu(\boldsymbol{q})$ at t_{in}, one finds $\xi_{k\alpha}^{(s)}(\boldsymbol{Q}) = \delta_{k\alpha}(2\pi)^{-3/2}\varphi_\mu(\boldsymbol{q})$ and $S_\alpha(\boldsymbol{Q})/\hbar = \vec{k}_\alpha \cdot \vec{R}$, while $\boldsymbol{\Phi}_a^{(0)}(\boldsymbol{X};\boldsymbol{Q}) = \left[\delta_{ka}\Phi_a^{A,B}(\boldsymbol{X};\boldsymbol{Q})\right]$ is a row matrix containing only the initial electronic state $\Phi_a^{A,B}$.

To initiate a calculation of trajectories for each set of initial phase space points $\{\vec{P}_s = \hbar\vec{k}_\alpha, \vec{R}_s, (\boldsymbol{p},\boldsymbol{q})_s\}$, one must choose, in a CM reference frame, a large relative distance Z_s outside the reach of interaction potentials, impact parameter b_s and azimuthal angle φ_s, and a collection of internal momenta $\boldsymbol{p}_s$ and positions $\boldsymbol{q}_s$ covering the initial vibrational–rotational phase space, with the momentum distribution coming from the Fourier-transformed internal states, $\tilde{\varphi}_\mu(\boldsymbol{p}_s)$, and with the distribution of internal positions in $\varphi_\mu(\boldsymbol{q}_s)$. This collection of positions and momenta generates a bundle $\mathcal{B}_\varphi$ of initial conditions labelled by the index $s \in \mathcal{B}_\varphi$, covering the starting phase-space of the collision dynamics, and this evolves as time proceeds.

The scattering amplitudes can be obtained equating the asymptotic form of the scattering wavefunction at positions $(\vec{R},\boldsymbol{q},\boldsymbol{X})$ to an eikonal wavefunction after collision, at location $\boldsymbol{Q}_s(t) = (\vec{R}_s, \boldsymbol{q}_s)$, $t \to t_{fn}$. In detail, comparing asymptotic forms in

$$\begin{aligned}\Psi_\alpha^{(+)}(\vec{R},\boldsymbol{q},\boldsymbol{X}) &\approx \Psi_\alpha^{(0)}(\vec{R},\boldsymbol{q},\boldsymbol{X}) + \sum_\gamma \psi_{\gamma,\alpha}^{(sc)}(\vec{R},\boldsymbol{q})\Phi_\gamma^{A,B}(\boldsymbol{X};\boldsymbol{q}) \\ &\approx \sum_s \Lambda_s \boldsymbol{\Phi}[\boldsymbol{X};\boldsymbol{Q}_s(t)]\boldsymbol{\xi}_\alpha^{(s)}[\boldsymbol{Q}_s(t)] \exp\left\{iS_\alpha^{(s)}[\boldsymbol{Q}_s(t)]/\hbar\right\}\end{aligned}$$

where $R = R_{fn} \to \infty$ on the first line corresponds with $t \to t_{fn}$ on the second line, we can extract the scattering amplitude $\psi_{\gamma,\alpha}^{(sc)} \approx (2\pi)^{-\frac{3}{2}} f_{\gamma\alpha}^{(+)}(\Omega)R^{-1}e^{ik_\gamma R}\varphi_\nu(\boldsymbol{q})$ into a final channel $\gamma = (\vec{k}_\gamma, \nu, c)$, with ν containing final rovibrational quantum numbers and with c standing for final electronic quantum numbers. This is obtained by integrating over electronic variables in $\left\langle \Phi_\gamma^{A,B} \middle| \Psi_\alpha^{(+)} \right\rangle_{\boldsymbol{q}}$ with $\Psi_\alpha^{(+)}$ from the eikonal sum over trajectories, and further integrating this with a final channel relative momentum and rovibrational distributions $\varphi_\nu(\boldsymbol{q})$ by adding over trajectory positions $\boldsymbol{q}_s$. This can be done selecting trajectories s which emerge at distance R along solid angle Ω and adding amplitudes over variables $\boldsymbol{q}_s$ as they appear in $\varphi_\nu(\boldsymbol{q})$ to relate the eikonal function to the state-to-state scattering amplitude $f_{\gamma\alpha}^{(+)}$. This also provides the final action function $S_\alpha^{(s)}[\boldsymbol{Q}(t_{fn})]/\hbar = k_\gamma R$ at solid angle Ω, evolved from its initial value $\vec{k}_\alpha \cdot \vec{R}$, and adding the bundle amplitudes gives

$$f_{\gamma\alpha}^{(+)}(\Omega)\varphi_\nu[\boldsymbol{q}] = (2\pi)^{\frac{3}{2}}\, R\, B_{\gamma\alpha}[\boldsymbol{Q}(t_{fn})],\ R \to \infty$$

$$B_{\gamma\alpha}[\boldsymbol{Q}(t_{fn})] = \sum_s \Lambda_s \xi_{\gamma\alpha}^{(s)}[\boldsymbol{Q}_s(t_{fn})] \exp\left\{iS_\alpha^{(s)}[\boldsymbol{Q}_s(t_{fn})]/\hbar\right\}$$

where the right-hand side position variables must be located at the final phase space consistent with the total energy E and final wavevector $\vec{k}_\gamma$ with

$k_\gamma^2 = 2M(E - E_\gamma)/\hbar^2$. We can use that $|\varphi_\nu(\boldsymbol{q})|^2$ integrates to unity over internal variables, so that multiplying both sides with $\varphi_\nu(\boldsymbol{q})^*$ and integrating over $\boldsymbol{q}$, we get

$$f_{\gamma\alpha}^{(+)}(\Omega) = (2\pi)^{\frac{3}{2}} R \int d\boldsymbol{q}|\varphi_\nu(\boldsymbol{q})| \exp[-i \arg[\varphi_\nu(\boldsymbol{q})]B_{\gamma\alpha}(\vec{R}, \boldsymbol{q}), \quad R \to \infty$$

where $\vec{R}$ is oriented along Ω. The integral can be done adding integrand values over the bundle in the phase space $\{\boldsymbol{P}(t), \boldsymbol{Q}(t)\}$ at the final time, for relative motion ending at the desired final $\vec{R}$. The numerical value of $B_{\gamma\alpha}$ at the final time is obtained as a sum over matrix elements in $\Lambda_s \boldsymbol{c}_\alpha^{(s)}(t_{fn}) \exp\left[\eta_\alpha^{(s)}(t_{fn})\right]$ for all trajectories s ending at $\{\boldsymbol{P}(t_{fn}), \boldsymbol{Q}(t_{fn})\}$, obtained from the coupled differential equations for $c_{c\alpha}^{(s)}(t)$ and from the scattering boundary conditions. The phase factor $\exp[\eta_\alpha(t_{fn})]$ contains a negative exponential with a divergence of streamlines which compensates for the large factor R. Finally, the scattering amplitude can be numerically extracted asymptotically for $R \to \infty$ by adding over a distribution of N'_{fn} final internal coordinates $\boldsymbol{q}_s$, weighted with the magnitudes $|\varphi_\nu(\boldsymbol{q}_s)|$ of the final internal states, as shown in the above integral.

The differential of state-to-state cross section $d\sigma_{\gamma\alpha}$ into the solid angle increment $d\Omega$ follows as before from the scattering flux as

$$d\sigma_{\gamma\alpha} = (\vec{J}^{(sc)} \cdot \vec{n}/|J^{(0)}|)R^2 d\Omega = (k_\gamma/k_\alpha)\left|f_{\gamma\alpha}^{(+)}(\Omega)\right|^2 d\Omega$$

which is proportional to the element of surface $dS = R^2 d\Omega$. Therefore, an increment $\Delta\sigma_{\gamma\alpha}$ can be obtained from a bundle of trajectories with all $\vec{R}$ relative positions exiting through an element of surface ΔS after a collision, within a scattering solid angle range $\Delta\Omega$.

Alternatively, the transition amplitude $f_{\gamma\alpha}^{(+)}(\Omega)$ can be obtained from a T-matrix expression involving scattering and free final eikonal functions $\Psi[\boldsymbol{X}, \boldsymbol{Q}(t)]$. We have previously written the state-to-state $\alpha \to \gamma$ differential cross section in terms of the transition integrals $T_{\gamma\alpha}$ so that now

$$\frac{d\sigma_{\gamma\alpha}}{d\Omega} = \frac{k_\gamma}{k_\alpha}\left(\frac{2\pi}{\hbar}\right)^4 M^2 |T_{\gamma\alpha}|^2$$

$$T_{\gamma\alpha} = \int d\boldsymbol{Q} \left\langle \chi_\gamma^0(t_{fn}) \middle| \hat{H}_{\boldsymbol{Q}}^{(fn)} \middle| \chi_\alpha(t_{fn}) \right\rangle \exp\left\{ \frac{i}{\hbar} \left[S_\alpha(t_{fn}) - S_\gamma^0(t_{fn})\right] \right\}.$$

Here the superscript 0 denotes a no-interaction state, and $\hat{H}_Q^{(fn)} = \hat{H}_Q - \hat{H}_{Q,fn}^{(0)}$ is the molecule–molecule coupling potential energy in the final collisional arrangement, with unperturbed Hamiltonian $\hat{H}_{Q,fn}^{(0)}$. The states χ_α and χ_γ^0 are given by expansions in the set of electronic states as shown above and by sums over trajectories with coefficient amplitudes in $\mathbf{B}_\gamma^{(0)}[\boldsymbol{Q}(t_{fn})]$ and $\mathbf{B}_\alpha[\boldsymbol{Q}(t_{fn})]$. The transition integrals $T_{\gamma\alpha}$ can be expressed as an average over N_{in} initial conditions, where each initial set $(\boldsymbol{P}, \boldsymbol{Q})_s$ generates an integral $(T_{\gamma\alpha})_s$ and gives the average $\overline{T_{\gamma\alpha}} = \sum_s (T_{\gamma\alpha})_s/N_{in}$. The eikonal action and amplitude functions depend on the initial coordinates and momenta in $(\boldsymbol{P}, \boldsymbol{Q})$ or alternatively on initial impact parameter, orientation angles, internal coordinates and momenta, and initial relative speed vector.

In a first-principles (or ab initio) dynamics treatment, electronic states are generated as needed over time along trajectories, instead of first generating analytical fits to PESs and their couplings for the dynamics. Here a set of orthonormal many-electron states $\Phi_k(\boldsymbol{X};\ \boldsymbol{Q})$, $k = 1$ to K, with $\langle \Phi_k \mid \Phi_l \rangle = \delta_{kl}$, for atomic positions $\boldsymbol{Q}_s(t)$, must be calculated at times t in a trajectory s at short intervals so that not only $\boldsymbol{H}_Q$ but also $-(i/\hbar)d\boldsymbol{Q}^\dagger/dt \cdot \langle \boldsymbol{\Phi} \mid \partial/\partial \boldsymbol{Q} \mid \boldsymbol{\Phi} \rangle = -i\hbar \langle \boldsymbol{\Phi} \mid \partial/\partial t \mid \boldsymbol{\Phi} \rangle$ and $-(\hbar^2/2M)\langle \boldsymbol{\Phi} \mid \partial^2/\partial \boldsymbol{Q}^2 \mid \boldsymbol{\Phi} \rangle$ can be generated along the trajectory. The Laplacian term can be obtained from $\langle \boldsymbol{\Phi} \mid \partial^2/\partial \boldsymbol{Q}^2 \mid \boldsymbol{\Phi} \rangle = -\langle \partial \boldsymbol{\Phi}/\partial \boldsymbol{Q} \mid \partial \boldsymbol{\Phi}/\partial \boldsymbol{Q} \rangle$ for real-valued wavefunctions. These two terms, containing gradients of wavefunctions, are small or null in diabatic electronic representations constructed to remove avoided PES crossings [5], in which case the first-principles (ab initio) treatment of the dynamics is greatly simplified.

5.4.2 Calculation of Nonadiabatic Dynamics for Coupled Electronic States

The previous semiclassical approach, as derived from the eikonal treatment for a general electronically diabatic collision in the phase space of trajectory variables $(\boldsymbol{P},\ \boldsymbol{Q})$, can be made more accurate and efficient for electronic PES transitions mediated by atomic momentum couplings, by introducing approximations to the atomic trajectories. The first stage in calculations of cross sections involves electronic structure calculations of the many-electron system and generation of PESs and their momentum couplings, in the matrices $\mathbf{H}_Q = \langle \boldsymbol{\Phi}(\boldsymbol{Q}) | \widehat{H}_Q | \boldsymbol{\Phi}(\boldsymbol{Q}) \rangle$ and $\mathbf{G}(\boldsymbol{Q}) = (\hbar/i)\langle \boldsymbol{\Phi}(\boldsymbol{Q}) \mid \nabla_Q \boldsymbol{\Phi}(\boldsymbol{Q}) \rangle$. They can be parametrized as functions of position variables, and this is followed by calculation of atomic trajectories over time, for the bundle generated by initial conditions.

Trajectories are governed by an effective potential energy $\mathcal{V}(\boldsymbol{P}, \boldsymbol{Q})$ and in particular by the quantum potential $V_{qu}(\boldsymbol{Q}) = \overline{V}(\boldsymbol{Q}) + V'(\boldsymbol{Q}) + V''(\boldsymbol{Q})$ where the leading term $\overline{V}(\boldsymbol{Q}) = \langle \chi | \widehat{H}_Q | \chi \rangle / \langle \chi | \chi \rangle$ starts at initial time t_{in} with a value given by the initial choice of an electronic state $k = a$ so that $\overline{V}[\boldsymbol{Q}(t_{in})] = \langle \Phi_a(\boldsymbol{Q}) | \widehat{H}_Q | \Phi_a(\boldsymbol{Q}) \rangle / \langle \Phi_a(\boldsymbol{Q}) | \Phi_a(\boldsymbol{Q}) \rangle = V_{aa}(\boldsymbol{Q})$. Initially $V'(\boldsymbol{Q}) + V''(\boldsymbol{Q}) = 0$ at large distances. The generated trajectories $[\boldsymbol{P}(t), \boldsymbol{Q}(t)]_s$ must be self-consistent with electronic transitions that appear in the above quantum potential [18] and are obtained from the coefficients $\mathbf{c}(t)$ in the equation

$$\left(-i\hbar\frac{d}{dt} + \frac{d\mathbf{Q}^\dagger}{dt} \cdot \mathbf{G}[\boldsymbol{Q}(t)] + \mathbf{H}_Q\right)\mathbf{c}(t) = \mathbf{0}$$

where the upright boldface $\mathbf{Q}$ is a $1 \times F$ column matrix of position variables, and $\mathbf{Q}^\dagger$ is a row matrix. The coupled differential equations in time for $[\boldsymbol{P}(t), \boldsymbol{Q}(t)]$ and $\mathbf{c}(t)$ can be solved advancing time in intervals $t_{n-1} \leq t \leq t_n$ from $t_0 = t_{in}$.

The aim of an efficient calculation of cross sections is to obtain accurate results with a manageable number of trajectories in the bundle. This can be done introducing physical approximations to the dynamics in the propagation of the trajectories and state amplitudes. Among choices of approximate procedures, some useful ones are: (**a**) a numerical procedure with self-consistently coupled trajectories and effective potential energy functions [10, 19] and (**b**) stochastic procedures using surface-hopping trajectories [36–38].

In (**a**), the trajectories in an effective potential $\mathcal{V}(\boldsymbol{Q})$ are generated over time while coupled to transition coefficients, for each set of initial conditions forming a bundle of trajectories. Frequently the full $\mathcal{V}$ can be approximated with the averaged potential $\overline{V}(\boldsymbol{Q})$, and trajectories can be separately obtained from it in each nth time interval and used in the equations for $\mathbf{c}^{(s)}(t)$. Results for the phase-space functions $S_s(\boldsymbol{Q})$ and electronic amplitudes $\boldsymbol{\xi}^{(s)}(\boldsymbol{Q})$ at the final time are collected from the final bundle to reconstruct the desired final rovibrational and electronic states. This is straightforward but generally requires propagation of a large number of trajectories, with results obtained from amplitude matrices $\boldsymbol{c}^{(s)}(t_{n-1}) = \boldsymbol{U}(t_n, t_{n-1})$ $\boldsymbol{c}^{(s)}(t_{n-1})$, and averages over all the initial conditions.

Transition amplitudes starting in channel α are obtained from the coefficients $c_{k\alpha}^{(s)}(t)$, which satisfy an equation of motion containing the interaction matrices in $\mathbf{H}_Q$ and $\mathbf{G}(\boldsymbol{Q})$, and are dependent on $\boldsymbol{Q}_s(t)$. This in turn makes the effective potential dependent on the electronic expansion coefficients and requires the simultaneous integration of the differential equations for the coefficients and the trajectory. An average cross section follows as $\overline{(d\sigma_{\lambda\alpha}/d\Omega)} = \sum_s (d\sigma_{\lambda\alpha}/d\Omega)_s/N_{in}$, calculated for a bundle of N_{in} trajectories given by a distribution of initial conditions, as detailed in Section 5.4.1.

In (**b**), involving a stochastic procedure, the time evolution of the system is divided into regions without induced electronic transitions and regions with localized transitions. In regions without transitions, the quantum potential is just the potential energy of the electronic state and trajectories move on the related single PES. In regions of electronic transitions or crossings at time intervals $t_C \pm \Delta t_C/2$, state populations are recalculated from transition probabilities between one electronic state and another, by a set of rules based on statistical and dynamics considerations.

State populations are built over time as given by a master-equation treatment with the state-to-state transition rates obtained from the molecular dynamics, in a stochastic procedure. The state transitions occur in a time interval $t_{C,in} \pm \Delta t_{C,in}$ for molecules in their approach and an interval $t_{C,out} \pm \Delta t_{C,out}$ in their way out. In between crossings, the electronic state remains unchanged. The total change for a single transition from initial to final time can be expressed as $\mathbf{c}^{(s)}(t_{fn}) = \prod_i \mathbf{U}^{(i)}(t_{fn}, t_{in})\boldsymbol{c}^{(s)}(t_{in})$ in terms of a time-evolution matrix $\mathbf{U}(t_{fn}, t_{in})$, which now can be obtained as a product of time-evolution matrices $\mathbf{U}^{(i)}(t, t')$

for each of five intervals i starting at times $t_{in} \le t_{C,in} - \Delta t_{C,in} \le t_{C,in} + \Delta t_{C,in} \le t_{C,out} - \Delta t_{C,out} \le t_{C,out} + \Delta t_{C,out} \le t_{fn}$, ending at a final t_{fn}. The second and fourth transitions are crossings given by matrix factors $\mathbf{U}^{(C,in)}(t, t')$ and $\mathbf{U}^{(C,out)}(t, t')$ which are non-diagonal in the electronic basis set, while the matrices for the other time-evolutions are diagonal, in the absence of transitions. These matrices are two-by-two dimensional if there is only one pair of coupled states. The total event may more generally happen among N_{St} electronic states and may include several such crossings, each involving two electronic states in separate regions, in a sequence. In this case, the time-evolution matrices have dimensions $N_{St} \times N_{St}$ and contain a 2×2 block for each pair transition in each time interval.

Useful prescriptions to construct the needed time-evolution matrices in the coupling regions are briefly described in what follows.

(a) Self-consistent trajectories

An alternative description of the dynamics is obtained approximating the effective or quantal potential with the time-dependent potential energy function

$$\overline{\mathcal{V}}(\boldsymbol{Q}, t) = \mathbf{c}^{\dagger}(t)[\overline{\boldsymbol{V}}(\boldsymbol{Q}) + \boldsymbol{V}'(\boldsymbol{Q})]\, \mathbf{c}(t)/[\mathbf{c}^{\dagger}(t)\mathbf{c}(t)]$$

$$\overline{\boldsymbol{V}}(\boldsymbol{Q}) = \langle \boldsymbol{\Phi} | \hat{H}_{\mathrm{Q}} | \boldsymbol{\Phi} \rangle,\ \boldsymbol{V}'(\boldsymbol{Q}) = \frac{1}{2}\left[\frac{d\mathbf{Q}^{\dagger}}{dt} \cdot \mathbf{G}(\boldsymbol{Q}) + \mathbf{G}^{\dagger}(\boldsymbol{Q}) \cdot \frac{d\mathbf{Q}}{dt}\right]$$

where $\mathbf{G}(\boldsymbol{Q}) = (\hbar/i)\langle \boldsymbol{\Phi} \,|\, \nabla_{\mathrm{Q}} \boldsymbol{\Phi} \rangle$, given by matrix elements over time-evolving electronic states, for each initial electronic state Φ_a and initial phase-space point set $\{\boldsymbol{P}_s, \boldsymbol{Q}_s, s = 1 \text{ to } N_{in}\}$, to generate trajectories $\boldsymbol{P}_s(t)$, $\boldsymbol{Q}_s(t)$ [18, 39]. This must be done while at the same time the coefficients $c_{k\alpha}^{(s)}(t)$ are generated solving the equation of motion for $\mathbf{c}(t)$, which contains the state interaction matrix $\mathbf{H}[\boldsymbol{Q}(t)]$ and momentum couplings $\mathbf{G}[\boldsymbol{Q}(t)]$, evolving from the initial time t_{in}. To allow for self-consistency, a useful procedure starts by generating relaxing coefficients $\mathbf{c}(t)$ in each interval $t_{n-1} \le t \le t_n$ for fixed trajectory values, followed by driving effects from perturbative corrections due to time-evolving trajectory variables. Cross sections $(d\sigma_{\lambda\alpha}/d\Omega)_s$ for each initial condition are obtained from coefficients in $\mathbf{c}_{\alpha}^{(s)}(t_{fn}) = \left[c_{k\alpha}^{(s)}(t_{fn})\right]$ at the final time after collision, while an average cross section follows as $\overline{(d\sigma_{\lambda\alpha}/d\Omega)} = \sum_s (d\sigma_{\lambda\alpha}/d\Omega)_s / N_{in}$, calculated for a bundle of trajectories given by the distribution of initial conditions, as detailed in Section 5.4.1.

Depending on what electronic representation (adiabatic or diabatic) is being used, momentum couplings $\mathbf{G}(\boldsymbol{Q})$ of electronic states may be large or small. The term $\boldsymbol{V}'(\boldsymbol{Q})$ in the effective potential is needed in the dynamics within the adiabatic representation. The additional potential energy matrix $\boldsymbol{V}''$ and $\nabla_{\mathrm{Q}}{}^{2}\boldsymbol{\xi}(\boldsymbol{Q})$ are expected to be small in a diabatic representation and negligible. The calculation of cross sections from self-consistent trajectories is a computationally straightforward task, but individual trajectories may be unphysical, and meaningful results usually require taking an average over a large number N_{in} of generated trajectories.

(b) Surface-hopping trajectories

The number of bundle trajectories needed for accurate cross section calculations can be reduced using statistical considerations to constrain allowed state-to-state crossings to physical events with large probabilities. Within the electronically adiabatic representation, nuclei can be assumed to move classically on a single electronically adiabatic PES $V_{kk}[\boldsymbol{Q}(t)]$ until a nonadiabatic coupling induces a localized transfer to another surface $V_{ll}[\boldsymbol{Q}(t)]$ within an interval $t_C \pm \Delta t_C/2$. The momentum-coupling transition energy

$$H'_{\boldsymbol{kl}} = -\frac{i}{\hbar}\frac{d\boldsymbol{Q}^{\dagger}}{dt}\cdot\left\langle \Phi_k \left| \frac{\partial}{\partial \boldsymbol{Q}} \right| \Phi_l \right\rangle = -i\hbar\left\langle \Phi_k \left| \frac{\partial}{\partial t} \right| \Phi_l \right\rangle$$

must be obtained in the region of the PESs crossing, where $H_{kk} = H_{ll}$, or at a pseudo-crossing where they approach and $\partial H_{kk}/\partial \boldsymbol{Q} = \partial H_{ll}/\partial \boldsymbol{Q}$. This gives an equation $\mathcal{S}(\boldsymbol{Q}) = 0$ with a multidimensional curvilinear solution $\boldsymbol{Q}_C$ linking the position variables along a PES seam where the potentials cross or pseudo-cross, with curvilinear variables (s, ζ) along and perpendicular to the seam.

In the surface hopping procedure [37], it is assumed that a transition $k \to l$ occurs only if and when a trajectory crosses the seam, at a time interval $t_C \pm \Delta t_C/2$ around $\boldsymbol{Q}(t_C) = \boldsymbol{Q}_C$. The momentum coupling is needed only at the crossing time t_C and involves only the velocity matrix $(d\boldsymbol{Q}/dt)_C$. This can be obtained as $(\Delta \boldsymbol{Q}/\Delta\zeta)\Delta\zeta/\Delta t$ from two parallel seam curves separated by a $\Delta\zeta$ with a slope $(\Delta \boldsymbol{Q}/\Delta\zeta)$, which is geometrically perpendicular to the direction of the seam, and from the ratio $\Delta\zeta/\Delta t$.

At such a time, the trajectory with population $P_k(t)$, $t \le t_C - \Delta t_C/2$ is promoted in a surface hopping to the new surface with population $P_l(t)$, $t \ge t_C + \Delta t_C/2$. During the transition interval Δt_C, these populations are related by a state-to-state transition probability $K^{(C)}_{k\to l}$ obtained from the differential equation for $\boldsymbol{c}(t)$ or are connected from a time-evolution 2×2 matrix as $\boldsymbol{c}_l(t + \Delta t/2) = \mathbf{U}^{(C)}(t + \Delta t/2, t - \Delta t/2)\boldsymbol{c}_k(t - \Delta t/2)$ in the crossing interval. The state-to-state transition probabilities $K_{k\to l}$ can frequently be approximated by the analytical values given by the Landau–Zener or the Rosen–Zener models (described in what follows) for crossing or pseudo-crossing situations, respectively.

Alternatively, the transition probabilities can be generated by a Monte Carlo stochastic procedure [37] developed to simplify the treatment, assuming that transitions involve the fewest needed switches at each crossing or pseudo-crossing C. This is done from $P_k(t) = |c_k(t)|^2$ and its time-derivative, with rate terms obtained from the equation for coefficients, leading to

$$\frac{dP_k}{dt} = \sum_{l\neq k}\frac{2}{\hbar}Im\left[\left(c_k^* H_{kl} c_l\right) + \left(c_k^* H'_{kl} c_l\right)\right] = \sum_{l\neq k} b_{k\leftarrow l}(t)$$

The state-to-state transition rates $b_{k\leftarrow l}$ can be used in a criterion based on the "fewest switches" needed to conserve the number N of trajectories going through C. For the special case of a single transition $1 \leftrightarrow 2$ with $P_1 + P_2 = 1$ and $P_k(t_C + \Delta t/2) - P_k(t_C - \Delta t/2) \cong \Delta t.\ (dP_k/dt)_C = \Delta t.\ b_{k \leftarrow l}(t_C)$, $k, l = 1, 2$, the criterion involves generation of a random number $0 \leq \zeta \leq 1$, and choosing for each of the N trajectories moving along state k: (i) A switch from state $k = 1$ to state 2 if $\Delta t.\ b_{2 \leftarrow 1}(t_C) > \zeta.\ P_1(t_C)$ and (ii) A switch from state $k = 2$ to 1 if $\Delta t.\ b_{1 \leftarrow 2}(t_C) > \zeta.\ P_2(t_C)$. The surface-hopping procedure (analytical or stochastic) is repeated as needed over time until the final scattering state is reached, with overall probabilities for each electronic state resulting from the accumulated pair transitions.

The stochastic procedure can be understood in terms of a master equation for populations,

$$\frac{dP_n}{dt} = \sum_m W_{m \to n} P_m(t),\ W_{m \to n} = \sum_C W_{m \to n}^{(C)}$$

where the transition rates $W_{m \to n}^{(C)} = K_{m \to n}^{(C)} / \Delta t_C$ add up from repeated crossings into net rates $W_{m \to n}$. At the end of the collisional event, the populations settled into stationary values $P_m(t_{fn})$ so that the relation $\sum_m W_{m \to n} P_m(t_{fn}) = 0$ must be satisfied.

The stochastic surface-hopping procedure, repeatedly applied to each pair transition from an initial electronic state a, gives final probability coefficients for each electronic state k after collisions and the related collisional cross sections $(d\sigma_{\lambda a}/d\Omega)$ [36, 37]. The surface hopping procedure has been tested for analytical models against accurate numerical results and has been applied to calculations of molecular collision cross sections to compare them with experimental measurements.

(c) Decoherence correction and recoherence

Treatments with trajectories, self-consistently generated for effective potentials or forced into surface hopping, have accuracy limitations relating to quantum decoherence in electronic state interactions. Full quantum coherence of electronic states in the collision dynamics is lost when state-to-state transitions are obtained from independently propagated trajectories, and this may result in temporary partial population of electronic states instead of full population after a transition. Ways to correct for limitations on how to describe transitions in extended (unlike localized) regions and for neglect of quantum coherence between states during transitions have been considered in a variety of extensions and modifications [38, 40, 41].

In one of the computational procedures [40, 42], decoherence due to transitions is corrected by adding to the trajectory force $d\boldsymbol{P}_s/dt$, where $\boldsymbol{P}_s(t)$ is as before the trajectory momentum, a new force $d\boldsymbol{P}_s^{(D)}/dt$ constructed to force suppression of decoherence. This returns electronic state populations to full values and appears to improve the accuracy of transition probabilities. Ways to construct the new forces have been detailed for a diabatic electronic representation, within a formulation with density matrices, imposing conservation of energy and angular momentum in transitions. It has been labelled a self-consistent decay of mixing, or SCDM, treatment. Here the decay (or suppression) of mixing due to decoherence follows from a force $d\boldsymbol{P}_s^{(D)}/dt$ obtained from a density matrix in a basis set of diabatic electronic states, with elements parametrized to decay exponentially.

5.4.3 Short-Wavelength Approximation and Two-State Models

We now introduce a short-wavelength approximation in the eikonal description, valid when the shape of the electronic wavefunction varies little over atomic distances a, while the nuclear motion wavefunctions may oscillate rapidly. This can be done for short wavelengths $\lambda = h/P \ll a$, where h is the Heisenberg constant, writing $V'(\boldsymbol{P}, \boldsymbol{Q}) = (\lambda/a)V^{(1)}$ and $V''(\boldsymbol{Q}) = (\lambda/a)^2 V^{(2)}$, with $V^{(1)}$ and $V^{(2)}$ electronically averaged functions changing little over distances λ and therefore not needed in this approximation. It therefore neglects the second and third terms in the quantum potential V_{qu}, which involve the electronically averaged gradients and Laplacians of the electronic wavefunctions. Similarly, the matrices $\mathbf{H}'_Q$ and $\mathbf{H}''_Q$ can be removed from the differential equation for $\boldsymbol{c}(t)$ in this approximation when using a diabatic electronic representation.

Corrections can be generated expanding the total potential energy $\mathcal{V}$ and $\boldsymbol{c}(t)$ in powers of a small λ/a [18]. This procedure is useful when the electronic representation has been chosen to eliminate avoided crossings of electronic potentials, in some diabatic electronic representation. Hence, in the eikonal approximation, we propagate the trajectory of the nuclei over time for each set s of initial conditions, under the influence of the diabatic potential $\overline{V}(\boldsymbol{Q})$. However, when using the adiabatic representation or one leading to avoided potential crossings, it is expected that at least the gradient of the electronic wavefunction becomes large around the crossings, so that $V'(\boldsymbol{P}, \boldsymbol{Q})$ and $\mathbf{H}'_Q$ must be included there.

It is expected that in calculations of molecular scattering the average potential $\overline{V}(\boldsymbol{Q})$ may differ from physically realistic potentials over time for a specific initial $(\boldsymbol{P}, \boldsymbol{Q})_s$, insofar trajectories move instead under specific electronic potential energies $V_{kk}[\mathbf{Q}(t)]$ while undergoing electronic transitions, and the average potential may therefore give unphysical collision dynamics and cross sections $\left(d\sigma_{j'j}/d\Omega\right)_s$ for that set of initial conditions. However, given a collection of N_{in} initial conditions $(\boldsymbol{P}, \boldsymbol{Q})_s$, then a superposition of the resulting collection of eikonal states can provide a physically meaningful result from the bundle of trajectories evolved over time.

Trajectory treatments of nonadiabatic electronic collisions can be illustrated with a simple model of atom–atom interactions involving only two electronic states $\Phi_j(\vec{R}), j = 1, 2$. A preliminary calculation of electronic states provides potential energies $H_{jj}(R)$ and momentum couplings $\vec{G}_{jk}(\vec{R}) = (\hbar/i)\langle\Phi_j(\vec{R})|\nabla_R\Phi_k(\vec{R})\rangle$. To relate the interatomic position to a time variable t, we again introduce the effective potential $\overline{H}(R) = w_1(R)H_{11}(R) + w_2(R)H_{22}(R)$ with chosen weights w_k. The corresponding local wavenumber $\overline{q}(R;E,b) = [K^2 - \overline{W}(R;b)]^{1/2} \approx \overline{k}$ is a function of the initial kinetic energy $E_{kin} = \hbar^2k_1^2/(2M)$ and impact parameter b, with $\overline{W}(R;b) = 2M[\overline{H}(R) + E_{kin}b^2/R^2]/\hbar^2]$, and a turning point $\overline{R}_T(v_1, b)$. The radial velocity is $\overline{v}(R;E,b) = \hbar\overline{q}(R;E,b)/M$, and a distance is obtained as a function of time, for trajectories going in (−) and out (+), from

$$\frac{dR_\pm}{dt} = \pm\hbar\overline{q}(R;b)/M = \pm(\hbar/M)[K^2 - \overline{W}(R;b)]^{\frac{1}{2}}$$

which can be integrated with initial condition $R(0) = \overline{R}_T$ for times $-\infty < t < \infty$ and $t = 0$ at the turning point. Solving for $R(t)$, positive and negative branches are designated as $R_\sigma(t; v_1, b)$, with $\sigma = \pm$, making explicit their dependence on initial velocity and impact parameter, and using the sign function $\sigma = sgn\,(t) = +1$ for $t > 0$ and $sgn(t) = -1$ for $t < 0$. At an asymptotic matching distance R_M, where interaction potentials vanish, the corresponding times coming in and going out are $-|t_M|$ and $+t_M$, respectively. The treatment follows along the lines of Section 5.3.4, where wavefunction amplitudes in channel j, originating in state a, are shown as $A_{ja}^{(\pm)}(R)$.

A change for differential equations in R to the time variable t can be done introducing coefficients $C_{ja}(t)$ obtained for the out- and in-branches of the average trajectory using the integrated function $R_\pm(t)$, so that in a matrix notation,

$$\mathbf{C}(t) = -s\mathbf{A}^{(\sigma)}[R_\sigma(t)][\mathbf{A}^{(-)}(R_M)]^{-1}$$

These coefficients are, more explicitly, functions $C_{ja}(t;\ v_1,\ b)$ and depend on the initial velocity and impact parameter, which are omitted for now. The continuity condition at the turning point becomes the turning-time condition $\mathbf{C}(0) = -\mathbf{A}^{(+)}(R_T)[\mathbf{A}^{(-)}(R_M)]^{-1} = \mathbf{A}^{(-)}(R_T)[\mathbf{A}^{(-)}(R_M)]^{-1}$. In particular, we have $C_{ja}(-|t_M|) = \delta_{ja}$ and obtain at a final time $\mathbf{C}(t_M) = -\mathbf{A}^{(+)}(R_M)[\mathbf{A}^{(-)}(R_M)]^{-1}$, from which, using the phases defined in Section 5.3.4,

$$S_{ja}^{(l)} = \exp[i(\eta_j + \eta_a)C_{ja}(\infty).$$

Defining also potential energy matrix elements

$$H_{jj'}^{(l)}(t) = \frac{\hbar^2}{2M}\frac{\overline{k}}{\left(k_jk_{j'}\right)^{\frac{1}{2}}}\left[W_{jj}^{(l)}[R_\sigma(t)]\left(1 - \frac{k_j}{\overline{k}}\right)\delta_{jj'} + W_{jj'}^{(l)}[R_\sigma(t)]\left(1 - \delta_{jj'}\right)\right]$$

where $W_{jj'}^{(l)}(R;b) = (2M/\hbar^2)H_{jj'}(R) + \delta_{jj'}k_a^2b^2/R^2$ from Section 5.3.4, we have in particular $H_{jj}^{(l)} \approx (\hbar^2/2M)W_{jj}^{(l)}(R)(\overline{k}/k_j - 1)$ for $R \to \infty$ and obtain the coupled differential equations

$$dC_{ja}/dt = -(i/\hbar)\sum_{j'}H_{jj'}^{(l)}(t)C_{j'a}(t)$$

to be integrated numerically or derived analytically in some special cases. To this effect, the coupled equations can be further simplified making a change of variables to remove the diagonal Hamiltonian terms.

Using a matrix notation, for given angular momentum quantum number l, the above coupled equations become

$$d\mathbf{C}/dt = -(i/\hbar)\mathbf{H}(t)\mathbf{C}(t).$$

Separating in the energy matrix diagonal and non-diagonal parts, $\mathbf{H} = \mathbf{H}_D + \mathbf{H}_N$, and defining new coefficients $\boldsymbol{b}(t)$ in

$$\mathbf{C}(t) = \exp\left[-(i/\hbar)\int_0^t dt'\mathbf{H}_D(t')\right]\boldsymbol{b}(t)$$

the new matrix equation is

$$d\boldsymbol{b}/dt = -(i/\hbar)\exp[-i\boldsymbol{\delta}(t)]\mathbf{H}_N(t)\boldsymbol{b}(t)\exp[i\boldsymbol{\delta}(t)]$$

$$\boldsymbol{\delta}(t) = -\int_0^t dt'\mathbf{H}_D\,(t')]/\hbar.$$

In particular, in the case of only two coupled equations,

$$db_1/dt = -i\Omega(t)\exp[-i\Delta(t)]b_2(t)$$

$$db_2/dt = -i\Omega^*(t)\exp[i\Delta(t)]b_1(t)$$

where $\Omega(t) = H_{12}(t)/\hbar$ and $\Delta(t) = \delta_1(t) - \delta_2(t)$, with $\delta_j(t) = -\int_0^t dt' H_{jj}(t')/\hbar$, must be solved with initial conditions $b_j(-\infty) = \delta_{j1}$ for an initial state $a = 1$. Solving formally for $b_2(t)$ in the second line, replacing it in the first line, and then solving for $b_1(t)$ by integration of its first-order equation, the scattering amplitudes are simply

$$S_{j1}^{(l)} = \exp\left\{-i\left[\eta_j + \int_0^\infty dt\, H_{jj}(t)/\hbar\right]\right\} b_j(\infty)\exp\left\{i\left[\eta_1 + \int_0^\infty dt\, H_{11}(t)/\hbar\right]\right\}.$$

Cross sections can be obtained from their expressions in terms of $t_{j'j}^{(l)} = \left(S_{j'j}^{(l)} - \delta_{j'j}\right)/(2i)$ and summations over orbital quantum numbers l. These can be transformed into integrals over $b = (l + 1/2)/k_1$ with the definition $S_{j'j}(v_1, b) = S_{j'j}^{(l)}$ and the asymptotic Legendre polynomial form valid for $|l\sin(\Theta)| \gg 1$,

$$P_l(\cos\Theta) \approx \left[\frac{2}{\pi l\sin(\Theta)}\right]^{\frac{1}{2}}\cos\left[\left(l + \frac{1}{2}\right) - \frac{\pi}{4}\right]$$

rewritten in terms of b. This gives, for initial channel $a = 1$,

$$\frac{d\sigma_{j1}}{d\Omega} = \frac{2k_1}{|\pi\sin\Theta|}\left|\int_0^\infty db\, b^{1/2}[S_{j1}(v_1, b) - \delta_{j1}]\cos\left(bk_1 - \frac{\pi}{4}\right)\right|^2$$

$$\sigma_{j1}(v_1) = 2\pi\int_0^\infty db\, b|S_{j1}(v_1, b) - \delta_{j1}|^2.$$

The integral over impact parameters for the differential cross section can be approximated by stationary phase methods by analogy with procedures for the semiclassical approximation to elastic scattering [4] to identify here the rainbow and glory impact parameters relevant to the differential cross sections in an inelastic collision.

Alternatively, transition integrals from Section 5.3.4 can be done over the time variable, as the collision proceeds between initial and final states, with definitions $dR = \hbar q^{(\sigma)}(t)dt/M, \overline{q}_{j'}[R_\sigma(t)] = q_{j'}^{(\sigma)}(t)$, $A_{ja}^{(\sigma)}[R_s(t)] = B_{ja}^{(\sigma)}(t)$, and $\overline{\alpha}_j[R_\sigma(t)] = \beta_j^{(\sigma)}(t)$, giving

$$I_{ja} = \left(\frac{\hbar}{M}\right) \int_{-\infty}^{\infty} dt\, q^{(\sigma)}(t) \sin\left[k_j R_\sigma(t) - \frac{l\pi}{2}\right]$$
$$\times \sum_{j'} H_{jj'}[R_\sigma(t)] \left[\frac{k_a}{q_{j'}^{(\sigma)}(t)}\right]^{\frac{1}{2}} B_{j'a}^{(\sigma)}(t) \exp\left[s\beta_{j'}^{(\sigma)}(t)\right]$$

where $\sigma = sgn\,(t)$. The summation over j' can be removed using the differential equation for $B_{ja}^{(\sigma)}(t)$, and the resulting integral can be further simplified integrating by parts. Again, these integrals can be combined with coefficients M_{a1} to obtain the transition amplitudes $t_{j1}^{(l)} = \sum_a I_{ja} M_{a1}$, which appear in the differential cross section.

Furthermore, given a set $(\boldsymbol{P}, \boldsymbol{Q})_s$ of initial positions and momenta, generating a bundle of N_{in} trajectories and related transition amplitudes, an average over the initial conditions can provide a physically meaningful average $\overline{(d\sigma_{j'j}/d\Omega)} = \sum_s \left(d\sigma_{j'j}/d\Omega\right)_s/N_{in}$.

Several two-state models have been developed that allow analytical solutions and provide insights on physical phenomena. They are briefly reviewed here, with details to be found in the literature [1]. In what follows, the potentials $H_{jj'}(R)$ are included in $W_{jj'}^{(l)}(R; b)$, which also contains the centrifugal potential for impact parameter b.

(A) In the *Landau–Zener* (or L–Z) model, the potentials in the coupling region $R_C \pm \Delta R_C$ are

$$H_{11}(R) = V_C - F_1 \cdot (R - R_C)$$
$$H_{22}(R) = V_C - F_2 \cdot (R - R_C)$$
$$H_{12}(R) = H_{21}(R) = U$$

where the F_j are constant forces and the coupling potential U is also a constant. The diagonal potential elements cross at R_C, and this is assumed to be larger than R_T, the turning point. At total energy E, the two wavefunctions $\psi_{j1}(R)$, $j = 1, 2$, describe scattering states. The trajectory is taken to be a straight line for impact parameter b and constant radial velocity $v_C = v_1(1-b^2/R^2)^{1/2}$, and crossing occurs at times $\pm|t_C|$, with $R - R_C = \pm v_C(t \mp |t_C|)$.

This model applies for example to charge transfer in the collisions $Na + Cl \rightarrow Na^+ + Cl^-$, $H + Mg^{2+} \rightarrow H^+ + Mg^+$, and $H + C^{2+} \rightarrow H^+ + C^+$.

An analytical solution for the coefficient $C_{j1}(t)$ can be obtained from its second-order differential equation in terms of the parabolic cylinder solution $D_\gamma(z)$ of the Weber differential equation [43], obtained by defining the constant $a = (F_1 - F_2)v_C/\hbar$ and making a change from t to a new independent variable $z = a^{1/2}\exp(i\pi/4)(t + |t_C|)$. The probability P_C of a single crossing from one potential to the other depends on the combination $U^2/(|a|\hbar^2)$ and takes the form [4]

$$P_C = 1 - \exp\left(-\frac{v_{LZ}}{v_C}\right), \quad v_{LZ} = 2\pi U^2/(\hbar|F_1 - F_2|).$$

In a simple description, the probability of transition $1 \rightarrow 2$ requires a crossing in the way in but not in the way out, or the reverse. This gives overall a probability $P_{1\rightarrow 2} = 2P_C(1 - P_C)$. A more detailed analysis of phase interference between the two transitions as the system evolves around $t = 0$ gives a result known as the *Landau–Zener–Stueckelberg* probability [4]:

$$P_{1\rightarrow 2} = 4P_C(1 - P_C)\sin(\phi)^2$$

$$\phi = \alpha + \beta - (2\hbar)^{-1}\int_{-|t_C|}^{|t_C|} dt\,(H_{22} - H_{11})_t$$

where α and β are numerical constants. Alternatively, phase interference may be described constructing the integrals I_{ja} for times from $-\infty$ to $+\infty$.

Estimates of the coupling potential U and of the L–Z velocity v_{LZ} have been given for several pairs of species [44, 45]. Examples of numerical results are shown in Figure 5.13, for the electron transfer between Li or Na, and I, where the L–Z dependence of cross section vs velocity gives a good parametrization of experimental results [46], and in its Table 2, where the right orders of magnitude for rates of electron transfer are found for several ion-neutral pairs [47].

(B) A second model is the *Rosen–Zener* one, with

$$H_{22} - H_{11} = \hbar\Omega$$

$$H_{12} = H_{21} = U\{\cosh[a(R - R_C)]\}^{-1}$$

with constant parameters Ω and a, where now the two potentials have a fixed energy difference but are strongly interacting, insofar the coupling peaks at R_C. The trajectory is again taken to be a straight line at a constant radial speed v_C with $R - R_C = \pm v_C(t \mp t_C)$ for outer or inner trajectories. These potentials apply to transitions such as $H^+ + He(1s^2) \rightarrow H(1s) + He^+(1s)$, where relevant potentials $(1\Sigma, 2\Sigma)$

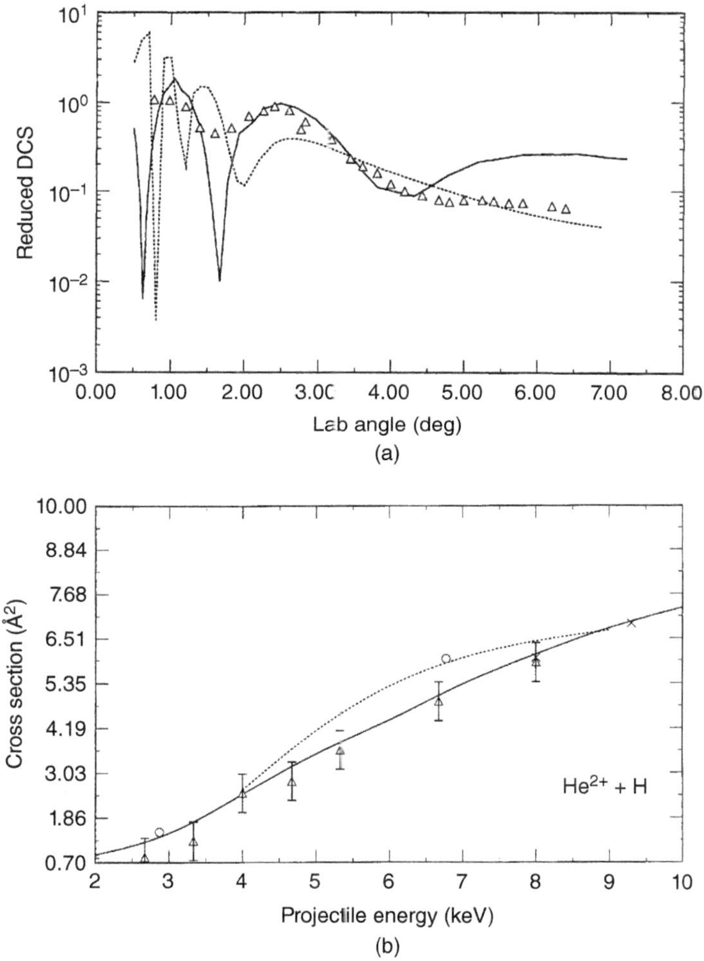

Figure 5.13 (a) Reduced elastic differential cross section (in degrees times area in au) vs the laboratory scattering angle for $H^+ + H$ collisions at 420 eV. Full lines from calculations in [39] / American Physical Society and triangles from experimental measurements. (b) Charge transfer integral cross section vs projectile energy for $He^{2+} + H$ collisions. Full line: Eik-TDHF; dotted line from Errea et al. (1992); points from experiments by Nutt et al. (1978) and by Shah et al (1978). Source: [65] / American Physical Society.

are nearly parallel. Analytical solutions are obtained with the change of variable $y = \{\tanh[av_C(t \mp t_C) - 1\}/2$ and relate to the hypergeometric function ${}_2F_1(y|.|.)$. The probability of transition in the coupling region is given by [1]

$$P_C = \sin^2[\pi|U|/(\hbar a v_C)]/\cosh[\pi\Omega/(2av_C)].$$

(C) A related, more general model is known as the *Nikitin–Demkov* model, [28] based on potentials with

$$H_{22} - H_{11} = \hbar\Omega[1 - \cos(\theta)\exp[-a(R - R_C)]\}$$

$$H_{12} = H_{21} = \hbar\Omega\ \sin(\theta)\exp[-a(R - R_C)]$$

where the parameters θ and α can be chosen to give a variety of potential energy shapes, including the Rosen–Zener ones. This model can be applied to transitions such as H^+ $Be^+(1s^2 2s) \rightarrow H(1s) + Be^{2+}(1s^2)$. Solutions are obtained with a change of variables as above, from the confluent hypergeometric function, and the transition probability in the coupling region is expressed in terms of the variable $\Omega/(av_C)$ [1].

5.4.4 First-Principles Dynamics Treatments

5.4.4.1 Nonadiabatic Multistate Dynamics

The previously described procedures (**a**) and (**b**) for semiclassical nonadiabatic dynamics, in Section 5.4.2, construct molecular wavefunctions from bundles of trajectories propagated in given PESs and their couplings, that satisfy scattering boundary conditions. They introduce expansions of the molecular wavefunction $\Psi(\boldsymbol{Q}, \boldsymbol{X})$ in the orthonormalized set of electronic states $\{\Phi_k(\boldsymbol{X}; \boldsymbol{Q})\}$ for fixed atomic positions, leading to Hamiltonian matrix equations. Usually the dynamics is obtained in two stages, starting first with the generation of PESs $\mathbf{H}(\boldsymbol{Q}) = \langle \boldsymbol{\Phi} | \hat{H}_Q | \boldsymbol{\Phi} \rangle$ and their momentum couplings in $\mathbf{G}(\boldsymbol{Q}) = (\hbar/i) \langle \boldsymbol{\Phi}(\boldsymbol{Q}) \mid \nabla_Q \boldsymbol{\Phi}(\boldsymbol{Q}) \rangle$ calculated from the electronic structure of the interacting molecules that are parametrized as functions of atomic position variables for all the molecular structures that participate in the dynamics. Energy surfaces and their couplings appear in the calculation of scattering cross sections in the second stage, after the introduction of classical trajectories $\boldsymbol{Q}(t)$ for nuclear motions. Accurate many-electron functions $\boldsymbol{\Phi}[\boldsymbol{Q}(t)]$, accounting for electronic correlation and excitation, must be generated in advance to include all involved position variables and to allow for correct electronic excitations and long-range interaction energies, and they are very demanding of computational resources.

Ab initio (or first-principles) molecular dynamics, also described as "direct dynamics" is done as an alternative, when feasible, to simultaneously generate the time-dependence of both electronic states and atomic motions. The expectation is that knowing the time evolution of atomic positions can decrease the number of electronic structure calculations needed, compared with the number needed for PES parametrization, to generate trajectories and transition probabilities. This is an active area of research relevant to nonadiabatic collisions and also to nonadiabatic molecular spectroscopy, as covered in [48–58] and several papers in [59].

Two alternative procedures are considered here for doing first-principles calculations. One is a general quantum-classical formulation using the eikonal representation of stationary scattering states $\Psi_{\alpha}^{(+)}(\boldsymbol{Q},\boldsymbol{X})$ and trajectories generated by Hamiltonian equations of motion for a bundle of trajectories with given initial conditions. The other procedure instead starts from the time-dependent Schrodinger equation for a wavefunction $\Psi(\boldsymbol{Q}, \boldsymbol{X}, t)$ and uses wavepackets to construct it for given initial scattering conditions; in both cases, one generates trajectories with positions $\boldsymbol{Q}(t)$ and momenta $\boldsymbol{P}(t)$, simultaneously with time-dependent electronic states $\boldsymbol{\Phi}[\boldsymbol{Q}(t)]$ and related time-dependent electronic density matrices, needed for calculations of $\mathbf{H}_{Q(t)}=\mathbf{H}[\boldsymbol{Q}(t)]$ and $\mathbf{G}[\boldsymbol{Q}(t)]$, to integrate time-dependent equations for scattering functions.

This requires advanced methods for many-electron structures, such as configuration interaction or multiconfiguration coupled cluster methods, suitable for treatments at all intermolecular distances and giving correct potential energies at all intermolecular distances [5], from large to very short distances for ground and excited electronic states. A general multiconfigurational self-consistent field treatment introduces super-molecular (including both collision partners) spin–orbitals (SOs) $\psi_j(\boldsymbol{x}_n;\boldsymbol{Q})=\sum_{\mu}c_{j\mu}(\boldsymbol{Q})\phi_{\mu}(\boldsymbol{x}_n;\boldsymbol{Q})$ as linear combinations of atomic orbitals $\phi_{\mu}(\boldsymbol{x}_n;\boldsymbol{Q})$, which are functions of spin–orbit electronic variables $\boldsymbol{x}_n$, $n = 1$ to N_{el} and of the position of all the nuclei at $\boldsymbol{Q}$. Atomic SOs and the coefficients building the molecular SOs change over time during a dynamical process. The molecular SOs are used to construct determinantal many-electron states $D_J(\boldsymbol{X};\boldsymbol{Q}) = N_{el}^{-1/2}\det[\psi_j(\boldsymbol{x}_n;\boldsymbol{Q})]$ and multiconfigurational electronic wavefunctions as their linear combinations $\Phi_k(\boldsymbol{X};\boldsymbol{Q}) = \sum_J C_{Jk}(\boldsymbol{Q})D_J(\boldsymbol{X};\boldsymbol{Q})$. The configuration coefficients $C_{Jk}(\boldsymbol{Q})$ depend also on atomic positions and therefore also change over time with the many-atom dynamics. Alternatively, the matrices $\mathbf{H}[\boldsymbol{Q}(t)]$ and $\mathbf{G}[\boldsymbol{Q}(t)]$ are given in terms of one- and two-electron density matrices like $\rho^{(1)}(\boldsymbol{x}_n,\boldsymbol{x}_{n'},t)$ for a time-dependent Hartree–Fock treatment [1, 60] or in terms of time-dependent Kohn–Sham densities $\rho^{(KS)}(\boldsymbol{x}_n,t)$ [61, 62] for a density functional appropriate to treat all needed collisional interactions, including short-range and long-range ones [5].

The eikonal treatment for stationary states in Section 5.4.1 provides the eikonal amplitude as $|\chi[\boldsymbol{Q}(t)]\rangle = |\boldsymbol{\Phi}(\boldsymbol{Q})\rangle\xi^{(s)}(\boldsymbol{Q})|\exp[\eta(t)]$ for each initial trajectory s. Introducing a time-dependent position variable $\boldsymbol{Q}(t)$, with $\xi^{(s)}[\boldsymbol{Q}(t)] = \mathbf{c}(t) = \left[c_k^{(s)}(t)\right]$, matrices $\mathbf{P}/M = d\mathbf{Q}/dt$, and $-(i\hbar/M)\mathbf{P}^{\dagger}(t).\ (\partial\xi^{(s)}/\partial\mathbf{Q}) = -i\hbar\, d\mathbf{c}/dt$, a simple differential equation for $\mathbf{c}(t)$ is obtained for the short-wavelength quantal dynamics of interest here. In a first-principles treatment, the equations of motion for positions $\boldsymbol{Q}(t)$ and momenta $\boldsymbol{P}(t)$ and for $\mathbf{c}(t)$ are integrated over time while many-electron functions $\Phi_k[\boldsymbol{X};\boldsymbol{Q}(t)]$ and matrices $\mathbf{H}[\boldsymbol{Q}(t)]$ and $\mathbf{G}[\boldsymbol{Q}(t)]$ are simultaneously generated. This can be done selecting electronically adiabatic or diabatic representation sets $\{\Phi_k\}$ and constructing matrix elements, as described in Section 5.2.

As a practical matter, it is convenient to work in the adiabatic representation, with electronic states $\Phi_\Gamma[\boldsymbol{Q}(t)]$ generated as eigenfunctions of the electronic Hamiltonian with eigenenergies $E_\Gamma[\boldsymbol{Q}(t)]$, using available efficient computational software. This provides the matrix $\langle\boldsymbol{\Phi}[\boldsymbol{Q}(t)]|\hat{H}_Q|\boldsymbol{\Phi}[\boldsymbol{Q}(t)]\rangle = \mathbf{E}[\boldsymbol{Q}(t)] = \left[E_\Gamma[\boldsymbol{Q}(t)]\delta_{\Gamma\Gamma'}\right]$ and momentum-coupling elements $G_{\Gamma\Gamma'}[\boldsymbol{Q}(t)] = (\hbar/i)\left\langle\Phi_\Gamma[\boldsymbol{Q}(t)]|\nabla_Q\Phi_{\Gamma'}[\boldsymbol{Q}(t)]\right\rangle$. The sequence of coupled nuclear-electronic equations to be repeatedly solved is then

$$\text{(A)}\ \frac{d\boldsymbol{Q}}{dt} = \frac{\partial\mathcal{H}}{\partial\boldsymbol{P}},\ \frac{d\boldsymbol{P}}{dt} = -\frac{\partial\mathcal{H}}{\partial\boldsymbol{Q}},$$

$$\text{(B)}\ \hat{H}_Q|\Phi_\Gamma[\boldsymbol{Q}(t)]\rangle = E_\Gamma[\boldsymbol{Q}(t)]|\Phi_\Gamma[\boldsymbol{Q}(t)]\rangle$$

$$\text{(C)}\ \left(-i\hbar\frac{d}{dt} + \left(\frac{d\mathbf{Q}}{dt}\right)^\dagger\cdot\mathbf{G}[\boldsymbol{Q}(t)] + \mathbf{E}[\boldsymbol{Q}(t)]\right)\boldsymbol{c}(t) = \mathbf{0}$$

with $\mathcal{H}(\boldsymbol{P},\boldsymbol{Q}) = \mathbf{P}\cdot\mathbf{P}/(2M) + \mathcal{V}(\boldsymbol{P},\boldsymbol{Q})$ and $\mathcal{V}(\boldsymbol{P},\boldsymbol{Q})$ a chosen effective potential energy function. To start the sequence with step (A), it is necessary to specify for each trajectory in a bundle, at an initial time t_{in}, the initial position and momenta $\boldsymbol{P}_s(t_{in}), \boldsymbol{Q}_s(t_{in})$, $s = 1$ to N_{in}, the initial coefficients in $\boldsymbol{c}(t_{in})$ corresponding to the initial scattering channel, and the initial force from the gradient $\partial\mathcal{V}/\partial\boldsymbol{Q}$ with $\mathcal{V}(\boldsymbol{Q}) = \boldsymbol{c}^\dagger\overline{\boldsymbol{V}}(\boldsymbol{Q})\boldsymbol{c}/(\boldsymbol{c}^\dagger\boldsymbol{c})$, insofar there is no momentum coupling at large distances. Propagation follows with $\mathcal{V}(\boldsymbol{P},\boldsymbol{Q}) \cong \overline{V}(\boldsymbol{Q}) + V'(\boldsymbol{Q})$ in the adiabatic representation, solving the Hamiltonian equations with the integration algorithms described in Chapter 2, Section 2.4, for classical trajectories. Step (B) must be done to generate the adiabatic electronic state $\Phi_\Gamma[\boldsymbol{Q}(t)]$ and eigenenergy $E_\Gamma[\boldsymbol{Q}(t)]$ and to construct the matrix elements $\langle\Phi_\Gamma[\boldsymbol{Q}(t)]\,|\,\nabla_Q\Phi_{\Gamma'}[\boldsymbol{Q}(t)]\rangle$, with accurate electronic wavefunctions allowing for treatment of all relevant interatomic distances during a collision. Step (C) is a set of coupled first-order ordinary differential equations with solutions that can be generated as described in Chapter 4, Section 4.2, using algorithms like time-step interpolation, or multistep prediction-correction methods, or as a Magnus cumulant expansion. Sequential integration in (A), (B), and (C) is complicated by the large disparity between scales of time oscillations, Δt_{el} for electronic and Δt_{nu} for nuclear motions, usually with $\Delta t_{el} \ll \Delta t_{nu}$. Propagation in time of $\boldsymbol{P}(t)$, $\boldsymbol{Q}(t)$, and the coefficients $\boldsymbol{c}(t)$ can be done with a common time-step $\Delta t_{el}/n$, with an integer n chosen for numerical accuracy and with the mentioned algorithms for solving systems of coupled first-order differential equations for nuclei and electronic states.

However, in the present treatment involving first-principles dynamics integration, it is important to introduce a procedure that minimizes the number of time-steps when the computationally demanding electronic structures are generated. A way to do this has been shown in our previous work on ion–atom collisions with a time-dependent Hartree–Fock description of the electronic states and trajectories obtained from an effective potential energy $\overline{V}(\boldsymbol{Q})$ [63], later on expanded to include an efficient treatment of time scales [39]. The same procedure is next developed here for a set of general $\Phi_\Gamma[\boldsymbol{Q}(t)]$ electronic states, to decrease the number of needed time-steps in the sequence (A)–(C). It constructs the matrix $\boldsymbol{c}(t)$ as a sum of two terms, in a so-called relax-and-drive procedure. This relies on the generation of a relaxing solutions $\boldsymbol{c}^{(0)}(t)$ within a large time interval $t_0 \leq t \leq t_1$, followed by a correction $\boldsymbol{c}^{(1)}(t)$ from driving forces in the interval, in a first-principles calculation of the cross sections.

Introducing the rate matrix $\mathbf{W}(t) = (d\boldsymbol{Q}/dt)^\dagger \cdot \mathbf{G}[\boldsymbol{Q}(t)] + \frac{\mathbf{E}[\boldsymbol{Q}(t)]}{\hbar}$, with $\mathbf{W_0} = \mathbf{W}(t_0)$ at a time t_0 and a known solution $\boldsymbol{c}_0 = \boldsymbol{c}(t_0)$, the transient solution $\boldsymbol{c}^{(0)}(t)$ in the interval $t_0 \leq t \leq t_1$ is obtained with a time-evolution operator $\mathbf{U}_0(t, t_0)$, as

$$i\, d\boldsymbol{c}^{(\mathbf{0})}/dt = \mathbf{W_0}\boldsymbol{c}^{(\mathbf{0})}(t)$$

$$\mathbf{W_0} = (d\mathbf{Q}/dt)^\dagger_{\mathbf{0}} \cdot \mathbf{G}[\boldsymbol{Q}(t_0)] + \mathbf{E}[\boldsymbol{Q}(t_0)]/\hbar$$

Here $\boldsymbol{c}^{(0)}(t)$ is a relaxation matrix, obtained from

$$\boldsymbol{c}^{(\mathbf{0})}(t) = \mathbf{U}_0(t, t_0)^\dagger \boldsymbol{c}_0 \mathbf{U}_0(t, t_0)$$

$$\mathbf{U}_0(t, t_0) = \exp[-i\mathbf{W_0}.(t - t_0)]$$

and it can be easily obtained in an interval where the momentum coupling $\mathbf{G}[\boldsymbol{Q}(t_0)]$ is negligible and $\mathbf{W_0} \cong \mathbf{E}[\boldsymbol{Q}(t_0)]/\hbar$ is a diagonal matrix. The time-evolution operator $\mathbf{U}_0(t, t_0)$ generates a local (labeled with a subindex L) interaction picture [64] for the full matrix $\boldsymbol{c}(t) = \mathbf{U}_0(t, t_0)^\dagger \boldsymbol{c}_L(t)\mathbf{U}_0(t, t_0)$ and for $\mathbf{W}(t) = \mathbf{U}_0(t, t_0)^\dagger \mathbf{W}_L(t)\mathbf{U}_0(t, t_0)$, so that $\Delta\mathbf{W}_L(t) = \mathbf{W}_L(t) - \mathbf{W}_0$ gives changes $\Delta\boldsymbol{c}_L(t) = \boldsymbol{c}_L(t) - \boldsymbol{c}_0$ within the interval $t_0 \leq t \leq t_1$. They are given by

$$i\, d\Delta\boldsymbol{c}_L/dt = \Delta\mathbf{W}_L(t)\boldsymbol{c}_L(t) = \Delta\mathbf{W}_L(t)\boldsymbol{c}_0 + \Delta\mathbf{W}_L(t)\Delta\boldsymbol{c}_L(t)$$

This can be rewritten as an integral equation and solved by iteration, as

$$\Delta\boldsymbol{c}_L(t) = \Delta\boldsymbol{c}_{\boldsymbol{L}}^{(1)}(t) + \Delta\boldsymbol{c}_{\boldsymbol{L}}^{(2)}(t) + \cdots$$

$$\Delta\boldsymbol{c}_L^{(1)}(t) = -i\int_{t_0}^{t} dt'\,\Delta\mathbf{W}_L(t')\boldsymbol{c}_0$$

$$\Delta\boldsymbol{c}_{\boldsymbol{L}}^{(2)}(t) = -i\int_{t_0}^{t} dt'\,\Delta\mathbf{W}_L(t')\Delta\boldsymbol{c}_{\boldsymbol{L}}^{(1)}(t')$$

Returning with $\mathbf{U}_0(t, t_0)$ to the starting matrices $\Delta\boldsymbol{c}(t)$ and $\Delta\mathbf{W}(t)$, the leading coefficient correction is

$$\Delta\boldsymbol{c}^{(1)}(t) = -i\int_{t_0}^{t} dt'\,\Delta\mathbf{W}(t')\boldsymbol{c}^{(\mathbf{0})}(t')$$

and provides the driven change $\Delta \boldsymbol{c}^{(1)}(t_1)$ to be added to the relaxed term $\boldsymbol{c}^{(0)}(t_1)$ for a corrected

$$\mathbf{c}(t_1) = \mathbf{c}^{(0)}(t_1) + \Delta \mathbf{c}^{(1)}(t_1)$$

which is accurate, provided that $\left\|\Delta \mathbf{c}_L^{(2)}(t)\right\| / \|\Delta \mathbf{c}^{(1)}(t)\| \ll 1$ in the time interval. This relax-and-drive procedure advances the dynamics by a time-interval $\Delta t = t_1 - t_0$ with a minimal requirement of electronic structure calculations that now are needed only at an intermediate time in each interval. In regions of small or null diabatic coupling of states, where PESs are smooth, it is possible to select time intervals $\Delta t \approx \Delta t_{nu} \gg \Delta t_{el}$. In configuration regions where diabatic couplings are large and PESs change rapidly with atomic distances, the time interval can be reduced to $\Delta t \approx \Delta t_{nu}/n \gg \Delta t_{el}$ with $n \approx 20$ or less, which is typically much less than the time scale Δt_{el} for electronic changes. The size of the time interval Δt can be chosen to make sure that $\left\|\Delta \mathbf{c}_L^{(2)}(t_1)\right\| / \|\Delta \mathbf{c}^{(1)}(t_1)\| \ll 1$ or the interval can be halved to a shorter $\Delta t/2$ and calculations repeated to satisfy this propagation condition.

The integral $\Delta \boldsymbol{c}^{(1)}(t_1)$ can be obtained by diagonalizing the matrix $\mathbf{W}_0$ to obtain its eigenvalues $w_k^{(0)}$, $k = 1$ to K, with the matrix $\boldsymbol{c}^{(0)}(t)$ containing exponentials with arguments $\left(w_k^{(0)} - w_{k'}^{(0)}\right)(t - t_0)$, and by interpolating $\Delta \mathbf{W}(t)$ between t_0 and t_1 as $\Delta \mathbf{W}(t) = \mathbf{W}_0 + (\mathbf{W}_1 - \mathbf{W}_0)(t - t_0)$, after $\Phi_\Gamma[\boldsymbol{Q}(t_1)]$ and $\mathbf{W}_1$ have been calculated. The integration in $\Delta \boldsymbol{c}^{(1)}(t_1)$ is then straightforward and provides $\boldsymbol{c}(t_1) = \boldsymbol{c}_1$ for the next time interval $t_1 \leq t \leq t_2$. In practice, as seen in calculations of ion–atom collisions [39], it is computationally more accurate to introduce an intermediate step at $t_{1/2} = t_0 + \Delta t/2$, which requires one additional electronic structure calculation at $t_{1/2}$. The advance through the whole time interval is summarized as

$$(\boldsymbol{P}, \boldsymbol{Q}, \mathcal{V}, \mathbf{c}, \mathbf{W})_0 \to (\boldsymbol{P}, \boldsymbol{Q})_{1/2} \to (\Phi_\Gamma, \mathcal{V}, \mathbf{c}, \mathbf{W})_{1/2} \to (\boldsymbol{P}, \boldsymbol{Q}, \mathcal{V}, \mathbf{c}, \mathbf{W})_1$$

by successive integration of the Hamilton equations between t_0 and $t_{1/2}$, generation of the electronic state effective potential and coefficients at $t_{1/2}$, and final integrations between $t_{1/2}$ and t_1 for $\boldsymbol{P}(t)$, $\boldsymbol{Q}(t)$ and for the coefficients in $\mathbf{c}(t)$.

The eigenenergies $\mathrm{E}_\Gamma[\boldsymbol{Q}(t)]$ provide ground and excited PESs in an adiabatic electronic representation, and the momentum couplings follow from the eigenfunction derivatives with respect to atomic displacements. Alternatively, it is possible to transform away the momentum couplings in a diabatic electronic representation, as described in Section 5.2, and solve then for the direct dynamics with a nondiagonal $\mathbf{H}[\boldsymbol{Q}(t)]$. The choice of an effective potential energy $\mathcal{V}(\boldsymbol{P}, \boldsymbol{Q})$ generates results of given accuracy for cross sections, with calculations done for selfconsistent-eikonal or surface hopping treatments, as described in Section 5.4.2. This is an active area of research, with variations and alternatives being explored, including a numerical treatment using the Verlet integration algorithm in an extension of the relax-and-drive procedure to ab initio dynamics of photoexcited molecules [50].

5.4.4.2 First-Principles Eik/TDHF Treatment

A simplified and computationally efficient treatment can be based on a description of the electronic dynamics by means of a time-dependent self-consistent (Hartree–Fock or HF) approximation, for processes where the electron dynamics involves only one-electron rearrangements [39]. A simple choice is then a Hartree–Fock determinantal wavefunction $D_J(\boldsymbol{X};\boldsymbol{Q}) = N_{el}{}^{-1/2}\det[\psi_j(\boldsymbol{x}_n;\boldsymbol{Q})]$ for N_{el} electrons constructed from a set of orthonormal electronic SOs $\{\psi_j, j=1 \text{ to } N_{el}\}$ with electron position and spin variables $x_n = (\vec{r}_n, \zeta_n)$, generated as the SOs change over time along a nuclear trajectory $\boldsymbol{Q}(t)$. This couples the nuclear motion and electronic dynamics and requires propagation of the time-dependent HF one-electron density matrix $\rho(\boldsymbol{x}_n, \boldsymbol{x}_{n'}, t) = \sum_{j\,occ}\psi_j(\boldsymbol{x}_n, t)\psi_j(\boldsymbol{x}_{n'}, t)^*$ along each trajectory, with $\psi_j(\boldsymbol{x}_n, t) = \psi_j[\boldsymbol{x}_n, \boldsymbol{Q}(t)]$. At an initial time, the occupied SOs can be constructed to localize electrons at each colliding molecule, for a given asymptotic electronic state Γ. The internal nuclear-motion states of the molecules and the relative motion of their center-of-mass involve, to start, a set of initial values for a bundle of trajectories with a distribution of parameters Λ_s.

The treatment can be based on the propagation of $\rho(\boldsymbol{x}_n, \boldsymbol{x}_{n'}, t)$ as a solution of an integro-differential equation in the time variable [39, 59]. In a bracket notation for electronic states with $\psi_j(\boldsymbol{x}_n, t) = \langle \boldsymbol{x}_n \mid \psi_j(t)\rangle$, the time evolution operator $\widehat{\rho}(t)$ is given by

$$\widehat{\rho}(t) = \sum_{j\,occ} \left|\psi_j(t)\right\rangle \left\langle\psi_j(t)\right|$$

$$\widehat{F}\widehat{\rho} - \widehat{\rho}\widehat{F} = i\hbar\partial\widehat{\rho}/\partial t$$

where $\widehat{F}(t) = \widehat{H}^{(1)} + \widehat{G}^{(1)}[\widehat{\rho}]$ is the Fock Hamiltonian with its one-electron term $\widehat{H}^{(1)}$ containing the electron kinetic energy plus its interaction with atomic ion cores, plus the sum of HF Coulomb and exchange energies in $\widehat{G}^{(1)}$ [60]. This equation must be solved coupled to a trajectory equation with the potential energy $V_S^{(HF)}[\boldsymbol{Q}_s(t)] = V_{cc} + tr\left[\widehat{\rho}_J^{(s)}(\widehat{H}^{(1)} + \widehat{F})/2\right]$, where V_{cc} comes from the interaction of atomic ion cores.

In the simple Eik/TDHF approximation [39], the coherence of trajectories is again assumed to average out and the relative motion of nuclei is obtained from a potential energy function $\overline{V}[\boldsymbol{Q}(t)]$ generated from an average density operator $\langle\widehat{\rho}_J\rangle = \sum_s w_s \widehat{\rho}_J^{(s)}$ with weights w_s chosen from an initial distribution of states. These equations can be further obtained in terms of matrices representing the density and HF operators, after introducing molecular SOs as combinations of atomic basis functions $\phi_p(\vec{r}, t)$ at each moving atomic center p [5]. Care however must be used to avoid spurious couplings arising from electron overlaps $\langle\phi_p \mid \partial\phi_q/\partial t\rangle$. This is described in [34, 39] and references there and is elaborated in [1].

The time-evolution of the TDHF density operator $\widehat{\rho}(t)$ is constructed as a sum of two terms, in a so-called relax-and-drive procedure, as described in the previous subsection, for a general many-electron state. This relies on the generation of a relaxing solutions $\hat{\rho}^{(0)}(t)$ within a large time interval $t_0 \leq t \leq t_1$, followed by a correction $\hat{\rho}^{(1)}(t) = \hat{\sigma}(t)$ from driving forces in the interval, as $\hat{\rho} = \hat{\rho}^{(0)}(t) + \hat{\sigma}(t)$. The relaxing term is obtained from $\hat{\rho}_0 = \hat{\rho}(t_0)$ and $\hat{F}_0 = \hat{F}(t_0)$ with a time-evolution operator $\hat{U}_0(t, t_0)$ as

$$\hat{\rho}^{(0)}(t) = \hat{U}_0(t, t_0)\hat{\rho}_0 \hat{U}_0(t, t_0)^{\dagger}, \hat{U}_0(t, t_0) = \exp[-i\hat{F}_0.(t - t_0)/\hbar]$$

As the atoms move along a trajectory, the Fock operator changes by $\Delta\hat{F}(t) = \hat{F}(t) - \hat{F}_0$, a perturbation driving the density operator, to add to first order [39]:

$$\hat{\sigma}(t) = (i\hbar)^{-1} \int_{t_0}^{t} dt' \, \hat{U}_0(t, t')\hat{D}^{(1)}(t')\hat{U}_0(t, t')^{\dagger},$$

$$\hat{D}^{(1)}(t) = \Delta\hat{F}(t)\hat{\rho}^{(0)}(t) - \hat{\rho}^{(0)}(t)\Delta\hat{F}(t)$$

The integral can be extended to time t_1, in a calculation that generates the TDHF SOs $\psi_i(\boldsymbol{x}_n, t_1)$ after a trajectory has been advanced, and can be obtained linearizing $\Delta\hat{F}(t)$ between t_0 and t_1. This procedure avoids the need to repeatedly generate $\hat{F}(t)$ in the interval, a computationally demanding task. It provides an ab initio (or first principles) treatment and a HF determinant evolving from an initial $D_I(\boldsymbol{X}, t_i) = D_I[\boldsymbol{X}; \boldsymbol{Q}(t_i)]$ into a final $D_J^0(\boldsymbol{X}, t_f)$ and from them a cross section for a diabatic $I \rightarrow J$ transition [18, 39].

The state-to-state differential cross section for going from state D_I containing SOs ψ_i to state D_J with SOs ψ_j is furthermore written as the product of the elastic differential cross section $(d\sigma/d\Omega)_I$ for the initial electronic state, multiplying the final transition probability from the evolved electronic state $D_I(t_f)$ to the free final state $D_J^0(t_f)$ after the interaction [39]. This is given in terms of electronic states represented by Slater determinants as [34]

$$\frac{d\sigma_{JI}}{d\Omega} = \left(\frac{d\sigma}{d\Omega}\right)_I \left|\left\langle D_J^0(t_f)\middle|D_I(t_f)\right\rangle\right|^2 = \left(\frac{d\sigma}{d\Omega}\right)_I \left|\det\left[\left\langle \psi_j^0(t_f)\middle|\psi_i(t_f)\right\rangle\right]\right|^2$$

where a property of overlap of determinants has been used to simplify the expression as a single determinant of a matrix.

In the special case of atom–atom collisions, with $(\boldsymbol{P}, \boldsymbol{Q}) = (\vec{P}, \vec{R})$, we can obtain state-to-state integral cross sections by integrating differential cross sections over solid angles or equivalently by integrating over the range of initial impact parameters (φ, b) leading to each deflection angle Ω, keeping in mind that the SOs depend on (E, φ, b) through the initial conditions for the trajectories, to obtain

$$\sigma_{JI}(E) = \int_0^{2\pi} d\varphi \int_0^{\infty} bdb \left|\det\left[\left\langle \psi_j^0(t_f)\middle|\psi_i(t_f)\right\rangle\right]\right|^2$$

where the time-evolved MOs are obtained from the time-dependent density matrix at the final time.

Results for the simple cases of elastic and electron transfer in $H^+ + H$ and $He^{2+} + H$ collisions at hyperthermal energies are shown in Figure 5.13(a) and (b).

The propagation procedure with the HF density matrix for a many-electron system has been used in calculations of electron transfer in ion–atom and ion–surface collisions. More recently a more elaborate treatment involving different time-steps for electronic and nuclear displacements has been computationally implemented and used to calculate intramolecular electronic transfer in a polyatomic compound [50].

5.4.4.3 Nuclear-Electronic Dynamics Treatment

It is possible to extend the previous treatments to include in the quantum dynamics not only the electronic system but also a collection of nuclei, which may undergo quantal dynamics [66, 67]. This is especially relevant to phenomena where rearrangement of electrons and protons are simultaneously present, as in proton tunneling over energy barriers and hydrogen bonding rearrangement in polyatomic molecules. In our notation, the quantal position and spin variables of electrons and protons are now grouped in $(\boldsymbol{X}^{(e)}, \boldsymbol{X}^{(p)})$, while position variables for other atoms, considered as undergoing classical-like motion, are designated here as a new set $\boldsymbol{Q}'(t)$.

The total wavefunction in the eikonal form is now $\Psi(\boldsymbol{Q}', \boldsymbol{X}^{(e)}, \boldsymbol{X}^{(p)}, t) = \chi(\boldsymbol{Q}', \boldsymbol{X}^{(e)}, \boldsymbol{X}^{(p)}, t) \exp[i\, S'(\boldsymbol{Q}', t)/\hbar]$. By analogy with the previous development for the coupled dynamics of quantal and classical variables, the time-dependent mechanical action $S'(\boldsymbol{Q}', t)$ satisfies a time-dependent Hamilton–Jacobi differential equation, or alternatively $\boldsymbol{Q}'(t)$ and $\boldsymbol{P}'(t) = \partial S'/\partial \boldsymbol{Q}'$ are positions and momenta obtained from classical Hamiltonian equations with initial conditions for a classical distribution of their initial values in a bundle of trajectories. The total eikonal amplitude along a trajectory is now $\chi^{(ep)}(\boldsymbol{X}^{(e)}, \boldsymbol{X}^{(p)}, t) = \chi[\boldsymbol{Q}'(t), \boldsymbol{X}^{(e)}, \boldsymbol{X}^{(p)}, t]$ and can be approximated for a collection of interacting electrons and protons.

A simple choice, again for each trajectory, is a self-consistent product of a determinantal wavefunction $D^{(e)}(\boldsymbol{X}^{(e)}, t)$ for electrons times a determinantal wavefunction $D^{(p)}(\boldsymbol{X}^{(p)}, t)$ for protons, as in $\chi^{(ep)}(\boldsymbol{X}^{(e)}, \boldsymbol{X}^{(p)}, t) = C^{(ep)} D^{(e)}(\boldsymbol{X}^{(e)}, t) D^{(p)}(\boldsymbol{X}^{(p)}, t)$. This factorized form can be replaced in a variational functional for the total energy, with a Hamiltonian operator $\hat{H} = \hat{H}^{(e)} + \hat{H}^{(p)} + \hat{H}^{(ep)}$, where $\hat{H}^{(e)}$ contains the kinetic energy of electrons and their Coulombic interaction with other electrons and atomic ions (excluding protons), $\hat{H}^{(p)}$ similarly contains proton kinetic energy and Coulombic interactions with other protons and atomic ions, and $\hat{H}^{(ep)}$ contains Coulombic interactions among electrons and protons. In some detail, the determinants are constructed from electronic and protonic orbitals as $D^{(\alpha)}(\boldsymbol{X}^{(\alpha)}, t) = N_\alpha^{-1/2} \det\left[\psi_{j_\alpha}^{(\alpha)}\left(\boldsymbol{x}_{n_\alpha}^{(\alpha)}, t\right)\right]$, $\alpha = e, p$, and $j_\alpha = 1$ to $N^{(\alpha)}$, while related Hartree–Fock-like density operators are $\hat{P}^{(\alpha)}(t) = \sum_{j_\alpha} \left|\psi_{j_\alpha}^{(\alpha)}(t)\right\rangle \left\langle \psi_{j_\alpha}^{(\alpha)}(t)\right|$.

These operators appear as variables in the energy variational expression and its minimization gives time-dependent H–F-like equations,

$$i\hbar\frac{d\widehat{P}^{(\alpha)}}{dt} = \left[\widehat{F}^{(\alpha)}(t), \widehat{P}^{(\alpha)}(t)\right]$$

$$\widehat{F}^{(\alpha)}(t) = \widehat{f}^{(\alpha)} + tr_{\beta}\left[\widehat{h}^{(\alpha\beta)}\widehat{P}^{(\beta)}(t)\right], \quad \beta = e, p\,(\neq \alpha)$$

where $\widehat{f}^{(\alpha)} = \widehat{h}^{(\alpha)} + \widehat{J}^{(\alpha)}[\widehat{P}^{(\alpha)}] + \widehat{K}^{(\alpha)}[\widehat{P}^{(\alpha)}]$, with the first term containing one-particle kinetic and atomic–ion interaction terms, and where $\widehat{J}^{(\alpha)}$ and $\widehat{K}^{(\alpha)}$ are Coulomb and exchange one-particle H–F operators. The operator $\widehat{F}^{(\alpha)}[\widehat{P}^{(\alpha)}, \widehat{P}^{(\beta)}]$ therefore contains both *e*- and *p*-density operators and their coupled equations must be solved simultaneously.

This nuclear-electronic orbital approach (or NEO approach) has been applied, among several examples, to proton transfer between the two oxygen atoms in *o*- hydroxybenzaldehyde (*o*-HBA) following a photoexcitation $S_0 \rightarrow S_1$. Comparisons of electronic spectra and excited conformations from TDHF and NEO–TDHF, as well as the related TDDFT versions, show clear differences in calculated values and good agreement with experimental results for the NEO results [67]. Versions of the NEO treatment can also be developed for first-principles dynamics with more accurate many-electron wavefunctions coming from electronic CI and CC treatments [66].

5.5 Electronic Rearrangement for Several Interatomic Variables

5.5.1 Potential Energy Surfaces and Their Couplings: Multistate Cases

Given the state $\Psi(\boldsymbol{Q},\boldsymbol{X}) = \sum_k \psi_k(\boldsymbol{Q})\Phi_k(\boldsymbol{X};\boldsymbol{Q})$ of a pair of molecules A and B as an expansion in a basis set of electronic wavefunctions $\Phi_k(\boldsymbol{X};\boldsymbol{Q})$ (not necessarily adiabatic ones), which vary with a set $\boldsymbol{Q} = (\vec{R}, \boldsymbol{q})$ of interatomic distances, its Schrodinger equation leads to a matrix equation for the coefficients ψ_k containing a potential energy matrix with elements $H_{kl}(\boldsymbol{Q})$, which are functions of distances, as well as a matrix of momentum coupling elements $G_{kl}(\boldsymbol{Q})$. It is usually possible to construct the wavefunction of the pair of molecules A and B so that at large intermolecular distances the internal molecular motions are independent and occur along well-defined separate electronic PESs $E_a^{(A)}(\boldsymbol{q}_A)$ and $E_b^{(B)}(\boldsymbol{q}_B)$.

However, as distances are decreased, the potential energy for each molecular electronic state, originating in the pair (a, b), becomes a function $E_k^{(AB)}(R, \boldsymbol{Q}_{int})$, which changes with the intermolecular distance and internal positions, excluding overall translations and rotations, and generates a PES showing as isocontours of energy in the multidimensional space of distances. Pairs of such PESs may display crossings or regions of minimal energy differences, which depend on what electronic representation is being used. They interact through nuclear momentum couplings, which also depend on the representation. Special points can be identified looking for minimal values of differences in each pair of potential surfaces, as one varies the internal coordinates, so that $\left|E_k^{(AB)}(R, \boldsymbol{Q}_{int}) - E_l^{(AB)}(R, \boldsymbol{Q}_{int})\right| = Min.$ This can be zero if there is a crossing along a surface in the hyperspace of internal coordinates, and its shape depends again on what electronic representation one is using. When dealing with a diatomic and a single internal distance, one may find a crossing or single intersection at a distance R_0; for two or more internal variables, one may find a conical intersection, crossing seams, or near crossings. All these are important in descriptions of electronic rearrangement.

The procedure for calculating intermolecular states by first fixing nuclear positions leads to the construction of many-electron states Φ_k dependent parametrically on all the nuclear position variables. Starting with the total number $3N$ of atomic position variables for a system with N atoms and indicating with $\boldsymbol{Q}_{RB}$ the set of rigid body variables (six for overall translations and rotations or five for a linear conformation), the remaining variables in a BF frame form a set $\boldsymbol{Q}_{BF} = \{Q_\nu, \nu = 1 \text{ to } N_{BF}\}$ of $N_{BF} = 3N - N_{RB}$ degrees of freedom. Generalizing the previous treatment for atom–atom collisions, matrix elements H_{kl} of the Hamiltonian operator for fixed nuclei depend on all the BF nuclear variables and as they vary the diagonal elements H_{kk} and H_{ll} may cross or show avoided crossings at specific points, lines, or multidimensional surfaces.

5.5.2 Crossings in Several Dimensions: Conical Intersections and Seams

The search for special intersections of PESs can be extended to systems with three or more atoms, involving more internal degrees of freedom.

Even for three atoms, there are $N_{int} = 3$ or 4 degrees of freedom for the atomic positions and enough variables to satisfy the two equations that indicate an intersection of PESs. This fixes only two internal degrees of freedom. However, there are many rearrangements of atoms during electronic excitation where most degrees of freedom remain inactive and only two are changing. The two changing ones can be identified by searching for the configurations, where $|H_{11}(\{Q_\nu\}) - H_{22}(\{Q_\nu\})|$ is a minimum for a set $\left\{Q_\nu^{(m)}\right\}$ and where distortion forces for two of them, called here Q_a and Q_b, are small around $Q_a^{(m)}$ and $Q_b^{(m)}$.

We follow the treatment in Chapter 5 of [5]. Recent detailed treatments of the resulting conical intersections of PESs and applications, also to spectroscopy and biomolecules, can be found in [68, 69].

We first treat this case, with displacement variables $x = Q_a - Q_a^{(m)}$ and $y = Q_b - Q_b^{(m)}$, and all other internal variables set at their values for the minimum and omitted from arguments for now. Consider again just two original states $|\Phi_j\rangle$ in a diabatic representation with diagonal Hamiltonian matrix elements $H_{11}(x, y)$ and $H_{22}(x, y)$ and a nondiagonal $H_{12}(x, y)$ real valued for now but taken to be complex later on. A transformation to the adiabatic representation states $\left|\Phi_k^{(a)}\right\rangle$ gives the PESs

$$E_{1,2}(x, y) = M(x, y) \pm N(x, y)$$

$$M(x, y) = \frac{1}{2}[H_{11}(x, y) + H_{22}(x, y)]$$

$$N(x, y) = \left\{[H_{11}(x, y) - H_{22}(x, y)]^2 + 4|H_{12}(x, y)|^2\right\}^{1/2}$$

The condition for intersection of PESs is $N(x, y) = 0$ or both $H_{11}(x, y) - H_{22}(x, y) = 0$ and $H_{12}(x, y) = 0$, which may be satisfied by roots (x_0, y_0), corresponding to variables $Q_v^{(0)}$ expected to be close to $Q_v^{(m)}$, $v = a, b$. When this happens for the original (diabatic set) matrix elements, the new energies $E_{1,2}(x, y)$ of the adiabatic states can be expanded around the roots up to second order in the displacements, so that

$$\begin{aligned} E_{1,2}(x, y) = {} & M^{(0,0)} + M^{(1,0)}x + M^{(0,1)}y + M^{(2,0)}x^2/2 + M^{(0,2)}y^2/2 + M^{(1,1)}xy \\ & \mp (N^{(1,0)}x + N^{(0,1)}y + N^{(2,0)}x^2/2 + N^{(0,2)}y^2/2 + N^{(1,1)}xy) \end{aligned}$$

Each of these two energy functions gives an isocontour surface for each constant energy in the space of E vs x, y. Since they are quadratic forms, the isosurfaces are conical forms and they touch at a degeneracy vertex, at the location of the roots (x_0, y_0) in a (x, y) plane. They can be opposing cones or collateral ones, depending on slopes along x and y. This in turn depends on what types of bonding surfaces are interacting. Figure 5.14 below illustrates conical intersections for a triatomic system ABC, as can be expected coming from (a) isomerization, that is bound-to-bound structure interaction, (b) dissociation in bound-to-unbound structure changes, or also (c) collisions as unbound-to-unbound structure events.

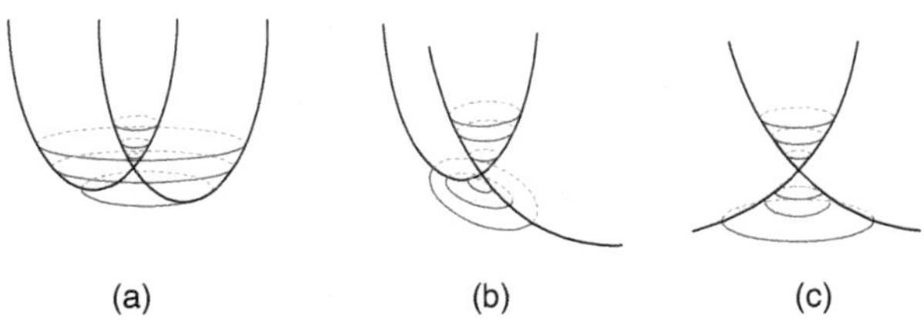

Figure 5.14 Illustration of three cases of conical intersections between two potential energy surfaces: (a) bound-to-bound, (b) bound-to-unbound, and (c) unbound-to-unbound.

Insights on the nature of these conical intersections can be gained constructing simple models of the intersecting conical surfaces (CISs) in a space E vs (x, y), where x and y are bond distances or angles, and the other $N_{int} - 2$ internal conformation variables are kept fixed. The intersection of two diabatic potential energies leads to CISs when coupled, and as the other variables are changed, the conical shapes can be pictured as moving along seams in a hyperspace of $N_{int} - 2$ dimensions. Examples of models are given as follows:

(a) Bound-to-bound CIS:

$$H_{11} = \frac{\omega_1^2}{2}\left(x + \frac{a}{2}\right)^2 + \frac{\omega_2^2}{2}y^2, \quad H_{22} = \frac{\omega_1^2}{2}\left(x - \frac{a}{2}\right)^2 + \frac{\omega_2^2}{2}y^2 - \Delta, \quad H_{12} = cy$$

This corresponds to two intersecting paraboloids and describes, for example, SH_2 of point group symmetry C_{2v}, in the excited states $1^1A''$ and $2^1A''$ [70], with x corresponding to the H–H distance r and y for the distance R from S to the CM of HH. The third variable is the angle between $\vec{r}$ and $\vec{R}$ orientations and is considered here to be chosen at a fixed value.

(b) Bound-to-unbound CIS:

$$H_{11} = \frac{\omega_1^2}{2}\left(x + \frac{a}{2}\right)^2 + \frac{\omega_2^2}{2}y^2, \quad H_{22} = A\ \exp[-\alpha(x - x_0)] + \frac{\omega_2^2}{2}y^2 - \Delta,$$

$$H_{12} = cy\ \exp\left[(x - x_C)^2/\lambda^2 + (y - y_0)^2/\mu^2\right]$$

where the unbound motion occurs as the x variable increases. This applies to the dissociation of excited HNO ($^1\Delta$) into H(^2S) + NO($^2\Pi s$) [71], with the two variables corresponding to the H to NO distance (x) and the HNO bond angle (y).

(c) Unbound-to-unbound CIS:

$$H_{11} = A\exp[-\alpha(x - x_0)], \quad H_{22} = B\exp[-\beta(y - y_0)] - \Delta$$

$$H_{12} = cy\ \exp\left[(x - x_0)^2/\lambda^2 + (y - y_0)^2/\mu^2\right]$$

This applies to a triatomic reaction A + BC → AB + C with a fixed ABC angle and x and y representing the bond distances of initial and final diatomics, such as in the collinear $H + H_2 \rightarrow H_2 + H$ reaction. Also, and more generally, a triatomic PES for three active electrons in an atom–diatomic reaction can be described by the LEPS surface, as mentioned in Chapter 4. That is the lower PES of a pair generated by the combination of two valence bond functions and the two PESs form a conical intersection when the radical in the LEPS expression is null, as can be expected to also occur for other monovalent atom combinations at certain interatomic distances.

The diabatic to adiabatic transformation of electronic states is generated as before by coefficients dependent now on the two variables x and y. For a complex valued coupling $H_{12}(x, y) = |H_{12}(x, y)| \exp[i\alpha(x, y)]$, the two new adiabatic states are as before

$$\left|\Phi_1^{(a)}\right\rangle = |\Phi_1\rangle \exp\left(-\frac{i\alpha}{2}\right)\cos(\beta) + |\Phi_2\rangle \exp\left(\frac{i\alpha}{2}\right)\sin(\beta)$$

$$\left|\Phi_2^{(a)}\right\rangle = -|\Phi_1\rangle \exp\left(-\frac{i\alpha}{2}\right)\sin(\beta) + |\Phi_2\rangle \exp\left(\frac{i\alpha}{2}\right)\cos(\beta)$$

but now with

$$\tan[2\beta(x,y)] = 2|H_{12}(x,y)|/[H_{11}(x,y) - H_{22}(x,y)]$$

which is well defined away from the intersection point.

More accurately, one must consider all the internal degrees of freedom and work with the displacements $\boldsymbol{q} = \boldsymbol{Q} - \boldsymbol{Q}^{(m)}$, to allow for couplings of the displacements in the two variables Q_ν, $\nu = a, b$, with the remaining degrees of freedom, as all atoms are displaced from the minimum conformation. Indicating with $\overline{\boldsymbol{Q}}$ the set of all internal coordinates other than Q_a or Q_b, the two PESs depend more generally also on the displacements $\overline{\boldsymbol{q}} = \overline{\boldsymbol{Q}} - \overline{\boldsymbol{Q}}^{(m)}$, and when these are not all zero, the PESs have form $E_{1or2}(x, y, \overline{\boldsymbol{q}}) = M(x, y, \overline{\boldsymbol{q}}) \pm N(x, y, \overline{\boldsymbol{q}})$. One finds in general that $N(x_0, y_0, \overline{\boldsymbol{q}}) \neq 0$ at the previous roots, and the energy degeneracy at the vertex is lifted. This is likely to distort the conical energy surfaces touching at the vertex, changing them into glancing paraboloids.

This avoided degeneracy is due to couplings to other variables within the adiabatic state description and therefore different in origin from the avoided crossing in the one-dimensional case, which arises from the two original electronic states having the same symmetry. The magnitude of the splitting at the vertex can be found doing perturbation theory with the vibronic perturbation energy operator generated by displacing internal variables from the reference values at the minimum conformation. Models can be presented in terms of adiabatic or nonadiabatic PESs [72]. Two adiabatic PESs are also subject to nonadiabatic momentum couplings $\vec{G}_{kl}^{(a)}$, with momentum components along the x and y directions, even if all other internal variables are assumed to be fixed.

Depending on the nature of the two variables, such as whether they describe distortions of a bond length or of a bond angle, the conical shapes may become flattened and in the limit of a breaking bond, the degeneracy may become a line, showing as a seam for two intersecting PESs along the direction of the breaking bond. This is what happens for example in the case of the PES for collinear $H^+ + H_2$ interacting with the PES for $H + H_2^+$ [36] and of the PES for $Ar^+ + H_2$ interacting with that of $Ar + H_2^+$, as shown in Figure 5.15, (a) for the adiabatic potential energies and (b) for their nonadiabatic momentum coupling $\tau = |\vec{G}_{kl}^{(a)}|/\hbar$ along the diatomic axis [73]. The crossing between the two lowest PESs can also be described as two intersecting potential energy lines in the plane of the energy E and the bond

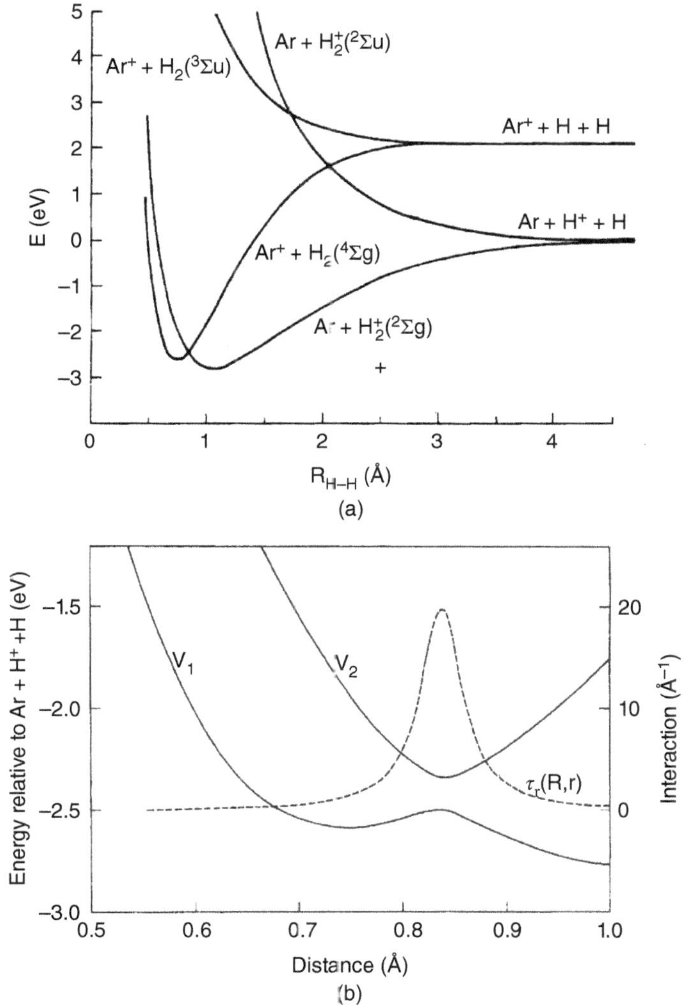

Figure 5.15 (a) Adiabatic potential energy surfaces for collinear $Ar^+ + H_2$ interacting with $Ar + H_2^+$ along the diatomic axis, as functions of the H to H distance and for large intermolecular Ar—H distances; (b) nonadiabatic momentum coupling between the two lowest PESs at a fixed intermolecular distance. Source: [73] / American Physical Society.

distance r in H_2 and in H_2^+, translated along a seam in a direction perpendicular to that plane, as the distance R between Ar^+ and H_2 is changing. The crossing in Figure 5.15 (a) below 1.0 Å (or 0.1 nm) is transformed due to the momentum coupling along R, into an avoided crossing, as shown in Figure 5.15 (b).

The two adiabatic PESs are also subject to nonadiabatic momentum couplings $\vec{G}_{kl}^{(a)}$, with momentum components along the x and y directions, even if all other internal variables are assumed to be fixed.

A special situation arises when the two original electronic states, 1 and 2, are degenerate with the same energy $H_{11}(x, y) = H_{22}(x, y)$ for all argument values. The degeneracy can be lifted due to a coupling $H_{12}(x, y) \neq 0$. Conditions under which this happens can be found assuming vibronic coupling and can be studied by expanding the electronic Hamiltonian operator of the system in powers of small displacements from the reference structure. Coupling matrix elements of the resulting force operator are different from zero if the direct product of the three representations of the point groups for the two original states and for any of the displacements contains the totally symmetric representation.

In this case, the degeneracy can be lifted with new states and PESs can be found from the solution of the two-state eigenvalue equations. This analysis has been followed to arrive at the Jahn–Teller theorem [74], stating that in any symmetric nonlinear molecule with electronically degenerate states, displacement of atoms lead to symmetry distortions that decrease the original symmetry.

If the energy of the two degenerate states is, to begin with, a paraboloid showing a single minimum vs x and y, the theorem indicates that this form will be distorted leading to a more complex form with two minima in the (E, x) plane. But now as a circle is traced in the (x, y) plane around the degeneracy point, there are energy maxima and minima corresponding to lower symmetry structures [75]. An example is provided by the radical C_5H_5 of symmetry D_{5h} to begin with and in doubly degenerate states of label ${}^2E_1''$. The coupling between the two states can be written in terms of the variables $x = \rho\cos(5\varphi)$ and $y = \rho\sin(5\varphi)$ and as the angle φ varies over a circle this shows energies of structures with the lower symmetry C_{2v}. The distortion may also arise from atomic momentum couplings between the two original states, involving derivatives in the x and y variables, again provided that symmetry arguments indicate that the coupling is not zero.

Conical intersections of PESs can be expected to appear when dealing with photoexcited polyatomic molecules and may show up in large numbers when the molecular system is complex, as in polymers and biomolecules. Recent reviews have covered the role that CISs play in molecular spectroscopy and photoinduced dynamics [76], in isomerization [77], organic photochemistry [78], and inorganic compounds with transition metal atoms.

The presence of CISs in the primary isomerization event in vision has been experimentally confirmed using ultrafast optical spectroscopy in the visible and near infrared [79]. It can be expected that many CISs will be present and linked through seams in photoinduced biomolecular dynamics.

5.5.3 Geometrical Phase and Generalizations

The electronic wavefunctions introduced so far are real valued, but more generally they can be constructed to also contain a phase factor making them complex valued. The phase factor can be obtained from the Schrodinger equation for the wavefunction but it is nonunique. The phase appears to affect theoretical results when adiabatic electronically excited states are involved and may be needed to reproduce experimental results. This can be determined by doing calculations of dynamical properties and comparing results with experimental measurements of molecular spectra or collisional cross sections. The phase of an electronic wavefunction has been considered and used in dynamics calculations under the various names geometric phase, or Berry phase, or molecular Aharonov–Bohm phase [71, 80–84].

The role of the geometric phase is relevant at the level of the adiabatic approximation. The electronic wavefunction can be allowed to take complex values from a given phase factor $\eta_k^{(a)}(\boldsymbol{Q}) = \exp[i\gamma_k(\boldsymbol{Q})]$ with γ_k a real-valued geometrical phase, so that now

$$\Psi^{(ad)}(\boldsymbol{Q},\boldsymbol{X}) = \eta_k^{(a)}(\boldsymbol{Q})\,\Phi_k^{(a)}(\boldsymbol{X};\boldsymbol{Q})\psi_k^{(a)}(\boldsymbol{Q}).$$

It appears as if a phase factor for the total molecular wavefunction should not have physical content. But here one must consider what happens when the nuclei undergo displacements through a hyperspace containing degenerate electronic wavefunction points, where other states with $l \neq k$ have the same energy $E_k(\boldsymbol{Q})$ and interfere quantum mechanically with the state above and also what happens due to momentum couplings of electronic states.

Even when degeneracies and couplings are absent, the phase may alter numerical values of PESs and the related dynamics. From the definition of the momentum $\boldsymbol{G}_{kk}^{(a)}(\boldsymbol{Q})$, now for electronic state $\eta_k^{(a)}\Phi_k^{(a)}$, one finds that it is not any longer null but that instead $\boldsymbol{G}_{kk}^{(a)}(\boldsymbol{Q}) = \hbar\partial\gamma_k/\partial\boldsymbol{Q}$, a hyper-momentum-like quantity in the $\boldsymbol{Q}$ hyperspace, and that by integration along a line integral from point $\boldsymbol{Q}_0$ to $\boldsymbol{Q}$, the phase function factor is

$$\eta_k^{(a)}(\boldsymbol{Q}) = \exp\left[\frac{i}{\hbar}\int_{\boldsymbol{Q}_0}^{\boldsymbol{Q}} d\boldsymbol{Q}''.\boldsymbol{G}_{kk}^{(a)}(\boldsymbol{Q}'').\right.$$

Further replacing the above wavefunction in the Schrodinger equation and projecting it on $\eta_k^{(a)}(\boldsymbol{Q})^*\left\langle\Phi_k^{(a)}(\boldsymbol{Q})\right|$ with integration over electronic variables, one obtains

$$\frac{1}{2M}\left[\frac{\hbar}{i}\nabla_Q + \boldsymbol{G}_{kk}^{(a)}(\boldsymbol{Q})\right]^2\psi_k^{(a)}(\boldsymbol{Q}) + \left[E_k^{(a)}(\boldsymbol{Q}) - E\right]\psi_k^{(a)}(\boldsymbol{Q}) = 0.$$

The effect of $G_{kk}^{(a)}(Q)$ on the nuclear wavefunction can be accounted for by writing it as $\psi_k^{(a)}(Q) = \varphi_k^{(a)}(Q)\eta_k^{(a)}(Q)^*$ with $\varphi_k^{(a)}$ real valued. Applying the momentum operator within the square bracket above, in sequence, one finds that

$$\left[-\frac{\hbar^2}{2M}\nabla_Q{}^2 + E_k^{(a)}(Q) - E\right] \varphi_k^{(a)}(Q) = 0.$$

Therefore, the introduction of the phase factor in the electronic state must be compensated by a phase factor in the nuclear motion wavefunction describing the system dynamics to be able to recover the form of the total wavefunction. If the electronic state is not degenerate, then the path in the line integral can be closed and deformed to disappear, corresponding to a choice of a null phase function. However, if the electronic state is degenerate with another one and they interact to create an intersection of PESs, such as the conical one, then there is a choice of paths around the intersection, which will give non-zero phases, and from these, it follows that there are related changes in the energies of each of the interacting states. This has been carefully studied in examples of adiabatic transformations of a state for three identical nuclei [71, 82] and in many-state models analogous to a spin in a magnetic field [83].

An expression for the geometrical phase can be obtained introducing an integration path parameter s labelling positions along a chosen path (s) and writing the equation for the electronic state $\Upsilon_k = \eta_k\Phi_k$ as

$$\hat{H}_{Q(s)}\,|\Upsilon_k[Q(s)]\rangle = E_k[Q(s)]\,|\Upsilon_k[Q(s)]\rangle$$

Differentiating it with respect to s, one can construct a state change at position $s_1 = s + \Delta s$ from

$$\left|\frac{\partial\Upsilon_k}{\partial s}\right\rangle_1 = \sum_{l\neq k}|\Upsilon_l\rangle_1 \frac{\langle\Upsilon_l|(\partial\hat{H}_{Q(s)}/\partial s)|\Upsilon_k\rangle_1}{E_k[Q(s_1)] - E_l[Q(s_1)]}$$

followed by a second step to $s_2 = s + 2\Delta s$ giving

$$\left\langle\Upsilon_k\left|\frac{\partial\Upsilon_k}{\partial s}\right.\right\rangle_2 = -\sum_{l\neq k}\frac{\langle\Upsilon_k|(\partial\hat{H}_{Q(s)}/\partial s)|\Upsilon_l\rangle_1\langle\Upsilon_l|(\partial\hat{H}_{Q(s)}/\partial s)|\Upsilon_k\rangle_2}{\{E_k[Q(s_1)] - E_l[Q(s_1)]\}^2} = \left(\frac{d\gamma_k}{ds}\right)_2.$$

This allows generation of the geometric phase in a step-wise procedure along a path, for a given electronic Hamiltonian and adiabatic basis set.

As a path moves around a CIS, the adiabatic state in the lowest surface will change with the parameter s and, after a closed loop in the x and y variables bringing them to their original values in $\tan(2\beta) = 2\,|H_{12}|/(H_{11} - H_{22})$, from Section 5.5.2, one finds that 2β will have changed by $\pm 2\pi$ and the adiabatic wavefunction coefficients $\cos(\beta)$ and $\sin(\beta)$ will have changed the wavefunction by a factor -1. This is an essential change indicating that one must incorporate the geometrical phase γ_k in electronic and nuclear wavefunctions in a treatment of the molecular dynamics so they compensate each other to make sure that results are invariant when the physical system is distorted along a path until it returns to its original conformation.

Inspection of the matrix equation given above for $\psi_k^{(a)}$ shows that there is a similarity with the form of the Schrodinger equation for a charged particle in an electromagnetic field described by vector and scalar potentials $\vec{A}$ and ϕ to be compared to $\boldsymbol{G}_{kk}^{(a)}(\boldsymbol{Q})$ and $E_k^{(a)}(\boldsymbol{Q})$. In the same way a gauge transformation of the electromagnetic potentials preserves the form of the particle equation of motion, now the transformation generated by the geometric phase γ_k plays a similar role for the equation of a single adiabatic state. More generally this can be extended for coupled adiabatic states by considering the previous coupled equations for moving nuclei and the column matrix $\boldsymbol{\psi}(\vec{R})$, which is also valid for a $\boldsymbol{\psi}(\boldsymbol{Q})$ in many variables, containing the square matrices $\boldsymbol{G}^{(a)}(\boldsymbol{Q})$ and $\boldsymbol{H}^{(a)}(\boldsymbol{Q})$. This suggests that the geometric phase in a molecular system can be generally treated using a unitary local gauge transformation $\boldsymbol{U}^{(a)}(\boldsymbol{Q})$ applied to the set of electronic states $\left\{\left|\Phi_k^{(a)}(\boldsymbol{Q})\right\rangle\right\}$. Corresponding transformations of $\boldsymbol{G}^{(a)}(\boldsymbol{Q})$ and $\boldsymbol{H}^{(a)}(\boldsymbol{Q})$ and of the nuclear motion states $\left\{\psi_k^{(a)}(\boldsymbol{Q})\right\}$ with the adjoint $\boldsymbol{U}^{(a)\dagger}$ and substitutions in the Schrodinger equation show that its form remains invariant under the new gauge transformation [85]. This generalizes the observation made above for a single adiabatic state and shows the way to construct geometric phase factors for many coupled states.

It can be expected that geometric phases in an adiabatic representation will play a quantitative role in the calculation of dynamical properties of electronically excited systems, such as optical spectra and collisional phenomena showing electronic rearrangement. This happens in the reaction $D + H_2 \rightarrow HD + H$, for which comparison of theory and experiment have shown the existence of a conical intersection and a need for a geometric phase in the calculation of the reaction cross section using adiabatic PESs [86].

References

1 B. H. Bransden and M. R. C. McDowell, *Charge Exchange and the Theory of Ion-Atom Collisions, Oxford*, England: Oxford University Press, 1992.

2 J. C. Slater, *Quantum Theory of Molecules and Solids*, Volume 1, New York: McGraw-Hill, 1963.

3 R. D. Piacentini and A. Salin, "Molecular treatment of the He_2+ -H collisions," *J. Phys. B: Atom. Molec. Phys.*, vol. 7, p. 1666, 1974.

4 M. Child, *Molecular Collision Theory*, New York: Academic Press, 1974.

5 D. A. Micha, *Molecular Interactions: Concepts and Methods*, Hoboken, NJ: John Wiley & Sons, Inc., 2020.

6 M. Born and K. Huang, *Dynamical Theory of Crystal Lattices, Appendix VIII, Oxford*, England: Oxford University Press, 1954.

7 J. O. Hirschfelder, C. F. Curtis and R. Byron Bird, *Molecular Theory of Gases and Liquids*, New York: Wiley, 1954.

8 J. O. Hirschfelder (Editor), *Intermolecular Forces (Adv. Chem. Phys. Vol. 12)*, New York: Wiley, 1967.

9 E. E. Nikitin, *Theory of Elementary Atomic and Molecular Processes in Gases*, London, England: Oxford University Press, 1974.

10 D. A. Micha, "Effective Hamiltonian methods for molecular collisions," *Adv. Quantum Chem.*, vol. 8, p. 231, 1974.

11 B. C. Garrett and D. G. Truhlar, "The coupling of electronically adiabatic states in atomic and molecular collisions," in *Theoretical Chemistry*, Volume 6, Part A (Editor: D. Henderson), New York: Academic Press, 1981, p. 216.

12 M. Baer, "The theory of electronic nonadiabatic transitions in chemical reactions," in *Theory of Chemical Reaction Dynamics*, Volume II (Editor: M. Baer), Boca Raton, FL: CRC Press, 1985, p. 219.

13 H. Nakamura, *Nonadiabatic Transition: Concepts, Basic Theories And Applications* (2nd edition), New York: World Scientific, 2012.

14 F. T. Smith, "Diabatic and adiabatic representations for atomic collision problems," *Phys. Rev.*, vol. 179, p. 111, 1969.

15 M. Baer, "Adiabatic and diabatic representationsfor atom-molecule collisions: Treatment of the collinear arrangement," *Chem. Phys. Lett.*, vol. 35, p. 112, 1975.

16 A. C. Mead and D. G. Truhlar, "Conditions for the definition of strictly diabatic electronic basis for molecular systems," *J. Chem. Phys.*, vol. 77, p. 6090, 1982.

17 J. A. Olson and D. A. Micha, "Electronic state representations at molecular potential pseudocrossings," *Intern. J. Quantum Chem.*, vol. 22, p. 971, 1982.

18 D. A. Micha, "A self-consistent eikonal treatment of electronic transitions in molecular collisions," *J. Chem. Phys.*, vol. 78, p. 7138, 1983.

19 J. M. Cohen and D. A. Micha, "Electronically diabatic atom-atom collisions: A self-consistent eikonal approximation," *J. Chem. Phys.*, vol. 97, p. 1038, 1992.

20 F. M. Fernandez and D. A. Micha, "Time-evolution of molecular states in electronically diabatic phenomena," *J. Chem. Phys.*, vol. 97, p. 8173, 1992.

21 R. J. Cave and M. D. Newton, "Calculation of electronic coupling matrix elements for ground and excited state electron transfer reactions: Comparison of the generalized Mulliken–Hush and block diagonalization methods," *J. Chem. Phys.*, vol. 106, p. 9213, 1997.

22 J. E. Subotnik, S. Yeganeh, R. J. Cave and M. A. Ratner, “Constructing diabatic states from adiabatic states: Extending generalized Mulliken–Hush to multiple charge centers with Boys localization,” *J. Chem. Phys.*, vol. 129, p. 244101, 2008.

23 P. J. Bruna and S. D. Peyerimhoff, “Excited-state potentials,” *Adv. Chem. Phys. Part I*, vol. 67, p. 1, 1987.

24 M. Tinkham, *Group Theory and Quantum Mechanics*, New York: McGraw-Hill, 1964.

25 K. K. Baeck and H. An, “Practical approximation of the non-adiabatic coupling termsfor same-symmetry interstate crossings by using adiabatic potential energies only,” *J. Chem. Phys.*, vol. 146, 064107, 2017.

26 Y. Shu, L. Zhang, X. Chen, S. Sun, Y. Huang and D. G. Truhlar, “Nonadiabatic dynamics algorithms with only potential energies and gradients: Curvature-driven coherent switching with decay of mixing and curvature-driven trajectory surface hopping,” *J. Chem. Theory Comput.*, vol. 18, p. 1320, 2022.

27 A. R. Edmonds, *Angular Momentum in Quantum Mechanics*, Princeton, NJ: Princeton University Press, 1960.

28 E. E. Nikitin and S. Y. Umanskii, *Theory of Slow Atomic Collisions*, New York: Springer-Verlag, 1984.

29 C. Cohen-Tanoudji, B. Diu and F. Laloe, *Quantum Mechanics*, Volume 1, Chap. IV, New York: Wiley, 1977.

30 R. E. Olson and F. T. Smith, “Collisional spectroscopy IV. Semiclassical theory of inelastic scattering with applications to He^+ + Ne,” *Phys. Rev A*, vol. 3, p. 1607, 1971.

31 H. Goldstein, C. P. Poole Jr., and J. L. Safko, *Classical Mechanics* (3rd edition), San Francisco, CA: Addison-Wesley, 2002.

32 A. Messiah, *Quantum Mechanics* vol. 1, Amsterdam: North-Holland, 1962.

33 J. C. Y. Chen and K. M. Watson, “Electronic transitions in slow collisions of atoms and molecules,” *Phys. Rev.*, vol. 174, p. 152, 1968.

34 D. A. Micha, “Time-dependent many-electron treatment of electronic energy and charge transfer in atomic collisions,” *J. Phys. Chem. A*, vol. 103, pp. 7562–7574, 1999.

35 J. B. Delos, W. R. Thorson and S. K. Knudson, “Semiclassical theory of inelastic collisions. I. Classical picture and semiclassical formulation,” *Phys. Rev. A*, vol. 6, p. 709, 1972.

36 J. C. Tully, “Nonadiabatic processes in molecular collisions,” in *Dynamics od Molecular Collisions Part B* (Editor: W. H. Miller), New York: Plenum Press, 1976, p. 217.

37 J. C. Tully, “Molecular dynamics with electronic transitions,” *J. Chem. Phys.*, vol. 93, p. 1061, 1990.

38 L. Wang, A. Akimov and O. V. Prezhdo, "Recent progress in surface hopping: 2011–2015," *J. Phys. Chem. Lett.*, vol. 7, pp. 2100–2112, 2016.

39 D. A. Micha and K. K. Runge, "Time-dependent many-electron approach to slow ion-atom collisions: The coupling of electronic and nuclear motions," *Phys. Rev. A*, vol. 50, p. 322, 1994.

40 C. Zhu, A. W. Jasper and D. G. Truhlar, "Non-Born–Oppenheimer trajectories with self-consistent decay of mixing," *J. Chem. Phys.*, vol. 120, p. 5543, 2004.

41 J. E. Subotnik, A. Jain, B. Landry, A. Petit, W. Ouyang and N. Bellonzi, "Understanding the surface hopping view of electronic transitions and decoherence," *Annu. Rev. Phys. Chem.*, vol. 67, p. 387, 2016.

42 D. G. Truhlar, "Decoherence in combined quantum mechanical and classical mechanical methods for dynamics," in *Quantum Dynamics of Complex Molecular Systems* (Editors: D. A. Micha and I. Burghardt), Berlin: Springer-Verlag, 2007, p. 227.

43 M. Abramowitz and I. A. Stegun, *Handbook of Mathematical Functions*, New York: Dover Publications, 1965.

44 R. E. Olson, F. T. Smith and E. Bauer, "Estimation of the coupling matrix elements for one-electron transfer systems," *Appl. Optics*, vol. 10, p. 1848, 1971.

45 R. Grice and D. Herschbach, "Long-range configuration interaction of ionic and covalent states," *Mol. Phys.*, vol. 27, p. 159, 1974.

46 A. P. Baede, "Charge transfer between neutrals at hyperthermal energies," *Adv. Chem. Phys.*, vol. 30, p. 462, 1975.

47 S. E. Butler and A. Dalgarno, "Charge transfer of multiply charged ions with hydrogen and helium Landau-Zener calculations," *Astr. J*, vol. 241, p. 838, 1980.

48 M. Baer and G. D. Billing (Editors), *The Role of Degenerate States in Chemistry, Volume 124*, New York: John Wiley & Sons, 2002.

49 G. A. Worth and M. A. Robb, "Applying direct molecylar dynamics to non-adiabatic systems," *Adv. Chem. Phys.*, vol. 124, p. 355, 2002.

50 X. Li, J. C. Tully, H. B. Schlegel and M. J. Frisch, "Ab initio ehrenfest dynamics," *J. Chem. Phys.*, vol. 123, 084106, 2005.

51 B. G. Levine, J. D. Coe, A. M. Virshup and T. J. Martinez, "Implementation of ab initio multiple spawning in the MOLPRO quantum chemistry package," *Chem. Phys.*, vol. 347, pp. 3–16, 2008.

52 D. Marx and J. Hutter, *Ab Initio Molecular Dynamics*, Cambridge, UK: Cambridge University Press, 2009.

53 K. Saita and D. V. Shalashilin, "On-the-fly ab initio molecular dynamics with multiconfigurational Ehrenfest method," *J. Chem. Phys.*, vol. 137, p. 22A506, 2012.

54 B. F. E. Curchod, U. Rothlisberger and I. Tavernelli, "Trajectory-based non-adiabatic dynamics with time-dependent density functional theory," *ChemPhysChem*, vol. 14, p. 1314, 2013.

55 B. F. E. Curchod and T. J. Martinez, "Ab initio nonadiabatic quantum molecular dynamics," *Chem. Rev.*, vol. 118, p. 3305, 2018.

56 S. Mai, P. Marquetand and L. Gonzalez, "Nonadiabatic dynamics: The SHRAC approach," *WIREs Comput Mol. Sci.*, vol. 8, p. e1370, 2018. DOI: 10.1002/wcms.1370

57 W. Zhou, A. Mandal and P. Huo, "Quasi-diabatic scheme for non-adiabatic on-the-fly simulations," *J. Phys. Chem. Lett.*, vol. 10, p. 7062, 2019.

58 T. R. Nelson, A. J. White, J. A. Bjorgaard, A. E. Sifain, Y. Zhang, B. Nebgen, S. Fernandez-Alberty, D. R. A. E. Mozyrsky and S. Tretiak, "Non-adiabatic excited state molecular dynamics: Theory and applications for photophysics and materials," *Chem. Rev.*, vol. 120, p. 2215, 2020.

59 J. Gao and P. J. Roosky, "The age of direct cemical dynamics," *Acc. Chem. Res.*, vol. 55, p. 471, 2022.

60 I. N. Levine, *Quantum Chemistry* (5th edition), Upper Saddle River, NJ: Prentice-Hall, 2000.

61 M. A. L. Marques and E. K. U. Gross, "Time-dependent density functional theory," *Ann. Rev. Phys. Chem.*, vol. 55, pp. 427–455, 2004.

62 M. E. Casida and M. Huix-Rotllant, "Progress in time-dependent density-functional theory," *Ann. Rev. Phys. Chem.*, vol. 63, pp. 287–323, 2012.

63 K. Runge, D. A. Micha and E. Q. Feng, "A time-dependent molecular orbital approach to electron transfer in ion-atom collisions," *Int. J. Quantum Chem. Symp.*, vol. 24, p. 781, 1990.

64 A. Messiah, *Quantum Mechanics* Volume 2, New York: North-Holland Publishing Company, 1962.

65 K. Runge and D. A. Micha, "Time-dependent approach to ion-atom collisions for systems with one active electron," *Phys. Rev. A*, vol. 53, p. 1388, 1996.

66 F. Pavosevic, T. Culpitt and S. Hammes-Schiffer, "Multicomponent quantum chemistry: Integrating electronic and nuclear quantum effects via the nuclear-electronic orbital method," *Chem. Rev.*, vol. 120, p. 4222, 2020.

67 L. Zhao, Z. Tao, F. Pavosevic, A. Wildman, S. Hammes-Schiffer and X. Li, "Real-time time dependent nuclear-electronic orbital approach: Dynamics beyond the Born-Oppenheimer approximation," *J. Phys. Chem. Lett.*, vol. 11, p. 4052, 2020.

68 S. Matsika and P. Krause, "Nonadiabatic events and conical intersections," *Ann. Rev. Phys. Chem.*, vol. 62, pp. 621–643, 2011.

69 W. Domcke, D. R. Yarkony and H. Köppel (Editors), *Conical Intersections: Electronic Structure, Dynamics and Spectroscopy*, Singapore: World Scientific Publishing, 2014.

70 N. Matsunaga and D. R. Yarkony, "Energies and derivative couplings in the vicinity of a conical intersection II. CH_2 and SH_2," *J. Chem. Phys.*, vol. 107, p. 7825, 1997.

71 G. Herzberg and H. C. Longuet-Higgins, "Intersection of potential energy surfaces in polyatomic molecules," *Disc. Faraday Soc.*, p. 77, 1963.

72 D. A. Worth and L. S. Cederbaum, "Beyond Born-Oppenheimer: Molecular dynamics through a conical intersection," *Ann. Rev. Phys. Chem.*, vol. 55, p. 127, 2004.

73 M. Baer and J. A. Beswick, "Incorporation of electronically nonadiabatic effects into bimolecular reactive systems. III. The collinear Ar + H_2+ system," *Phys. Rev. A*, vol. 19, p. 1559, 1979.

74 H. A. Jahn and E. Teller, "Stability of polyatomic molecules in degenerate electronic states. I. Orbital degeneracy," *Proc. R. Soc. A*, vol. 161, p. 220, 1937.

75 G. Herzberg, *Molecular Spectra and Molecular Structure III. Polyatomic Molecules*, New York: Van Nostrand Reinhold, 1966.

76 W. Domcke and D. R. Yarkony, "Role of conical intersections in molecular spectroscopy and photoinduced chemical dynamics," *Annu. Rev. Phys. Chem.*, vol. 63, p. 325, 2012.

77 B. G. Levine and T. J. Martinez, "Isomerization through conical intersections," *Annu. Rev. Phys. Chem.*, vol. 58, p. 613, 2007.

78 M. A. Robb, "Conical intersections in organic photochemistry," in *Conical Intersections: Theory, Computation, and Experiment* (Editors: W. Domcke, D. R. Yarkony and H. Koeppel), Singapore: World Scientific, 2011, p. 3.

79 D. Polli, P. Altoe, O. Weingart, K. M. Spillane, C. Manzoni, D. Brida, G. Tomasello, G. Orlandi, P. Kukura, R. A. Mathies, M. Garavelli and G. Cerullo, "Conical intersection dynamics of the primary photoisomerization event in vision," *Nature*, vol. 467, p. 440, 2010.

80 H. C. Longuet-Higgins, "The intersection of potential energy surfaces in polyatomic molecules," *Proc. R. Soc. Lond. A*, vol. 344, p. 147, 1975.

81 A. J. Stone, "Spin-orbit coupling and the intersection of potential energy surfaces in polyatomic molecules," *Proc. R. Soc. Lond. A*, vol. 351, p. 141, 1976.

82 C. A. Mead and D. G. Truhlar, "On the determination of Born-Oppenheimer nuclear motion wavefunctions including complications due to conical intersections and identical nuclei," *J. Chem. Phys.*, vol. 70, p. 2284, 1979.

83 M. V. Berry, "Quantal phase factors accompanying adiabatic changes," *Proc. R. Soc. Lon. A*, vol. 392, p. 45, 1984.

84 Y. Aharonov and D. Bohm, "Significance of the electromagnetic potential in the quantum theory," *Phys. Rev.*, vol. 115, p. 485, 1959.

85 T. Pacher, C. A. Mead, L. S. Cederbaum and H. Koeppel, "Gauge theory and quasidiabatic states in molecular physics," *J. Chem. Phys.*, vol. 91, p. 7057, 1989.

86 A. Kuppermann and Y.-S. M. Wu, "The geometric phase effect shows up in chemical reactions," *Chem. Phys. Lett.*, vol. 205, p. 577, 1993.

6

Reactive Collisions

CONTENTS

Quantum Molecular Dynamics: Energy Transfer and Reactivity, First Edition. David A. Micha.

6.1 Arrangement Channels and Coordinate Transformations

6.1.1 A Three-Atom System

A treatment of reactive molecular collisions, where some bonds are broken and others are formed, requires careful descriptions of many-atom arrangements for reactants and products and choices of coordinate frames and space variables in scattering functions. Following early publications on atom–diatomic reactive collisions using wavefunction descriptions and calculations of cross sections [1, 2], general quantal treatments have been presented in numerous reviews over following years, describing progress on the theoretical treatment of many-atom rearrangements without and with electronic excitations, and on higher accuracy calculations of cross sections. There is a vast literature on electronically adiabatic and nonadiabatic (or diabatic) reactions, and references here provide only a small selection of contributions, chosen to give expanded coverage as needed. Electronically adiabatic reactions involve a single potential energy function for the atomic motions, which can be treated as quantal or classical [3–16]. This has been expanded to treatments of nonadiabatic (or diabatic) reactions where several electronic states of the whole system are involved, requiring both potential energy functions and atomic momentum couplings among them [6, 17–20]. More recently, direct (or ab initio) molecular dynamics methods have been combined with treatments of nonadiabatic phenomena. These are very demanding of computational resources and are being developed in the context of machine learning and artificial intelligence procedures as described in the Section 6.7.3. This chapter covers first adiabatic reactions and then follows with nonadiabatic reactions using the concepts and methods of the chapter on diabatic collisions.

To simplify, consider reactions among three atoms A, B, and C. Starting with the system A + BC with bonding between B and C, for the three-atom system in bonding arrangement 1 with internal quantum states α (labelled by electronic quantum number I, vibrational–rotational quantum numbers v_I, and relative motion wavevector $\vec{k}$), a reaction may lead to the following arrangements, labelled by $n = 0, 1, 2, 3$, and 4 in quantum states λ:

(0) A + B + C, in arrangement 0 and states λ_0
(1) A + BC, in arrangement 1 and states α
(2) B + CA, in arrangement 2 and states β
(3) C + AB, in arrangement 3 and states γ
(4) ABC, in arrangement 4 and states λ_4.

A state-to-state reaction is then labelled as $(1\alpha) \to (n\lambda)$. An example is the $1 \to 3$ rearrangement reaction

$$\mathrm{F}\left({}^2P_M\right) + \mathrm{H}_2\left({}^1\Sigma, vjm_j\right) \to \mathrm{FH}\left({}^1\Sigma, v'j'm_j'\right) + \mathrm{H}({}^2S)$$

with diatomic vibrational quantum number v and rotational numbers (j, m_j) and all species in their ground electronic state. The quantum scattering state must also specify, in addition to α, the relative momentum $\vec{P}_1 = \hbar\vec{k}_1$ between F and H_2 and must similarly give $\vec{P}_3$ in final state γ of arrangement 3.

Convenient relative position vectors for an atom–diatomic system can be constructed from the atomic position vectors $(\vec{r}_A, \vec{r}_B, \vec{r}_C)$ locating the centers of atomic masses m_A, m_B, m_C in a space-fixed (SF) reference frame. For arrangement 1, the positions of the total center of mass (CM) $\vec{R}_o$, internal relative position $\vec{r}_1$ in the diatomic BC, and relative position $\vec{R}_1$ of atom A to the CM of BC are

$$\vec{R}_o = \frac{m_A\vec{r}_A + m_B\vec{r}_B + m_C\vec{r}_C}{m_A + m_B + m_C}$$

$$\vec{r}_1 = \vec{r}_C - \vec{r}_B$$

$$\vec{R}_1 = \vec{r}_A - \frac{m_B\vec{r}_B + m_C\vec{r}_C}{m_B + m_C}$$

and similarly for arrangements 2 and 3, with position vectors obtained by permutations of A, B, and C. They are shown in Figure 6.1. These are called Jacobi position vectors, and corresponding masses are the total, internal reduced, and relative reduced masses

$$M_o = m_A + m_B + m_C, \quad m_1 = \frac{m_B m_C}{m_B + m_C}, \quad M_1 = \frac{m_A(m_B + m_C)}{m_A + m_B + m_C}$$

needed in a treatment of their dynamics.

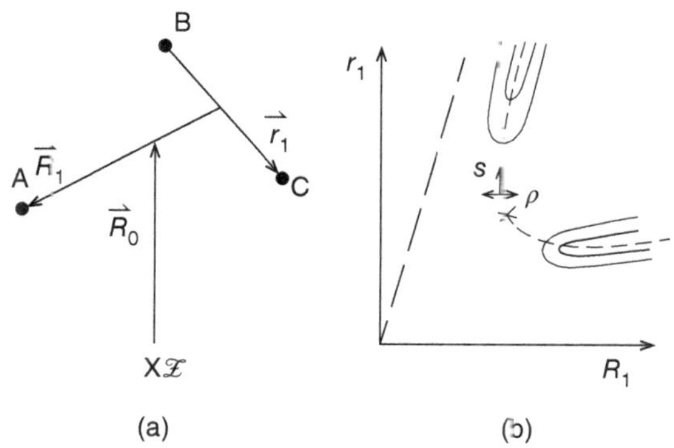

Figure 6.1 (a) Center-of-mass, relative, and internal position vectors for channel 1. (b) Reaction curvilinear coordinates s and ρ.

A compact matrix notation helps to relate positions in different arrangements. Defining a 3×3 transformation matrix $\mathbf{M}_{jo}$ so that for $n = 1$

$$\begin{pmatrix} \vec{R}_o \\ \vec{R}_1 \\ \vec{r}_1 \end{pmatrix} = \mathbf{M}_{1o} \begin{pmatrix} \vec{r}_A \\ \vec{r}_B \\ \vec{r}_C \end{pmatrix}, \quad \mathbf{M}_{1o} = \begin{pmatrix} \dfrac{m_A}{M_o} & \dfrac{m_B}{M_o} & \dfrac{m_C}{M_o} \\ 1 & -\dfrac{m_B}{m_B + m_C} & -\dfrac{m_C}{m_B + m_C} \\ 0 & -1 & 1 \end{pmatrix}.$$

Writing the left equality as $\mathbf{Q}_1 = \mathbf{M}_{1o}\mathbf{Q}_o$ with $\mathbf{Q}$ a 3×1 matrix, introducing $\mathbf{M}_{no}$ for arrangement n and the inverse of $\mathbf{M}_{1o}$, the transformation between arrangements 1 and n is then

$$\mathbf{Q}_n = \mathbf{M}_{no}(\mathbf{M}_{1o})^{-1}\mathbf{Q}_1.$$

6.1.2 Curvilinear and Hyper-spherical Coordinates

Rearrangement in a chemical reaction can be described by trajectories on one or more potential energy surfaces (PESs), which are functions of the atomic positions. The reaction can be treated in a CM reference frame obtained by a space translation of the SF frame to the CM frame of the whole molecular system, with $\vec{R}_o = 0$, which takes care of three degrees of freedom. Furthermore, the axes of the SF–CM frame can be rotated as convenient by choosing three orientation or Euler angles (α, β, γ) for the new rotated body-fixed-CM (BF–CM) frame. For a reaction $A + BC \rightarrow AB + C$, one of the BF–CM axes can be taken along the relative position vector $\vec{R}_j, j = 1, 2, 3$, for reactants or for products, and then the CM frame can be rotated around it to get the three atoms on a (XZ)-plane. This means that the number of needed scalar coordinates decreases from nine to only three variable degrees of freedom for each planar triatomic arrangement in its frame.

Taking the intrinsic reaction coordinates to be the two bond lengths r_1 and r_3 of BC and AB and the angle ϑ between them, a reaction occurs on a triatomic PES depending on all three variables. Reaction arrangements are described by relative and internal bond vectors $\vec{R}_j, \vec{r}_j$, with angle γ_j between them. Changes proceed from coordinates (R_1, r_1, γ_1) to (R_3, r_3, γ_3) going through intermediate conformations along a reaction path with curvilinear coordinates (s, ρ, γ), where s is a linear displacement along the path and ρ is a displacement perpendicular to the path, so that

$$(R_1, r_1, \gamma_1) \rightarrow (s, \rho, \gamma) \rightarrow (R_3, r_3, \gamma_3)$$

with (s, ρ) functions of r_1 and r_3, as shown in Figure 6.1(b). In particular, curvilinear coordinate values $(s^{\ddagger}, \rho^{\ddagger}, \gamma^{\ddagger})$, defined at the location of a transition conformation from reactant to product, usually a potential energy barrier

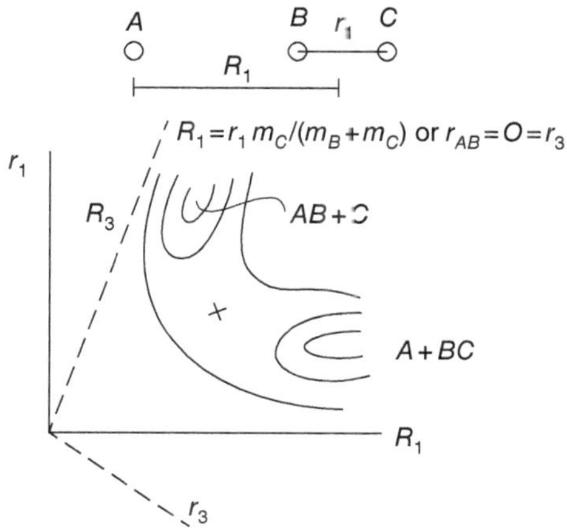

Figure 6.2 Isocontours vs reaction coordinates for channels 1 and 3, in a collinear reaction.

maximum along s, play a role in the treatment of activation-complex reaction rates. The special case of a colinear reaction (with $\gamma_1 = \gamma_3 = \pi$) is shown in Figure 6.2.

Another useful set of coordinates are the so-called hyperspherical coordinates (HSC), which are common to all A + BC, B + CA, and C + AB arrangements and allow a symmetrized expansion of a scattering wavefunction in arrangement functions of the same coordinates, formally similar to what appears in inelastic collisions. The coordinate set in a CM frame contains a hyper-radius ϱ, two arrangement conformation angles η_j (or ω_j) and ϕ_j, and three Euler angles (α, β, γ) specifying a rotation of frames from an SF to a BF one. The hyper-radius is given in terms of the Jacobi mass-scaled position vectors

$$\vec{\underline{R}}_j = \lambda_j \vec{R}_j, \vec{\underline{r}}_j = \lambda_j^{-1} \vec{r}_j$$

with the scale factor $\lambda_j = \sqrt{M_j/\mu} = \sqrt{\mu/m_j}$, where $\mu = \sqrt{m_A m_B m_C/M_o}$, as $\varrho = \sqrt{\underline{R}_j^{\,2} + \underline{r}_j^{\,2}}$. A hyper-angle has been chosen in two different ways in the literature, with $\eta_j = \tan^{-1}(\underline{r}_j/\underline{R}_j)$ generating the Delves HSC or with $\omega_j = 2\tan^{-1}(\underline{r}_j/\underline{R}_j)$ generating a symmetrized HSC. A second angle is given by $\gamma_j = \cos^{-1}[\vec{R}_j \cdot \vec{r}_j/(R_j\, r_j)]$, with both angles ω_j and γ_j within $[0, \pi]$, in the choice of symmetrized HSC. A picture of the PES isocontours for arrangement channels in the plane of the two hyper-angles is shown in Figure 6.3 [9]. Changes in the angle η_j for fixed $\underline{R}_j$ account for variations in the diatomic vibrational coordinate in arrangement j, while changes in γ_j can be related to orientation of the diatomic.

6.1.3 Arrangements and Potential Energies

The equations of motion for the triatomic system can be obtained introducing first the Hamiltonian function of all momenta and positions, $H(\vec{p}_A, \vec{p}_B, \vec{p}_C, \vec{r}_A, \vec{r}_B, \vec{r}_C) = p_A^2/(2m_A) + p_B^2/(2m_B) + p_C^2/(2m_C) + V(\vec{r}_A, \vec{r}_B, \vec{r}_C)$, a sum of atomic kinetic energies and a potential energy V assumed here to be a function of atomic positions only,

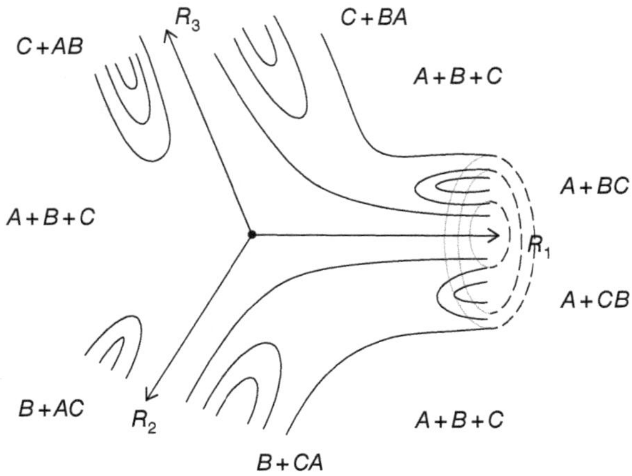

Figure 6.3 PES isocontours for the three arrangement channels showing the location of constant values of the potential energy V vs the two symmetrized hyper-angles (ω_j, φ_j).

in the SF–CM frame. This potential energy follows from an assumption that the dynamics occurs in an adiabatic electronic state changing smoothly from reactants A + BC to products AB + C. Cases where more than one electronic state are involved (a nonadiabatic process) due to couplings of electronic states induced by nuclear motions can be treated as shown in the following section.

The analytical form of the total potential energy depends on what variables one uses and is the sum of the internal potential energy $v_j(\vec{r}_j)$ of the diatomic plus the coupling potential energy $V^{(j)}(\vec{r}_j, \vec{R}_j)$ between atom and diatomic in arrangement $j = 1, 2, 3$, so that $V(\vec{r}_j, \vec{R}_j) = v_j(\vec{r}_j) + V^{(j)}(\vec{r}_j, \vec{R}_j)$. The following table lists the scattering channel, arrangement, internal and coupling potentials, including the arrangements for three free atoms and three interaction atoms (Table 6.1).

If the reaction is not electronically adiabatic, then it is necessary to include in the treatment the potential energy functions V_Γ of nuclear positions for all the

Table 6.1 Scattering channel, arrangement, internal and coupling potentials, including the arrangements for three free atoms and three interaction atoms.

Channel	Arrangement	Internal potential	Coupling potential
0	A, B, C	$v_0 = 0$	$V^{(0)} = V$
1	A, (B, C)	$v_1 = V_{BC}$	$V^{(1)} = V - v_1$
2	B, (C, A)	$v_2 = V_{CA}$	$V^{(2)} = V - v_2$
3	C, (A, B)	$v_3 = V_{AB}$	$V^{(3)} = V - v_3$
4	(A, B, C)	$v_4 = V_{ABC}$	$V^{(4)} = 0$

involved electronic states Γ and also their state coupling functions $G_{\Gamma\Gamma'}$ due to displacements of the nuclei during rearrangement.

6.2 Classical Reaction Dynamics

6.2.1 Atomic Rearrangements

Starting again with a three-atom system in a SF–CM reference frame with fixed axes orientations and considering reactions involving atom–diatomic arrangements $j = 1$ to 3, or $A + BC \rightarrow AB + C$, a classical treatment involves the positions and momenta of the three atoms. In a SF frame, the Jacobi positions are $\vec{R}_o$, $\vec{r}_j$, and $\vec{R}_j$, with corresponding CM, internal and relative momenta $\vec{P}_o$, $\vec{p}_j$, and $\vec{P}_j$, and corresponding masses M_o, m_j, and M_j. In a CM frame, $\vec{P}_o = 0$, $\vec{R}_o = 0$, and the treatment involves only six position variables instead of nine. Here again we consider an electronically adiabatic reaction where the interaction potentials are functions only of the atomic positions for a given electronic state.

Simplifying by working in the CM frame, the Hamiltonian is $H_{CM} = P_j^2/(2M_j) + p_j^2/(2m_j) + V(\vec{r}_j, \vec{R}_j)$, with $V(\vec{r}_j, \vec{R}_j) = v_j(\vec{r}_j) + V^{(j)}(\vec{r}_j, \vec{R}_j)$, as needed in arrangement j. Equations of motion follow by differentiating this Hamiltonian with respect to position and momentum variables, to obtain forces and generalized momenta, for given initial conditions $(\vec{r}_j, \vec{R}_j)_{in}$ and $(\vec{p}_j, \vec{P}_j)_{in}$ at a starting time t_{in} when the collision partners are far away and their coupling potential energy $V^{(j)}$ is vanishing. What follows is a summary of the classical treatment, which has been covered in great detail in several places [4, 21, 22].

Generally, a set of $F = 3N$ generalized coordinates $\boldsymbol{Q} = (Q_1, Q_2, \dots, Q_F)$ can be introduced for a system with N atoms, with their atomic (Cartesian, or Jacobi, or curvilinear) coordinates dependent on the time t. In a CM frame, $F = 3N-3$. Corresponding generalized velocities are $\dot{Q}_n = dQ_n/dt$, $n = 1$ to F.

The potential energy function, taken here to depend only on coordinates, is $\mathcal{V}(Q_1, Q_2, \dots, Q_F)$, while the kinetic energy function is generally of form $\mathcal{K}(Q_1, Q_2, \dots, Q_F, \dot{Q}_1, \dot{Q}_2, \dots, \dot{Q}_F)$. The Lagrangian function of the system is $\mathcal{L} = \mathcal{K} - \mathcal{V}$ and the generalized momenta are given by $P_n = \partial\mathcal{L}/\partial\dot{Q}_n$, which are alternatively used instead of velocities in a treatment with a Hamiltonian function

$$\mathcal{H} = \mathcal{K}(Q_1, Q_2, \dots, Q_F, P_1, P_2, \dots, P_F) + \mathcal{V}(Q_1, Q_2, \dots, Q_F).$$

Equations of motion for positions and momenta follow from

$$\dot{P}_n = -\partial\mathcal{H}/\partial Q_n = -\partial\mathcal{V}/\partial Q_n \text{ and } \dot{Q}_n = \partial\mathcal{H}/\partial P_n = \partial\mathcal{K}/\partial P_n$$

to be supplemented with initial conditions by the sets $\boldsymbol{Q}_{in} = \boldsymbol{Q}(t_{in})$ and $\boldsymbol{P}_{in} = \boldsymbol{P}(t_{in})$. These sets of initial conditions in the $2F$-dimensional phase space $(\boldsymbol{P}, \boldsymbol{Q})$

must be constructed in accordance with experimental conditions, coming from a molecular-beams laboratory arrangement or from a thermodynamical environment.

6.2.2 Reaction Probabilities

In a molecular beams experiment, for example, for crossed beams of A and BC, an initial intermolecular relative translational energy E_T and initial molecular internal rotational and vibrational energies E_R and E_V can be selected in a SF–CM frame. These provide the relative velocity υ_T from $E_T = M_1\, \upsilon_T^2/2$ and internal angular momentum $L_{int} = L_{BC}$ from $E_R = L_{BC}{}^2/(2I_{BC})$. A statistical distribution of initial conditions in the phase space of variables ($\boldsymbol{P}$, $\boldsymbol{Q}$) includes in addition the impact parameter b of the collision, assuming that incoming trajectories are along a Z-axis in CM and the relative angular momentum orientation L_Z (perpendicular to the initial plane of collision) or alternatively the angle η between the initial collision plane and the XZ-plane. A calculation of reaction probabilities must include a sampling of initial intermolecular distances R_1, molecular orientation angles $(\vartheta, \varphi)_1$, planar angle η_1, and internal bond distances r_1.

Given a distribution of initial parameters, $W_{in}(\Lambda_{in})$, with $\Lambda_{in} = (R_1, \eta_1, \vartheta_1, \varphi_1, r_1;\ \upsilon_T, E_R, E_V, b)$ and propagating over time, the bundle of classical trajectories originating on its set of values, for chosen υ_T, E_R, E_V, b, leads to a final distribution $W_{fn}(\boldsymbol{P}, \boldsymbol{Q};\ \upsilon_T, E_R, E_V, b)$ of points in the phase space of variables ($\boldsymbol{P}$, $\boldsymbol{Q}$) after reactions. This can be analyzed to extract the fraction of trajectories leading to the final products AB + C at large R_3. For the phase space with final parameters Λ_{fn} after a reaction with final fixed values υ'_T, E'_R, E'_V and with a distribution of (η_3, ϑ_3, φ_3, r_3) values for the AB structure, the $1 \to 3$ partial reaction probability is, in a compact notation,

$$P^{(R)}_{1\to 3}\left(\upsilon'_T, E'_R, E'_V; \upsilon_T, E_R, E_V, b\right) = \int_{\Lambda_{fn}} d\boldsymbol{P}\, d\boldsymbol{Q}\; W_{fn}(\boldsymbol{P}, \boldsymbol{Q}; \upsilon_T, E_R, E_V, b)$$

from which it is possible to construct differential cross sections $d\sigma_{1\to 3}/d\Omega$ into scattering solid angle $d\Omega = 2\pi b\; db/R_3^2$ and integral cross sections $\sigma_{1\to 3}$ as

$$\sigma_{1\to 3} = 2\pi \int b\, db\, P^{(R)}_{1\to 3}\left(\upsilon'_T, E'_R, E'_V; \upsilon_T, E_R, E_V, b\right).$$

Continuing with the example of A + BC → AB + C for three atoms, a connection can be established between the shapes of PESs and the distributions of final T, R, and V energies, given the initial ones [23, 24]. Considering the PES $V(r_{AB}, r_{CB}, \vartheta_{ABC})$ for ABC with a fixed interbond angle ϑ_{ABC} but as a function of bond distances r_{AB}, r_{BC}, the surface shows a reactant entrance valley for large r_{AB}, a transition region with an activation energy, and an exit product valley as r_{BC} becomes large.

Table 6.2 Conditions favored with large reaction probabilities $P^{(R)}_{1\to 3}$ in terms of energy fractions $f_X = E_X/(E_T + E_R + E_V)$ for three typical shapes of potential energy surfaces. Early and late reaction refer to the location of the activation energy barrier being near the reactant entrance valley and near product exit valley potential energy, respectively.

PES	Initial energy	Final energy
Early reaction	High f_T	High f'_V
Late reaction	High f_V	High f'_T
Local minimum	High f_T	All f_X

Depending on the location and shape of the activation region, the PES describes an early reaction, a late reaction or, if the activation region shows a local well, a trapping reaction. Energies are schematically shown in Figure 6.2 as isocontours of V vs r_{AB}, r_{BC} and along a reaction coordinate s.

General qualitative relations can be deduced between a distribution of an initial energy fraction $f_X = E_X/(E_T + E_R + E_V)$, $X = T, R, V$ and a final one $f_X{}'$ for each type of PES [23]. For example, following a bundle of trajectories, it is found that a reaction with an early activation energy has a larger probability for a high initial f_T and leads to a high final $f_V{}'$. A brief table is next shown giving conditions leading to high reaction probabilities $P^{(R)}_{1\to 3}$ (Table 6.2).

Pictorial descriptions of PESs for a simple collinear A + BC to AB + C reaction with different activation energies are shown in Figure 6.4. Isocontours with entrance and exit valleys correspond to lines of constant potential energy. Panels are arranged from bottom to top for increasing entrance valley energy (or attraction energy, labeled A_T) down from the activation barrier location. Percentages of attraction (A_T), mixed (M_T), and repulsion (R_T, for the exit valley) are shown to the right. Model calculations done for a light incoming atom show that reactions for larger R_T give more highly excited vibrational products AB. From [23–25].

Conceptual and computational simplifications for the trajectories arise when one introduces mass-scaled position variables. To simplify, consider a triatomic system ABC held to a linear conformation by their potential energy, so that $r_{AB} + r_{BC} = r_{AC}$. The kinetic energy in arrangement 1 is

$$K = \frac{1}{2} m_{A,BC} \left(\dot{R}_{A,BC}\right)^2 + \frac{1}{2} m_{BC} \left(\dot{r}_{BC}\right)^2$$

where $m_{A,BC} = M_1$ and $\dot{R}_{A,BC} = \dot{R}_1$ are the reduced mass and relative speed, and $m_{BC} = m_1$ and $\dot{r}_{BC} = \dot{r}_1$ are the diatomic reduced mass and internal speed, respectively, for A + BC. The equivalent expression for arrangement 3, AB + C,

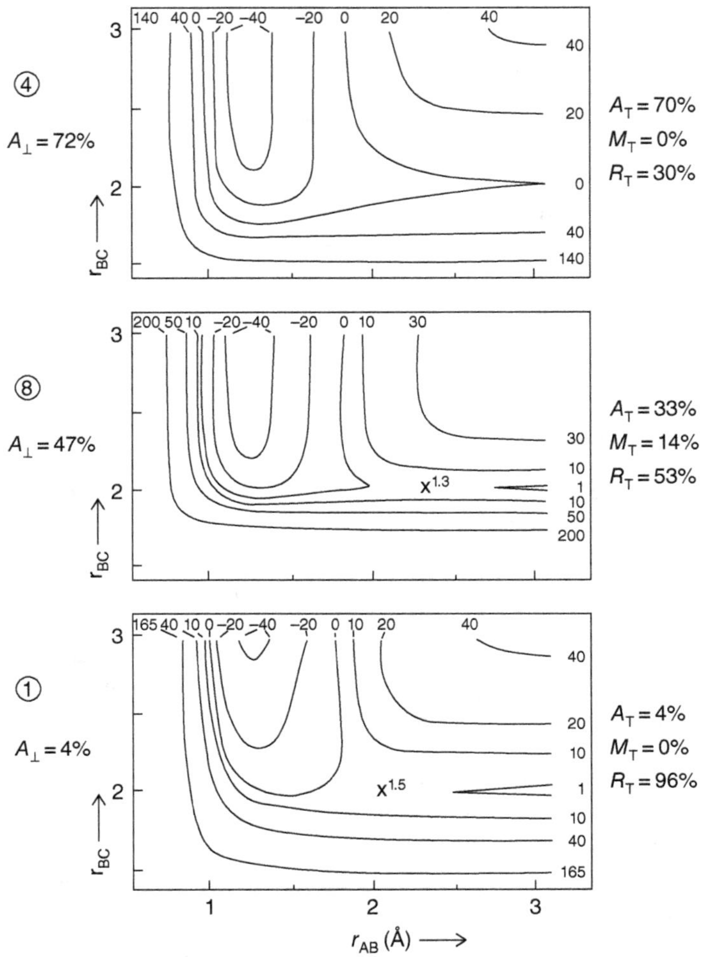

Figure 6.4 Potential energy surfaces for A + BC → AB + C with varying activation energy points and varying attraction valley, mixed, and repulsion valley energies. Source: [23] / John Wiley & Sons.

follows from a cyclic change of atomic labels. In terms of the mass ratio $a_1 = (m_{A,BC}/m_{BC})^{1/4}$, new variables are

$$\underline{R}_1 = a_1 R_1, \underline{r}_1 = (a_1)^{-1} r_1$$

with similar definitions for $\underline{R}_3$ and $\underline{r}_3$. Using these, one finds

$$K = \frac{1}{2}\mu\ [(\dot{\underline{R}}_1)^2 + (\dot{\underline{r}}_1)^2] = \frac{1}{2}\mu\ [(\dot{\underline{R}}_3)^2 + (\dot{\underline{r}}_3)^2]$$

with $\mu = [(m_A m_B m_C)/M]^{1/2}$. Therefore, the motion in channel 1 or 3 can be pictured as a point of mass μ moving in a two-dimensional space with orthogonal axes.

Furthermore, the two sets of variables are simply related by an angle transformation between the two space frames [4]. Starting from $R_1 = (m_C/m_{BC})r_1 + r_3$, the new variables satisfy

$$\begin{pmatrix} \underline{R}_1 \\ \underline{r}_1 \end{pmatrix} = \begin{pmatrix} a_1{}^2(m_C/m_{BC}) & a_1 a_3 \\ 1 & 0 \end{pmatrix} \begin{pmatrix} \underline{r}_1 \\ \underline{r}_3 \end{pmatrix} = \mathbf{T}_1 \begin{pmatrix} \underline{r}_1 \\ \underline{r}_3 \end{pmatrix}$$

and the similar expression relating $(\underline{R}_3, \underline{r}_3)$ to the same 1×2 matrix to the right,

$$\begin{pmatrix} \underline{R}_3 \\ \underline{r}_3 \end{pmatrix} = \mathbf{T}_3 \begin{pmatrix} \underline{r}_1 \\ \underline{r}_3 \end{pmatrix} = \mathbf{T}_3 \mathbf{T}_1{}^{-1} \begin{pmatrix} \underline{R}_1 \\ \underline{r}_1 \end{pmatrix} = \begin{pmatrix} -\cos\beta & -\sin\beta \\ -\sin\beta & \cos\beta \end{pmatrix} \begin{pmatrix} \underline{R}_1 \\ \underline{r}_1 \end{pmatrix}$$

where $\beta = \tan^{-1}[M_o m_B/(m_A m_C)]$ is a rotation angle between the frames.

Another qualitative aspect relates the shapes of PESs to mass ratios in the exchange of species B between A and C. Masses appear in the kinetic energy

$$K = \frac{1}{2} m_{A,BC} (\dot{R}_{A,BC})^2 + \frac{1}{2} m_{BC} (\dot{r}_{BC})^2$$

where $m_{A,BC} = M_1$ and $\dot{R}_{A,BC} = \dot{R}_1$ are the relative reduced mass and speed, respectively, for A + BC. It can be re-expressed in terms of mass-scaled variables $X = (m_{A,BC})^{1/2} R_{A,BC}$ and $x = (m_{BC})^{1/2} r_{BC}$ as $K = (\dot{X}^2 + \dot{x}^2)/2$. Similarly, for AB + C and after cyclic changes of atomic labels, $K = (\dot{Y}^2 + \dot{y}^2)/2$. The PES $V = V_1(X, x) = V_3(Y, y)$ in terms of these variables, for a fixed interbond angle, changes shape as the mass m_B of the transferred atom changes values compared to the other two [24]. In a plot of two frames with orthogonal axes, for (X, x) and (Y, y), with X along R_1 and Y along R_3, indicating incoming and outgoing PES valleys, the angle β between valley directions becomes small as the transferred mass m_B becomes smaller than m_A and m_C. This is shown in Figure 6.1. Incoming trajectories must traverse a PES along a reaction path s of large curvature into the outgoing valley, and this leads to a product diatomic in a high vibrational state. Labelling each atom as H for heavy or L for light, it is possible to analyze different combinations, such as HLH for Cl H I, and to extract qualitative understanding about the distribution of T, R, and V energies of the product molecule [25].

6.3 Quantal Theory of Adiabatic Reactions

6.3.1 Wavemechanical Treatment

A well-established treatment of reactivity introduces steady-state scattering wavefunctions like $\Psi_{1\alpha}^{(+)}$ with outgoing boundary conditions and incoming waves in initial channel 1α for A + BC. This is the solution of a Schrodinger equation,

$$\hat{H}\Psi_{1\alpha}^{(+)} = E\Psi_{1\alpha}^{(+)}$$

with a Hamiltonian operator $\hat{H} = \hat{H}_0 + \hat{V}$, the sum of kinetic and potential energies for the interacting molecular species. For a three-atom reaction starting with A + BC, $\hat{H}_0 = \hat{K}_A + \hat{K}_B + \hat{K}_C$ and $\hat{V} = \hat{V}_{AB} + \hat{V}_{BC} + \hat{V}_{CA} + \hat{V}_{ABC}$, written in terms of atomic kinetic energy operators $\hat{K}_J$ and of atom-pair V_{JK} potential interactions plus a triatomic reminder V_{ABC}. To treat interactions in each collision arrangement j, with an internal potential energy $\hat{V}_j$, it is convenient to decompose the potential into internal and intermolecular interactions as $\hat{V} = \hat{V}_j + \hat{V}^{(j)}$, and the Hamiltonian as

$$\hat{H} = \hat{H}_j + \hat{V}^{(j)}, \quad \hat{H}_j = \hat{K}_{j,rel} + \hat{H}_{j,int}, \quad \hat{H}_{j,int} = \hat{K}_{j,int} + \hat{V}_j$$

containing relative motion and internal motion kinetic energy operators. All these terms are, in general, functions of atomic positions and electronic variables. To simplify here, we consider again electronically adiabatic reactions and postpone nonadiabatic phenomena to the following section.

The total wavefunction has components in all arrangements and is given by

$$\Psi_{1\alpha}^{(+)}(\vec{R}, \vec{r}) = \sum_j \Psi_{1\alpha}^{(j)}(\vec{R}_j, \vec{r}_j).$$

For example in A + BC for arrangement $j = 1$, $\hat{V}_1 = \hat{V}_{BC}$, $\hat{K}_{1,int} = \hat{K}_B + \hat{K}_C$, and $\hat{K}_{1,rel}$ is the relative kinetic energy of A with respect to BC, with reduced mass M_1. Internal motions (rotational, vibrational, and electronic) of the diatomic in arrangement j, in the absence of interactions with the atom, are given by eigenstates $\phi_{j\nu}$, with $\nu = (J, v_J)$ internal quantum numbers, of the internal Hamiltonian $\hat{H}_{j,int} = \hat{K}_{j,int} + \hat{V}_j$ as in

$$\left(\hat{H}_{j,int} - E_{j\nu}\right)\left|\phi_{j\nu}\right\rangle = 0.$$

The total energy is then $E = E_{j\nu} + E_j(k)$, with $E_j(k) = \hbar^2k^2/(2M_j)$ the kinetic energy of relative motion in arrangement j, where the reduced mass is M_j. In general, the total scattering wavefunction $\Psi_{1\alpha}^{(+)}$ has components in all arrangement channels including dissociated molecular states and temporarily bound total states. For A + BC, this involves all the $n = 0, 1, 2, 3, 4$ arrangement channels.

A frequent special case involves only atom–diatom states and only arrangements $j = 1, 2, 3$. We start here with this situation, in the absence of electronic excitations or diatomic dissociation. Indicating relative and internal position variables with $\vec{R}_j$ and $\vec{r}_j$, the total scattering wavefunction can be expanded as

$$\Psi_{1\alpha}^{(+)}(\vec{R}, \vec{r}) = \sum_j \sum_\nu \psi_{j\nu,1\alpha}^{(+)}(\vec{R}_j)\phi_{j\nu}(\vec{r}_j)$$

with the index $\nu = 1$ to N_j for a basis set of N_j pairs of rovibrational states in arrangement j. A free-motion initial state is given by

$$\Psi_{1\alpha}^{(0)}(\vec{R}_1, \vec{r}_1) = \psi_{1\alpha}^{(0)}(\vec{R}_1)\phi_{1\alpha}(\vec{r}_1),$$

where the relative motion is a plane-wave free function for relative momentum $\hbar\vec{k}$ with a chosen normalization as in $\psi_{1\alpha}^{(0)}(\vec{R}_1) = \exp(i\vec{k}.\vec{R}_1)/(2\pi)^{3/2}$.

For a given total energy $E = E_{j\nu} + \hbar^2 k_{j\nu}{}^2/(2M_j)$, the incoming relative flux vector is $\vec{J}_{1\alpha}^{(0)} = (\hbar\vec{k}_{1\alpha}/M_1)(2\pi)^{-3}$. The scattering functions $\psi_{j\nu,1\alpha}^{(+)}(\vec{R}_j)$ satisfy asymptotic boundary conditions at large distances $R_j \to \infty$ for collisions into a solid angle Ω_j, with outgoing waves in all collision channels, so that

$$\Psi_{1\alpha}^{(+)}(\vec{R}_1,\vec{r}_1) \simeq \Psi_{1\alpha}^{(0)}(\vec{R}_1,\vec{r}_1) - \sum_{\alpha'} f_{1\alpha',1\alpha}^{(+)}(\Omega_1)\frac{\exp(ik_{1\alpha'}R_1)}{(2\pi)^{3/2}R_1}\phi_{1\alpha'}(\vec{r}_1), \quad \text{for } R_1 \to \infty$$

$$\Psi_{1\alpha}^{(+)}(\vec{R}_j,\vec{r}_j) \simeq \sum_{\nu} f_{j\nu,1\alpha}^{(+)}(\Omega_j)\frac{1}{(2\pi)^{\frac{3}{2}}}\frac{\exp(ik_{j\nu}R_j)}{R_j}\phi_{j\nu}(\vec{r}_j) \quad \text{for } R_j \to \infty (j \neq 1).$$

Replacing the scattering components in the general expression for the flux of particles, given by

$$\vec{J}_{j\nu,1\alpha}^{(+)}(\vec{R}_j) = \frac{\hbar}{2Mi}\left\{\psi_{j\nu,1\alpha}^{(+)}\left(\vec{R}_j\right)^* \nabla_{\vec{R}}\psi_{j\nu,1\alpha}^{(+)}\left(\vec{R}_j\right) - \left[\nabla_{\vec{R}}\psi_{j\nu,1\alpha}^{(+)}\left(\vec{R}_j\right)^*\right]\psi_{j\nu,1\alpha}^{(+)}\left(\vec{R}_j\right)\right\},$$

the flux into each channel is expressed in terms of the state-to-state scattering amplitudes $f_{j\nu,1\alpha}^{(+)}(\Omega_j)$, as

$$\vec{J}_{j\nu,1\alpha}^{(+)} = \frac{\hbar k_{j\nu}}{M_j}(2\pi)^{-3}\left|f_{j\nu,1\alpha}^{(+)}(\Omega_j)\right|^2\frac{\vec{n}_j}{R_j^2}$$

where $\vec{n}_j = \vec{R}_j/R_j$ is a unit vector in the direction of the emerging scattering wave.

The related state-to-state differential cross section is then

$$\frac{d\sigma_{j\nu,1\alpha}}{d\Omega_j} = \frac{\left(\vec{J}_{j\nu,1\alpha}^{(+)}.\vec{n}_j\right)R_j^2}{J_{1\alpha}^{(0)}} = \frac{k_{j\nu}M_1}{k_{1\alpha}M_j}\left|f_{j\nu,1\alpha}^{(+)}(\Omega_j)\right|^2.$$

Writing the Hamiltonian equation for $\Psi_{1\alpha}^{(+)}$ as $(E - \hat{H}_j)\Psi_{1\alpha}^{(+)} = \hat{V}^{(j)}\Psi_{1\alpha}^{(+)}$, with $E = E_{j\nu} + \hbar^2 k_{j\nu}{}^2/(2M_j)$, the partial scattering functions $\psi_{j\nu,1\alpha}^{(+)}$ are given by projecting both sides of the equation on the internal states $\langle\phi_{j\nu}|$ as

$$\frac{\hbar^2}{2M_j}\left(k_{j\nu}{}^2 + \nabla_{Rj}^2\right)\psi_{j\nu,1\alpha}^{(+)}\left(\vec{R}_j\right) = \sum_{l,\mu}\left\langle\phi_{j\nu}\middle|\hat{V}^{(j)}\left(\vec{R}_j\right)\middle|\varphi_{l\mu}\right\rangle\psi_{l\mu,1\alpha}^{(+)}\left(\vec{R}_j\right).$$

This can be written in a compact matrix notation as

$$\Psi_{1\alpha}^{(+)}\left(\vec{R}_j,\vec{r}_j\right) = \boldsymbol{\phi}(\vec{r}_j)\boldsymbol{\psi}^{(j)}\left(\vec{R}_j\right)$$

introducing the row $N_j \times 1$ matrix $\boldsymbol{\phi}(\vec{r}_j)$ and column $1 \times N_j$ matrix $\boldsymbol{\psi}^{(j)} = \left[\psi_{j\nu,1\alpha}^{(+)}\right]$ and square $N_j \times N_j$ matrices $[\delta_{j'\nu',j\nu}\hbar^2 k_{j\nu}{}^2/(2M_j)] = \epsilon_j$, $\hbar^2\nabla_{Rj}^2/(2M_j)\mathbf{I}$ and $[\langle\phi_{j\nu}|\hat{V}^{(j)}|\phi_{l\mu}\rangle] = \hat{\boldsymbol{V}}^{(j)}$ giving

$$\left[\epsilon_j + \hbar^2\nabla_{Rj}^2/(2M_j)\mathbf{I}\right]\boldsymbol{\psi}^{(j)}\left(\vec{R}_j\right) = \hat{\boldsymbol{V}}^{(j)}\left(\vec{R}_j\right)\boldsymbol{\psi}^{(j)}\left(\vec{R}_j\right).$$

It is possible to obtain the wavefunction $\Psi^{(+)}_{1\alpha}(\vec{R}_j, \vec{r}_j)$ as a solution of this set of coupled partial differential equations, with reaction scattering asymptotic conditions, by constructing local solutions in each arrangement j with coordinates $(\vec{R}_j, \vec{r}_j)$, similar to solutions for inelastic scattering, and then by matching wavefunctions and their derivatives at the boundaries of each pair of arrangements. This was done for the collinear reactive collision in $H + H_2 \rightarrow H_2 + H$, where the change between channel coordinates is simple [26], and was followed by a full treatment of the reaction in three-dimensional space [27], adding expansions in basis sets of orbital angular momentum functions in each arrangement. Details of numerical computations have been reviewed, also for other systems such as $F + H_2 \rightarrow FH + H$ [9, 10].

In some detail, a wavefunction in each arrangement is expanded as done for inelastic scattering, adding here an index j for each reaction arrangement, so that

$$\Psi^{(+)}_{1\alpha}(\vec{R}, \vec{r}; \vec{k}_\alpha) = \sum_{j\lambda} \psi^{(+)}_{j\lambda,1\alpha}(R_j; \vec{k}_{1\alpha}) \Phi_{j\lambda}(\Omega_j, \vec{r})$$

$$\Phi_{j\lambda}(\Omega_j, \vec{r}) = Y_{lm}(\Omega_j)\phi_{j\nu}(\vec{r})$$

$$\psi^{(+)}_{j\lambda,1\alpha}(R_j; \vec{k}_\alpha) = (2\pi)^{-\frac{3}{2}} \sum_{l,m} 4\pi i^l (k_\alpha R_j)^{-1} \psi_{j\lambda,1\alpha}(R_j; k_\alpha) Y_{lm}(\Omega_\alpha)^*$$

where $\lambda = (lm\nu)$. The indices $(j\lambda)$, called here partial-wave reaction channel indices, run over $N_o^{(j)}$ values for open channels and $N_c^{(j)}$ values for closed channels. Expanding Hamiltonian and states in the $\{\Phi_{j\lambda}\}$ basis set give matrices of total order $N^{(j)} = N_o^{(j)} + N_c^{(j)}$. Defining column matrices of states as $\boldsymbol{\psi}^{(j)}_{1\alpha}(R) = [\psi_{j\lambda,1\alpha}(R)]$, they satisfy a set of coupled ordinary differential equations given in matrix form as $\widehat{\mathbf{D}}_R \boldsymbol{\psi}^{(j)}_{1\alpha}(R) = \mathbf{U}(R)\boldsymbol{\psi}^{(j)}_{1\alpha}(R)$, where $\widehat{\mathbf{D}}_R$ is a matrix differential operator derived from the Schrodinger equation. Asymptotic forms of the wavefunctions contain for each arrangement scattering matrices $\boldsymbol{S}^{(j)}$, which can be extracted from numerical solutions of the equations by matching functions and derivatives to scattering functions, at convenient large distances for each arrangement.

6.3.2 Integro-differential Equations for Reaction Amplitudes

The equation for the scattering wavefunction and its boundary condition can be combined in an integral equation as previously done for elastic collisions, introducing the free scattering Green function (or free propagator) for outgoing waves $G_0^{(+)}(\vec{R}, \vec{R}')$, which satisfies

$$\frac{\hbar^2}{2M_j}\left(k_j^2 + \nabla_R^2\right) G_0^{(+)}(\vec{R}, \vec{R}') = \delta(\vec{R} - \vec{R}')$$

and is given in channel j by

$$G_0^{(+)}(\vec{R}, \vec{R}'; k_j) = -\frac{1}{4\pi}\frac{2M_j}{\hbar^2}\frac{\exp(ik_j|\vec{R} - \vec{R}'|)}{|\vec{R} - \vec{R}'|}.$$

Writing the Hamiltonian equation for $\Psi^{(+)}_{1\alpha}$ as $(E - \widehat{H}_j)\Psi^{(+)}_{1\alpha} = \widehat{V}^{(j)}\Psi^{(+)}_{1\alpha}$, with $E = E_{j\nu} + \hbar^2 k_{j\nu}{}^2/(2M_j)$, the partial scattering functions $\psi^{(+)}_{j\nu,1\alpha}$ are again given by projecting both sides of the equation on the internal states $\langle\phi_{j\nu}|$ as

$$\left(\frac{\hbar^2 k_{j\nu}{}^2}{2M_j} + \nabla^2_{Rj}\right)\psi^{(+)}_{j\nu,1\alpha}(\vec{R}_j) = \sum_{l,\mu}\langle\phi_{j\nu}|\widehat{V}^{(j)}(\vec{R}_j)|\phi_{l\mu}\rangle\psi^{(+)}_{l\mu,1\alpha}(\vec{R}_j)$$

where the bracket involves integration only over the internal coordinates in $\phi_{j\nu}$. It follows that the formal solution of these coupled differential equations, in integral form, is

$$\psi^{(+)}_{j\nu,1\alpha}(\vec{R}_j) = \psi^{(0)}_{1\alpha}(\vec{R}_j)\delta_{j\nu,1\alpha} + \int d^3R_j{}'\, G^{(+)}_0\left(\vec{R}_j, \vec{R}'_j; k_{j\nu}\right)$$
$$\times \sum_{l,\mu}\left\langle\phi_{j\nu}|\widehat{V}^{(j)}\left(\vec{R}'_j\right)|\phi_{l\mu}\right\rangle\psi^{(+)}_{l\mu,1\alpha}\left(\vec{R}'_j\right)$$

as can be verified by differentiation of both sides with the left-operator of the previous equation. Here $\delta_{j\nu,1\alpha} = 1$ or 0 for $j\nu = 1\alpha$ or $j\nu \neq 1\alpha$, and $\langle\phi_{j\nu}|\widehat{V}^{(j)}(\vec{R}_j)|\phi_{l\mu}\rangle = \int d^3r\phi_{j\nu}(\vec{r}_j)^* V^{(j)}(\vec{R}_j, \vec{r}_j)\phi_{l\mu}(\vec{r}_l) = V^{(j)}_{j\nu,l\mu}(\vec{R}_j)$ is a state-to-state coupling potential from channel l into exit channel j.

At large distances $R_j \to \infty$ and using the previously given asymptotic form of the free propagator, the amplitude multiplying an outgoing spherical wave gives, with brackets including integration over $\vec{R}_j$,

$$f^{(+)}_{j\nu,1\alpha}(\Omega_j) = -\frac{1}{4\pi}\frac{2M_j}{\hbar^2}(2\pi)^{3/2}\left\langle\vec{k}_{j\nu}|\widehat{T}^{(j)}_{\nu,\alpha}|\vec{k}_{1\alpha}\right\rangle,$$

$$\left\langle\vec{k}_{j\nu}|\widehat{T}^{(j)}_{\nu,\alpha}|\vec{k}_{1\alpha}\right\rangle = \sum_{l,\mu}\left\langle\psi^{(0)}_{j\nu}\phi_{j\nu}|\widehat{V}^{(j)}|\psi^{(+)}_{l\mu,1\alpha}\phi_{l\mu}\right\rangle$$

where $\left\langle\vec{k}_{j\nu}\middle|\widehat{T}^{(j)}_{\nu,\alpha}\middle|\vec{k}_{1\alpha}\right\rangle$ is a scattering transition matrix element between two states with the shown initial and final wavevectors, and the brackets involve all (relative position and internal) variables. This expression, containing the wave $\psi^{(+)}_{l\mu,1\alpha}$ incoming in channel 1α and the final arrangement interaction potential $\widehat{V}^{(j)}$, is referred as a final-channel expression. An alternative form with a scattering state $\psi^{(-)}_{j\nu,l\mu}\phi_{l\mu}$ incoming in channel $j\nu$ and with initial interaction $\widehat{V}^{(1)}$ from state $\psi^{(0)}_{1\alpha}\phi_{1\alpha}$ is called an initial channel expression.

In compact form, these equations can be written in terms of the free motion propagator $\widehat{G}^{(+)}_{0,j\nu}$ with matrix elements $G^{(+)}_0\left(\vec{R}_j, \vec{R}'_j; k_{j\nu}\right) = \left\langle\vec{R}_j\middle|\widehat{G}^{(+)}_{0,j}(k_{j\nu})\middle|\vec{R}'_j\right\rangle$ or more generally in terms of the propagator for channel j, formally defined by $\widehat{G}^{(+)}_{0,j}(E) = (E + i\eta - \widehat{H}_j)^{-1}, \eta \to 0+$. This satisfies the operator equation $(E - \widehat{H}_j)\widehat{G}^{(+)}_{0,j} = \widehat{I}$, the identity operator. Working in the function space with basis set $\left\{|\phi_{j\nu}\rangle\psi^{(0)}_{j\nu}\right\}$, the scattering amplitude $\Psi^{(+)}_{1\alpha}$ satisfies the operator integral equation

$$\Psi^{(+)}_{1\alpha} = \hat{I}_1\Psi^{(0)}_{1\alpha} + \hat{G}^{(+)}_{0,j}\hat{V}^{(j)}\Psi^{(+)}_{1\alpha} = \Psi^{(0)}_{1\alpha} + \hat{G}^{(+)}_{0,j}\hat{T}^{(j)}\Psi^{(0)}_{1\alpha}$$

where $\hat{I}_1$ is a projection operator on channel 1.

This derivation leads to an expression which is formally correct but not useful in calculations of rearrangement cross sections because the propagator $\hat{G}^{(+)}_{0,j}$ generates the correct asymptotic form of the total wavefunction only for channel j as $R_j \to \infty$ but not for the other asymptotic distances $R_k \to \infty$ when $j \neq k$. Instead, it is more general to rewrite the integral equation in terms of the total propagator $\hat{G}^{(+)}$, which satisfies $(E - \hat{H})\hat{G}^{(+)} = \hat{I}$. This follows by multiplying the above equation on both sides times $\left(\hat{G}^{(+)}_{0,j}\right)^{-1} = E - \hat{H}_j$ and regrouping terms, using $= V_j + \hat{V}^{(j)}$, to obtain

$$\Psi^{(+)}_{1\alpha} = \hat{I}_1\Psi^{(0)}_{1\alpha} + \hat{G}^{(+)}\hat{V}\Psi^{(0)}_{1\alpha}.$$

This integral equation allows equivalent descriptions of asymptotic conditions for all channels. However, this integral equation must be carefully applied for rearrangement collisions to avoid introduction of an unphysical integral-equation kernel, as has been analyzed in seminal work by Faddeev [28], who provided a rigorous solution in the special case when the total potential energy is a sum of pair potentials. This was followed by generalizations of the theory for coupled rearrangement channels [5, 8, 29] and for more general many-atom interactions [16].

In the present treatment and also in what follows on coupled arrangement channels (CAC), calculations involve integrals over functions of space variables for the different arrangements. Some integrals, describing transitions among states of the same chemical arrangement k, are similar to those in inelastic scattering and can be obtained as direct integrals in the variables $(\vec{R}_k, \boldsymbol{q}_k)$ of the arrangement, with $\boldsymbol{q}_k$ a set of internal variables of the pair of colliding molecules. They have been described in the chapter on inelastic collisions, where it has been shown that for an atom–diatomic arrangement in the variables $(\vec{R}_k, \vec{r}_k)$, multiple integrals can be expressed in terms of products of one-dimensional integrals in R_k and r_k, introducing orientation-rotational functions of the angles in $(\vec{R}_k, \vec{r}_k)$ as states $Y_{lm_l}(\Omega_k)Y_{jm_j}(\omega_k)$, a product of spherical harmonic functions.

Other more complicated integrals involve functions in two different arrangements k and j and require transformation between corresponding variables and functions, in so-called exchange integrals [12, 27, 30]. These can be calculated in a SF reference frame with Cartesian axes (X, Y, Z) common to all the arrangements or can be treated in a BF-k reference frame (X_k, Y_k, Z_k) for each arrangement k, with the new Z_k axis of orientation vector $\vec{n}_k$ pointing along the vector $\vec{R}_k$. Integrands in SF contain products such as $F^{(k)}_{k\mu}(\vec{R}_k)\phi_{k\mu}(\boldsymbol{q}_k)F^{(j)}_{j\nu}(\vec{R}_j)\phi_{j\nu}(\boldsymbol{q}_j)$ involving two different sets of space variables for arrangements j and k, and it is necessary in calculations to transform variables to a single common set. For this purpose, it helps to choose an origin of coordinates for the SF frame and to introduce the BF frames of j and k arrangements and a rotational operator $\hat{\mathcal{R}}(\alpha, \beta, \gamma)$, function of the Euler angles (α, β, γ) [31], bringing states in frame BF-k to their form for BF-j. Choosing to do calculations in the BF-j frame, the new variables are $(\alpha, \beta, \gamma = 0)$, $\vec{R}_j' = R_j \vec{n}_j$, $\boldsymbol{q}_j'$, and $\vec{R}_k(\vec{R}_j', \boldsymbol{q}_j')$, $\boldsymbol{q}_k(\vec{R}_j', \boldsymbol{q}_j')$. Integrations are done over space volume $R_j' \, d\alpha \sin(\beta) d\beta \, d\boldsymbol{q}_j'$. Transformations of orientation–rotation functions coupled to total angular momenta with quantum numbers (J, K), between two BF frames, can readily be done using the corresponding rotational matrix of elements $D_{K'K}^{(J)}(\alpha, \beta, \gamma)$.

In detail for atom–diatomic systems using BF frames, integrand factors in the six-dimensional integrals (in the variables $(\vec{R}_k, \vec{r}_k)$ or $(\vec{R}_j, \vec{r}_j)$) can be transformed by a frame rotation with three Euler angles $(\alpha, \beta, \gamma)_{kj}$ to obtain overlap integrals between total angular momentum states in a representation $(ljJK)$ given in BF-j by

$$\mathcal{Y}_{l,j}^{JK}(\Omega_j, \omega_j) = \sum_{m_l, m_j} \langle l m_l j m_j | JK \rangle Y_{l m_l}(\Omega_j) Y_{j m_j}(\omega_j)$$

in arrangement j, with orbital lm_l and rotational jm_j quantum numbers, using Clebsch–Gordan coefficients [31], and a similar state $\mathcal{Y}_{l',j'}^{JK'}(\Omega_k, \omega_k)$ constructed from $Y_{l'm_{l'}}(\Omega_k) Y_{j'm_{j'}}(\omega_k)$ in arrangement k. The BF frames can be chosen so that the plane of the three atoms is (XZ)-j and (XZ)-k, both perpendicular to a common Y-axis, and so that the y-coordinate of each atom is zero. Transformation from BF-k to BF-j requires only a rotation by the angle β_{jk} bringing $\vec{R}_k$ into the direction of $\vec{R}_j$, and coordinates transform as

$$x_j = x_k \cos(\beta_{jk}) + z_k \sin(\beta_{jk}), \quad z_j = -x_k \sin(\beta_{jk}) + z_k \cos(\beta_{jk}).$$

The orientation and rotational states change as

$$Y_{l m_{l'}}(\Omega_k) = \sum_{m_l} D_{m_{l'} m_l}^{(l)}(0, -\beta_{jk}, 0) \, Y_{l m_l}(\Omega_j)$$

$$Y_{j m_{j'}}(\omega_k) = \sum_{m_j} D_{m_{j'} m_j}^{(j)}\left(0, -\beta_{jk}, 0\right) \, Y_{j m_j}(\omega_j)$$

and from them one gets

$$\mathcal{Y}_{l',j'}^{JK'}(\Omega_k, \omega_k) = \sum_{K,l,j} C_{l'j',lj}^{JKK'} \, \mathcal{Y}_{l,j}^{JK}(\Omega_j, \omega_j)$$

where the coefficients $C_{l'j',lj}^{JKK'}$ can be expressed in terms of $D_{m_{p'} m_p}^{(p)}(0, -\beta_{jk}, 0)$, with $|l-j| \leq p \leq l+j$, $m_p' = m_l' + m_j'$, and $m_p = m_l + m_j$, using Clebsch–Gordan (or 3-j) coefficients [31].

The exchange integrals for the (A, B, C) system in the overall SF frame are six-dimensional and can be done in the BF-j frames, specifying the three Euler angles that provide the orientation of the plane formed by the three atoms.

Integration can be done over the three orientation angles using properties of spherical harmonics, and the remaining three-dimensional integrals, in the variables that give the triangle of three atoms, can be chosen for example to be r_k, r_j, and ϑ_{kj}, the angle between the two bonds, and can be calculated numerically.

6.3.3 Coupled Arrangement Channels Method

A more general treatment of the collision event, which includes, in addition to atom–diatom arrangements $j = 1, 2, 3$, a channel with three free atoms $j = 0$ and also a channel $j = 4$ with three bound atoms, can be developed within an operator formalism that starts with the decomposition

$$\Psi_{1\alpha}^{(+)} = \sum_{j=0}^{4} \Psi_{1\alpha}^{(j)}.$$

Here the term with $j = 4$ is included to allow for atomic motion correlations at short distances for all atoms, and it is null for asymptotic distances $R_j \to \infty$ in all channels. The remaining channels must show asymptotic forms with scattering amplitudes corresponding to an atom-plus-diatomic, or three free atoms. For collisions starting in channel 1 with total energy E below dissociation into three free atoms, the asymptotic conditions are

$$\Psi_{1\alpha}^{(j)}(\vec{R}_j, \vec{r}_j) \simeq \sum_{v} \left[\delta_{jv,1\alpha} \psi_{1\alpha}^{(0)}(\vec{R}_1)\phi_{1\alpha}(\vec{r}_1) + f_{jv,1\alpha}^{(+)}(\Omega_j) \frac{1}{(2\pi)^{3/2}} \frac{\exp(ik_{jv} R_j)}{R_j} \phi_{jv}(\vec{r}_j) \right]$$

for $R_j \to \infty$, and with the summation over internal states ϕ_{jv} limited by $E_{jv} \le E$.

Proceeding again from $\hat{H}\Psi_{1\alpha}^{(+)} = E\Psi_{1\alpha}^{(+)}$ and with $\hat{H} = \hat{H}_j + \hat{V}^{(j)}$ for each arrangement, let

$$(E - \hat{H}_j)\Psi_{1\alpha}^{(j)} = \hat{V}^{(j)}\Psi_{1\alpha}^{(j)} + \sum_{k \neq j} (\hat{H} - E)\Psi_{1\alpha}^{(k)} = \sum_{k} V^{(jk)}\Psi_{1\alpha}^{(k)}$$

$$V^{(jk)} = \hat{V}^{(j)}\delta_{jk} + (\hat{H} - E)(1 - \delta_{jk})$$

where $V^{(jk)}$ is a new channel-to-channel (or re-coupling) coupling potential energy, which must satisfy $\sum_j V^{(jk)} = \hat{V}^{(j)}$ so that when one adds over index j the first and last side of the first line above, one can recover the original equation for $\Psi_{1\alpha}^{(+)}$. An alternative expression for the recoupling potential is $V^{(jk)} = \hat{H} - E + \delta_{jk}(E - \hat{H}_j)$.

This channel-to-channel coupling can be done in a variety of ways, which however must contain physical meaning for the reaction process. An acceptable selection for a reaction starting in $j = 1$ and with no channel 0 present is $V^{(11)} = 0$, $V^{(12)} = \hat{V}^{(2)}$, $V^{(13)} = V^{(14)} = 0$, and cyclically in channel labels, for $j = 2$ to 4. This selection couples channel 1 with 2, channel 2 with 3, channel 3 with 4, and channel 4 with 1 and satisfies the summation requirement to reconstruct the first line above. A counter-example would be $V^{(jj)} = \widehat{V}^{(j)}$ and $V^{(jk)} = 0$ for $j \neq k$, which would not couple rearrangement channels even though it satisfies the summation rule.

For $j = 1$, the described choice gives $(E - \widehat{H}_1)\Psi^{(1)}_{1\alpha} = \widehat{V}^{(2)}\Psi^{(2)}_{1\alpha}$. When $R_1 \to \infty$, the left side goes to zero but $\widehat{V}^{(2)}$ does not, so that $\Psi^{(2)}_{1\alpha}$ must be null asymptotically in channel 1. And when $R_2 \to \infty$, then $\widehat{V}^{(2)} \approx 0$, and this requires that $\Psi^{(1)}_{1\alpha} \approx 0$ asymptotically in channel 2. Using the same arguments cyclically for the other channels, it is concluded that $\Psi^{(j)}_{1\alpha} \approx 0$ in all channels except in j, so that this component is localized into the region of space variables of channel j, as shaped by the interaction potential energy $\widehat{V}(\vec{r}_A, \vec{r}_B, \vec{r}_C)$ there. This localization property can be used to improve the efficiency of calculations of the state components. Other acceptable choices of channel-to-channel couplings are possible and have been used in calculations [8].

The differential equations for the channel components, with their asymptotic conditions, can be formally solved introducing the propagator $\widehat{G}^{(+)}_{0,j} = (E + i\eta - \widehat{H}_j)^{-1}$ for free motion in channel j, with $\eta \to 0+$. Its asymptotic form for $R_j \to \infty$, as it operates on a function $F(\vec{R}_j, \vec{r}_j)$, is

$$\left\langle R_j \Phi_{j\nu} \middle| \widehat{G}^{(+)}_{0,j} \middle| F \right\rangle \approx -\frac{1}{4\pi}\frac{2M_j}{\hbar^2}\frac{\exp(i\,k_{j\nu}R_j)}{R_j}(2\pi)^{\frac{3}{2}} \left\langle \psi^{(0)}_{j\nu}\Phi_{j\nu} \middle| F \right\rangle .$$

The formal solution for the equation with coupled amplitudes is

$$\Psi^{(j)}_{1\alpha} = \delta_{j1}\Psi^{(0)}_{1\alpha} + \widehat{G}^{(+)}_{0,j}\sum_{k=1}^{4} V^{(jk)}\Psi^{(k)}_{1\alpha}$$

which generalizes the Lippmann–Schwinger equation to make it applicable to rearrangement collisions and gives asymptotically

$$f^{(+)}_{j\nu,1\alpha}(\Omega_j) = -\frac{1}{4\pi}\frac{2M_j}{\hbar^2}(2\pi)^3\sum_{k=1}^{4}\left\langle \Psi^{(0)}_{j\nu} \middle| V^{(jk)} \middle| \Psi^{(k)}_{1\alpha} \right\rangle$$

from which the state-to-state rearrangement differential cross section $d\sigma_{j\nu,1\alpha}/d\Omega_j$ is obtained.

6.3.4 Rearrangements in a Three-atom System

A special treatment of rearrangement scattering can be developed for a triatomic system with potential energy

$$V = V_{AB} + V_{BC} + V_{CA} + V_{ABC}$$

where atom-pair potentials $V_{IJ}(\vec{r}_I, \vec{r}_J)$ depending on atomic positions are large and the remaining intrinsic triatomic term V_{ABC} is small. This is the case when the three atoms have closed atomic shells, as in He + Ar_2. Such decomposition can also be extended to some pairs of open-shell atoms with doublet states like H(^2S) where diatomic states are singlet or triplet like $^{1,3}\Sigma$ for H_2. This is a general procedure extensively developed in the diatomics-in-molecules treatment [25]. In these cases, the atom-pair decomposition can be extended using atomic spin variables $\xi = \pm 1/2$ and pair potentials $V_{IJ}(\vec{r}_I, \xi_I; \vec{r}_J, \xi_J)$.

The rearrangement scattering equations can now be rewritten in terms of atom-pair transition amplitudes and arrangement channels can be identified as before with $j = 1$ for A + BC, $j = 2$ for B + CA, $j = 3$ for C + AB, and $j = 4$ for ABC, each containing now only a pair potential in each channel, as in $V_1 = V_{BC}$, $V^{(1)} = 0$, $V_2 = V_{CA}$, $V_3 = V_{AB}$, and $V_4 = V_{ABC}$. The free-motion Hamiltonian is a sum of atomic kinetic energies as in $\hat{H}_0 = \hat{K}_A + \hat{K}_B + \hat{K}_C$. Inter-channel couplings can be chosen as

$$V^{(jk)} = V_j(1 - \delta_{jk}).$$

The calculation of reaction cross sections can be done solving the CAC differential (or integral) equations just described, or the cross sections can be obtained within an operator formalism that introduces two-atom transition operators, as shown in more detail in Chapter 7.

The coupled differential equations are simplified when two or the three atoms are identical, meaning that they have identical nuclei and electronic shells. It is then possible to combine arrangement amplitudes to satisfy nuclear exchange symmetry, for either Boson or Fermion nuclei, to decrease the number of coupled reaction equations. This has been developed in the literature for reaction differential equations [32, 33] and also for the coupled operator equations that arise in many-atom treatments [16, 34].

6.4 Calculation of Adiabatic Reaction Cross Sections

6.4.1 Approximation of Distorted Waves for Rearrangement Collisions

A straightforward approximation for the scattering amplitude $f^{(+)}_{j\nu,1\alpha}(\Omega_j)$ follows by replacing $\Psi^{(k)}_{1\alpha}$ with the known unperturbed $\Psi^{(0)}_{1\alpha}$, which provides a Born-type approximation,

$$\frac{d\sigma_{j\nu,1\alpha}{}^{(B)}}{d\Omega_j} = \frac{k_{j\nu}\,M_1}{k_{1\alpha}M_j}\left(\frac{M_j}{2\pi\hbar^2}\right)^2 \left|\sum_{k=1}^{4}\left\langle \Psi^{(0)}_{j\nu}\middle| V^{(jk)} \middle| \Psi^{(0)}_{1\alpha}\right\rangle\right|^2.$$

However, this is of limited use for molecular collisions because the short-range repulsive forces of atomic cores lead to large distortions of the incoming waves and exclude them from short-range interaction regions. The Born approximation may work for reactions happening only at large impact parameters, where repulsion by cores is minimal. The effects of short-range repulsive potentials can be accounted for by means of a perturbation treatment, which introduces distortion potentials and distorted wavefunctions, in extensions of treatments for inelastic collisions [1, 35].

A better alternative to the Born approximation is to replace the wavefunction $\Psi^{(k)}_{1\alpha}$ with a distorted-wave function $\Phi^{(k)}_{1\alpha}$ generated by a distortion potential $U^{(k)}(R_k, \boldsymbol{Q}_k)$ in channel k, containing a repulsive potential energy in the short-range region along intermolecular distance R_k and dependent on other orientation and internal coordinates in $\boldsymbol{Q}_k$. Returning to the coupled equations for the components $\Psi^{(j)}_{1\alpha}$ and subtracting on both sides $U^{(j)}\Psi^{(j)}_{1\alpha}$, one obtains

$$(E - \widehat{H}_{Dj})\Psi^{(j)}_{1\alpha} = \sum_k U^{(jk)}\Psi^{(k)}_{1\alpha}$$

with the distortion Hamiltonian $\widehat{H}_{Dj} = \widehat{H}_j + U^{(j)}$ and the new recouplings $U^{(jk)} = V^{(jk)} - \delta_{jk}U^{(j)}$. These coupled equations can be solved introducing the distorted-waves propagator $\widehat{G}^{(+)}_{Dj} = (E + i\eta - \widehat{H}_{Dj})^{-1}$ and the eigenstates in $(E - \widehat{H}_{Dj})\Phi^{(\pm)}_{1\alpha} = 0$ with outgoing (+) or incoming (−) waves, writing for scattering starting in arrangement 1,

$$\Phi^{(+)}_{1\alpha}(\vec{R}_1, \vec{r}_1) \approx \psi^{(0)}_{1\alpha}(\vec{R}_1)\phi_{1c}(\vec{r}_1) + \sum_{\alpha'} f^{(D)}_{1\alpha',1\alpha}(\Omega_1)\frac{1}{(2\pi)^{3/2}}\frac{\exp\left(ik_{1\alpha'}R_1\right)}{R_1}\phi_{1\alpha'}(\vec{r}_1) \text{ for } R_1 \to \infty$$

asymptotically for $R_j \to \infty$. Similarly, distorted functions $\Phi^{(-)}_{j\nu}$ have incoming-wave boundary conditions in arrangement j.

From the scattering components starting in arrangement 1,

$$\Psi^{(j)}_{1\alpha} = \delta_{j1}\Phi^{(+)}_{1\alpha} + \widehat{G}^{(+)}_{Dj}\sum_k U^{(jk)}\Psi^{(k)}_{1\alpha}$$

it follows that

$$f^{(+)}_{j\nu,1\alpha}(\Omega_j) = \delta_{j\nu,1\alpha}f^{(D)}_{1\alpha,1\alpha}(\Omega_j) + f^{(C)}_{j\nu,1\alpha}(\Omega_j)$$

$$f^{(D)}_{1\alpha,1\alpha}(\Omega_j) = -\frac{1}{4\pi}\frac{2M_1}{\hbar^2}(2\pi)^3\left\langle \Phi^{(-)}_{1\alpha}\middle| U^{(1)}\middle|\Phi^{(+)}_{1\alpha}\right\rangle$$

$$f^{(C)}_{j\nu,1\alpha}(\Omega_j) = -\frac{1}{4\pi}\frac{2M_j}{\hbar^2}(2\pi)^3\sum_{k=1}^{4}\left\langle \Phi^{(-)}_{j\nu}\middle| U^{(jk)}\middle|\Psi^{(k)}_{1\alpha}\right\rangle$$

where the first term gives the distorted-wave scattering amplitude of the initial arrangement, and the second term gives all the reaction amplitudes arising from arrangement couplings. Alternatively,

$$f_{j\nu,1\alpha}^{(+)}(\Omega_j) = -\frac{1}{4\pi}\frac{2M_j}{\hbar^2}(2\pi)^3 \left\langle \vec{k}_{j\nu}\middle|\hat{T}_{\nu,\alpha}^{(j)}\middle|\vec{k}_{1\alpha}\right\rangle$$

$$\left\langle \vec{k}_{j\nu}\middle|\hat{T}_{\nu,\alpha}^{(j)}\middle|\vec{k}_{1\alpha}\right\rangle = \delta_{j\nu,1\alpha}T_{1\alpha,1\alpha}^{(D)} + T_{j\nu,1\alpha}^{(C)}$$

with $T_{1\alpha,1\alpha}^{(D)}$ a T-matrix term for elastic scattering in channel 1 and $T_{j\nu,1\alpha}^{(C)}$ a T-matrix reaction amplitude.

The distorted-wave approximation (DWA) approximates $\Psi_{1\alpha}^{(k)}$ with $\delta_{k1}\Phi_{1\alpha}^{(+)}$ in the reaction amplitude to obtain

$$\frac{d\sigma_{j\nu,1\alpha}{}^{(DWA)}}{d\Omega_j} = \frac{k_{j\nu}\,M_1}{k_{1\alpha}M_j}\left(\frac{M_j}{2\pi\hbar^2}\right)^2 \left|\left\langle \Phi_{j\nu}^{(-)}\middle|U^{(j1)}\middle|\Phi_{1\alpha}^{(+)}\right\rangle\right|^2$$

which allows calculations requiring only single-channel (but possibly multistate) scattering functions, by space integrations.

Calculation of cross sections for an A + BC → AB + C reaction starts within the DWA with a choice of a realistic PES and proceeds to introduce related distortion potentials for initial and final arrangements. The distortion potentials $U^{(j)}$ are in general functions of the intermolecular distance R_j and also of orientation and internal variables of the diatomic in arrangement j, in which cases the needed states $\Phi_{j\nu}^{(\pm)}$ must be obtained as solutions of inelastic scattering events in the j-arrangement.

The DWA is effectively an approximation of the scattering components $\Psi_{1\alpha}^{(j)}$ to first order in the recoupling potentials $U^{(jk)} = V^{(jk)} - \delta_{jk}U^{(j)}$. Therefore, the choice of the distortion potential $U^{(j)}(R_j, \boldsymbol{Q}_j)$ is very relevant to the accuracy of the DWA. This choice provides distorted waves and must be followed by integration in the state-to-state transition matrix elements. This procedure was followed in early applications of the DWA before introduction of the CAC formalism with coupled arrangements, using instead a final-channel potential coupling equivalent to the choice $V^{(j1)} = \hat{V}^{(j)}$.

The CAC formalism is more general, but it yet requires a correct choice for the distortion potential. For a simple radial potential $U^{(j)}(R_j)$, the distorted waves can be calculated from elastic scattering equations in arrangement j but do not account for the coupling of internal states $\phi_{j\nu}$, and the potential generates a state-distorted approximation (SDWA). This potential can be generalized to include state-coupling (or inelastic events) in each arrangement, in which case many internal states may be needed to describe the effects of rearrangement.

At low collision energies (thermal or hyperthermal), a useful alternative choice is obtained allowing for the adiabatic adaptation of states of each collision partner to the presence of the other one at a fixed distance R_j, generating adiabatic states $\eta_\nu^{(j)}(\boldsymbol{q}_j; R_j)$ to be used in a spatial average of the full potential energy function, with $\boldsymbol{q}_j$ the set of internal and orientation variables changing during collisions in arrangement j. For a reaction A + BC → AB + C, the adiabatic states in the entrance channels are $\eta_\alpha^{(1)}(\boldsymbol{q}_1; R_1)$, and they are $\eta_\gamma^{(3)}(\boldsymbol{q}_3; R_3)$ in the exit channels.

The adiabatic states satisfy the equation

$$[\widehat{K}_{Q_j} + V(R_j, \boldsymbol{Q}_j)]\eta_\nu^{(j)}(\boldsymbol{Q}_j; R_j) = \epsilon_\nu^{(j)}(R_j)\eta_\nu^{(j)}(\boldsymbol{Q}_j; R_j)$$

where $\widehat{K}_{Q_j}$ is a kinetic energy operator, $V(R_j, \boldsymbol{Q}_j) = V_j(\boldsymbol{Q}_j) + V^{(j)}(R_j, \boldsymbol{Q}_j)$ is the full PES, with V_j indicating the internal potential energy function of molecules in arrangement j, and $\varepsilon_\nu^{(j)}$ is an adiabatic energy changing only with R_j. For small values of R_j, the interaction potential $V^{(j)}(R_j, \boldsymbol{Q}_j)$ is highly repulsive and the adiabatic states are negligibly small. The distortion potential is

$$U^{(j)}(R_j) = \sum_\nu \left|\eta_\nu^{(j)}(R_j)\right\rangle \Delta\epsilon_\nu^{(j)}(R_j) \left\langle \eta_\nu^{(j)}(R_j)\right|$$

using the bracket notation so that $\eta_\nu^{(j)}(\boldsymbol{Q}_j; R_j) = \left\langle \boldsymbol{Q}_j \middle| \eta_\nu^{(j)}(R_j)\right\rangle$, with $\Delta\epsilon_\nu^{(j)}(R_j) = \epsilon_\nu^{(j)}(R_j) - \epsilon_\nu^{(j)}(\infty)$. It already includes the internal bonding potential V_j. The distortion Hamiltonian is then $\widehat{H}_{Dj} = \widehat{K}_{R_j} + \widehat{K}_{Q_j} + U^{(j)}$. A scattering distorted state in arrangement j takes the form $\Phi_{\alpha}^{(j)} = \sum_\nu F_{j\nu,1\alpha}(R_j)\eta_\nu^{(j)}(\boldsymbol{Q}_j; R_j)$ and its intermolecular factors $F_{j\nu,1\alpha}(R_j)$ are given, after projection of the CAC equations on $\left\langle \eta_\nu^{(j)}(R_j)\right|$, by

$$\left[E - \left(\widehat{K}_{R_j} + \Delta\epsilon_\nu^{(j)}\right)\right] F_{j\nu,1\alpha}(R_j) = \sum_{k,\mu} \left\langle \eta_\nu^{(j)}(R_j)\middle|(V^{(jk)} - \delta_{jk}U^{(j)})\middle|\eta_\mu^{(k)}(R_j)\right\rangle F_{k\mu,1\alpha}(R_j)$$

which is a relatively simple set of coupled radial differential equations in the intermolecular distance parameter R_j. The adiabatic states $\eta_\nu^{(j)}(\boldsymbol{Q}_j; R_j)$ are however quite complicated in their most general form, which involves coupled vibrational and hindered rotation motions. A variety of simpler choices of adiabatic states has been explored in calculations [36]. One such choice for atom–diatomic reactions factors out the rotational state for $\vec{r}_j$ so that $\eta_\nu^{(j)}(\vec{r}_j; R_j) = \eta_\nu^{(j)}(r_j; R_j)Y_{lm}(\hat{r}_j)$ and calculates the vibrational functions $\eta_\nu^{(j)}(r_j; R_j)$ numerically. This gives an adiabatic vibrational basis set leading to the VADW approximation, which has been found to provide good distributions P_{v_3} of final vibrational states in some applications [36].

However, more accurate adiabatic states are needed to account for vibrational distortions due to bond rearrangement and their coupling to hindered rotations during the reaction, as required to obtain accurate magnitudes of reaction cross sections [37]. This has been shown for the $H + H_2 \to H_2 + H$ reaction with the choice of adiabatic states

$$\eta_\nu^{(j)}(\vec{r}_j; R_j) = \frac{1}{\sqrt{4\pi}} \sum_\lambda (2\lambda + 1)^{\frac{1}{2}} r_j^{-1} F_\lambda(r_j; R_j) P_\lambda(\cos\gamma_j)$$

where $\cos\gamma_j = \vec{r}_j \cdot \vec{R}_j/(r_jR_j)$ and an expansion in the P_λ Legendre polynomials has been introduced to describe hindered rotations. The state $F_\lambda(r_j; R_j)$ is a solution

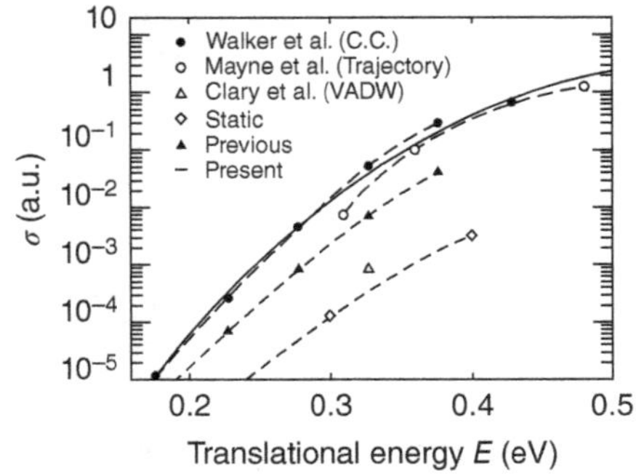

Figure 6.5 Integral reaction cross section vs collision energy for $H + H_2$ from an adiabatic DWA with a distortion potential energy surface containing reactant and product energy wells [37], compared to accurate coupled-channel results [38].

of a radial Schrodinger equation with the potential energy varying around $V(r_j, R_j, \gamma_j = 0)$, suitable for a reaction proceeding through a linear activated complex, and giving eigenenergies $\varepsilon_n(R_j)$. This PES shows at a large distance R_j two wells vs the distance r_j, one for the incoming energy valley and another for the product valley, with both merging at shorter R_j distances. Calculated numerically and provided that $\varepsilon_n(R_j)$ varies smoothly with R_j, the distorted scattering waves are expanded as $\Phi_{1\alpha}^{(j)} = \sum_v F_{jv,1\alpha}(R_j)\eta_v^{(j)}(\boldsymbol{Q}_j; R_j)$ in an improved DWA, also called an adiabatic T-matrix approximation [37]. Results are shown in Figure 6.5 and compare very well with accurate coupled-channel values for integral cross sections obtained with the LSTH PES and the R-matrix propagation procedure [38, 39].

Other adiabatic DWAs tend to give correct final relative probabilities of product states but give reaction cross sections which are quite smaller, because their distortion potentials do not allow for needed wavefunction amplitudes in the reaction region. Conditions under which the different DWAs are reliable have been reviewed in the literature [1, 10, 11, 36, 37].

In the DWAs and also in what follows using normalizable functions, calculations involve integrals over functions of space variables for the different arrangements. As previously mentioned, some integrals, describing transitions among states of the same chemical arrangement k, are similar to those in inelastic scattering and can be directly obtained as integrals in the variables $(\vec{R}_k, \boldsymbol{Q}_k)$ of the arrangement. Other more complicated integrals involve functions in two different arrangements k and l and require transformation between corresponding variables, in so-called atomic exchange integrals [12, 27, 30].

6.4.2 Variational Calculations with Expansions in Normalizable Functions

Even though scattering states are not normalizable (or L^2) functions, it is possible to develop a computational solution of the reaction equations, introducing expansions in basis sets of L^2 functions in each arrangement [10, 14]. Considering the density amplitude

$$\zeta_{1\alpha}^{(j)} = \sum_k U^{(jk)} \Psi_{1\alpha}^{(k)}$$

in channel j as a function of variables $(\vec{R}_j, \vec{r}_j)$ for relative and internal positions, it is seen that the amplitude vanishes as $R_j \to \infty$, by construction of $U^{(jk)}(\vec{R}_j, \vec{r}_j)$, and is also null for large $\vec{r}_j$ values insofar these are variables in bound internal state (excluding here dissociative states). Therefore one can introduce an orthogonalized basis set $\left\{ u_n^{(j)}(\vec{R}_j, \vec{r}_j), n = 1 \text{ to } N_j \right\}$, with $\left\langle u_m^{(i)} \middle| u_n^{(j)} \right\rangle = \delta_{ij}\delta_{mn}$, and expand the amplitude as

$$\zeta_{1\alpha}^{(j)}(\vec{R}_j, \vec{r}_j) = \sum_n u_n^{(j)}(\vec{R}_j, \vec{r}_j) c_{n,1\alpha}^{(j)}$$

with coefficients $c_{n,1\alpha}^{(j)}$ to be determined as solutions from the coupled equations for the channel functions $\Psi_{1\alpha}^{(j)}$.

The channel functions and an accurate form of the transition amplitude can be derived using the S-matrix form of the Kohn variational treatment for scattering with outgoing and incoming asymptotic conditions for wavefunctions or equivalently the T-matrix form [40], in a version expanded to include coupled arrangements, and expansions of the wavefunctions in basis sets of L^2 functions. A Kohn variational functional

$$\tilde{J}_{j\nu,1\alpha}^{(K)} \left[\tilde{\Psi}_{j\nu}^{(-)}, \tilde{\Psi}_{1\alpha}^{(+)} \right] = \left\langle \tilde{\Psi}_{j\nu}^{(-)} \middle| (E - \hat{H}) \middle| \tilde{\Psi}_{1\alpha}^{(+)} \right\rangle$$

of trial functions $\tilde{\Psi}_{j\nu}^{(\pm)}$ is constructed, with each trial function chosen as a sum of a multichannel distorted wave with trial transition amplitudes $\tilde{T}_{j\nu,1\alpha}$ in its asymptotic form plus a function describing short-range structures, as previously described. The Kohn variational theorem states that for a first-order variation (indicated by a delta increment) one has that

$$\delta \left\{ \tilde{J}_{j\nu,1\alpha}^{(K)} \left[\tilde{\Psi}_{j\nu}^{(-)}, \tilde{\Psi}_{1\alpha}^{(+)} \right] - \tilde{T}_{j\nu,1\alpha} \right\} = 0.$$

The functional minus the chosen trial T-matrix is, as shown, stationary for first-order variations of parameters. The stationary value (or *stat*) of the functional is an approximation to the exact T-matrix value with respect to variations of the trial functions and provides the scattering matrix $\tilde{S}_{j\nu,1\alpha} = \delta_{j\nu,1\alpha} + \tilde{T}_{j\nu,1\alpha}$.

Alternatively letting $\tilde{T}_{j\nu,1\alpha} = stat(T_{j\nu,1\alpha}) - \tilde{T}_{j\nu,1\alpha}$, and insofar $stat\left(J_{j\nu,1\alpha}^{(K)} \left[\tilde{\Psi}_{j\nu}^{(-)}, \tilde{\Psi}_{1\alpha}^{(+)} \right]\right) = 0$, it follows that $stat(T_{j\nu,1\alpha}) = \tilde{T}_{j\nu,1\alpha} + \tilde{J}_{j\nu,1\alpha}^{(K)} \left[\tilde{\Psi}_{j\nu}^{(-)}, \tilde{\Psi}_{1\alpha}^{(+)} \right]$ gives the transition amplitude after calculation of the stationary values for the functional parameters.

To implement its use, the short-range term in the trial functions is given as an expansion in L^2 basis functions with a set of trial coefficients $\left\{\tilde{c}^{(j)}_{n,jv}, n = 1 \text{ to } N_j\right\}$ to be determined by variation of the functional. A trial function for arrangement j with scattering originating in state 1α is

$$\Psi^{(j)}_{1\alpha}(\vec{R}_j, \vec{r}_j) = \delta_{j,1}\Phi^{(+)}_{1\alpha}(\vec{R}_1, \vec{r}_1) + \sum_n u^{(j)}_n(\vec{R}_j, \vec{r}_j)\tilde{c}^{(j)}_{n,1\alpha}$$

$$\Phi^{(+)}_{1\alpha}(\vec{R}_1, \vec{r}_1) \approx \psi^{(0)}_{1\alpha}(\vec{R}_1)\phi_{1\alpha}(\vec{r}_1) - \frac{(2\pi)^2 M_1}{\hbar^2}\sum_{\alpha'}\tilde{T}_{1\alpha',1\alpha}\frac{1}{(2\pi)^{3/2}}\frac{\exp(ik_{1\alpha'}R_1)}{R_1}\phi_{1\alpha'}(\vec{r}_1) \quad \text{for } R_1 \to \infty$$

and the Kohn functional becomes a function of the coefficients in $\left\{\tilde{c}^{(j)}_{n,jv}\right\}$ and of $\tilde{T}_{jv,1\alpha}$.

For reactions with CAC, the trial function is $\Psi^{(+)}_{1\alpha} = \sum_{j=0}^{4}\Psi^{(j)}_{1\alpha}$ and the variational expression becomes

$$\delta\left\{\tilde{J}^{(K)}_{jv,1\alpha}\left[\left\{\tilde{c}^{(j)}_{n,jv}, j = 1 \text{ to } 4\right\}\right] - \tilde{T}_{jv,1\alpha}\right\} = 0$$

with parameters $\tilde{c}^{(j)}_{n,jv}$ and $\tilde{T}_{jv,1\alpha}$ to be determined. This procedure provides accurate and stable solutions for the coefficients and properly symmetrized state-to-state transition amplitudes [10, 15].

The expansion coefficients must be obtained for each total energy E. For a single energy, the efficiency of the expansion depends on the size of the chosen basis set. This set can be constructed in a variety of ways. For example, for an atom–diatomic reaction, the basis functions can be products of spherical harmonics for the orientation angles of $\vec{R}_j$ and $\vec{r}_j$, times harmonic oscillator functions in the distance variables R_j and r_j. Basis functions with variable parameters can be optimized, again using the S-matrix form of the Kohn variational treatment for reactive scattering.

A compact derivation of equations satisfied by the trial coefficients can be derived in matrix form, writing for the incoming trial arrangement component i

$$\tilde{\Psi}^{(i)}_{1\alpha} = \delta_{i,1}\tilde{\Phi}^{(+)}_{1\alpha} + \boldsymbol{u}^{(i)}\tilde{\boldsymbol{c}}^{(i)}_{1\alpha}$$

after introducing the row $1 \times N_i$ matrix $\boldsymbol{u}^{(i)}(\vec{R}_i, \vec{r}_i)$ with elements $u^{(i)}_m$ and the $N_i \times 1$ column matrix of coefficients $\boldsymbol{c}^{(i)}_{1\alpha} = \left[c^{(i)}_{m,1\alpha}\right]$. Further introducing the $N_C \times 1$ column supermatrix (a matrix with column matrix elements) $\boldsymbol{C}^{(+)}_{1\alpha} = \left[\boldsymbol{c}^{(1)}_{1\alpha}, \boldsymbol{c}^{(2)}_{1\alpha}, \boldsymbol{c}^{(3)}_{1\alpha}, \boldsymbol{c}^{(4)}_{1\alpha}\right]$ for all the arrangements, where $N_C = \sum_j N_j$ is the total number of expansion functions, and the $1 \times N_C$ row supermatrix $\boldsymbol{U}$ with components $\boldsymbol{u}^{(k)}$, $k = 1$ to 4, row matrices, the total outgoing trial wavefunction starting in state 1α is simply

$$\tilde{\Psi}_{1\alpha}^{(+)} = \sum_{j=0}^{4} \tilde{\Psi}_{1\alpha}^{(j)} = \tilde{\Phi}_{1\alpha}^{(+)} + \boldsymbol{U}\,\tilde{\boldsymbol{C}}_{1\alpha}^{(+)}.$$

Similarly, for the total incoming trial wavefunction leading to state $j\nu$,

$$\tilde{\Psi}_{j\nu}^{(-)} = \tilde{\Phi}_{j\nu}^{(-)} + \boldsymbol{U}\,\tilde{\boldsymbol{C}}_{j\nu}^{(-)}$$

with

$$\tilde{\Phi}_{j\nu}^{(-)}(\vec{R}_j, \vec{r}_j) \approx -\frac{(2\pi)^2 M_j}{\hbar^2} \sum_{\nu'} \tilde{T}_{j\nu', j\nu} \frac{1}{(2\pi)^{\frac{3}{2}}} \frac{\exp\left(i - k_{j\nu'} R_j\right)}{R_j} \phi_{j\nu'}(\vec{r}_j) \quad \text{for } R_j \to \infty.$$

Further introducing the super-matrices $\tilde{\boldsymbol{A}}_{1\alpha}^{(+)} = \left\langle \boldsymbol{U}^T \middle| (E - \widehat{H}) \middle| \tilde{\Phi}_{1\alpha}^{(+)} \right\rangle$, $\tilde{\boldsymbol{A}}_{j\nu}^{(-)} = \left\langle \tilde{\Phi}_{j\nu}^{(-)} \middle| (E - \widehat{H}) \middle| \boldsymbol{U} \right\rangle$ of dimensions $N_C \times 1$ and $1 \times N_C$, respectively, and $\boldsymbol{B} = \langle \boldsymbol{U}^T | (E - \widehat{H}) | \boldsymbol{U} \rangle$ of dimensions $N_C \times N_C$, containing $(\boldsymbol{u}^{(k)})^T$ transpose $N_k \times 1$ column matrices, the Kohn variational functional takes the form

$$\tilde{J}_{j\nu,1\alpha}^{(K)} \left[\tilde{\Psi}_{j\nu}^{(-)}, \tilde{\Psi}_{1\alpha}^{(+)} \right] = \tilde{J}_{j\nu\,1\alpha}^{(DW)} \left[\tilde{\Phi}_{j\nu}^{(-)}, \tilde{\Phi}_{1\alpha}^{(+)} \right] + \left(\tilde{\boldsymbol{C}}_{j\nu}^{(-)} \right)^{\boldsymbol{T}} \tilde{\boldsymbol{A}}_{1\alpha}^{(+)} + \tilde{\boldsymbol{A}}_{j\nu}^{(-)} \tilde{\boldsymbol{C}}_{1\alpha}^{(+)} + \left(\tilde{\boldsymbol{C}}_{j\nu}^{(-)} \right)^{\boldsymbol{T}} \boldsymbol{B}\, \tilde{\boldsymbol{C}}_{1\alpha}^{(+)}$$

Here $\tilde{J}_{j\nu,1\alpha}^{(DW)} \left[\tilde{\Phi}_{j\nu}^{(-)}, \tilde{\Phi}_{1\alpha}^{(+)} \right] = \left\langle \tilde{\Phi}_{j\nu}^{(-)} \middle| (E - \widehat{H}) \middle| \tilde{\Phi}_{1\alpha}^{(+)} \right\rangle = S_{j\nu,1\alpha}^{(DW)} \left(\tilde{T}_{j\nu,1\alpha} \right)$ is the functional formed from the known trial distorted waves.

Now $\tilde{J}_{j\nu,1\alpha}^{(K)}$ is a function of trial expansion coefficients in $\left\{ \tilde{c}_{1\alpha}^{(i)} \right\}, \left\{ \tilde{c}_{j\nu}^{(k)} \right\}$ and of the transition amplitude $\tilde{T}_{j\nu,1\alpha}$ in the distorted waves. Variationally optimized values for all these are obtained solving the equations from $\delta \left(\tilde{J}_{j\nu,1\alpha}^{(K)} - S_{j\nu,1\alpha}^{(DW)} \right) = 0$, which are derived taking partial derivatives with respect to the trial parameters,

$$\frac{\partial \tilde{J}_{j\nu,1\alpha}^{(K)}}{\partial \tilde{c}_{1\alpha}^{(i)}} = \tilde{\boldsymbol{A}}_{j\nu}^{(-)} + \left(\tilde{\boldsymbol{C}}_{j\nu}^{(-)} \right)^{\boldsymbol{T}} \boldsymbol{B} = 0, \quad \frac{\partial \tilde{J}_{j\nu,1\alpha}^{(K)}}{\partial \tilde{c}_{j\nu}^{(k)}} = \tilde{\boldsymbol{A}}_{1\alpha}^{(+)} + \boldsymbol{B}\, \tilde{\boldsymbol{C}}_{1\alpha}^{(+)} = 0$$

and $\partial S_{j\nu,1\alpha}^{(DW)} / \partial \tilde{T}_{j\nu,1\alpha} = 0$, which simply gives $stat(\tilde{T}_{j\nu,1\alpha}) = T_{j\nu,1\alpha}^{(DW)}$. Solving the above linear equations for $\tilde{\boldsymbol{C}}_{1\alpha}^{(+)}$ and $\tilde{\boldsymbol{C}}_{j\nu}^{(-)}$ and replacing in the functional finally leads to

$$stat(T_{j\nu,1\alpha}) = T_{j\nu,1\alpha}^{(DW)} - \tilde{\boldsymbol{A}}_{j\nu}^{(-)} \boldsymbol{B}^{-1} \tilde{\boldsymbol{A}}_{1\alpha}^{(+)}$$

which gives the optimized result for the transition amplitude in a compact way.

As an example for atom–diatomic reactions, calculations involve choices of functions factored into internal vibrational states $\chi_v^{(j)}(r_j)$, internal-relative rotational states $\mathcal{Y}_\Gamma^{(j)}(\hat{r}_j, \widehat{R}_j)$ and localized (usually Gaussian) L^2 functions $g_p^{(j)}(R_j)$ for relative distances, so that

$$u_n^{(j)}(\vec{R}_j, \vec{r}_j) = g_p^{(j)}(R_j) \chi_v^{(j)}(r_j) \mathcal{Y}_\Gamma^{(j)}(\hat{r}_j, \widehat{R}_j)$$

while the trial distorted wave $\tilde{\Phi}_{1\alpha}^{(+)}$ can be chosen as a free wave plus the form $f_q^{(j)}(R_j)\chi_v^{(j)}(r_j)\mathcal{Y}_\Gamma^{(j)}(\hat{r}_j, \hat{R}_j)$, with $f_q^{(j)}(R_j) = f_{cut}^{(j)}(R_j)\exp(i\, k_{1\alpha}R_j)/R_j$ as an outgoing wave in channel 1α, and with a cut-off pre-factor so that $f_{cut}^{(j)}(0) = 0$. Calculations involve direct (single-arrangement) integrals over variables $(R_j,\, r_j)$ similar to the ones found in inelastic collisions, as well as atom exchange (rearrangement) integrals, which can be best done in mixed variables like $(r_1,\, r_3,\, \vartheta_{13})$ with bond distances and the angle between them.

This computational procedure has been followed to generate integral reaction cross section vs final rotational quantum number j for $D + H_2 \rightarrow HD + H$ from a variational treatment, expanding scattering states with a basis set of normalized functions [30]. Figure 6.6 shows panels for two selected total collision energies (a) $E = 0.93$ eV and (b) 1.086 eV, compared with experimental values [41]. A panel (c) gives integral cross sections for $D + H_2\ (v = j = 0) \rightarrow HD\ (v' = 0,\, 1,$ all $j') + H$ vs total energy compared to experimental values [41]. Figure 6.7 gives state-to-state differential reaction cross sections $\sigma_{11\leftarrow 00}(\theta)$ vs collision energy E for $H + H_2(v = 0, j = 0) \rightarrow H_2(v', j') + H$ from a variational treatment, expanding scattering states with a basis set of normalized functions: (i) for $(v' = 1, j' = 1)$ and (ii) for $(v' = 1, j' = 3)$. Parameter values give variations with CM deflection angles θ. These results are from [10].

6.4.3 Treatments Using Matching of Wavefunctions

Rearrangement where a bond is broken and another one is formed, such as in atom plus diatom reactions, requires introduction of different space variables for each arrangement. This can complicate calculations, insofar space integrals require transformations of functions of different variable sets into a common set. They can be avoided in a treatment where wavefunctions are locally constructed, numerically propagated and matched at selected boundaries. This can be done for arrangements of reactants and products, matching their functions at boundaries of an internal region of close interactions described by internal wavefunctions.

This was first developed in detail for the reaction $H + H_2 \rightarrow H_2 + H$ in a model of planar scattering [42] and later on done in a three-dimensional description [27, 43]. Here, the PES for each arrangement was divided into four regions, called asymptotic (a), weak interaction (w), strong interaction (s), and internal (i) regions. Wavefunctions were described in the BF frame of each arrangement and were expanded in each region in suitable vibrational–rotational states. Indicating each region with the index $\lambda = a, w, s, i$, it is convenient to transform from a SF frame to a BF frame with a space rotation of Euler angles $\alpha = \phi_\lambda, \beta = \theta_\lambda, \gamma = \psi_\lambda$. In terms of the vectors $\vec{R}_\lambda, \vec{r}_\lambda$, and the angle γ_λ between them, the wavefunction $\Psi_{JM}^{(\lambda)}$ in region λ with total angular momentum quantum numbers $(J\, M)$ in the SF

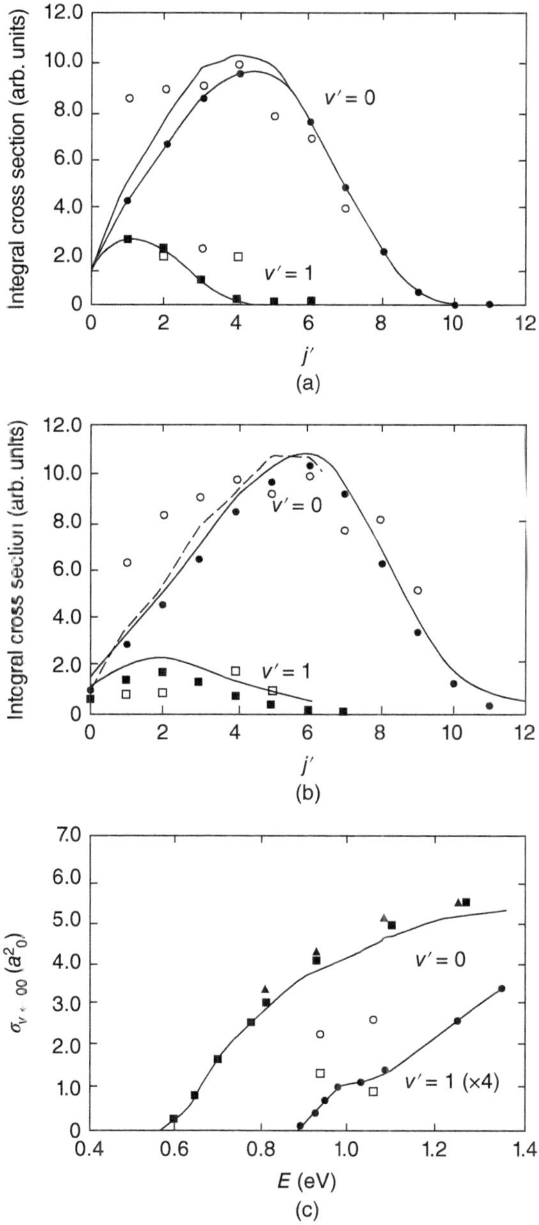

Figure 6.6 Integral reaction cross section vs final rotational quantum number j for $D + H_2 \rightarrow HD + H$ from a variational treatment, expanding scattering states with a basis set of normalized functions [30], for two selected total collision energies (a) $E = 0.93$ eV and (b) 1.086 eV, compared with experimental values [41]. Panel (c) gives integral cross sections for $D + H_2$ $(v = j = 0) \rightarrow HD\ (v' = 0, 1,$ all $j') + H$ vs total energy compared to experimental values [41].

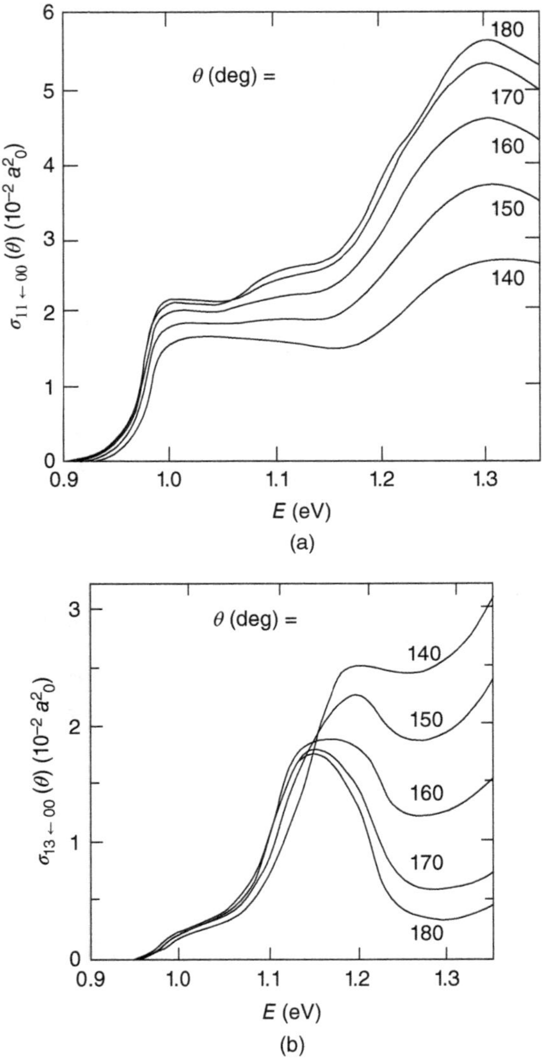

Figure 6.7 State-to-state differential reaction cross section $\sigma_{11\leftarrow 00}(\theta)$ vs collision energy E for $H + H_2(v = 0, j = 0) \rightarrow H_2(v', j') + H$ from a variational treatment, expanding scattering states with a basis set of normalized functions: (a) for ($v' = 1$, $j' = 1$) and (b) for ($v' = 1$, $j' = 3$). Parameter values give variations with CM deflection angles θ. Source: [10] / Annual Reviews.

frame can be transformed to $\Psi_{JK}^{(\lambda)}$ in the BF frame and expanded as

$$\Psi_{JM}^{(\lambda)}(\vec{R}_\lambda, \vec{r}_\lambda) = \sum_{K=-J}^{J} D_{MK}^{J}(\phi_\lambda, \theta_\lambda, 0)\Psi_{JK}^{(\lambda)}(R_\lambda, r_\lambda, \gamma_\lambda, \psi_\lambda)$$

$$\Psi_{JK}^{(\lambda)}(R_\lambda, r_\lambda, \gamma_\lambda, \psi_\lambda) = \sum_{L\geq|K|} Y_{LK}(\gamma_\lambda, \psi_\lambda) \mathrm{w}_{JLK}^{(\lambda)}(R_\lambda, r_\lambda).$$

Replacing this double expansion in the differential equation for $\Psi_{JM}^{(\lambda)}$ and projecting both sides on $\langle D_{MK}^{J} Y_{LK} |$ to extract the coupled equations for the radial functions $w_{JLK}^{(\lambda)}(R_\lambda, r_\lambda)$, one arrives at the matrix differential equation $\left(\boldsymbol{T}_J^{(\lambda)} + \boldsymbol{V}^{(\lambda)} \right) \boldsymbol{w}_J^{(\lambda)} = E\boldsymbol{w}_J^{(\lambda)}$, where $\boldsymbol{T}_J^{(\lambda)}$ and $\boldsymbol{V}^{(\lambda)}$ are square matrix representations with matrix indices (LK) of kinetic and potential energy operators in the variables (R_λ, r_λ) and $\boldsymbol{w}_J^{(\lambda)}$ is a column matrix with the same indices [27]. In each region, the radial functions are expanded again, this time in terms of a vibrational basis set $\left\{ \varphi_{vL}^{(\lambda)} \right\}$ so that

$$w_{JLK}^{(\lambda)}(R_\lambda, r_\lambda) = \sum_{v} f_{JLKv}^{(\lambda)}(R_\lambda) \varphi_{vL}^{(\lambda)}(r_\lambda)$$

Finally, the functions $f_{JLKv}^{(\lambda)}$ and their derivatives $df_{JLKv}^{(\lambda)}/dR_\lambda$ must be matched at the boundary surface of two contiguous space regions λ and λ'. This can be done in a compact notation for square matrices $\boldsymbol{f}_J^{(\lambda)}$ and $d\boldsymbol{f}_J^{(\lambda)}/dR_\lambda$, for elements with the indices LKv.

Matching of wavefunction values and their derivatives at inter-region surfaces assure continuity of the wavefunctions. Reaction scattering matrices $\mathbf{S} = (\mathbf{I} + i\mathbf{R})/(\mathbf{I} - i\mathbf{R})$ and related $\mathbf{R}$-matrices are extracted from the asymptotic forms of wavefunctions at large intermolecular distances. In the strong and internal regions, it was necessary in applications to subdivide the regions into subregions and to propagate solutions by expanding them in subregion local basis sets in each one, and by matching in between, a computationally demanding procedure.

A stable numerical procedure for propagation of wavefunctions, through a finite region of space and matching at chosen interregion surfaces, is based on the R-matrix propagation treatment for solving partial differential equations as they appear in collisional dynamics and applications to molecular reactions [38, 39, 44, 45].

This can be elaborated in the simple case of an atom–diatomic reaction governed by a potential energy $V^{(\lambda)}(R_\lambda, r_\lambda, \gamma_\lambda)$ in region λ. A constant function $S_i^{(\lambda)}(R_\lambda, r_\lambda, \gamma_\lambda) = C_i$ generates a two-dimensional surface (or isocontour) bordering a volume or sector i between surfaces i and i–1, with a unit normal vector $\vec{n}_i^{(\lambda)}$ (directed outward of a closed surface region) at each point, in the direction of the wavefunction flow. A wavefunction $F^{(\lambda)}(R_\lambda, r_\lambda, \gamma_\lambda)$ with a vector gradient $\vec{\nabla}_\lambda F^{(\lambda)}(R_\lambda, r_\lambda, \gamma_\lambda)$ in the chosen variables has a component perpendicular to the surface given by the scalar product $\vec{n}_i^{(\lambda)} \cdot \vec{\nabla}_\lambda F_i^{(\lambda)}$. The wavefunction and its normal gradient are related by an R-operator $\widehat{R}_{i-1}^{(\lambda)}$ at surface i–1 so that $\left| F_{i-1}^{(\lambda)} \right\rangle = \widehat{R}_{i-1}^{(\lambda)} \left(\vec{n}_{i-1}^{(\lambda)} \cdot \left| \vec{\nabla}_\lambda F_{i-1}^{(\lambda)} \right\rangle \right)$. Propagation between two nearby surfaces $S_i^{(\lambda)}$ and $S_{i-1}^{(\lambda)}$ can be done by a recursion relation between $\widehat{R}_i^{(\lambda)}$ and $\widehat{R}_{i-1}^{(\lambda)}$.

In a simple notation omitting the region label and with $\vec{n}_i \cdot \left|\vec{\nabla}_\lambda F_i\right\rangle = \left|\overline{F}_i\right\rangle$, the relation between functions at surfaces i and $i-1$ can be written with step operators $\widehat{A}^{(i)}, \ldots, \widehat{D}^{(i)}$ as $\left|F_i\right\rangle = \widehat{A}^{(i)}\left|F_{i-1}\right\rangle + \widehat{B}^{(i)}\left|\overline{F}_{i-1}\right\rangle, \left|\overline{F}_i\right\rangle = \widehat{C}^{(i)}\left|F_{i-1}\right\rangle + \widehat{D}^{(i)}\left|\overline{F}_{i-1}\right\rangle$, and these can be reorganized to express (in the present notation) functions in terms of gradients as

$$\left|F_{i-1}\right\rangle = \widehat{r}^{(i)}_{00}\left|\overline{F}_{i-1}\right\rangle - \widehat{r}^{(i)}_{01}\left|\overline{F}_i\right\rangle, \left|F_i\right\rangle = \widehat{r}^{(i)}_{10}\left|\overline{F}_{i-1}\right\rangle + \widehat{r}^{(i)}_{11}\left|\overline{F}_i\right\rangle.$$

Replacing here $\left|F_{i-1}\right\rangle = \widehat{R}_{i-1}\left|\overline{F}_{i-1}\right\rangle$ and solving for $\left|\overline{F}_{i-1}\right\rangle$, with $\left|F_i\right\rangle = \widehat{R}_i\left|\overline{F}_i\right\rangle$, the following recurrence relation appears,

$$\widehat{R}_i = \widehat{r}^{(i)}_{11} - \widehat{r}^{(i)}_{10}\left(\widehat{r}^{(i)}_{00} + \widehat{R}_{i-1}\right)^{-1}\widehat{r}^{(i)}_{01}$$

providing a relation between sequential R-operators. This can be implemented numerically by expansion of wavefunctions and r-operators in a basis sets of normalizable functions in the sector bound by $S_i^{(\lambda)}$ and $S_{i-1}^{(\lambda)}$ [39].

In computational work, matrix expressions of the recursion operators $\widehat{r}^{(i)}_{uv}$ follow from a treatment introducing a basis set to expand the wavefunction in each sector i, to generate a row matrix $\boldsymbol{f}^{(i)}$ of expansion coefficients and square matrices $\boldsymbol{r}^{(i)}_{uv}$ and $\boldsymbol{R}_i$. These can be explicitly obtained from the matrix elements of a propagator (or Green function) for the Schrodinger partial differential equation with zero-gradient (Dirichlet) conditions at the boundary surfaces between sectors [45]. In reactive collisions, R-matrix propagation involves its matching at sector surfaces in each arrangement λ and also between two different arrangements λ and λ', so that for reactions one needs to propagate all the matrices $\boldsymbol{R}_i^{(\lambda\lambda')}$.

6.4.4 Treatments Using Hyperspherical Coordinates

An alternative procedure that avoids transformations between arrangement variables is based on the introduction of global hyperspherical coordinates (or HSC) valid for all arrangements and the related partial differential equations for reactive collisions [9]. Insofar HSC are common to reactant and product wavefunctions, their differential equations appear similar to the ones solved for inelastic collisions. They have been applied to the reactions $H + H_2 \rightarrow H_2 + H$ and $F + H_2 \rightarrow HF + H$ and compared with experimental results when available [46–48].

To describe the procedure for a generic $A + BC \rightarrow AB + C$ reaction involving three arrangements labelled here with the index λ, the wavefunction is given first in a CM–SF frame. It can be transformed by rotation to a BF frame specifying Euler angles $(\phi, \theta, \psi)_\lambda$ for the frame orientation, with the remaining three variables given in the Delves HSC by the hyper-radius ρ and the two hyper-angles

η_λ and γ_λ related, respectively, to vibrational and rotational coordinates of the diatomic in arrangement λ. An expansion basis set is constructed for fixed ρ, as a function of the set $\boldsymbol{X} = \{(\phi, \theta, \psi)_\lambda, \gamma_\lambda, \eta_\lambda\}$ of the form [46]

$$B^{(JM)}_{vjk}(\boldsymbol{X};\rho)_\lambda = \left(\rho^{\frac{5}{2}} \sin\eta_\lambda \cos\gamma_\lambda\right)^{-1} D^{(J)}_{Mk}(\phi,\theta,0)_\lambda Y_{jk}(\gamma_\lambda,\psi_\lambda)\phi_v(\eta_\lambda)$$

where $D^{(J)}_{Mk}$ is a Wigner rotational function, Y_{jk} is a spherical harmonic with the diatomic orientation, and ϕ_v is a vibrational function for quantum number v. The total wavefunction, a sum including all the arrangements, is expanded as

$$\Psi_{JM}(\rho,\boldsymbol{X}) = \sum_{\lambda vjk} B^{(JM)}_{vjk}(\boldsymbol{X};\rho)_\lambda\, g_{\lambda vjk}(\rho)_\lambda$$

with the coefficients $g_{\lambda vjk}$ as functions of only the hyper-radius common to all arrangements. These coefficients form a column supermatrix $\mathbf{g}(\rho)$ with indices running over all the arrangements and satisfying a set of coupled ordinary differential equations

$$d^2\mathbf{g}/d\rho^2 = \boldsymbol{\Delta}(\rho)^{-1}\mathbf{U}(\rho)\mathbf{g}(\rho)$$

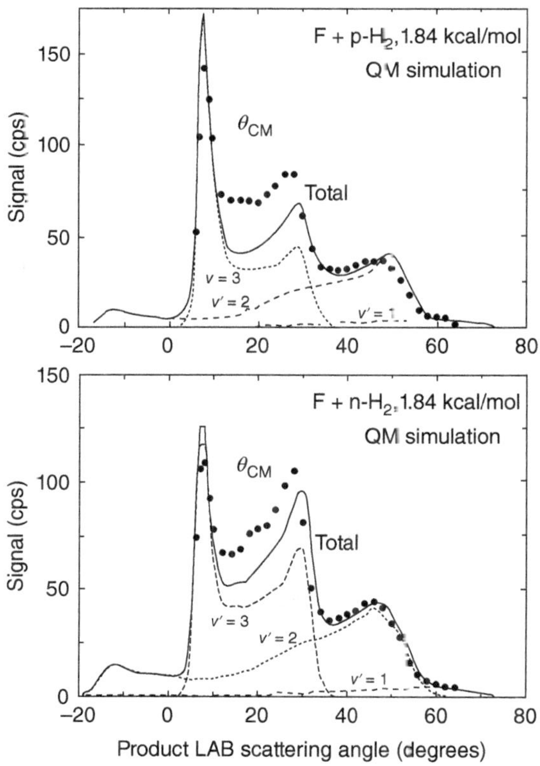

Figure 6.8 Comparisons of experimental (solid points) and theoretical (solid line) differential cross sections in the LAB frame for $F + H_2 \rightarrow HF + H$ at collision energy of 1.84 kcal/mol with resolved distributions for the vibrational quantum number v' of HF (Source: [48] / American Chemical Society). Results are shown for the two reactant beams of p-H_2 and n-H_2. Quantum dynamics calculations were done with a potential energy surface generated by Stark and Werner.

derived from the Schroedinger equation, with overlap integrals in $\mathbf{\Delta}$ and matrix elements of the Hamiltonian in $\mathbf{U}$.

Figure 6.8 shows comparisons of experimental (solid points) and theoretical (solid line) differential cross sections in the LAB frame for $F + H_2 \rightarrow HF + H$ at a collision energy of 1.84 kcal/mol with resolved distributions for the vibrational quantum number v' of HF [48]. Results are shown for the two reactant beams of p-H_2 and n-H_2. Quantum dynamics calculations were done with a PES generated by Stark and Werner.

This figure shows remarkably good agreement between experimental measurements and quantal theoretical calculations. Comparisons with results from classical trajectories show that the large peak for $v' = 3$ is absent in the classical modelling but appears in the quantal results. This can be related to quantal tunneling over the reaction activation barrier and to quantal state-couplings between vibrational and high rotational states of HF, which appear only in a quantal description.

6.5 Electronically Diabatic Reactions

6.5.1 Quantal Treatment

Many reactive collisions involve ground electronic states coupled to electronically excited ones, and the dynamics occurs on all the related PESs [17, 19, 49]. Electronic rearrangements during a reaction are labelled as electronically nonadiabatic if the reaction involves transitions among excited and ground electronic states and are also alternatively called electronically diabatic reactions. Their quantal treatment can be done quite generally expanding on the treatment in the previous chapter on diabatic collisions, differentiating here variables and states for the different reaction arrangements j. This is done introducing scattering state functions $\Psi^{(+)}_{1\alpha}(\boldsymbol{Q}^{(j)}, \boldsymbol{X}^{(j)})$ of both sets of electronic (positions and spin) variables $\boldsymbol{X}^{(j)}$ and atomic position variables $\boldsymbol{Q}^{(j)} = \{Q^{(j)}_n, n = 1 \text{ to } F\}$, for an initial quantum state 1α, $\alpha = (I, \Gamma)$, where I refers to atomic rovibrational quantum numbers and Γ to electronic quantum numbers. We also consider as special cases atom–diatomic reactions, where $\boldsymbol{Q}^{(j)} = (\vec{R}_j, \vec{r}_j)$ in each arrangement j.

The equation of motion for the scattering states involve a Hamiltonian operator $\widehat{H} = \widehat{K}_{at} + \widehat{H}_Q$, which contains the relative kinetic energy operator $\widehat{K}_{at}$ of atomic ions, and the remaining total energy for fixed atomic ion positions in $\widehat{H}_Q$, for a given arrangement. In more detail, in each arrangement j,

$$\widehat{H}_Q = \widehat{K}_{el} + \widehat{V}_{el,el} + \widehat{V}_{at,el} + \widehat{V}_{at,at}$$

where the potential energies involve interactions among electrons and ions and possibly spin–orbit couplings. The atomic kinetic energy depends on the velocities of nuclei (or atomic ions) at position variables q_n, each one with a mass m_n, $n = 1$ to $3N_{at}$. These variables can be mass-scaled to simplify the equations, using instead variables $Q_n = (m_n/M)^{1/2} q_n$, in each arrangement, so that $\hat{K}_{at} = -\hbar^2(\nabla_Q)^2/2M = -(\hbar^2/2M)\sum_n \partial^2/\partial Q_n^{\,2}$, is given by the sum of squares of a ∇_Q multidimensional vector gradient operator. The potential energies $\hat{V}_{at,at}$ and $\hat{V}_{at,el}$ are functions of $\boldsymbol{Q}$ and of electronic variables and contain short-range interactions from the closed electronic shells of atomic ions [50].

For atomic collisions at relative velocities low compared with electron velocities, the usual situation for thermal and hyperthermal collisions, it is convenient to expand the scattering state in a basis of N_{ES} molecular electronic states $\Phi_\Gamma(\boldsymbol{X}; \boldsymbol{Q})$, each a function of electronic variables for fixed nuclei, defined in a SF reference frame with axes (X, Y, Z). The choice of an electronic basis set of eigenfunctions of $\hat{H}_Q$ provides the adiabatic electronic representation, but other representations in terms of diabatic states are also convenient for collisional phenomena.

A scattering state between species A and B in arrangement j, starting in arrangement 1, can be expanded in an electronic basis set, at relative distances R_j, as

$$\Psi_{1\alpha}^{(j)}(\boldsymbol{Q}^{(j)}, \boldsymbol{X}^{(j)}) = \sum_{\Gamma} \psi_{\Gamma,1\alpha}^{(+j)}(\boldsymbol{Q}^{(j)}) \Phi_\Gamma(\boldsymbol{X}^{(j)}; \boldsymbol{Q}^{(j)})$$

$$\approx \Psi_{1\alpha}^{(0)}(\boldsymbol{Q}, \boldsymbol{X}) + \sum_{\gamma} \psi_{\gamma,1\alpha}^{(sc,j)}(\vec{R}_j, \boldsymbol{q}^{(j)}) \Phi_\gamma^{A,B}(\boldsymbol{X}^{(j)}; \boldsymbol{Q}^{(j)}), \quad \text{as } R_j \to \infty$$

with expansion coefficients $\psi_{\Gamma 1\alpha}^{(+)}$ dependent on all the internuclear position vectors, and of asymptotic form

$$\psi_{\gamma,1\alpha}^{(sc,j)}(\vec{R}_j, \boldsymbol{q}^{(j)}) = (2\pi)^{-\frac{3}{2}} f_{\gamma 1c}^{(+j)}(\Omega_j) R^{-1} e^{ik_\gamma R_j} \varphi_{\gamma_A}^{A}(\boldsymbol{q}_A) \varphi_{\gamma_B}^{B}(\boldsymbol{q}_B)$$

with $\boldsymbol{q}^{(j)} = (\boldsymbol{q}_A, \boldsymbol{q}_B)$, while $\Phi_\gamma^{A,B}(\boldsymbol{X}^{(j)}; \boldsymbol{Q}^{(j)}) \approx \Phi_{\Gamma_A}^{A}(\boldsymbol{X}_A; \boldsymbol{q}_A) \Phi_{\Gamma_B}^{B}(\boldsymbol{X}_B; \boldsymbol{q}_B)$, as $R_j \to \infty$.

The total wavefunction originating in arrangement 1 has components in all arrangements and is given by

$$\Psi_{1\alpha}^{(+)}(\boldsymbol{Q}, \boldsymbol{X}) = \sum_j \Psi_{1\alpha}^{(j)}(\boldsymbol{Q}^{(j)}, \boldsymbol{X}^{(j)})$$

Expanding on the treatment for diabatic collisions, introducing brackets for integration over only electronic variables for fixed atomic positions, and a matrix notation, we write $\Phi_\Gamma(\boldsymbol{X}; \boldsymbol{Q}) = \langle \boldsymbol{X} | \boldsymbol{\Phi}(\boldsymbol{Q}) \rangle$ and, for given initial state 1α,

$$\left| \Psi_{1\alpha}^{(j)}(\boldsymbol{Q}^{(j)}) \right\rangle = \left| \boldsymbol{\Phi}(\boldsymbol{Q}^{(j)}) \right\rangle \boldsymbol{\psi}^{(j)}(\boldsymbol{Q}^{(j)})$$

where the first factor is a $1 \times N_{ES}$ row of states Φ_Γ, $\Gamma = 1$ to N_{ES}, for arrangement j and the second factor is a $N_{ES} \times 1$ column of arrangement amplitudes $\psi_\Gamma^{(j)}(\boldsymbol{Q}^{(j)})$. Operating on $\Psi_{1\alpha}^{(j)}$ with the momentum operator $\hat{P}_n = (\hbar/i)\nabla_n$, where $\nabla_n = \partial/\partial Q_n^{(j)}$ for short, we find two terms in

$$\frac{\hbar}{i}\nabla_n \left| \Psi(\boldsymbol{Q}^{(j)}) \right\rangle = \left| \boldsymbol{\Phi}(\boldsymbol{Q}^{(j)}) \right\rangle \frac{\hbar}{i} \nabla_n \boldsymbol{\psi}^{(j)}(\boldsymbol{Q}^{(j)}) + \left[\frac{\hbar}{i} \nabla_n \left| \boldsymbol{\Phi}(\boldsymbol{Q}^{(j)}) \right\rangle \right] \boldsymbol{\psi}^{(j)}(\boldsymbol{Q}^{(j)})$$

and using the completeness relation of the electronic basis in front of the second term, we have

$$\frac{\hbar}{i}\nabla_n\left|\Psi(\boldsymbol{Q}^{(j)})\right\rangle = \left|\boldsymbol{\Phi}(\boldsymbol{Q}^{(j)})\right\rangle\left[\mathbf{I}\frac{\hbar}{i}\nabla_n + \mathbf{G}_n(\boldsymbol{Q}^{(j)})\right]\boldsymbol{\psi}^{(j)}(\boldsymbol{Q}^{(j)})$$

$$\mathbf{G}_n(\boldsymbol{Q}^{(j)}) = (\hbar/i)\langle\boldsymbol{\Phi}(\boldsymbol{Q}^{(j)})|\nabla_n\boldsymbol{\Phi}(\boldsymbol{Q}^{(j)})\rangle$$

where $\mathbf{I}$ is the $N_{ES}\times N_{ES}$ identity matrix. The symbol $\langle\boldsymbol{\Phi}(\boldsymbol{Q}^{(j)})|$ represents a $N_{ES}\times 1$ column matrix containing the complex conjugate elements $\Phi_\Gamma{}^*$. The $N_{ES}\times N_{ES}$ matrix $\mathbf{G}_n$ is vector-like with components where each one is a n-momentum-coupling matrix $\left[\left(G_n^{(j)}\right)_{\Gamma'\Gamma}\right]$. These components have the meaning of momentum fluctuations, resulting from the electronic transitions induced by atomic displacements, as shown in detail in the chapter on diabatic collisions. The $\left[\left(G_n^{(j)}\right)_{\Gamma'\Gamma}\right]$ matrix is Hermitian, as can be seen by taking the gradient of the normalization condition for the electronic basis. Its diagonal elements are null when the electronic functions are real-valued, as follows from $\nabla_n\langle\Phi_\Gamma(\boldsymbol{Q}^{(j)})\mid\Phi_\Gamma(\boldsymbol{Q}^{(j)})\rangle = 0$, but they are not necessarily zero when the electronic functions are complex-valued.

Operating again on $\nabla_n|\Psi(\mathbf{Q}^{(j)})\rangle$ with the gradient ∇_n, one ends up with the scalar product of a vector with $F = 3\text{N}-3$ components $[\mathbf{I}(\hbar/i)\nabla_n + \mathbf{G}_n]$ times itself. The treatment in Section 6.3 for an adiabatic reaction in a single electronic state $\Gamma = \text{g}$, using an expansion in atomic rovibrational states $\phi_{j\text{v}}(\mathbf{q}^{(j)})$, can be carried through in parallel fashion for a diabatic reaction here in terms of the expanded electronic-nuclear basis sets $\Phi_\Gamma(\mathbf{X}^{(j)};\ \mathbf{Q}^{(j)})$ for each arrangement. The previous matrix equation in the adiabatic basis set of N_j internal (rovibrational) states, given there as

$$\left[\epsilon_j + \mathbf{I}\hbar^2/(2M_j)\sum_n\nabla_n^2\right]\boldsymbol{\psi}^{(j)}(\boldsymbol{Q}^{(j)}) = \hat{\boldsymbol{V}}^{(j)}(\boldsymbol{Q}^{(j)})\boldsymbol{\psi}^{(j)}(\boldsymbol{Q}^{(j)})$$

for arrangement amplitudes $\psi^{(+)}_{j\text{v},1\alpha}(\boldsymbol{Q}^{(j)})$ in the $N_j\times 1$ column matrix $\boldsymbol{\psi}^{(j)}$, takes here a matrix form in a larger basis set of $N_{ES}N_j = N^{(j)}$ electronic-atomic functions with the kinetic energy operator replaced by the form including momentum couplings so that now

$$\left\{\epsilon_j - \frac{1}{2M_j}\sum_n\left[\mathbf{I}\frac{\hbar}{i}\nabla_n + \mathbf{G}_n(\boldsymbol{Q}^{(j)})\right]^2\right\}\boldsymbol{\psi}^{(j)}(\boldsymbol{Q}^{(j)}) = \hat{\boldsymbol{V}}^{(j)}(\boldsymbol{Q}^{(j)})\boldsymbol{\psi}^{(j)}(\boldsymbol{Q}^{(j)}).$$

with $\boldsymbol{\psi}^{(j)}$ a $N^{(j)}\times 1$ matrix, ε_j, $\hat{\boldsymbol{V}}^{(j)}$ and $\mathbf{I}$ are $N^{(j)}\times N^{(j)}$ matrices, and where the second power of the square bracket is a scalar product of a vector times itself with each element given by a $N^{(j)}\times N^{(j)}$ matrix. This is a set of $N^{(j)}$ coupled differential equations that must be solved with the appropriate boundary conditions for the $\boldsymbol{\psi}^{(j)}$ amplitudes, for bound or unbound (decaying or scattering) states in arrangement j.

Equations given in Sections 6.3–6.5 for adiabatic reactions can similarly be written in matrix form for N_j adiabatic rovibrational basis functions and a single electronic state. They can be generalized to new matrix equations for nonadiabatic collisions involving N_{ES} electronic states in the larger basis sets of $N_{ES}N_j = N^{(j)}$, replacing the adiabatic kinetic energy $N_j \times N_j$ matrix operator $\hbar^2\nabla_n^2/(2M_j)\mathbf{I}$ with the $N^{(j)} \times N^{(j)}$ larger $(1/2M_j)[\mathbf{I}(\hbar/i)\nabla_n + \mathbf{G}_n(\boldsymbol{Q}^{(j)})]^2$ matrix operator, to include momentum couplings of electronic states.

In the context of the CAC general treatment for three-atom rearrangements, excluding cases with three free atoms, the total scattering wavefunction is again

$$\Psi_{1\alpha}^{(+)} = \sum_{j=1}^{4} \Psi_{1\alpha}^{(j)}$$

with arrangement amplitudes $\Psi_{1\alpha}^{(j)}$ related as before by

$$\Psi_{1\alpha}^{(j)} = \delta_{j1}\Psi_{1\alpha}^{(0)} + \widehat{G}_{0,j}^{(+)} \sum_{k=1}^{4} V^{(jk)}\Psi_{1\alpha}^{(k)}$$

but now with matrix expansions in an electronic basis set with rows times columns as $\Psi_{1\alpha}^{(j)} = \boldsymbol{\Phi}\boldsymbol{\psi}^{(j)}$, leading to coupled CAC equations for amplitudes $\boldsymbol{\psi}^{(j)}$.

Nonadiabatic state-to-state reaction amplitudes are given as before from asymptotic expressions containing $\left\langle \Psi_{j\nu}^{(0)} | V^{(jk)} | \Psi_{1\alpha}^{(k)} \right\rangle$, but here the bracket notation signifies integrations over both atomic $\boldsymbol{Q}^{(j)}$ and electronic $\boldsymbol{X}^{(j)}$ variables. Integration over electronic variables involve electronic states usually written in terms of many-electron wavefunctions composed of atomic orbital functions of electronic position and spin variables $(\vec{r}_a, \zeta_c)$, which are common to all arrangements. Doing the integrations over electronic variables provide inter-arrangement matrices $\boldsymbol{V}^{(jk)}(\boldsymbol{Q}) = \langle \boldsymbol{\Phi}(\boldsymbol{Q}) | V^{(jk)} | \boldsymbol{\Phi}(\boldsymbol{Q}) \rangle_Q$ dependent on atomic variables. Integration over atomic variables are both direct for $j = k$ and of atom-exchange type for $j \neq k$ and can be done in one of the sets $\boldsymbol{Q}^{(j)}$ or $\boldsymbol{Q}^{(k)}$ of atomic variables. The state-to-state amplitudes are then, restoring indices for initial scattering states,

$$f_{j\nu,1\alpha}^{(+)}(\Omega_j) = -\frac{1}{4\pi}\frac{2M_j}{\hbar^2}(2\pi)^3 \sum_{k=1}^{4} \int dQ\, \boldsymbol{\psi}_{j\nu}^{(j0)}(\boldsymbol{Q})^\dagger \boldsymbol{V}^{(jk)}(\boldsymbol{Q})\boldsymbol{\psi}_{1\alpha}^{(k+)}(\boldsymbol{Q})$$

obtained from the j-arrangement wavefunction in the limit $R_j \to \infty$. Nonadiabatic state-to-state differential reaction cross section are given as before by

$$\frac{d\sigma_{j\nu,1\alpha}}{d\Omega_j} = \frac{k_{j\nu}\, M_1}{k_{1\alpha}M_j}\left| f_{j\nu,1\alpha}^{(+)}(\Omega_j)\right|^2$$

with components from all the $k = 1$–4 arrangements and their cross terms.

As described in the chapter on diabatic collisions, it is possible to choose a convenient electronic representation to calculate the collision dynamics. The adiabatic representation requires momentum couplings $\left(G_n^{(j)}\right)_{\Gamma'\Gamma}$ between electronic states Γ' and Γ, with atomic motions in adiabatic PESs $E_\Gamma^{(a)}(\boldsymbol{Q}^{(j)})$ and undergoing momentum-induced transitions. Alternatively, a diabatic electronic representation can be constructed so that $\left(G_n^{(j)}\right)_{\Gamma'\Gamma} = 0$, with atomic motions governed by Hamiltonian matrix elements $H_{\Gamma'\Gamma}^{(d)}(\boldsymbol{Q}^{(j)})$.

Applications of this treatment include reactions of open-shell atoms in different spin-orbital states. Examples are the reactions of $F(^2P_{1/2,3/2})$ with H_2 and D_2, forming products $HF + H$ and $DF + D$. These have been extensively studied both theoretically and experimentally, with compared results showing excellent agreement, provided that correct PESs are used. Calculations of accurate PESs and couplings have been done with correlated-electron functions (at the MRCI and CASSCF levels) [51], providing $F + H_2$ adiabatic PESs of axial Σ, $\Pi_{\pm 1}$ symmetries around the CM intermolecular $\vec{R}$ axis (or 1A′, 2A′, 1A″ symmetries on reflection at the three-atom plane). Adding spin–orbit coupling energies, they give six PESs [52]. Calculations included the nonadiabatic coupling between states 1A′ and 2A′, for a total of four functions of interatomic distances. The Hamiltonian matrix in the adiabatic basis set was transformed to a quasi-diabatic matrix form, and diabatic PESs were fit to functions of the three bond distances with sums of single-, double-, and triple-atom terms. The same fitting was done for the diabatic coupling element in the diabatic Hamiltonian matrix. The followed procedure is similar to what was described in the previous chapter on diabatic collisions. These results also apply to the electronic structure of isoelectronic $F(^2P_{1/2,3/2}) + D_2$. A sketch of the changing energies going from reactants to products is given in Figure 6.9(a) for (F, D, D), showing a reaction barrier in the ground electronic state of about 1.2 kcal/mol for linear (F, D, D) after subtraction of zero-point vibrational energies.

Reaction cross sections were obtained with stationary scattering wavefunctions of HSC and with interactions obtained in a quasi-diabatic electronic representation. Experiments with crossed molecular beams and time-of-flight detection methods have provided measured cross sections, with product D atoms emerging in the backward direction relative to an incoming F beam. Comparisons of measured and calculated cross sections give excellent agreement [53]. Figure 6.9(b) for $F + D_2(v, j = 0) \rightarrow DF(v', j') + D$ show integral reaction cross sections vs collision energies in the CM frame. Reactions of the higher electronic energy $F(^2P_{3/2})$ species are larger at very small collision energies, where the internal (spin–orbit) F energy contributes to overcome a reaction barrier to the final ground electronic states HF or DF, and the reaction events are nonadiabatic. However, at larger collision energies, when the reaction barrier can be overcome with available

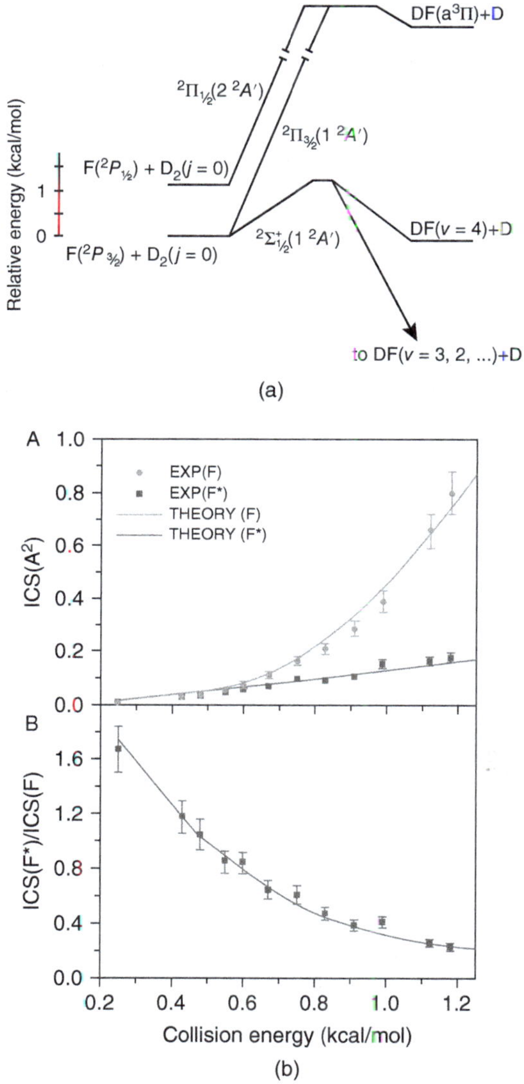

Figure 6.9 (a) Sketch of changing triatomic energy levels going from reactants to products for (F, D, D), showing a reaction barrier in the ground electronic state of about 1.2 kcal/mol for linear (F, D, D) after subtraction of zero-point vibrational energies. (b) $F + D_2$ $(v, j = 0) \rightarrow DF(v', j') + D$ integral reaction cross sections vs collision energies in the CM frame. Panel (A) shows cross sections summed over product vibrational and rotational levels, for collision energies $0.25 \leq E_c \leq 1.2$ kcal/mol. Panel (B) shows ratios of the cross sections in (A). Solid curves are theoretical results, whereas the points indicate experimental results. Source: [53] / American Association for the Advancement of Science.

kinetic energy, the adiabatic reaction events are more probable. Panel (A) shows cross sections summed over product vibrational and rotational levels, for collision energies $0.25 \leq E_c \leq 1.2$ kcal/mol. Panel (B) shows ratios of the cross sections in (A). Solid curves are theoretical results, whereas the points indicate experimental results.

6.5.2 Eikonal and Semiclassical Treatments

Nonadiabatic reactive collisions are frequently described in eikonal and semiclassical formulations, where atomic motions in bonds undergoing rearrangement are treated as classical-like. This can be done expanding on the developments in the chapter on diabatic collisions.

An alternative semiclassical treatment makes use of a bundle of time-dependent atomic trajectories with $\boldsymbol{Q}(t) = \{Q_n(t)\}$ positions and $\boldsymbol{P}(t) = \{P_n(t)\}$ momenta, to construct eikonal phases $S[\boldsymbol{Q}(t)]$ and amplitudes $|\chi[\boldsymbol{Q}(t)]\rangle = |\boldsymbol{\Phi}[\boldsymbol{Q}(t)]\rangle \boldsymbol{c}(t) \exp[\gamma(t)]$, with $\boldsymbol{c}(t) = [c_{jv,1\alpha}(t)]$, as they evolve from reactants to products. This is similar to the treatment of diabatic dynamics of the previous chapter, with reaction rates extracted statistically by collecting final distributions of the desired C + D products. Trajectories are obtained for a Hamiltonian $\mathcal{H}(\boldsymbol{P}, \boldsymbol{Q}) = \boldsymbol{P} \cdot \boldsymbol{P}/(2M) + \mathcal{V}(\boldsymbol{P}, \boldsymbol{Q})$, with an effective potential energy $\mathcal{V}(\boldsymbol{P}, \boldsymbol{Q}) = \mathrm{tr}[\widehat{w}(\boldsymbol{P}, \boldsymbol{Q})\widehat{V}(\boldsymbol{Q})]$, where $\widehat{V}(\boldsymbol{Q}) = \widehat{V}_{el,el} + \widehat{V}_{at,el} + \widehat{V}_{at,at}$ is an operator over electronic variables, and $\widehat{w}(\boldsymbol{P}, \boldsymbol{Q})$ is a chosen density distribution, also an operator over electronic variables, over which the trace is taken.

In a calculation of trajectories starting in initial arrangement 1, for each set of initial phase space points $\{\vec{P}_s = \hbar\vec{k}_{1\alpha}, \vec{R}_s, (\boldsymbol{p}, \boldsymbol{q})_s\}, s = 1$ to N_{in}, one must choose (in a CM reference frame) a large relative distance Z_s outside the reach of interaction potentials, impact parameter b_s and azimuthal angle φ_s, and a collection of internal momenta p_n and positions q_n covering the initial vibrational–rotational phase space of reactants, with the distribution of internal positions in $\phi_{1\alpha}(\boldsymbol{q}_s)$, and the momentum distribution coming from the Fourier-transformed internal states $\tilde{\phi}_{1\alpha}(\boldsymbol{p}_s)$. This collection of positions and momenta generates a bundle $\mathcal{B}_\varphi$ of initial conditions labelled by the index $s \in \mathcal{B}_\varphi$, covering the starting phase-space of the collision dynamics, and this phase-space evolves over time. As the trajectories evolve leading to products C + D, they can be collected in a final arrangement j, where a set of internal positions and momenta can reconstruct the final product internal states $\phi_{jv}(\boldsymbol{q})$ and $\tilde{\phi}_{jv}(\boldsymbol{p})$.

Reaction rates follow then from the time-dependent differential equation giving coefficients $c_{jv,1\alpha}(t)$ in the bundle and from calculation of statistical averages of reaction cross sections. This allows calculation of a reaction rate increment $\Delta\dot{N}_{jv,1\alpha}(t; \vec{P}_s, \vec{R}_s) = J_{0,1\alpha}\Delta\sigma_{jv,1\alpha}(t; \vec{P}_s, \vec{R}_s)$, with trajectories generated by the effective potential energy $\mathcal{V}(\boldsymbol{P}, \boldsymbol{Q})$. It provides statistical values at a final time from $\sum_s \Delta\sigma_{jv,1\alpha}(t_{fn}; \vec{P}_s, \vec{R}_s)/N_{in} = \overline{\Delta\sigma_{jv,1\alpha}}$ as differential reaction cross sections $I_{jv,1\alpha}(\Omega, v) = \overline{\Delta\sigma_{jv,1\alpha}}/\Delta\Omega$.

Within an eikonal treatment for coupled arrangements in the CAC method, each arrangement wavefunction $\Psi_{1\alpha}^{(j)}$ can be written in terms of eikonal phase and amplitude, using the formalism on coupled-arrangement channels in this chapter, for reactive collisions. State-to-state reaction amplitudes $f_{jv,1\alpha}^{(+)}(\Omega_j)$ follow from the transition integrals $\left\langle \Psi^{(0)}_{j\nu} \middle| V^{(jk)} \middle| \Psi^{(k)}_{1\alpha} \right\rangle$. In the eikonal treatment, each arrangement has again an expansion in the eikonal form

$$\Psi^{(j)}_{1\alpha}(\mathbf{Q}^{(j)}, \mathbf{X}^{(j)}) = \chi^{(j)}_{1\alpha}(\mathbf{Q}^{(j)}, \mathbf{X}^{(j)}) \exp\left[iS^{(j)}_{1\alpha}(\mathbf{Q}^{(j)})/\hbar\right]$$

with pre-exponential amplitudes expanded in the basis set of electronic states to yield

$$\chi^{(j)}_{1\alpha}(\mathbf{Q}^{(j)}, \mathbf{X}^{(j)}) = \sum_{\Gamma} \xi^{(+j)}_{\Gamma,1\alpha}(\mathbf{Q}^{(j)})\Phi_{\Gamma}(\mathbf{X}^{(j)}; \mathbf{Q}^{(j)}) = \mathbf{\Phi}(\mathbf{X}^{(j)}; \mathbf{Q}^{(j)})\boldsymbol{\xi}^{(+j)}_{1\alpha}(\mathbf{Q}^{(j)}).$$

The electronic state-specific amplitudes $\xi^{(+j)}_{\Gamma,1\alpha}$ and the action phases $S^{(j)}_{1\alpha}$ change in a classical-like treatment along trajectories, as described in the chapter on diabatic energy transfer phenomena.

The set of momentum and position variables can be arranged into column matrices $\mathbf{P}^{(j)} = \partial S^{(j)}/\partial \mathbf{Q}^{(j)}, \mathbf{Q}^{(j)}$ in arrangement j and can be obtained for a fixed energy E and for a general effective potential energy $\mathcal{V}^{(j)}(\mathbf{P}^{(j)}, \mathbf{Q}^{(j)})$ similarly to what was done for nonreactive diabatic collisions. Position and momentum functions of a time variable t, satisfying here Hamiltonian equations for $\mathcal{H}^{(j)} = \mathbf{P}^{(j)\dagger} \cdot \mathbf{P}^{(j)}/(2M_j) + \mathcal{V}^{(j)}$ for an initial set of conditions $\{(\mathbf{P}^{(1)}, \mathbf{Q}^{(1)})_s, s = 1 \text{ to } N_{in}\}$, can be obtained from the initial relative position $\vec{R}_s$ and momentum $\vec{P}_s$ in a collision, and from the distributions of internal positions and momenta $(\mathbf{p}^{(j)}, \mathbf{q}^{(j)})_s$ in the vibrational–rotational states of the colliding molecules in arrangement j. The set of all these trajectories form a bundle in phase space, constructed to satisfy the initial conditions, and the trajectories give the action function over time from the line integral

$$S^{(j)}_s(t) = S^{(j)}(\mathbf{Q}_s) + \int_{\mathbf{Q}^{(j)}_s}^{\mathbf{Q}^{(j)}(t)} d\mathbf{Q}^{(j)\dagger} \cdot \mathbf{P}^{(j)}$$

for the total mechanical action $S_s(t)$ along each trajectory. Pre-exponential amplitudes $\chi^{(j)}_{1\alpha}(\mathbf{Q}^{(j)}, \mathbf{X}^{(j)})$ can be obtained from the coupled differential equations for the scattering functions $\boldsymbol{\xi}^{(+j)}_{1\alpha}(\mathbf{Q}^{(j)})$, as shown in the chapter on diabatic phenomena.

6.6 First-Principles (Ab initio) Treatments of Reactive Collisions

6.6.1 Time-dependent Eikonal Method

The eikonal procedure has been developed in Chapter 5 for stationary states undergoing diabatic collisions, introducing asymptotic conditions for scattering in Section 5.4 there, and a treatment with stationary states has been extended here in Section 6.5 within the coupled-arrangements-channels formalism of reactive collisions. An alternative procedure can be based on a time-dependent eikonal method, suitable for treatments of reactive collisions in terms of evolving wavepackets with suitable initial conditions. This alternative can be helpful in computational treatments of reactions involving polyatomic molecules and multiple arrangement channels. Starting from the time-dependent Schrodinger equation for a temporal wavefunction written in the eikonal form $\Psi(\boldsymbol{Q},\boldsymbol{X},t)=\chi(\boldsymbol{Q},\boldsymbol{X},t)\exp[i\,S(\boldsymbol{Q},t)/\hbar]$ [54], this satisfies the partial differential equation

$$(\hat{K}_{\mathbf{Q}}+\hat{H}_{\mathbf{Q}})|\Psi(\boldsymbol{Q},t)\rangle = \mathrm{i}\hbar\,\partial|\Psi(\boldsymbol{Q},t)\rangle/\partial t$$

with an initial wavefunction $|\Psi(\mathbf{Q},t_{in})\rangle = |\Psi^{(in)}(\mathbf{Q})\rangle = \chi^{(0)}(\boldsymbol{Q},\boldsymbol{X},t_{in})\exp[i\,S^{(0)}(\boldsymbol{Q},t_{in})/\hbar]$ involving wavepackets $\chi^{(I)}(\vec{R},\boldsymbol{q}_I,\boldsymbol{X}_I,t)$, I = A, B, for each molecule A and B so that $\chi^{(0)}(\boldsymbol{Q},\boldsymbol{X},t_{in})=\chi^{(0)}_{rel}(\vec{R},t_{in})\chi^{(A)}(\vec{R},\boldsymbol{q}_A,\boldsymbol{X}_A,t_{in})\chi^{(B)}(\vec{R},\boldsymbol{q}_B,\boldsymbol{X}_B,t_{in})$ at large distances before collision and with CM relative kinetic energy E_k. The time-dependent eikonal amplitude χ satisfies, following Section 5.4,

$$(\hat{L}_{\mathbf{Q}}+\hat{H}_{\mathbf{Q}}-\partial S(\boldsymbol{Q},t)/\partial t)\,|\chi(\boldsymbol{Q},t)\rangle = \mathrm{i}\hbar\,\partial\,|\chi(\boldsymbol{Q},t)\rangle\,/\partial t$$

and as before $P_n = \partial S/\partial Q_n$, however with the action function $S(\boldsymbol{Q},t)$ satisfying a time-dependent Hamilton–Jacobi equation, or equivalently with variables $\boldsymbol{P}_s(t),\boldsymbol{Q}_s(t)$ satisfying classical Hamiltonian equations to be solved for each initial set of s-trajectory values, as described above.

For a given initial state χ_α of the (A,B) pair, a state-to-state $\alpha\to\beta$ transition amplitude at a time $t>t_{in}$ is given by

$$A_{\beta\alpha}(t)=\int d\mathbf{Q}\,\left\langle\chi^{(0)}_\beta(\boldsymbol{Q},t)|\chi_\alpha(\boldsymbol{Q},t)\right\rangle\exp\left\{i\left[S_\alpha(\boldsymbol{Q},t)-S^{(0)}_\beta(\boldsymbol{Q},t)\right]/\hbar\right\}$$

from which a transition probability rate $R_{\beta\alpha}(t)=d|A_{\beta\alpha}|^2/dt$ is obtained.

The collisional amplitude can be expanded in the basis set of electronic states, giving in the bracket notation for electronic states $|\chi_\alpha(\boldsymbol{Q},t)\rangle = |\boldsymbol{\Phi}(\boldsymbol{Q})\rangle\xi_\alpha(\boldsymbol{Q},t)\exp[\gamma_\alpha(t)]$, where the first factor is a row matrix of kets and the second is a column matrix of state amplitude functions, $\xi_\alpha(\boldsymbol{Q},t)=[\xi_{k\alpha}(\boldsymbol{Q},t)]$. The exponential function of time is

$$\gamma_\alpha(t)=-\frac{i}{\hbar}\int_{t_{in}}^{t}dt'\left\{-\frac{i\hbar}{2M}\sum_n\frac{\partial P_n}{\partial Q_n}(t')+\frac{1}{2M}\sum_n P_n[\mathbf{Q}(t')]^2+\partial S(\boldsymbol{Q},t')/\partial t'\right\}_\alpha$$

and has been introduced to eliminate scalar terms in the equation for χ_α. This gives in matrix notation

$$\left\{\frac{1}{2M}\sum_n\left[\mathbf{I}\frac{\hbar}{i}\nabla_n+\mathbf{G}_n(\mathbf{Q})\right]^2+\frac{1}{M}\sum_n P_n(\mathbf{Q}).\left[\mathbf{I}\frac{\hbar}{i}\nabla_n+\mathbf{G}_n(\mathbf{Q})\right]+\mathbf{H}_{\mathbf{Q}}\right\}$$
$$\times\,\xi_\alpha(\boldsymbol{Q},t)=\mathrm{i}\hbar\,\partial\xi_\alpha(\boldsymbol{Q},t)/\partial t$$

which is a set of coupled partial differential equations for the unknowns $\xi_{k\alpha}(\boldsymbol{Q},t)$, to be solved with initial conditions for scattering.

6.6.2 Wavepacket-spawning Transitions

The scattering coefficients $\xi_{k\alpha}(\boldsymbol{Q}, t)$, for each electronic state k, starting in the collisional state α, can be expanded in a set of time-dependent Gaussian functions for $F = N_{df}$ degrees of freedom [55],

$$g_s[\boldsymbol{Q}; \boldsymbol{P}_s(t), \boldsymbol{Q}_s(t), \alpha_s, \beta_s] = (2\alpha_s/\pi)^{N_{df}/4} \times \exp\left[-\alpha_s(\boldsymbol{Q} - \boldsymbol{Q}_s)^2 + i\boldsymbol{P}_s.(\boldsymbol{Q} - \boldsymbol{Q}_s)/\hbar + i\beta_s(t)/\hbar\right]$$

giving time-dependent coefficients $b_{k\alpha}^{(s)}$ [56] in

$$\xi_{k\alpha}(\boldsymbol{Q}, t) = \sum_{s \in B_\mu} b_{k\alpha}^{(s)}(t) g_s[\boldsymbol{Q}; \boldsymbol{P}_s(t), \boldsymbol{Q}_s(t)].$$

The collisional state $\chi_\alpha = \sum_k \Phi_k(\boldsymbol{X}; \boldsymbol{Q})\xi_{k\alpha}(\boldsymbol{Q}, t)$ is a superposition of trajectory functions $\chi_\alpha^{(s)}$ in a bundle $\mathcal{B}^{(\mu)}$ constructed to include at the initial time the distribution of initial internal position and momentum variables in $\phi_\mu(\boldsymbol{q})$ and $\tilde{\varphi}_\mu(\boldsymbol{p})$ and to include relative position and momentum variables for the centers of mass of the colliding wavepackets, to begin with.

The set of phase space parameters $\{\boldsymbol{P}_s(t), \boldsymbol{Q}_s(t)\}$ can be obtained in principle from equations of motion with an effective potential energy $\mathcal{V}(\boldsymbol{P}, \boldsymbol{Q})$ as introduced for the time-independent eikonals and containing gradients and Laplacians of eikonal amplitudes, but this is usually approximated by the simpler potential energy $\overline{V}_\alpha = \langle\chi_\alpha|\hat{H}_Q|\chi_\alpha\rangle/\langle\chi_\alpha|\chi_\alpha\rangle$, assuming that effects on transition rates from gradient and Laplacian terms in $\mathcal{V} = V_{qu}$ average to zero, for sums over a bundle of trajectories The parameters α_s and β_s can be chosen as convenient for computational treatments, with usual choices being a single parameter α independent of trajectory index and a $\beta_s(t)$ given by the time evolution of the Gaussian wavepacket along a trajectory in a locally harmonic PES [56, 57].

The distribution of initial values in phase space at time t_{in}, to be used in the Gaussian basis functions, must be chosen at large distances before interaction of A and B, in accordance with the momentum distribution coming from the Fourier-transformed internal states, $\tilde{\varphi}_\mu(\boldsymbol{p}_s)$, and the distribution of internal positions in $\varphi_\mu(\boldsymbol{q}_s)$. The initial set of phase space points $\{\boldsymbol{P}_s(t_{in}), \boldsymbol{Q}_s(t_{in})\}$ must cover the ranges of these distributions, and collisional rates are obtained from sums of Gaussian wavepackets with each phase-space point weighted by the magnitudes of the distributions.

The set of expansion coefficients $\left\{b_{k\alpha}^{(s)}(t)\right\}$ follow from the equation of motion of $|\chi(\boldsymbol{Q}, t)\rangle$ for a given initial state α, with this state and its operators expanded in the double basis set of electronic states Φ_k and trajectory Gaussian wavepackets g_s [56]. Generally, this means that each coefficient $b_k^{(s)}$ is coupled to all other ones $b_{k'}^{(s')}$ so that over time both electronic states k and k' are coupled, and also trajectories s and s' are coupled. This means that the trajectories (or paths) evolve coherently, and the time-dependent differential equations for the coefficients $\left\{b_k^{(s)}(t)\right\}$ must be solved simultaneously with the equations for the trajectory variables $\{\boldsymbol{P}_s(t), \boldsymbol{Q}_s(t)\}$.

However, in many earlier applications, the quantum coherence of trajectories s and s' has been ignored, assuming that effects of their coupling on transition rates will average out when adding over a bundle of trajectories. Each independent trajectory s has then been obtained from a potential $\overline{V}_\alpha^{(s)} = \left\langle \chi_\alpha^{(s)} \middle| \widehat{H}_Q \middle| \chi_\alpha^{(s)} \right\rangle / \left\langle \chi_\alpha^{(s)} \middle| \chi_\alpha^{(s)} \right\rangle$. Also, the number of Gaussian functions in a bundle $\mathcal{B}_k(t)$ of independent trajectories, changing over time, has been adjusted for each electronic state k as a transition occurs to a new electronic state l at a time t_C. At a crossings or avoided crossing into the new electronic PES, new Gaussian functions have been spawned after the transition to form a larger bundle $\mathcal{B}_l$, as needed to proceed with the dynamics, so that $\mathcal{B}_k^{(\mu)} \subset \mathcal{B}_l^{(\mu)}$ in a so-called spawning procedure [56]. This has been implemented in a computational package which furthermore provides a first-principles treatment by simultaneously generating electronic states along the trajectories [58].

6.6.2.1 Coherent-paths Treatment

More accurately, it is possible to also allow for the coupling of trajectories s and s' in a procedure, which has been called a multiconfigurational Ehrenfest method, however with the understanding that these are atomic-motion configurations and not electronic configurations [59]. A similar procedure is called here a coherent-paths treatment to emphasize that trajectories are coupled during propagation of wavepackets. Consistent with the present focus on addition over paths in bundles of trajectories, a coherent-paths procedure allows for both electronic and trajectory couplings in the equation of motion for $\chi_\alpha(\boldsymbol{Q}, \boldsymbol{X}, t)$, now after expansion in the double basis set $\{g_s(\boldsymbol{Q}, t)\Phi_k(\boldsymbol{X}; \boldsymbol{Q})\}$ of Gaussians (for each path s) times electronic-state functions. The expansion coefficients $a_k^{(s)}(t)$ are provided by the set of coupled time-dependent ordinary differential equations [55]

$$
\begin{aligned}
&i\hbar \sum_{s' \in \mathcal{B}_\mu} \left(\Delta_{ss'} \frac{db_k^{(s')}}{dt} + \dot{\Delta}_{ss'} b_k^{(s')} \right) \\
&\quad = \sum_{s' \in \mathcal{B}_\mu} \sum_{k'} \left\langle g_s \Phi_k \middle| \left[-\frac{\hbar^2}{2M} \frac{\partial^2}{\partial \boldsymbol{Q}^2} - \frac{i\hbar}{M} \mathbf{P}(t).\frac{\partial}{\partial \boldsymbol{Q}} + \widehat{\mathbf{H}}_Q \right] \middle| g_{s'} \Phi_{k'} \right\rangle b_{k'}^{(s')}
\end{aligned}
$$

where $\Delta_{ss'} = \int d\boldsymbol{Q}\, g_s(\boldsymbol{Q})^* g_{s'}(\boldsymbol{Q})$ and $\dot{\Delta}_{ss'} = \int d\boldsymbol{Q}\, g_s(\boldsymbol{Q})^* dg_{s'}(\boldsymbol{Q})/dt$. To the right, $(\mathbf{H}_Q)_{sk,s'k'} = \langle g_s \Phi_k | \widehat{\mathbf{H}}_Q | g_{s'} \Phi_{k'} \rangle$ is a matrix element in the double expansion basis sets, with similar matrix elements for first and second space derivatives. These coupled equations display the quantal coherence coupling of trajectories (or paths) s and s'.

6.7 Reduced Dimensionality, Optical-potential, and Machine-learning Treatments

6.7.1 Reduced Dimensionality Models

Similarly to what has been done earlier in the chapter on inelastic collisions, it is possible to simplify the treatment of electronically adiabatic reactive collisions by decreasing the number of relevant atomic degrees of freedom in the dynamics modelling. This can be done for reactive collisions by selecting certain degrees of freedom, such as active (breaking or forming) bond distances, or a bending angle in a reaction activated complex, to be treated in detail while constraining the dynamics of other degrees of freedom in the reaction. This means that reactions can be described as happening on PESs of reduced dimensionality. An example is given in atom–diatom reactions $A + BC \rightarrow AB + C$ when one fixes the angle ϑ between bonds A–B and B–C at its value $\vartheta^{\ddagger}$ for the activated complex $ABC^{\ddagger}$ between reactants and products and treats the reaction as happening on a two-dimensional PES dependent only on the breaking and forming diatomic bond distances. This can be improved by treating ϑ as a fast bend variable vibrating around $\vartheta^{\ddagger}$ while the bond distances change adiabatically.

Specifically for an atom–diatomic reaction, in terms of Jacobi coordinates $(\vec{R}_j, \vec{r}_j)$ for arrangement j, with γ_j the angle between these vectors and abscissa angle ψ_j for the triatomic plane in a BF frame, scattering can be treated with the previously introduced state $\Psi_{JK}^{(j)}(R_j, r_j, \gamma_j, \psi_j)$ constructed under the assumption that distances (R_j, r_j) are slowly varying, while angles (γ_j, ψ_j) change rapidly around reference values $(\gamma_j, \psi_j)^{\ddagger}$. An adiabatic factorization of the total wavefunction, omitting the label j for now, can be chosen as [60]

$$\Psi_{JK}(R, r, \gamma, \psi) = \exp(iK\psi)\xi_{nK}(\gamma; r, R)U_{nJK}(r, R)/(rR\sqrt{2\pi})$$

where ξ_{nK} is an adiabatic triatomic bend state dependent only parametrically on (r, R) and is labelled by a bending quantum number n. Replacing Ψ_{JK} in its Schrodinger equation, one can derive the differential equation satisfied by ξ_{nK} when (r, R) are fixed and find

$$\left\{\hat{L}_{\gamma}^2/[2I(r, R)] + V(r, R, \gamma) - V(r, R, \gamma = 0) - \epsilon_{nK}(r, R)\right\} \xi_{nK}(\gamma; r, R) = 0$$

where $\hat{L}_{\gamma}$ is the angular momentum operator for bending with the moment of inertia $I(r, R)$, and $\epsilon_{nK}(r, R)$ is the energy eigenvalue of the adiabatic bend state. The remaining wavefunction factor $U_{nJK}(r, R)$ is obtained from its differential equation for scattering with the adiabatic potential $V(r, R, \gamma = 0) + \epsilon_{nK}(r, R)$, with only two variables, for each quantum number n. Its asymptotic form in arrangement j contains the diatomic vibrational states $\phi_{v_j}^{(j)}(r_j)$ and it provides the transition amplitudes $T_{v_1 \rightarrow v_j}^{(JnK)}$ for a transition between arrangements 1 and j.

The $v_1 \to v_j$ state-to-state transition probability is obtained from the sum over bending states as $P^{(J)}_{v_1 \to v_j} = \sum_{nK} \left| T^{(JnK)}_{v_1 \to v_j} \right|^2$. It has been calculated for reactions of $H + H_2$ and $O(^3P) + H_2$, among others [60].

This special case illustrates how it is possible more generally to reduce the dimensionality of a reaction process by selecting certain degrees of freedom as active and by considering whether the related potential energy functions change smoothly or steeply in those variables. This can be followed by a treatment where some states are assumed to change adiabatically, to simplify the reaction dynamics. Given the set of $F = 3N$ generalized coordinates $\boldsymbol{Q} = (Q_1, Q_2, \ldots, Q_F)$ for a system with N atoms, its potential energy function in a path from reactants to products may allow separation of the coordinates into a set $\boldsymbol{R} = (Q_1, Q_2, \ldots, Q_f)$, $f < F$, of slow varying variables and a set $\boldsymbol{S}$ of fast varying ones. A wavefunction in a BF frame can be factored in the adiabatic form $\Psi_{JK}(\boldsymbol{Q}) = \xi_{nK}(\boldsymbol{S}; \boldsymbol{R}) U_{nJK}(\boldsymbol{R})$, with the adiabatic factor ξ_{nK} satisfying an equation

$$\{\hat{K}_{\boldsymbol{S}} + V(\boldsymbol{R}, \boldsymbol{S}) - V(\boldsymbol{R}, \boldsymbol{S}_0) - \epsilon_{nK}(\boldsymbol{R})\}\xi_{nK}(\boldsymbol{S}; \boldsymbol{R}) = 0$$

where $\hat{K}_{\boldsymbol{S}}$ is the kinetic energy operator of fast variables and $\boldsymbol{S}_0$ is a chosen set of reference values. Scattering is described by $U_{nJK}(\boldsymbol{R})$ in the reduced dimensionality potential $V(\boldsymbol{R}, \boldsymbol{S}_0) + \epsilon_{nK}(\boldsymbol{R})$ with kinetic energy $\hat{K}_{\boldsymbol{R}}$. This can be done in a treatment that also allows for corrections due to nonadiabatic momentum couplings of states involving gradients $\hat{\nabla}_{\boldsymbol{R}}$ of ξ_{nK} and U_{nJK} similar to what was shown in the chapter on nonadiabatic collisions, here with the momentum coupling matrix $\vec{\mathbf{G}}_{\boldsymbol{R}} = -i\hbar[\langle U_{nJK} | \hat{\nabla}_{\boldsymbol{R}} U_{n'JK}\rangle]$ leading to nonadiabatic transitions in a reduced dimensionality.

6.7.2 Optical Potentials Treatment

Alternatively, instead of simplifying the reaction dynamics by classifying molecular degrees of freedom (spatial variables) into active and inactive ones, it is instead possible to simplify the dynamics by classifying molecular states in each reaction arrangement into two sets: an explicitly active set involved in the dynamics and a second set of intermediate states to be treated by means of an optical potential in the arrangement. This can be done by generalizing to reactions the treatment with optical potentials for inelastic collisions in Chapter 3. They are mathematical constructs obtained from a partitioning of the space of molecular states $|\phi_{j\nu}\rangle$ in arrangement j into a subspace to be explicitly treated, and a complimentary subspace of the remaining (intermediate) states, which lead to flux loss. The partition provides formally complex-valued interaction PESs for the reaction. The treatments can be based on physical consideration of quantal flux absorption [61] or more generally can be based on state-partitioning methods with projection operators, which lead to effective Hamiltonians [35]. This is developed in Chapter 7.

6.7.3 Machine-learning Procedures

The calculation of electronically diabatic reaction cross sections, and also of photodissociation rates, requires very extensive computational efforts in a system of several interacting atoms. They involve calculations of PESs for ground and excited electronic states and their nonadiabatic couplings. In a full quantal treatment, this requires calculation in advance of the PESs and couplings for all relevant atomic positions in the collision or dissociation and their parametrization to be used in the quantal dynamics with all the coupled translational and rovibrational states. A treatment where atomic motions are instead treated classically requires PESs, their nonadiabatic momentum couplings, and forces only along each of many trajectories. The computational demands in both cases are multiplied as the number of atoms increases in the whole system.

Ways to deal with these demands have been recently devised using artificial intelligence (AI) to generate PESs and couplings with machine-learning (ML) procedures applied to neural networks (NNs). Here the treatment is presented in its basic form, in terms of the ML training of a NN, as shown in Figure 6.10. Details on the NN procedure for generation of ground electronic state PESs can be found in [50]. This has been recently generalized to deal with electronically excited systems, with emphasis on photoexcitation of polyatomics.

The ML procedures require access to large sets of input data and choices of input descriptors to the NNs. The input can be provided in the form of a collection of atom-a embedding energies $E_a^{(emb)}$ containing information about distances to neighbor atoms and what atomic orbital shells are needed for electronic excitation. These are entered into a first layer of the NN with nodes m and are

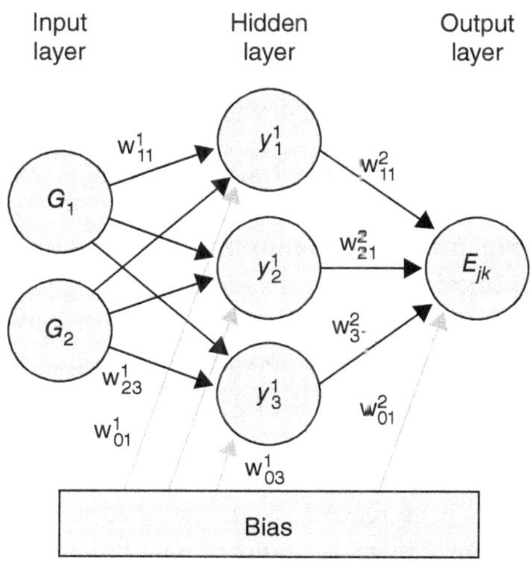

Figure 6.10 Simple neural network with two nodes G_1, G_2 in an input layer containing initial assigned values of input variables G_a. A hidden layer $n = 1$ with $m = 1, 2, 3$ nodes in states $y_{a,m}^{(1)}$ gives outputs biased by an activation function. The outputs of the hidden layer nodes are added to generate the PESs and their coupling $E_{jk}(Q)$ in a final output. Source: [62] / American Physical Society.

transferred to subsequent layers with chosen connections between nodes and chosen biases, designed to train the NN and guide it to desired outputs. Here the output of the NN can be guided to extract PESs and their coupling $E_{jk}(\boldsymbol{Q})$.

A frequently used AI-NN procedure [63] involves at least an input NN layer with nodes μ containing each $G_a^{(\mu)}$ value, outputting their values into a hidden layer (number 1) of nodes m (as many as needed for training) coding partial energy components $x_{a,m}^{(1)}$ given by

$$x_{a,m}^{(1)} = b_m^{(1)} + \sum_{\mu} G_a^{(\mu)} w_{\mu,m}^{(1)}.$$

Here the weights $w_{\mu,m}^{(1)}$ are parameters to be found, connecting the global variable in the input layer to node m in the hidden layer. This partial energy component is accepted with an activation (or acceptance) probability biased by the magnitude of $b_m^{(1)}$. The bias parameter is introduced to allow for an acceptance somewhere between zero and one for the output of node m, given by a sigmoid-shape (activation) function $f^{(1)}\left(x_{a,m}^{(1)}\right) = y_{a,m}^{(1)}$ varying from zero at minus infinity to one at plus infinity. This function can be chosen for example as $[1 + \tanh(x)]/2$. The values of the bias $b_m^{(1)}$ and of the weights $w_{\mu,m}^{(1)}$ are obtained in a training process involving minimization of the standard deviation of sets of output values generated by the NN as compared with precalculated output values selected to test the NN.

In some detail, the embedded atom energies are

$$E_a^{(emb)} = \sum_m y_{a,m}^{(1)} \left(\left\{ G_a^{(\mu)} \right\}; b_m^{(1)}, \left\{ w_{\mu,m}^{(01)} \right\} \right),$$

where the set $\left\{ G_a^{(\mu)} \right\}$ gives a global atomic environment, and are a sum of the outputs from all the nodes of the hidden layer. The parameters $b_m^{(1)}$ and $\left\{ w_{\mu,m}^{(01)} \right\}$ in the acceptance function follow from a NN training that optimizes the fitting of a set of reference PES values $E_{a,ref}^{(emb)}$ for given input atomic coordinates, using usually the mean absolute deviation (MAD)

$$MAD = \frac{1}{N_{at}} \sum_a \left| E_{a,ref}^{(emb)} - E_{a,NN}^{(emb)} \right|$$

which must be minimized by the selection of parameters. An alternative is to minimize the root-mean-square error. Once the parameters are known, the interpolation of values to new atomic positions must be checked, and this is done using a validation set of previously generated PES values different from the ones used for training the NN. The total PES is given by the sum of embedded atom energies, $V = \sum_a E_a^{(emb)}$.

The earlier procedures using embedded atom energies for ground electronic PESs have recently been extended to deal with electronically excited states, particularly in calculations of populations of photoexcited polyatomics vs relaxation times in internal conversion [64] and isomerization [65]. High-quality multireference many-electron calculations were used to generate diabatic PESs displaying crossings and including their diabatic couplings.

This AI-NN procedure has also been followed in a study of the two lowest quasi-diabatic PESs and nonadiabatic couplings of LiFH [66], relevant to the reaction $Li + HF \rightarrow LiF + H$, using previously generated data from extensive multiconfiguration electronic structure calculations with single and double excitations [67]. Here the input to a multilayer NN consisted of a collection of potential energy, forces, and momentum couplings at selected values of the LiF and HF bond distances, and the LiFH bond angle. Training of the NN was done extracting bias and interlayer parameters by minimization of standard deviations for energies, forces, and momentum couplings. This showed high accuracy in the AI-NN procedure from comparisons with the original results for isocountours of the PESs and for the momentum couplings at the avoided crossing of the surfaces.

An alternative procedure also using AI-NNs would be applicable to a quantum-classical treatment of reactions, with cross sections and rates obtained from a bundle of trajectories $\{\mathbf{Q}(t_s), s = 1 \text{ to } N_{in}\}$. In this treatment, the PESs and couplings need not be generated in advance or parametrized, but they are needed at each time step along each trajectory. A large number of trajectories are usually needed for accurate calculations of cross sections or rates, with time steps shortened in regions of PESs couplings and transitions. Here it is again desirable to consider using an AI-NN procedure to decrease the number of required trajectories. It should be possible to input into the NN a large selection of initial conditions and train the NN to generate desired distributions of internal state populations for products of a reaction or in photodissociation.

This is a very active and evolving area of research, and several collections of papers on AI-NN for molecular dynamics have recently become available [68–70], with many new results likely to appear in the near future.

References

1 M. Karplus, "Special results of theory: Distorted waves," in *Molecular Beams and reaction Kinetics*, (Editor: C. Schlier), New York: Academic Press, 1970, p. 407.

2 D. A. Micha, "Quantum mechanical model for simple molecular reactions," *Arkiv Fysik*, vol. 30, pp. 411–423, 1965.

3 C. Schlier, *Molecular Beams and Reaction Kinetics*, New York: Academic Press, 1970.

4 M. Child, *Molecular Collision Theory*, New York: Academic Press, 1974.

5 A. Micha, "Quantum theory of reactive molecular collisions," *Adv. Chem. Phys.*, vol. 30, pp. 7–75, 1975.

6 W. H. Miller (Editor), *Dynamics of Molecular Collisions Parts A and B*, New York: Plenum Press, 1976.

7 R. B. Bernstein (Editor), *Atom-Molecule Collision Theory: A Guide to the Experimentalist*, New York: Plenum Press, 1979.

8 M. Baer, "General theory of reactive scattering: The differential equation approach (Chapter 3)," in *Theory of Chemical Reaction Dynamics*, Volume I, (Editor: M. Baer), Boca Raton, FL: CRC Press, 1985, pp. 91–162.

9 A. Kuppermann, "Accurate quantum calculations of reactive systems," in *Theoretical Chemistry: Advances and Perspectives*, Volume 6A, New York: Academic Press, 1981, pp. 79–164.

10 W. H. Miller, "Recent advances in quantum mechanical reactive scattering theory," *Annu. Rev. Phys. Chem.*, vol. 41, pp. 245–281, 1990.

11 R. B. Walker and J. C. Light, "Reactive molecular collisions," *Annu. Rev. Phys. Chem.*, vol. 31, pp. 401–433, 1980.

12 R. E. Wyatt, "Direct-mode chemical reactions I: Methodology for accurate quantal calculations," in *Atom-molecule Collision Theory*, (Editor: R. B. Bernstein), New York: Plenum Press, 1979, pp. 567–594.

13 E. Deumens and Y. N. Ohrn, "Electron-nuclear dynamics with diabatic and adiabatic wave packets," *J. Phys. Chem.*, vol. 92, p. 3181, 1988.

14 D. G. Truhlar, D. W. Schwenke and D. Kouri, "Quantum dynamics of chemical reactions by converged algebraic variational calculations," *J. Phys. Chem.*, vol. 94, pp. 7346–7352, 1990.

15 J. Z. H. Zhang, *Theory and Application of Quantum Molecular Dynamics*, Singapore: World Scientific, 1999.

16 D. A. Micha, "Rearrangement in molecular collisions: A many-body approach," in *Theory of Chemical Reaction Dynamics*, Volume II, (Editor: M. Baer), Boca Raton, FL: CRC Press, 1985, pp. 181–218.

17 J. C. Tully, "Nonadiabatic processes in molecular collisions," in *Dynamics od Molecular Collisions Part B*, (Editor: W. H. Miller), New York: Plenum Press, 1976, p. 217.

18 J. C. Tully, "Mixed quantum-classical dynamics," *Faraday Discuss.*, vol. 110, p. 407, 1998.

19 D. H. Zhang and H. Guo, "Recent advances in quantum dynamics of bimolecular reactions," *Annu. Rev. Phys. Chem.*, vol. 67, pp. 135–158, 2016.

20 L. J. Butler, "Chemical reaction dynamics beyond the Born-Oppenheimer approximation," *Annu. Rev. Phys. Chem.*, vol. 49, p. 125, 1998.

21 R. N. Porter and L. M. Raff, "Classical trajectory methods in molecular collisions," in *Dynamics of Molecular Collisions Part B*, (Editor: W. H. Miller), New York: Plenum Press, 1976, pp. 1–52.

22 D. G. Truhlar and J. T. Muckerman, "Reactive scattering cross sections III. Quasiclassical and semiclassical merthods (Chapter 16)," in *Atom-Molecule Collision Theory*, (Editor: R. B. Bernstein), New York: Plenum Press, 1979, p. 505.

23 J. C. Polanyi, "Some concepts in reaction dynamics (Nobel lecture)," *Angew. Chem. Int. Engl.*, vol. 26, pp. 952–971, 1987.

24 I. W. M. Smith, *Kinetics and Dynamics of Elementary Gas Reactions*, London: Butterworths, 1980.

25 P. J. Kuntz, "Features of potential energy surfaces and their effects on collisions," in *Dynamics of Molecular Collisions Part B*, (Editor: W. H. Miller), New York: Plenum Press, 1976, pp. 53–120.

26 D. J. Truhlar and A. Kuppermann, "Exact and approximate quantum mechanical reaction probabilities and rate constants for the collinear H + H_2 reaction," *J. Chem. Phys.*, vol. 56, p. 2232, 1972.

27 G. C. Schatz and A. Kuppermann, "Quantum mechanical reactive scattering for three-dimensional atom plus diatom systems. I. Theory," *J. Chem. Phys.*, vol. 65, p. 4642, 1976.

28 L. D. Faddeev, *Mathematical Aspects of the Three-Body Problem in Quantum Scattering Theory*, New York: Davey, 1965.

29 D. J. Kouri, "General theory of reactive scattering: The integral equation approach (Chapter 4)," in *Theory of Chemical Reaction Dynamics, volume I (Editor* M. Baer*)*, Boca Raton, FL, CRC Press, 1985, pp. 163–226.

30 J. Z. H. Zhang and W. H. Miller, "Quantum reactive scattering via the S-matrix version of the Kohn variational principle: Differential and integral cross sections for D + H_2 → HD + H," *J. Chem. Phys.*, vol. 91, p. 1528, 1989.

31 A. R. Edmonds, *Angular Momentum in Quantum Mechanics*, Princeton NJ: Princeton University Press, 1960.

32 W. H. Miller, "Coupled equations and the minimum principle for collisions of an atom and a diatomic molecule, including rearrangement," *J. Chem. Phys.*, vol. 50, p. 407, 1969.

33 B. H. Choi, R. T. Poe, J. C. Sun and K. T. Tang, "Atom-molecule reactive scattering and symmetrized cross sections for a system containing identical nuclei," *J. Chem. Phys.*, vol. 78, p. 5590, 1983.

34 D. A. Micha, "Collision dynamics of three interacting atoms: Permutational symmetry for identical nuclei," *J. Chem. Phys.*, vol. 60, p. 2480, 1974.

35 D. A. Micha, "Effective hamiltonian methods for molecular collisions," in *Advances in Quantum Chemistry*, Volume 8, New York: Academic Press, 1974, p. 211.

36 J. N. L. Connor, "The distorted wave theory of chemical reactions," in *The Theory of Chemical Reaction Dynamics*, (Editor: D. C. Clary), Dordrecht, Holland: Reidel, 1985, p. 247.

37 J. C. Sun, B. H. Choi, R. T. Poe and K. T. Tang, "Adiabatic T-matrix theory for three-dimensional reactive scattering: Application to the ($H + H_2$) system," *J. Chem. Phys.*, vol. 78, pp. 4523–4532, 1983.

38 R. B. Walker, E. B. Stechel and J. C. Light, "Accurate H3 dynamics on an accurate H3 potential surface," *J. Chem. Phys.*, vol. 69, p. 2922, 1978.

39 J. C. Light, R. B. Walker, E. B. Stechel and T. G. Schmalz, "R-matrix propagation methods in inelastic and reactive collisions," *Comput. Phys. Commun.*, vol. 17, pp. 89–97, 1979.

40 B. H. Bransden, "Chapter 2: Special methods in potential scattering," in *Atomic Collision Theory*, New York: Benjamin, 1970, p. 54.

41 D. L. Phillips, H. B. Levene and J. J. Valentini *J. Chem. Phys.*, vol. 79, p. 5202, 1984.

42 A. Kuppermann, D. C. Schatz and M. Baer, "Quantum mechanical reactive scattering for planar atom plus diatom systems. I. Theory," *J. Chem. Phys.*, vol. 65, p. 4596, 1976.

43 G. C. Schatz and A. Kuppermann, "Quantum mechanical reactive scattering. II. Accurate cross sections for $H + H_2$," *J. Chem. Phys.*, vol. 65, p. 4668, 1976.

44 J. Light and R. Walker, "An R-matrix approach to the solution of coupled equations for atom-molecule reactive scattering," *J. Chem. Phys.*, vol. 65, p. 4272, 1976.

45 B. R. Schneider and R. B. Walker, "The coupled channel R-matrix propagation method," *J. Chem. Phys.*, vol. 70, p. 2466, 1979.

46 G. C. Schatz, "Quantum reactive scattering using hypersherical coordinates: Results for $H + H_2$ and Cl + HCl," *Chem. Phys. Lett.*, vol. 150, p. 92, 1988.

47 G. A. G. A. Parker and R. T. Pack, "Quantum reactive scattering in three dimensions using hyperspherical (APH) coordinates. VI. Analytic basis method for surface functions," *J. Chem. Phys.*, vol. 98, p. 6883, 1993.

48 F. J. Aoiz, L. Banares, B. Martınez-Haya, J. F. J. F. Castillo, D. E. Manolopoulos, K. Stark and H.-J. Werner, "Ab initio simulation of molecular beam experiments for the $F + H_2$ to HF + H reaction," *J. Phys. Chem. A*, vol. 101, pp. 6403–6414, 1997.

49 M. Baer and G. D. Billing (Editors), *The Role of Degenerate States in Chemistry*, New York: John Wiley & Sons, 2002.

50 D. A. Micha, *Molecular Interactions: Concepts and Methods*, Hoboken, NJ: John Wiley & Sons, 2020.

51 M. H. Alexander, D. E. Manolopoulo and H.-J. Werner, "An investigation of the $F + H_2$ reaction based on a full ab initio description of the open-shell character of the F(2P) atom," *J. Chem. Phys.*, vol. 113, p. 11084, 2000.

52 G. Li, H.-J. Werner, F. Lique and M. H. Alexander, "New ab initio potential energy surfaces for the $F + H_2$ reaction," *J. Chem. Phys.*, vol. 127, 174302, 2007.

53 L. Che, Z. Ren, X. Wang, W. Dong, D. Dai, X. Wang, D. Zhang, X. Yang, L. Sheng, G. Li, H.-J. Werner, F. Lique and M. H. Alexander, "Breakdown of the Born-Oppenheimer approximation in the F + o-D_2 → DF + D reaction," *Science*, vol. 317, p. 1061, 2007.

54 D. A. Micha, "Time-dependent many-electron treatment of electronic energy and charge transfer in atomic collisions," *J. Phys. Chem. A*, vol. 103, pp. 7562–7574, 1999.

55 B. F. E. Curchod and T. J. Martinez, "Ab initio nonadiabatic quantum molecular dynamics," *Chem. Rev.*, vol. 118, p. 3305, 2018.

56 M. Ben-Nun and T. J. Martinez, "Nonadiabatic molecular dynamics: Validation of the multiple spawning," *J. Chem. Phys.*, vol. 108, p. 7244, 1998.

57 E. J. Heller, "Time-dependent approach to semiclassical dynamics," *J. Chem. Phys.*, vol. 62, p. 1544, 1975.

58 B. G. Levine, J. D. Coe, A. M. Virshup and T. J. Martinez, "Implementation of ab initio multiple spawning in the MOLPRO quantum chemistry package," *Chem. Phys.*, vol. 347, pp. 3–16, 2008.

59 K. Salta and D. V. Shalashilin, "On-the-fly ab initio molecular dynamics with multiconfigurational Ehrenfest method," *J. Chem. Phys.*, vol. 137, p. 22A506, 2012.

60 J. Bowman, "Reduced dimensionality theory of quantum reactive scattering," in *Advances in Chemical Physics, Volume 61*, (Editors: S. A. Rice and I. Prigogine), New York: Wiley, 1985, p. 115.

61 D. A. Micha, "Optical models in molecular collision theory," in *Dynamics of Molecular Collisions, Part A*, (Editor: W. H. Miller), New York: Plenum Press, 1976, pp. 81–129.

62 J. Behler and M. Parrinello, "Generalized neural network representation of high-dimensional potential-energy surfaces," *Phys. Rev. Lett.*, vol. 98, p. 146401, 2007.

63 J. Behler, "Neural network potential energy surfaces in chemistry: A tool for large scale simulations," *Phys. Chem. Chem. Phys.*, vol. 13, p. 17930, 2011.

64 J. Westermayr, M. Gastegger, M. F. S. J. Menger, S. Mai, L. Gonzalez and P. Marquetand, "Machine learning enables long-time scale molecular photodynamics simulations," *Chem. Sci.*, vol. 10, p. 8100, 2019.

65 W.-K. Chen, X.-Y. Liu, W.-H. Fang, P. O. Dral and G. Cui, "Deep learning for nonadiabatic excited-state dynamics," *J. Phys. Chem. Lett.*, vol. 9, p. 6702, 2018

66 Y. Guan, D. H. Zhang, H. Guo and D. H. Yarkony, "Representation of coupled adiabatic potential energy surfaces using neural network based quasi-diabatic Hamiltonians: 1,2 2A′ states of LiFH", *Phys. Chem. Chem. Phys.*, vol. 21, p. 14205, 2019.

67 A. W. Jasper, M. D. Hack and D. G. Truhlar, "Coupled quasidiabatic potential energy surfaces for LiFH," *J. Chem. Phys.*, vol. 116, p. 8353, 2002.

68 O. V. Prezhdo, "Advancing physical chemistry with machine learning," *J. Phys. Chem. Lett.*, vol. 11, 2020.

69 N. E. Jackson, B. M. Savoie, A. Statt and M. A. Webb, "Introduction to machine learning for molecular simulation," *J. Chem. Theory Comput.*, vol. 19, p. 4335, 2023.

70 A. L. Ferguson and J. Pfaendtner, "Virtual special issue on machine learning in physical chemistry," *J. Phys. Chem. A*, vol. 128, p. 5225, 2024.

7

The Quantum Scattering Operator and the Statistical Density Operator

CONTENTS

7.1 Scattering Operators and Transition Rates

7.1.1 Time-dependent Treatment

Previous chapters have emphasized the wave mechanical treatment of collisions, which provides physical images in terms of scattering waves [1]. It leads to partial differential equations or integral equations for wavefunctions, which can be solved with well-tested numerical methods to provide cross sections and rates. Alternatively, a general and compact treatment of molecular scattering can be

Quantum Molecular Dynamics: Energy Transfer and Reactivity, First Edition. David A. Micha.

developed in terms of scattering operators [2–6]. It is developed in what follows for time-dependent and time-independent (stationary) collisional events and leads to operator relations that can be implemented in position or momentum representations [5] for computational purposes. We follow here our formulation in [7].

The interaction of two colliding molecules A and B can be described within a time-dependent formulation introducing a quantum state $\Psi(t)$ which can be constructed as a function $\Psi(\boldsymbol{X}, \boldsymbol{Q}, t) = \langle \boldsymbol{X}, \boldsymbol{Q} \mid \Psi(t) \rangle$ of all electronic and nuclear position and spin variables. The wavefunctions must satisfy over time the operator equation of motion

$$\hat{H}\Psi(t) = i\hbar\partial\Psi/\partial t$$

which must be solved with boundary conditions $\langle \Psi \mid \Psi \rangle = 1$ and the initial condition $\Psi(t_{in}) \approx \Psi_0(t_{in}) = \Psi^{(0)}(\lambda)$ at an initial time before interactions, where λ signifies a set of parameters with a given initial distribution of values. The total Hamiltonian operator can be written as $\hat{H} = \hat{H}_0 + \hat{V}$, a sum of a time-independent Hamiltonian $\hat{H}_0 = \hat{H}_{rel} + \hat{H}_{int}$ describing a reference (unperturbed) system plus a perturbation $\hat{V}$, coupling relative and internal motions, which in a general case can depend on time. Here, we consider collisions in an isolated system so that $\hat{V}$ is time independent. A formal solution for the state between times t_0 and t is given by a time-evolution operator $\hat{U}$ as

$$\Psi(t) = \hat{U}(t, t_0)\Psi(t_0)$$

$$\hat{U}(t, t_0) = \exp[-i\hat{H}(t - t_0)/\hbar]$$

in an expression valid for events at times t after or before t_0. The time evolution can be made explicit introducing propagators (or Green functions) $\hat{G}^{(\pm)}(t, t_0)$ for advanced or retarded events satisfying, with $\delta(t)$ a Dirac delta function, the equation

$$(i\hbar\partial/\partial t - \hat{H})\hat{G}^{(\pm)}(t, t_0) = \delta(t - t_0),$$

$$\hat{G}^{(+)}(t, t_0) = 0 \ \text{ for } t < t_0, \text{“forward propagation”}$$

$$\hat{G}^{(-)}(t, t_0) = 0 \ \text{ for } t > t_0, \text{“backward propagation”}$$

so that $\hat{G}^{(+)}(t, t_0)$ describes events only for $t > t_0$.

The equation for $\hat{G}^{(\pm)}(t, t_0)$ reduces to that for $\hat{U}(t, t_0)$ when $t \neq t_0$. Introducing the step (or Heaviside) function $\theta(\tau) = 1$ or 0, for $\tau \geq 0$ or < 0, satisfying $d\theta/d\tau = \delta(\tau)$, they are related by

$$\hat{G}^{(\pm)}(t, t_0) = \mp(i/\hbar)\,\theta[\pm(t - t_0)]\hat{U}(t, t_0)$$

so that $\Psi(t) = \pm i\hbar\hat{G}^{(\pm)}(t, t_0)\Psi(t_0)$ for $t >$ or $< t_0$. Given the unperturbed solution $\hat{G}_0^{(\pm)}(t, t_0)$ to the above differential equation with the unperturbed Hamiltonian $\hat{H}_0$, it follows that

$$\widehat{G}^{(\pm)}(t,t_0) = \widehat{G}_0^{(\pm)}(t,t_0) + \int_{-\infty}^{\infty} dt' \widehat{G}_0^{(\pm)}(t,t')\widehat{V}\widehat{G}^{(\pm)}(t',t_0)$$

as can be verified differentiating both sides with respect to time. Therefore, for $t \geq t_0$

$$\Psi^{(+)}(t) = i\hbar\widehat{G}_0^{(+)}(t,t_0)\Psi(t_0) + \int_{-\infty}^{\infty} dt' \widehat{G}_0^{(+)}(t,t')\widehat{V}\,\Psi^{(+)}(t')$$

is a state evolving forward in time. Taking $t_0 \to -\infty$ as an early time before collisional interactions take place between A and B, with $\Psi_0^{(+)}(t) = i\hbar\widehat{G}_0^{(+)}(t,t_0)\Psi(t_0)$, for $t_0 \to -\infty$, a free motion wavepacket, the general integral equation for a time-dependent forward scattering state is

$$\Psi^{(+)}(t) = \Psi_0^{(+)}(t) + \int_{-\infty}^{\infty} dt' \widehat{G}_0^{(+)}(t,t')\widehat{V}\,\Psi^{(+)}(t').$$

Given the free motion wavepacket $\Psi_0^{(+)}(t)$ before interactions, this integral equation can be solved, for example, by iteration or with the numerical methods described in Chapter 4, to generate the full forward scattering state over times $t \geq t_0$ from an initial known state at $t_0 \to -\infty$. Similarly, a treatment can be developed for cases where a free moving state $\Psi_0^{(-)}(t)$ of A plus B is instead known at a final time $t_0 \to +\infty$ and is given at earlier times $t \leq t_0$ by a scattering state

$$\Psi^{(-)}(t) = \Psi_0^{(-)}(t) + \int_{-\infty}^{\infty} dt' \widehat{G}_0^{(-)}(t,t')\widehat{V}\,\Psi^{(-)}(t')$$

in terms of the backward propagator $\widehat{G}_0^{(-)}$.

7.1.2 Time-independent Treatment

An alternative general formulation can be given changing variables from the time t to an energy variable s with dimensions of energy. This provides equations closely related to the ones for stationary states developed in previous chapters and is applicable also to scattering of wavepackets by time-independent interaction potential energies between colliders A and B. Introducing Fourier transforms [5, 8]

$$\tilde{f}(s) = \int_{-\infty}^{+\infty} dt \exp(ist/\hbar) f(t)$$

$$f(t) = (2\pi\hbar)^{-1} \int_{-\infty}^{+\infty} ds \exp(ist/\hbar)\tilde{f}(s)$$

the equations for $\Psi^{(\pm)}(t)$ can be transformed to the new variable using the convolution relation

$$\int_{-\infty}^{+\infty} dt\, e^{ist/\hbar} \int_{-\infty}^{+\infty} dt'\, f(t-t')g(t') = \tilde{f}(s)\tilde{g}(s)$$

and that the functions $\widehat{G}^{(\pm)}(t, t_0) = \widehat{G}^{(\pm)}(\tau)$, and $\theta(\tau) = \exp(-\epsilon\tau/\hbar)$, for $\lim \epsilon \to 0+$, and $\tau = t - t_0 \geq 0$ can be Fourier transformed to the s variable. Using that the transform of $\Psi^{(\pm)}(t)$ is $\tilde{\Psi}^{(\pm)}(s) = \Psi^{(\pm)}(s)$, to simplify notations, the convolution relation gives

$$\Psi^{(\pm)}(s) = \Psi_0^{(\pm)}(s) + \widehat{G}_0^{(\pm)}(s)\widehat{V}\Psi^{(\pm)}(s)$$

$$\widehat{G}_0^{(\pm)}(s) = -\frac{i}{\hbar}\int_{-\infty}^{+\infty} dt\, e^{ist/\hbar}\theta(t)e^{-i\widehat{H}_0 t/\hbar} = \lim_{\eta\to 0+}(s + i\eta - \widehat{H}_0)^{-1}.$$

Using the notation $\lim_{\eta\to 0+} s \pm i\eta = s^{(\pm)}$, a compact form is $\widehat{G}_0^{(\pm)}(s) = (s^{(\pm)} - \widehat{H}_0)^{-1}$ and similarly for the full propagator transform $\widehat{G}^{(\pm)}(s) = (s^{(\pm)} - \widehat{H})^{-1}$. The above equation for $\Psi^{(\pm)}(s)$ is the Lippmann–Schwinger equation [2] for a general wavepacket scattering.

Applying the convolution relation to the previous integral equation for $\widehat{G}^{(\pm)}(t, t_0)$ leads to the transformed equation

$$\widehat{G}^{(\pm)}(s) = \widehat{G}_0^{(\pm)}(s) + \widehat{G}_0^{(\pm)}(s)\widehat{V}\widehat{G}^{(\pm)}(s) = \widehat{G}_0^{(\pm)}(s) + \widehat{G}^{(\pm)}(s)\widehat{V}\widehat{G}_0^{(\pm)}(s)$$

with the second equality obtained from the operator identity $(\widehat{A} - \widehat{B})^{-1} = \widehat{A}^{-1} + (\widehat{A} - \widehat{B})^{-1}\widehat{B}\widehat{A}$. It can be formally solved for

$$\widehat{W}^{(\pm)}(s) = \widehat{G}^{(\pm)}(s)\left[\widehat{G}_0^{(\pm)}(s)\right]^{-1} = \widehat{I} + \widehat{G}^{(\pm)}(s)\widehat{V}$$

and provides a new expression for the scattering states,

$$\Psi^{(\pm)}(s) = \widehat{W}^{(\pm)}(s)\Psi_0^{(\pm)}(s)$$

where $\widehat{W}^{(\pm)}(s)$ is a wave operator. This in turn gives

$$\Psi^{(\pm)}(s) = \Psi_0^{(\pm)}(s) + \widehat{G}_0^{(\pm)}(s)\widehat{T}^{(\pm)}(s)\Psi_0^{(\pm)}(s)$$

which defines the transition operator

$$\widehat{T}^{(\pm)}(s) = \widehat{V}\widehat{W}^{(\pm)}(s) = \widehat{V} + \widehat{V}\widehat{G}^{(\pm)}(s)\widehat{V}$$

central to calculations of transition probabilities. Furthermore, from $\widehat{G}^{(\pm)}(s) = \widehat{G}_0^{(\pm)}(s)[\widehat{I} + \widehat{V}\widehat{G}^{(\pm)}(s)]$, it follows that

$$\widehat{W}^{(\pm)}(s) = \widehat{I} + \widehat{G}_0^{(\pm)}(s)\widehat{V}\widehat{W}^{(\pm)}(s)$$

$$\widehat{T}^{(\pm)}(s) = \widehat{V} + \widehat{V}\widehat{G}_0^{(\pm)}(s)\widehat{T}^{(\pm)}(s)$$

with the second line known as the Lippman–Schwinger equation for the T-operator [2]. These equations can be iterated on the right side to provide expansions in powers of $\widehat{V}$ for wavefunctions and for transition amplitudes.

The connection with the previous treatments of stationary scattering wavefunctions starts with the expansion of a free-motion $\Psi_0^{(\pm)}(t)$ in a basis set of stationary (time-independent) wavefunctions $\Psi_\nu^{(0)}$ with $\nu = (\vec{k}, n)$ for total (kinetic plus internal) energies $E = (\hbar^2 k^2/2M) + E_p^{(A)} + E_q^{(B)}$ and quantum numbers $n = (p, q)$ for internal electronic-nuclear states, as

$$\Psi_0^{(\pm)}(t) = \int d\nu\, c_\nu^{(\pm)} \Psi_\nu^{(0)} \exp(iEt/\hbar)$$

where expansion coefficients $c_\nu^{(\pm)}$ are chosen to construct given wavepackets for A + B before or after their interaction. Its Fourier transform is

$$\Psi_0^{(\pm)}(s) = \int d\nu\, c_\nu^{(\pm)} \Psi_\nu^{(0)}\, 2\pi\hbar\delta(s - E)$$

Applying to this the wave-operator $\widehat{W}^{(\pm)}$, it constructs $\Psi^{(\pm)}(s)$ as a superposition of stationary states $\Psi_\nu^{(\pm)} = \widehat{W}^{(\pm)}(E)\Psi_\nu^{(0)}$ and the wavepacket

$$\Psi^{(\pm)}(t) = \int d\nu\, c_\nu^{(\pm)} \Psi_\nu^{(\pm)} \exp(iEt/\hbar)$$

where $\Psi_\nu^{(\pm)} = \Psi_\nu^{(0)} + \widehat{G}_0^{(\pm)}(s)\widehat{T}^{(\pm)}(s)\Psi_\nu^{(0)}$ is a stationary wavefunction, given in terms of the transition operator $\widehat{T}^{(\pm)}(s)$.

Using the Dirac bracket notation, normalization and completeness relations are given by

$$\left\langle \Psi_\nu^{(\pm)} \middle| \Psi_{\nu'}^{(\pm)} \right\rangle = \delta_{\nu\nu'} = \delta^3(\vec{k} - \vec{k}')\delta_{nn'}$$

$$\sum_b \left|\Psi_b\right\rangle \left\langle \Psi_b\right| + \int d\nu\, \left|\Psi_\nu^{(\pm)}\right\rangle \left\langle \Psi_\nu^{(\pm)}\right| = \hat{1}$$

where Ψ_b is a bound state possibly present during interactions, with the first line established using properties of the propagator $\widehat{G}^{(\pm)}(s)$ [7].

Wavefunctions are obtained in the coordinate representation, as $\Psi_\nu^{(\pm)}(\boldsymbol{Q}, \boldsymbol{X}) = \left\langle \boldsymbol{Q}, \boldsymbol{X} \middle| \Psi_\nu^{(\pm)} \right\rangle$ for atomic coordinates $\boldsymbol{Q} = (\vec{R}, \boldsymbol{q})$ and electronic variables $\boldsymbol{X}$. Introducing a basis set of K electronic wavefunctions $\Phi_k(\boldsymbol{X}; \boldsymbol{Q})$ for chosen atomic coordinates, omitting here the subindex ν,

$$\Psi^{(\pm)}(\boldsymbol{Q}, \boldsymbol{X}) = \sum_{k=1}^{K} \Phi_k(\boldsymbol{X}; \boldsymbol{Q})\, \psi_k^{(\pm)}(\boldsymbol{Q}) = \boldsymbol{\Phi}(\boldsymbol{X}; \boldsymbol{Q})\, \boldsymbol{\psi}^{(\pm)}(\boldsymbol{Q})$$

written in terms of the row $1 \times K$ matrix $\boldsymbol{\Phi}$ and the column $K \times 1$ matrix $\boldsymbol{\psi}^{(\pm)}$. The atomic motion amplitudes $\boldsymbol{\psi}^{(\pm)}(\boldsymbol{Q})$, with components for all the electronic states, can be further expanded in a basis set of J internal rovibrational states $\phi_j(\boldsymbol{q})$, giving $\boldsymbol{\psi}^{(\pm)}(\boldsymbol{Q}) = \sum_j \phi_j(\boldsymbol{q}) \boldsymbol{\psi}_j^{(\pm)}(\vec{R}), j = 1$ to J, with relative motion amplitudes forming a $(K.J) \times 1$ row of rows $\boldsymbol{\psi}^{(\pm)}(\vec{R}) = \left[\psi_\gamma^{(\pm)}(\vec{R}; \vec{k})\right]$, $\gamma = (k, j)$, with γ labelling an electronic-nuclear channel state. The combined electronic-nuclear basis set of $K.J$ functions $\{\Phi_k(\boldsymbol{X}; \boldsymbol{Q})\phi_j(\boldsymbol{q}) = \Upsilon_\gamma(\boldsymbol{q}, \boldsymbol{X})\}$ provide a total wavefunction expansion

$\Psi^{(\pm)}(\boldsymbol{Q},\boldsymbol{X}) = \sum_\gamma \Upsilon_\gamma(\boldsymbol{q},\boldsymbol{X})\psi_\gamma^{(\pm)}(\vec{R})$. In this basis set, the operator expressions for scattering equations are given by matrix expressions in terms of propagator and transition matrices, like

$$\boldsymbol{\Psi}_\gamma^{(\pm)}(\vec{R}) = \boldsymbol{\Psi}_\gamma^{(0)}(\vec{R}) + \int d^3R'\widehat{\boldsymbol{G}}_0^{(\pm)}(\vec{R},\vec{R}')\left[\widehat{\boldsymbol{T}}^{(\pm)}(s)\boldsymbol{\Psi}_\gamma^{(0)}\right](\vec{R}'),$$

$$\boldsymbol{\Psi}_\gamma^{(0)}(\vec{R}) = \left[\psi_\gamma^{(0)}(\vec{R};\vec{k})\delta_{\gamma'\gamma}\right]$$

$$\widehat{\boldsymbol{G}}_0^{(\pm)}(\vec{R},\vec{R}') = \left\langle \vec{R}\middle|(s^{(\pm)} - \widehat{\boldsymbol{H}}_0)^{-1}\middle|\vec{R}'\right\rangle = -\frac{2M}{4\pi\hbar^2}\left[\frac{e^{ik_\gamma\left|\vec{R}-\vec{R}'\right|}}{\left|\vec{R}-\vec{R}'\right|}\delta_{\gamma\gamma'}\right]$$

where the form of the diagonal matrix $\widehat{\boldsymbol{G}}_0^{(\pm)}(\vec{R},\vec{R}')$ elements has been obtained from integration in the complex s energy plane [8]. In the limit of $R \to \infty$ after a collision, one finds the scattering amplitude at solid angle Ω for the $\gamma \to \gamma'$ transition given by

$$f_{\gamma'\gamma}(\omega) = -\frac{2M}{4\pi\hbar^2}(2\pi\hbar)^3\left\langle \boldsymbol{\Psi}_{\gamma'}^{(0)}\middle|\widehat{\boldsymbol{T}}^{(\pm)}\middle|\boldsymbol{\Psi}_\gamma^{(0)}\right\rangle$$

where $\left|\boldsymbol{\Psi}_\gamma^{(0)}\right\rangle$ is a column matrix in the basis set $\{\Upsilon_{\gamma'}\}$ and $\widehat{\boldsymbol{T}}^{(\pm)}$ is a square matrix in the basis set. Here, the bracket indicates integration over $\vec{R}$, and ω is the solid angle from $\vec{k}$ to $\vec{k}'$.

7.1.3 Rates of Change of Observables

The scattering operator formalism provides ways to calculate rates of change of observables, like the density or the momentum flux of a species as a result of collisions. Let an operator $\widehat{A}$ be the property and the system Hamiltonian be given by $\widehat{H} = \widehat{H}_0 + \widehat{V}$ as the sum of an unperturbed motion Hamiltonian plus a potential energy of interaction. We consider a property that is a constant over time in the absence of interactions so that it satisfies the commutator relation

$$i\hbar(d\widehat{A}/dt) = \widehat{H}_0\widehat{A} - \widehat{A}\widehat{H}_0 = [\widehat{H}_0,\widehat{A}] = 0$$

During an interaction described by a time-dependent state $\Psi^{(+)}(t)$, the time evolution of the averaged property is given by [5]

$$d\left\langle \Psi^{(+)}(t)\middle|\widehat{A}\middle|\Psi^{(+)}(t)\right\rangle/dt = (i\hbar)^{-1}\langle\Psi^{(+)}(t)|[\widehat{A},\widehat{V}]|\Psi^{(+)}(t)\rangle$$

which applies to a wavepacket or to a stationary state. For a stationary state of energy E so that $\Psi^{(+)}(t) = \Psi_\nu^{(+)}\exp(iEt/\hbar)$, with $\nu = (\vec{k}, n)$, this gives before taking the limit $\lim(\eta \to 0+)$,

$$d\left\langle \Psi_\nu^{(+)}\middle|\widehat{A}\middle|\Psi_\nu^{(+)}\right\rangle/dt = (i\hbar)^{-1}\left\langle \Psi_\nu^{(+)}\middle|[\widehat{A},\widehat{V}]\middle|\Psi_\nu^{(+)}\right\rangle = (2/\hbar)Im\left\langle \Psi_\nu^{(+)}\middle|\widehat{A}\widehat{V}\middle|\Psi_\nu^{(+)}\right\rangle$$

Using $\left\langle \Psi_\nu^{(+)} \right| = \left\langle \Psi_\nu^{(0)} \right| + \left\langle \Psi_\nu^{(+)} \right| \widehat{V}\, \widehat{G}_0^{(-)}$, together with

$$\widehat{G}_0^{(\pm)}(E) = \mathcal{P}(E - \widehat{H}_0)^{-1} \mp i\pi\delta(E - \widehat{H}_0)$$

for $\lim \eta \rightarrow 0+$, where $\mathcal{P}$ indicates taking the principal value of the required integral, one arrives to the useful relation

$$d\left\langle \Psi_\nu^{(+)} \middle| \widehat{A} \middle| \Psi_\nu^{(+)} \right\rangle /dt = (2/\hbar) Im \left\langle \Psi_\nu^{(0)} \middle| \widehat{A}\widehat{V} \middle| \Psi_\nu^{(+)} \right\rangle$$
$$+ (2/\hbar) \left\langle \Psi_\nu^{(+)} \middle| \widehat{V}\delta(E - \widehat{H}_0)\widehat{A}\widehat{V} \middle| \Psi_\nu^{(+)} \right\rangle .$$

An important application is to the choice $\widehat{A} = \left| \Psi_{\nu'}^{(0)} \right\rangle \left\langle \Psi_{\nu'}^{(0)} \right|$, a projection operator which selects the population of state $\Psi_{\nu'}^{(0)}$, with total energy E' after interactions. The first term above is null, while the second one gives the rate $R_{\nu'\nu}$ of transition $\nu \rightarrow \nu'$ per unit time as

$$R_{\nu'\nu} = \frac{2\pi}{\hbar} \left| \left\langle \Psi_{\nu'}^{(0)} \middle| \widehat{V} \middle| \Psi_\nu^{(+)} \right\rangle \right| \delta(E - E') = \frac{2\pi}{\hbar} \left| \left\langle \Psi_{\nu'}^{(0)} \middle| \widehat{T}^{(+)}(E) \middle| \Psi_\nu^{(0)} \right\rangle \right|^2 \delta(E - E')$$

Specifically, a state-to-state collisional cross section between state $\nu = (\vec{k}, n)$ and ν' is given by the rate of the transition per unit of incoming flux $(P/M)\,(2\pi\hbar)^{-3}$, with $\vec{P} = \hbar\vec{k}$. For a final momentum $\vec{P'}$ at solid angle $\Delta\omega$ relative to $\vec{P}$, the cross section within that solid angle is given by

$$\int_{\Delta\omega} d^3P'\, R_{n'n}(\vec{P'}, \vec{P}) = (P/M)\,(2\pi\hbar)^{-3} \Delta\sigma_{n'n}(\vec{P'}, \vec{P})$$

and after a change of variable in the delta function from E' to P',

$$d\sigma_{n'n}(\vec{P'}, \vec{P})/d\omega = (2\pi)^4 \hbar^2 M^2 (P'/P) \left| \left\langle \Psi_{\nu'}^{(0)} \middle| \widehat{T}^{(+)}(E) \middle| \Psi_\nu^{(0)} \right\rangle \right|^2$$

where $(P')^2/(2M) = E - E_{n'}$.

7.2 Partitioning the Space of State Wavefunctions

7.2.1 Effective Hamiltonian Operators

Effective Hamiltonians can be constructed quite generally to describe scattering phenomena involving two molecules, where some states of the whole system are active in the dynamics and must be treated explicitly, while all other states are secondary to the treatment and are involved only as intermediaries in the couplings among the active states. This classification of states and its consequences for scattering can be quantified introducing general projection operators, useful in scattering [9] and quasi-bound or resonance phenomena [10]. Some of this has already been covered in the chapter on inelastic collisions, where a projection operator was constructed from the internal states of the involved molecules. Here, we present a more general treatment with projection operators which can be applied to distortion, adiabatic, and sudden collisions, and to resonance phenomena, following the treatment in [7].

Hermitian projection operators (or projectors) $\widehat{P}$ of interest in scattering satisfy self-adjoint $\widehat{P} = \widehat{P}^{\dagger}$ and idempotency $\widehat{P} = \widehat{P}^2$ relations and are chosen to also satisfy the asymptotic condition at large distances between the molecules so that $\lim[\widehat{H}, \widehat{P}] \cong 0$, as $R \to \infty$, with $\widehat{H}$ being the full Hamiltonian. This projector is constructed to include asymptotically all the active states and, in particular, satisfies the relation $\widehat{P}\Psi_{0\nu} = \Psi_{0\nu}$, where $\Psi_{0\nu}$ is an initial unperturbed total state and is an eigenfunction of the asymptotic Hamiltonian $\widehat{H}_0$ so that $\widehat{H}_0\Psi_{0\nu} = E\Psi_{0\nu}$ for total energy E. A complementary projector is $\widehat{Q} = \widehat{I} - \widehat{P}$, also self-adjoint and idempotent, and satisfies $\widehat{Q}\widehat{P} = \widehat{0}$. Given an operator $\widehat{A}$ for the whole system, it can be decomposed into block diagonal and block non-diagonal parts as $\widehat{A} = \widehat{A}_D + \widehat{A}_N$ with

$$\widehat{A}_D = \widehat{P}\widehat{A}\widehat{P} + \widehat{Q}\widehat{A}\widehat{Q}, \quad \widehat{A}_N = \widehat{P}\widehat{A}\widehat{Q} + \widehat{Q}\widehat{A}\widehat{P}$$

Given two operators $\widehat{A}$ and $\widehat{B}$, we find

$$(\widehat{A}\,\widehat{B})_D = \widehat{A}_D\widehat{B}_D + \widehat{A}_N\widehat{B}_N$$
$$(\widehat{A}\,\widehat{B})_N = \widehat{A}_D\widehat{B}_N + \widehat{A}_N\widehat{B}_D.$$

This decomposition is applicable to the equation for the resolvent $\widehat{G}^{(\pm)}(E) = (E^{(\pm)} - \widehat{H})^{-1}$ in $(E^{(\pm)} - \widehat{H})\widehat{G}^{(\pm)}(E) = \widehat{I}$. It leads to

$$(E^{(\pm)} - \widehat{H}_D)\widehat{G}_D^{(\pm)}(E) - \widehat{H}_N\widehat{G}_N^{(\pm)}(E) = \widehat{I}$$
$$(E^{(\pm)} - \widehat{H}_D)\widehat{G}_N^{(\pm)}(E) - \widehat{H}_N\widehat{G}_D^{(\pm)}(E) = \widehat{0}$$

The second line can be formally solved for $\widehat{G}_N^{(\pm)}$ with this replaced in the first line to obtain

$$\widehat{G}_D^{(\pm)} = (E^{(\pm)} - \widehat{H}_D)^{-1} + (E^{(\pm)} - \widehat{H}_D)^{-1}\widehat{H}_N\widehat{G}_D^{(\pm)}\widehat{H}_N(E^{(\pm)} - \widehat{H}_D)^{-1}$$

relating $\widehat{G}_D^{(\pm)}$ to the simpler resolvent $(E^{(\pm)} - \widehat{H}_D)^{-1}$.

From the second line, $\widehat{G}_N^{(\pm)}(E) = (E^{(\pm)} - \widehat{H}_D)^{-1}\widehat{H}_N\widehat{G}_D^{(\pm)}(E)$ and replacing this in the first line,

$$(E^{(\pm)} - \widehat{\mathcal{H}}_D)\widehat{G}_D^{(\pm)}(E) = \widehat{I}$$
$$\widehat{\mathcal{H}}_D = \widehat{H}_D + \widehat{H}_N(E^{(\pm)} - \widehat{H}_D)^{-1}\widehat{H}_N$$

where $\widehat{\mathcal{H}}_D$ is an effective Hamiltonian operator which describes events where the dynamics involving states selected by $\widehat{P}$ and $\widehat{Q}$ are kept separate, but it also accounts in each set for the indirect effects of the other set of states.

Scattering states can be obtained from $\Psi_{\nu}^{(\pm)} = \widehat{W}^{(\pm)}(E)\Psi_{\nu}^{(0)} = \pm i\eta\widehat{G}^{(\pm)}(E)\Psi_{\nu}^{(0)}$ and

$$\widehat{G}^{(\pm)}(E) = [\widehat{I} + (E^{(\pm)} - \widehat{H}_D)^{-1}\widehat{H}_N]\widehat{G}_D^{(\pm)}(E)$$

so that

$$\Psi_{\nu}^{(\pm)} = [\widehat{I} + (E^{(\pm)} - \widehat{\mathcal{H}}_D)^{-1}\widehat{H}_N]\widehat{W}_D^{(\pm)}(E)\Psi_{\nu}^{(0)},$$

$$\widehat{W}_D^{(\pm)}(E) = \pm i\eta\widehat{G}_D^{(\pm)}(E) = \widehat{I} + \widehat{G}_D^{(\pm)}(E)(\widehat{\mathcal{H}}_D - E).$$

where to the right the wave-operator $\widehat{W}_D^{(\pm)}$ describes scattering in the separate sets of states, followed to its left by an operator coupling the two sets of states.

7.2.2 Optical Potentials and Scattering Resonances

The formalism providing the $\widehat{\mathcal{H}}_D$ effective Hamiltonian operator can give the flux loss in inelastic or reactive collisions by means of an optical potential defined in the P-subspace, or it can describe scattering resonances in the Q-subspace.

The effective Hamiltonian $\widehat{\mathcal{H}}_D$ can be related to an effective optical potential operator in the P-subspace, writing

$$\widehat{\mathcal{H}}_P = \widehat{P}\widehat{H}_0\widehat{P} + \widehat{\mathcal{V}}_P^{(\pm)}(E)$$

$$\widehat{\mathcal{V}}_P^{(\pm)}(E) = \widehat{P}\widehat{V}\widehat{P} + \widehat{P}\widehat{H}\widehat{Q}(E^{(\pm)} - \widehat{Q}\widehat{H}\widehat{Q})^{-1}\widehat{Q}\widehat{H}\widehat{P}$$

Further using the relation for special functions $(E^{(\pm)} - X)^{-1} \approx \mathcal{P}[(E - X)^{-1}] \mp i\pi\delta(E - X)$ where X is an integration variable, with $\mathcal{P}$ indicating the principal value of the inverse in the integral [5], and when the limit $\lim_{\eta\to 0+} E \pm i\eta = E^{(\pm)}$ is taken after integration,

$$\widehat{\mathcal{V}}_P^{(\pm)}(E) \approx \widehat{\mathcal{V}}_{P,r}^{(\pm)}(E) \mp i\widehat{\mathcal{V}}_{P,i}^{(\pm)}(E)$$

$$\widehat{\mathcal{V}}_{P,r}^{(\pm)}(E) = \widehat{P}\widehat{V}\widehat{P} + \widehat{P}\widehat{H}\widehat{Q}\mathcal{P}[(E - \widehat{Q}\widehat{H}\widehat{Q})^{-1}]\widehat{Q}\widehat{H}\widehat{P},\ \widehat{\mathcal{V}}_{P,i}^{(\pm)}(E)$$
$$= \pi\widehat{P}\widehat{H}\widehat{Q}\delta(E - \widehat{Q}\widehat{H}\widehat{Q})\widehat{Q}\widehat{H}\widehat{P}$$

where $\widehat{\mathcal{V}}_{P,r}^{(\pm)}(E)$ and $\widehat{\mathcal{V}}_{P,i}^{(\pm)}(E)$ are real-value operators giving real- and imaginary-parts of the effective optical operator, corresponding to the two parts of a model optical potential for inelastic or reactive collisions. Furthermore, different choices of states in the P-subspace lead to several useful approximations already described in previous chapters. This includes (i) choosing asymptotic internal molecular states $\phi_n(\boldsymbol{q})$ to construct $\widehat{P}$ and an optical potential for distorted-wave scattering; (ii) choosing adiabatic internal states $z_\alpha(\boldsymbol{q};\vec{R})$ that adjust to intermolecular distances in slow collisions, giving an optical potential in an adiabatic approximation; and (iii) choosing sudden-collision states to construct $\widehat{P}(\boldsymbol{q})$ for fixed internal variables, as a projector on wavefunctions $\phi_n(\boldsymbol{q})\chi^{(+)}_{\vec{k}}(\vec{R};\boldsymbol{q})$, where $\chi^{(+)}_{\vec{k}}(\vec{R};\boldsymbol{q})$ is a scattering eigenfunction of a Hamiltonian where internal variables are fixed. State expansions resulting from operating with $\hat{P}(\boldsymbol{q})$ instead involve integration over the relative motion wavevector $\vec{k}$, in a sudden approximation. Details on these approximations can be found in [7].

Resonance scattering involving quasi-bound states, with their energy levels and widths arising from interaction with scattering states, can be described starting with the effective Hamiltonian $\hat{\mathcal{H}}_Q$. They originate in a molecule–molecule interaction potential energy function with a localized minimum at a relative position $\vec{R}_m$ that holds the two molecules in nearly bound states, although coupled to scattering states. In the limit where this coupling can be ignored, to start with, the states are a combination of a set of K bound orthonormalized states Φ_κ and we choose $\hat{Q} = \sum_{\kappa=1}^{K} |\Phi_\kappa\rangle\langle\Phi_\kappa|$. Further constructing, from those bound states, new eigenstates $|\xi_\kappa\rangle$ of $\hat{Q}\hat{H}\hat{Q}$ with eigenenergy $\mathcal{E}_\kappa$ in

$$(\mathcal{E}_\kappa - \hat{Q}\hat{H}\hat{Q})\,|\xi_\kappa\rangle = 0$$

the resolvent of $\hat{Q}\hat{H}\hat{Q}$ in Q-subspace, replaced in the effective Hamiltonian in P-subspace, gives

$$\hat{\mathcal{H}}_P = \hat{P}\hat{H}\hat{P} + \hat{P}\hat{H}\hat{Q}\sum_{\kappa=1}^{K}\frac{|\xi_\kappa\rangle\langle\xi_\kappa|}{E^{(\pm)} - \mathcal{E}_\kappa}\hat{Q}\hat{H}\hat{P}$$

so that the effective Hamiltonian varies rapidly with energy E around $\mathcal{E}_\kappa$.

The behavior of the scattering wavefunction in the neighborhood of a resonance energy can be obtained from $\hat{P}\Psi^{(\pm)}_\nu = \hat{W}^{(\pm)}_P(E)\hat{P}\Psi^{(0)}_\nu$, where the wave operator is $\hat{W}^{(\pm)}_P(E) = \hat{P} + \hat{G}^{(\pm)}_P(E)(\hat{\mathcal{H}}_P - E)$. The resolvent $\hat{G}^{(\pm)}_P = \hat{P}\hat{G}^{(\pm)}_D\hat{P}$ involves a term from direct transitions in the P-subspace and also an indirect transitions term containing $\hat{G}^{(\pm)}_Q = (\hat{Q}E^{(\pm)} - \hat{\mathcal{H}}_Q)^{-1}$ as previously shown. The energy dependence of the latter term in $\hat{P}\Psi^{(\pm)}_\nu$ must be obtained from the eigenfunctions and complex-valued eigenenergies $f_\kappa(E)$ in $[f_\kappa(E) - \hat{\mathcal{H}}_Q]\,|\zeta_\kappa\rangle = 0$. Relevant resonance energy features follow from the complex-valued roots $E_\kappa = E'_\kappa - i\Gamma_\kappa/2$ of the equation $f_\kappa(E) = E$. The new real-value energy E'_κ is usually close to the previous $\mathcal{E}_\kappa$, and in addition one finds a resonance energy width Γ_κ. The time dependence of the resonance component of a wavepacket follow from the superposition of resonance stationary states from $\nu = \kappa$ in

$$\hat{P}\Psi_{res}(\mathrm{t}) = \sum_\kappa c_\kappa\hat{P}\Psi^{(\pm)}_\kappa\exp\left[-i\left(E'_\kappa - i\Gamma_\kappa/2\right)t/\hbar\right]$$

Additional details can be found in [7].

7.3 Many-Atom Scattering Operators

7.3.1 Reactions in Three-atom Systems

The formulation of CAC for reactive collisions described in Chapter 6, Section 6.3, can be implemented introducing two-body transition operators. Here we consider electronically adiabatic interactions in a treatment that can be extended to diabatic interactions, as done in Chapter 5. The potential energy for a triatomic (A, B, C) system, where here A, B, and C are atoms, can frequently be written as

$$V = V_{AB} + V_{BC} + V_{CA} + V_{ABC}$$

where atom-pair potentials $V_{IJ}(\vec{r}_I, \xi_I; \vec{r}_J, \xi_J)$ can depend on atomic positions and atomic spin variables $\xi = \pm 1/2$, and the remaining intrinsic triatomic term V_{ABC} is usually small [11]. Atom-pair potential energies are available for closed shell atoms, and also for open-shell atoms with singlet and triplet pair interactions [12]. A general treatment of three-atom interactions in terms of atom-pair scattering operators has been developed in detail [13, 14], based on the Faddeev–AGS equations [15, 16] for three interacting particles, generalized to apply to particles with structure, such as atoms with their electronic structure, incorporating computational methods for atom-pair interaction potentials.

The rearrangement scattering equations can now be rewritten in terms of atom-pair transition amplitudes, and arrangement channels can be identified as before with $j = 1$ for A + BC, $j = 2$ for B + CA, $j = 3$ for C + AB, and $j = 4$ for ABC, each containing now only a pair potential in each channel, as in $V_1 = V_{BC}$, $V^{(1)} = 0$, $V_2 = V_{CA}$, $V_3 = V_{AB}$, and $V_4 = V_{ABC}$. The free-motion Hamiltonian is a sum of atomic kinetic energies as in $\hat{H}_0 = \hat{K}_A + \hat{K}_B + \hat{K}_C$. Inter-channel couplings $V^{(jk)}$ in the CAC formulation of Chapter 6 can be chosen as

$$V^{(jk)} = V_j(1 - \delta_{jk})$$

Further introducing a two-atom transition operator as

$$\hat{T}_j = V_j + V_j \hat{G}_0^{(+)}(E) \hat{T}_j$$

where $\hat{G}_0^{(+)}(E) = (E + i\eta - \hat{H}_0)^{-1}$, the CAC equation for each arrangement amplitude $\Psi_{1\alpha}^{(j)}, j = 1$ to 4, can be written in terms of the two-atom transition operators. They satisfy the CAC equations in Section 6.3.3

$$\Psi_{1\alpha}^{(j)} = \delta_{j1} \Psi_{1\alpha}^{(0)} + \hat{G}_j^{(+)} \sum_{k=1}^{4} V^{(jk)} \Psi_{1\alpha}^{(k)}$$

with the propagator $\widehat{G}_j^{(+)} = (E + i\eta - \widehat{H}_j)^{-1}$ for free motion in arrangement j, with $\eta \to 0+$. It contains within $\widehat{G}_j^{(+)} V^{(jk)}$ the factor $\widehat{G}_j^{(+)} V_j = \widehat{G}_0^{(+)} \widehat{T}_j$ and this leads to [17]

$$\Psi_{1\alpha}^{(j)} = \delta_{j1} \Psi_{1\alpha}^{(0)} + \widehat{G}_0^{(+)} \widehat{T}_j \sum_{k=1}^{4} (1 - \delta_{jk}) \Psi_{1\alpha}^{(k)}$$

as the new coupled amplitude equations.

In this way, the three-atom problem has been related to calculations of two-atom transition operators for pairs in $j = 1, 2$, and 3 arrangements. These, however, must be obtained for the total energy E of the three atoms, or off-energy shell, which differs from the collisional energies for each atom pair, since one must add the kinetic energy of the third atom. The amplitude for $j = 4$ can be obtained from the approximation $\widehat{T}_4 \cong V_4$ when the three-atom potential term is small. Ways to calculate the atom-pair transition operators have been developed using expansions in diatomic basis sets [18, 19] and the numerical integration of differential equations [20].

The equation for $\Psi_{1\alpha}^{(j)}$ provides insight on collisions at high atomic velocities where the magnitude $\left\| \widehat{G}_0^{(+)} \widehat{T}_j \right\|$ is small, and it is then acceptable to expand the solution as a series by iteration, replacing $\Psi_{1\alpha}^{(k)}$ with $\delta_{k1} \Psi_{1\alpha}^{(0)}$ to the right. Scattering amplitudes follow, in the notation of Chapter 6, from

$$f_{j\nu,1\alpha}^{(+)}(\Omega_j) = -\frac{1}{4\pi} \frac{2M_j}{\hbar^2} (2\pi)^3 \left\langle \vec{k}_{j\nu} \middle| \widehat{T}_{\nu,\alpha}^{(j)} \middle| \vec{k}_{1\alpha} \right\rangle,$$

$$\left\langle \vec{k}_{j\nu} \middle| \widehat{T}_{\nu,\alpha}^{(j)} \middle| \vec{k}_{1\alpha} \right\rangle = \sum_{k=1}^{4} \left\langle \Psi_{j\nu}^{(0)} \middle| V^{(jk)} \middle| \Psi_{1\alpha}^{(k)} \right\rangle$$

For inelastic collisions in channel 1, this gives for single interactions

$$T_{1\alpha',1\alpha} = \left\langle \Psi_{1\alpha'}^{(0)} \middle| (T_2 + T_3 + T_4) \middle| \Psi_{1\alpha}^{(0)} \right\rangle$$

showing a superposition of events mediated by atom-pair interactions of A with C and with B plus the intrinsic three-atom interaction. This expression can also be used for collisional dissociation in $A + BC \to A + (B + C)$ when $\Psi_{1\alpha'}^{(0)}$ contains the scattering interaction of B with C. For the reaction $1 \to 3$ in $A + BC \to AB + C$, iteration leads to

$$T_{3\gamma,1\alpha} = \left\langle \Psi_{3\gamma}^{(0)} \middle| (E - \widehat{H}_0 + T_2 + T_4) \middle| \Psi_{1\alpha}^{(0)} \right\rangle$$

Here the term with $(E - \widehat{H}_0)$ corresponds to a glancing event leading to stripping rearrangement, and the other two come from atom–atom interactions. The glancing contribution is found to correctly describe some reactions such as in $Ar^+ + H_2$, $K + I_2$ and $K + Br_2$ [21, 22].

A treatment applicable to calculations of atom-diatomic scattering cross sections at low collision energies using the Faddeev–AGS formalism has also been developed and has been applied to three-atom scattering resonances in $H+H_2$ collisions [14, 23, 24].

7.3.2 Multiple Scattering

The operator formalism is useful for scattering involving extended systems, such as polyatomic molecules, where multiple collisions may arise among subsystems. Here we again consider electronically adiabatic interactions in a treatment that can be extended to diabatic interactions, as done in Chapter 5. The interaction potential energy operator between polyatomics A and B can be expressed as a sum of long-range plus short-range terms dependent on the relative position $\vec{R}$ and on all the internal (vibrational and rotational) variables $\boldsymbol{q}=(\boldsymbol{q}_A, \boldsymbol{q}_B)$, as

$$\hat{V}(\vec{R},\boldsymbol{q}) = \hat{V}_{LR}(\vec{R},\boldsymbol{q}) + \hat{V}_{SR}(\vec{R},\boldsymbol{q})$$

with $\hat{V}_{LR}$ constructed to contain weak long-range (Van der Waals) attractive interactions between A and B and a smooth cut-off at short distances, while $\hat{V}_{SR}$ contains strongly repulsive short-range interactions and intermediate-range energy minima. The first term can usually be treated as a perturbative distortion potential, while the second term requires a detailed computational treatment of strongly coupled scattering states.

Using the derivation with distorted waves in Chapter 3, for $\hat{V} = \hat{V}^{(1)} + \hat{V}^{(2)}$ where $\hat{V}^{(1)}$ is a distorting potential giving distorted waves $\Psi^{(+)}_{1,\alpha}(\vec{R},\boldsymbol{q}) = \chi^{(+)}_{1,\alpha}(\vec{R})\phi_\alpha(\boldsymbol{q})$ and is included in a distortion resolvent operator $\hat{G}^{(+)}_1$, the total wavefunction was given as $\Psi^{(+)}_\alpha = \Psi^{(+)}_{1,\alpha} + \hat{G}^{(+)}_1\hat{V}^{(2)}\Psi^{(+)}_\alpha$. Here $\Psi^{(+)}_{1,\alpha}$ can be rewritten with the distorted-wave operator $\widehat{W}^{(+)}_1(s)$ or the distorted-transition operator $\hat{T}^{(+)}_1(s)$. Identifying LR and SR potentials as $\hat{V}_{LR} = \hat{V}^{(1)}$ and $\hat{V}_{SR} = \hat{V}^{(2)}$, the distorted-wave treatment in Chapter 3 shows that the total transition operator is [25, 26]

$$\hat{T}^{(+)} = \hat{T}^{(+)}_{LR} + \left[\widehat{W}^{(-)}_{LR}\right]^\dagger (\hat{V}_{SR} + \hat{V}_{SR}\hat{G}^{(+)}\hat{V}_{SR})\widehat{W}^{(+)}_{LR}$$

with $\hat{G}^{(+)} = (E^{(+)} - \hat{H})^{-1}$ the resolvent for the total event. The first term describes transitions induced by the LR interactions, while the second term describes events starting, to the right, with LR distortion from $\widehat{W}^{(+)}_{LR}$, followed by events affected by SR interactions modified by the LR interaction in $\hat{G}^{(+)}$, and finally ending with LR interactions, from $\widehat{W}^{(-)}_{LR}$.

Alternatively, the interaction potential energy between polyatomics A and B can be expressed in terms of atomic position vectors $\vec{r}_j$ in sets $\boldsymbol{q}_A = \{\vec{r}_a\}$ and $\boldsymbol{q}_B = \{\vec{r}_b\}$. It is then

$$\hat{V}(\vec{R},\boldsymbol{q}) = \sum_{a,b} v_{ab} + \sum_{a,b,c} v_{abc} + \cdots$$

a sum over atom-pair potential energies v_{ab}, triplets, and so on.

The two previous examples of potential energy decompositions for multiple scattering can be covered by a treatment where the potential energy of interaction between A and B is

$$\hat{V} = \sum_{C} \hat{V}_C$$

and C is an atom cluster. The corresponding transition operator can be written as a sum of terms constructed from the cluster transition operator $\hat{T}_C = \hat{V}_C + \hat{V}_C \hat{G}_0^{(+)} \hat{T}_C$ which involves interactions only within cluster C, with $\hat{G}_0^{(+)}$ being the resolvent for the free motion Hamiltonian $\hat{H}_0 = \hat{H}_{rel} + \hat{H}_A + \hat{H}_B$, as [25, 26]

$$\hat{T} = \sum_{C} \hat{T}^{(C)}$$

$$\hat{T}^{(C)} = \hat{T}_C + \hat{T}_C \hat{G}_0^{(+)} \sum_{C' \neq C} \hat{T}^{(C)}$$

where iteration of $\hat{T}^{(C)}$ on the right side generates a sequence of cluster–cluster collision terms. It can be verified that replacing the second line into the first recovers the usual expression for the total transition operator $\hat{T}$.

In applications to a many-atom system, it is possible to express $\hat{V}$ as a sum of atom-pair components, in the so-called diatomic-in-molecules procedure [14]. In this case, the above equation for $\hat{T}^{(C)}$ can be obtained from two-atom transition operators, which are suitable for computational work.

7.3.3 Coupled Arrangement-channel Effective-Hamiltonian Equations

Arrangement amplitudes $\Psi_{1\alpha}^{(k)}$ in Chapter 6 were decomposed into a distorted-wave component plus an additional term due to couplings, with the distorted wave generated by a distortion potential energy $U^{(k)}(R_k, \mathbf{Q}_k)$. The utility of optical potentials for rearrangements can be displayed replacing the previously introduced distortion potentials $U^{(k)}(R_k, \mathbf{Q}_k)$ in arrangement channel k (see Section 6.4.1) with more general non-local optical potential operators $\widehat{\mathcal{U}}^{(k)}(R_k)$. These are obtained using projection operators $\hat{P}^{(k)} = \sum_{\nu}^{(k)} |\phi_{k\nu}\rangle \langle \phi_{k\nu}|$, which select a subset S_k of internal states $\phi_{k\nu}$ active in that channel during reaction, and using the complementary projection operators $\hat{Q}^{(k)} = \hat{I} - \hat{P}^{(k)}$ (with $\hat{I}$ the identity operator) to select inactive states. The original arrangement coupling $V^{(k)}$ can be decomposed as $V^{(k)} = \hat{U}_P^{(k)} + \overline{V}^{(k)}$, with $\hat{U}_P^{(k)} = \hat{P}^{(k)} V^{(k)} \hat{P}^{(k)}$ now a potential coupling only active states, plus a reminder coupling potential $\overline{V}^{(k)}$. The reaction dynamics can be described within the wavefunction space of active internal states in all the arrangements, by introducing effective Hamiltonians $\widehat{\mathcal{H}}_P^{(k)}$ which follow from the formal elimination of the inactive states in the $\widehat{Q}^{(k)}$ [27].

Rearrangement is treated introducing the optical potential (OP) Hamiltonian $\widehat{H}_{OP,j} = \widehat{H}_{P,j} + \widehat{\mathcal{V}}_P^{(j)}$, where $\widehat{\mathcal{V}}_P^{(j)}$ is in general non-Hermitian and energy dependent, as the $\widehat{\mathcal{V}}_P^{(\pm)}(E)$ in Section 7.2.1, and also introducing new recoupling potentials among active states of arrangements j and k,

$$\mathcal{V}_P^{(jk)} = \widehat{P}^{(k)} V^{(jk)} \widehat{P}^{(k)} - \delta_{jk} \widehat{\mathcal{V}}_P^{(j)}$$

An additional aspect here is that the scattering states for the Hamiltonian $\widehat{H}_{OP,j}$ allow for absorption or emission of scattering flux.

The effective Hamiltonian and related optical potential can be derived within the more general coupled-arrangement channel formulation of Section 6.3.3. The arrangement components are partitioned as $\Psi_{1\alpha}^{(k)} = (\widehat{P}^{(k)} + \widehat{Q}^{(k)})\Psi_{1\alpha}^{(k)} = \Psi_{P,1\alpha}^{(k)} + \Psi_{Q,1\alpha}^{(k)}$ and replaced in the coupled arrangement equation,

$$(E - \widehat{H}_j)\left(\Psi_{P,1\alpha}^{(j)} + \Psi_{Q,1\alpha}^{(j)}\right) = \sum_k V^{(jk)}\left(\Psi_{P,1\alpha}^{(k)} + \Psi_{Q,1\alpha}^{(k)}\right).$$

Both sides are multiplied to the left, firstly times $\widehat{P}^{(j)}$ and secondly times $\widehat{Q}^{(j)}$ to obtain

$$\widehat{P}^{(j)}(E - \widehat{H}_j)\Psi_{P,1\alpha}^{(j)} - \widehat{P}^{(j)}\widehat{H}_j\Psi_{Q,1\alpha}^{(j)} = \sum_k \widehat{P}^{(j)} V^{(jk)}\left(\Psi_{P,1\alpha}^{(k)} + \Psi_{Q,1\alpha}^{(k)}\right)$$

$$\widehat{Q}^{(j)}(E - \widehat{H}_j)\Psi_{Q,1\alpha}^{(j)} - \widehat{Q}^{(j)}\widehat{H}_j\Psi_{P,1\alpha}^{(j)} = \sum_k \widehat{Q}^{(j)} V^{(jk)}\left(\Psi_{P,1\alpha}^{(k)} + \Psi_{Q,1\alpha}^{(k)}\right)$$

Formally solving for $\Psi_{Q,1\alpha}^{(j)}$ in the second line and substituting it in the first line, the result gives

$$\left(E - \widehat{\mathcal{H}}_P^{(j)}\right)\Psi_{P,1\alpha}^{(j)} = \sum_k \mathcal{V}_P^{(jk)}\Psi_{P,1\alpha}^{(k)}$$

$$\widehat{\mathcal{H}}_P^{(j)} = \widehat{P}^{(j)}\widehat{H}_j\widehat{P}^{(j)} + \widehat{\mathcal{V}}_P^{(j)},$$

$$\widehat{\mathcal{V}}_P^{(j)} = \widehat{P}^{(j)}V^{(j)}\widehat{P}^{(j)} + \widehat{P}^{(j)}\,\widehat{H}_j\widehat{Q}^{(j)}\left(E + i\eta - \widehat{Q}^{(j)}\widehat{H}\widehat{Q}^{(j)}\right)^{-1}\widehat{Q}^{(j)}\widehat{H}_j\widehat{P}^{(j)}$$

$$\widehat{\mathcal{V}}_P^{(jk)} = \widehat{P}^{(j)}V^{(jk)}\widehat{P}^{(k)} + \widehat{P}^{(j)}V^{(jk)}\widehat{Q}^{(k)}\left(E + i\eta - \widehat{Q}^{(k)}\widehat{H}\widehat{Q}^{(k)}\right)^{-1}\widehat{Q}_k V^{(jk)}\widehat{P}^{(k)}$$

with $\widehat{\mathcal{V}}_P^{(j)}$ being an effective potential energy operator in arrangement j, showing how the chosen active states in $\widehat{P}^{(j)}$ are directly coupled among themselves and also indirectly into and out of intermediate states in the complementary subspace of $\widehat{Q}^{(j)}$, which leads to flux changes in the subspace of $\widehat{P}^{(j)}$. Similarly, $\widehat{\mathcal{V}}_P^{(jk)}$ describes arrangement changes due to couplings between active states in arrangement channels j and k, directly and indirectly with propagation in arrangement k.

7.4 Density Operator Treatments

7.4.1 Equation of Motion for the Density Operator

Collisional phenomena where two molecules interact in isolation from their surroundings, as treated in the preceding chapters, can be rigorously described in terms of quantal states or wavefunctions that satisfy equations of motion with well-defined initial and boundary conditions. Some of the theoretical models introduce in addition statistical considerations about initial conditions, such as a distribution of collision velocities and of initial internal (rovibrational) states to account for laboratory conditions, for example, in cross molecular beam experiments. Also, it is usual to introduce a distribution of internal position and velocity variables that reconstruct rotational–vibrational wavefunctions of the molecules, in quantal–classical models. These models combine quantal and statistical concepts and are best treated introducing statistical density operators [28, 29] to generalize wavefunction treatments.

Statistical density operators provide a general description of quantal coherence in collisional dynamics, and how coherence evolves and can disappear during molecular interactions. Decoherence arises from couplings between a set of quantum states of interest in a collisional measurement, and others states present during the event. Decoherence can be treated using partitioning methods where states are classified as being in a set $\mathcal{P}$ of active quantal states of interest or in a set $\mathcal{Q}$ of remaining intermediate states [30–32]. Quantal decoherence also arises from the interaction between quantum states and classical motions in quantal–classical models of molecular collision dynamics. Statistical density operators can be represented in a basis set of position–coordinate states for these models, classifying degrees of freedom as quantal or classical like. Further introducing partial Wigner transforms (or PWT) [33, 34] of the density functions, involving the classical-like degrees of freedom, leads to related equations of motion for the PWT density operator in phase–space coupled to the equation for quantal time evolution.

In addition, statistical density operators can incorporate, in their equation of motion and initial conditions, a distribution present in experimental setups with molecular beams, such as a distribution of molecular velocities and internal states, or of random external forces due to interactions with solvents like gasses or solid surfaces. External forces create a thermal environment with its thermal distributions of properties, and statistical densities can provide a link between state-to-state cross sections and kinetic rates obtained from an average over an initial thermal distribution of states and a sum over final states [29].

A theoretical treatment introduces a time-dependent statistical density operator (or DOp) $\hat{\Gamma}(t)$ for the quantum dynamics of a complex molecular system, including electronically diabatic molecular interactions [30, 32]. It allows for a derivation of equations of motion for the DOp describing quantum coherence of collisional states and for a consistent treatment of initial and boundary conditions. We follow here the treatments in [35] however with some changes in notation to be consistent with previous chapters.

The time-evolution of $\widehat{\Gamma}(t)$ is obtained from the Liouville–von Neuman (or L–vN) equation of motion [29] for a Hamiltonian operator $\widehat{\mathrm{H}}(t)$, which depends on time and on variables $\xi = (\boldsymbol{Q}, \boldsymbol{X})$ for atomic position $\boldsymbol{Q}$ and electronic $\boldsymbol{X}$ variables, as

$$i\hbar\frac{\partial\widehat{\Gamma}}{\partial t} = \widehat{\mathrm{H}}(t)\widehat{\Gamma}(t) - \widehat{\Gamma}(t)\widehat{\mathrm{H}}(t) = [\widehat{\mathrm{H}}(t), \widehat{\Gamma}(t)]$$

with initial conditions $\widehat{\Gamma}(t_{in}) = \widehat{\Gamma}_{in}$ dictated by an experimental setup or by a theoretical model. It is normalized by taking its trace over all its variables, as $tr[\widehat{\Gamma}(t)] = 1$, and provides the expectation value of a property operator $\widehat{\mathrm{A}}(t)$ as $\widehat{\mathrm{A}}(t) = tr[\widehat{\mathrm{A}}(t)\widehat{\Gamma}(t)]$. Calculations of the DOp require introduction of a quantal representation. It can be a done with a basis set of scattering states $\left\{|\nu\rangle = \left|\vec{k}_\nu, n\right\rangle\right\}$, where $|n\rangle$ is an electronic-nuclear state of the whole system and $\hbar\vec{k}_\nu$ is a relative motion momentum, to transform the operator equation of motion into a set of coupled ordinary differential equations in time for elements in a matrix $\boldsymbol{\Gamma}(t) = \left[\Gamma_{\nu\nu'}(t)\right]$. Or the DOp can be given in a coordinate representation with quantal variables in a ket $|\xi\rangle = |\boldsymbol{Q}, \boldsymbol{X}\rangle$, as $\widehat{\Gamma}(t) = \int d\xi \int d\xi' \; |\xi\rangle\,\Gamma(\xi, \xi', t)\left\langle\xi'\right|$, leading to coupled partial differential equations for amplitude functions $\Gamma(\xi, \xi', t)$.

State-to-state $\alpha \to \beta$ transition probabilities $P_{\beta\alpha}$ and rates $R_{\beta\alpha}$ can be related to the density operator starting in an initial state $\widehat{\Gamma}^{(\alpha)}(t_{in}) = \left|\vec{k}_\alpha, a\right\rangle\left\langle\vec{k}_\alpha, a\right| = \widehat{\mathrm{P}}^{(\alpha)}$ and evolving as $\widehat{\Gamma}^{(\alpha)}(t)$ to give at a final time t_{fn} a probability of finding state β selected with a projection operator $\widehat{\mathrm{P}}^{(\beta)} = \left|\vec{k}_\beta, b\right\rangle\left\langle\vec{k}_\beta, b\right|$. Probabilities are obtained from the trace over states in $P_{\beta\alpha}(t) = tr\,[\widehat{\mathrm{P}}^{(\beta)}\widehat{\Gamma}^{(\alpha)}(t)]$. A state-to-state transition rate after interaction follows from $R_{\beta\alpha} = \lim(dP_{\beta\alpha}/dt)$, $t \to \infty$, or

$$R_{\beta\alpha} = (i\hbar)^{-1} tr\left\{\widehat{\mathrm{P}}^{(\beta)}\left[\widehat{\mathrm{H}}(t), \widehat{\Gamma}^{(\alpha)}(t)\right]\right\}, t \to \infty$$

$$R_{\beta\alpha} = (i\hbar)^{-1} tr\left\{\left[\widehat{\mathrm{P}}^{(\beta)}, \widehat{\mathrm{H}}(t_{fn})\right]\widehat{\Gamma}^{(\alpha)}(t_{fn})\right\}$$

with the second line obtained from the permutational property of the trace of a product and with $t_{fn} = \lim t \to \infty$. This can be evaluated in an isolated collision, where $\widehat{\mathrm{H}} = \widehat{\mathrm{H}}_0 + \widehat{V}$ is time-independent, with $\widehat{\mathrm{H}}_0\left|\vec{k}_\beta, b\right\rangle = E_\beta\left|\vec{k}_\beta, b\right\rangle$, and with $[\widehat{P}^{(\beta)}, \widehat{\mathrm{H}}(t_{fn})] = [\widehat{P}^{(\beta)}, \widehat{\mathrm{V}}]$ insofar $[\widehat{P}^{(\beta)}, \widehat{\mathrm{H}}_0] = 0$.

The density operator $\widehat{\Gamma}^{(\alpha)}(t)$ can be obtained by integration over time of the L-vN equation. This can be conveniently done in the interaction picture for operators $\widehat{\mathrm{V}}_I(t)$ and $\widehat{\Gamma}_I^{(\alpha)}(t)$ generated by the time-evolution operator

$\hat{U}_0(t) = \exp[-i\hat{H}_0\,(t - t_{in})/\hbar]$ as in

$$\hat{\Gamma}^{(\alpha)}(t) = \hat{U}_0(t)\hat{\Gamma}_I^{(\alpha)}(t)\hat{U}_0(t)^\dagger$$

which substituted in the L–vN equation yields

$$i\hbar\frac{\partial\hat{\Gamma}_I^{(\alpha)}}{\partial t} = \hat{V}_I(t)\hat{\Gamma}_I^{(\alpha)}(t) - \hat{\Gamma}_I^{(\alpha)}(t)\hat{V}_I(t) = \left[\hat{V}_I(t), \hat{\Gamma}_I^{(\alpha)}(t)\right]$$

$$\hat{\Gamma}_I^{(\alpha)}(t) = \hat{\Gamma}_I^{(\alpha)}(t_{in}) + (i\hbar)^{-1}\int_{t_{in}}^{t} dt'\left[\hat{V}_I(t'), \hat{\Gamma}_I^{(\alpha)}(t')\right]$$

$$\hat{\Gamma}^{(\alpha)}(t) = \hat{P}^{(\alpha)} + (i\hbar)^{-1}\int_{t_{in}}^{t} dt'\,\hat{U}_0(t-t')\left[\hat{V}, \hat{\Gamma}^{(\alpha)}(t')\right]\hat{U}_0(t-t')^\dagger$$

in an expression suitable for expansion by iteration in powers of the interaction potential $\hat{V}$. Here we have used $\hat{\Gamma}^{(\alpha)}(t_{in}) = \hat{P}^{(\alpha)}$, $t_{in} = \lim t \to -\infty$, and $\hat{U}_0(t-t') = \hat{U}_0(t)\hat{U}_0(t')^\dagger$. The state-to-state rate follows from the above expression for $R_{\beta\alpha}$ as

$$R_{\beta\alpha} = -\hbar^{-2}\left(tr\left\{\left[\hat{P}^{(\beta)}, \hat{V}\right]\int_{-\infty}^{t} dt'\hat{U}_0(t-t')\left[\hat{V}, \hat{\Gamma}^{(\alpha)}(t')\right]\hat{U}_0(t-t')^\dagger\right\}\right), \quad t \to \infty,$$

As a special case to lowest (second) order in the potential interaction, using $\hat{\Gamma}^{(\alpha)}(t') \cong \hat{P}^{(\alpha)}$ in the integral, the rate is

$$R_{\beta\alpha}^{(2)} = -\hbar^{-2}\left(tr\left\{\int_{-\infty}^{t} dt'\left[\hat{P}^{(\beta)}, \hat{V}\right]\hat{U}_0(t-t')\left[\hat{V}, \hat{P}^{(\alpha)}\right]\hat{U}_0(t-t')^\dagger\right\}\right), \quad t \to \infty.$$

Finally, collisional cross sections follow as $\sigma_{\beta\alpha} = R_{\beta\alpha}/J_\alpha$, with J_α an incoming flux of particles per unit time and unit area. These results are directly generalized to statistical transition rates R_{BA} between a set of states, $A = \{\alpha\}$ with initial probabilities w_α and a final set $B = \{\beta\}$. These equations can also be expanded to account for transitions between coherent superpositions of states, replacing in previous equations a single projector $\hat{P}^{(\nu)}$ with a multiprojector $\hat{P}^{(N)} = \sum_{\nu\in N}\hat{P}^{(\nu)}$.

Two especially useful results are obtained by (i) partitioning the density operator into active and nonactive contributions or by (ii) following its time-evolution for coupled quantal and classical-like variables, as follows.

7.4.2 Partitioning of the Density Operator

A state-space partitioning method was described in Chapter 3 while introducing the formalism of optical potentials [7, 27]. The partitioning there was done for wavefunctions, but a similar procedure can also be followed for the density operator [36–38]. Section 7.2 introduced the block diagonal and nondiagonal partition of an operator, with projection operators $\hat{P}$ and its complementary projector $\hat{Q} = \hat{I} - \hat{P}$, both self-adjoint and idempotent, and satisfying $\hat{Q}\hat{P} = \hat{0}$. Here $\hat{P}$ can select active states during scattering, while $\hat{Q}$ selects intermediate states in a collision. Furthermore, the statistical density operator contains an initial distribution of all those states, consistent with experimental conditions for colliding beams or for a collisional environment in a gas. Given the density operator $\widehat{\Gamma}$ for the whole system, it can be decomposed into block diagonal $\widehat{\Gamma}_D$ and block nondiagonal $\widehat{\Gamma}_N$ parts as $\widehat{\Gamma} = \widehat{\Gamma}_D + \widehat{\Gamma}_N$ with

$$\widehat{\Gamma}_D = \widehat{P}\widehat{\Gamma}\widehat{P} + \widehat{Q}\widehat{\Gamma}\widehat{Q}, \widehat{\Gamma}_N = \widehat{P}\widehat{\Gamma}\widehat{Q} + \widehat{Q}\widehat{\Gamma}\widehat{P}.$$

Similarly the time-independent Hamiltonian operator is partitioned as $\widehat{\mathrm{H}} = \widehat{\mathrm{H}}_D + \widehat{\mathrm{H}}_N$. Given two operators $\widehat{A}$ and $\widehat{B}$, and the relations $(\widehat{A}\,\widehat{B})_D = \widehat{A}_D\widehat{B}_D + \widehat{A}_N\widehat{B}_N$ and $(\widehat{A}\,\widehat{B})_N = \widehat{A}_D\widehat{B}_N + \widehat{A}_N\widehat{B}_D$, and applying these to the L–vN equation for $\widehat{\Gamma}(t)$ written in terms of the Liouvillian superoperator $\widehat{\mathcal{L}}$ defined by the commutator in $\widehat{\mathcal{L}}\widehat{\Gamma} = [\widehat{\mathrm{H}}, \widehat{\Gamma}]$, so that $i\hbar\partial\widehat{\Gamma}/\partial t = \widehat{\mathcal{L}}\widehat{\Gamma}(t)$, gives

$$i\hbar\frac{\partial\widehat{\Gamma}_D}{\partial t} = \widehat{\mathcal{L}}_D\widehat{\Gamma}_D(t) + \widehat{\mathcal{L}}_N\widehat{\Gamma}_N(t)$$

$$i\hbar\frac{\partial\widehat{\Gamma}_N}{\partial t} = \widehat{\mathcal{L}}_D\widehat{\Gamma}_N(t) + \widehat{\mathcal{L}}_N\widehat{\Gamma}_D(t)$$

with $\widehat{\mathcal{L}}_D$ and $\widehat{\mathcal{L}}_N$ generated by $\widehat{\mathrm{H}}_D$ and $\widehat{\mathrm{H}}_N$.

The block-diagonal part $\widehat{\Gamma}_D$ can be formally obtained from an integral equation, introducing a time-evolution superoperator $\widehat{\mathcal{U}_D}(t, t') = (-i/\hbar)\widehat{\mathcal{L}}_D \cdot (t - t')$. This can be used to solve for $\widehat{\Gamma}_N(t)$ from the second line above, giving

$$\widehat{\Gamma}_N(t) = \widehat{\Gamma}_N(t_{in}) + (i\hbar)^{-1}\int_{t_{in}}^{t} dt'\, \widehat{\mathcal{U}_D}(t, t')\widehat{\mathcal{L}}_N\widehat{\Gamma}_D(t')$$

as can be verified differentiating both sides with respect to time. Replacing this on the first line leads to

$$i\hbar\frac{\partial\widehat{\Gamma}_D}{\partial t} = \widehat{\mathcal{L}}_D\widehat{\Gamma}_D(t) + \widehat{\mathcal{L}}_N\widehat{\Gamma}_N(t_{in}) + (i\hbar)^{-1}\int_{t_{in}}^{t} dt'\, \widehat{\mathcal{L}}_N\widehat{\mathcal{U}_D}(t, t')\widehat{\mathcal{L}}_N\widehat{\Gamma}_D(t')$$

where the second term to the right describes the effect of an initial statistical coupling of active states in $\widehat{P}$ and intermediate states in $\widehat{Q}$, while the third term describes the dynamical delay due to coupling from active states into intermediate ones and back to active states.

Calculation of a block diagonal density operator including initial statistical distributions and dynamical delay can be done as mentioned before introducing a quantal representation. It can be done with a basis set of scattering states $\left\{|\nu\rangle = \left|\vec{k}_\nu, n\right\rangle\right\}$, separated into active and intermediate states, to transform the operator equation of motion for $\widehat{\Gamma}_D(t)$ into a set of coupled ordinary differential equations in time for elements in a matrix $\mathbf{\Gamma}_D(t) = \left[\Gamma^{(D)}_{\nu\nu'}(t)\right]$ or the DOp can be given in a coordinate representation leading to coupled partial differential equations for amplitude functions $\Gamma_D(\xi, \xi', t)$.

7.4.3 Quantum-Classical Density Operator

7.4.3.1 Eikonal Representation

The statistical density operator is useful for nonadiabatic collisions, where initial conditions or models involve distributions of variables like atomic velocities or positions. In a coordinate representation, with states $|\xi\rangle = |\, \boldsymbol{Q}, \boldsymbol{X}\rangle$, it depends on both atomic $\boldsymbol{Q}$ and electronic $\boldsymbol{X}$ variables in the function $\Gamma(\xi, \xi', t)$. As done in Section 7.1, the dependence on electronic variables can be treated expanding operators in a basis set of K electronic wavefunctions $\Phi_k(\boldsymbol{X}; \boldsymbol{Q})$ for chosen values of atomic coordinates. This gives for the density function

$$\Gamma(\xi, \xi', t) = \sum_k^K \sum_{k'}^K \Phi_k(\boldsymbol{X}; \boldsymbol{Q})\Gamma_{kk'}(\boldsymbol{Q}, \boldsymbol{Q}', t)\Phi_{k'}(\boldsymbol{X}; \boldsymbol{Q}')$$

Here we want to further separate among the atomic variables $\boldsymbol{Q}$ a subset $\boldsymbol{q} = (q_1, q_2, \ldots, q_F)$ to be treated as quantal variables for internal vibrational–rotational motions, and another subset of classical-like time-dependent atomic position vectors called here $\boldsymbol{r} = (\vec{R}, \vec{r}_1, \vec{r}_2, \ldots, \vec{r}_M)$, with $\vec{R}$ the relative position vector of molecules A and B in a collision involving M atoms, introducing states $|\boldsymbol{Q}\rangle = |\, \boldsymbol{q}, \boldsymbol{r}\rangle$.

The Hamiltonian $\hat{H} = \hat{K}_{at} + \hat{H}_Q$ contains the relative kinetic energy operator $\hat{K}_{at}$ of atomic ions and the remaining total energy for fixed nuclei in $\hat{H}_Q = \hat{K}_{el} + \hat{V}_{el,el} + \hat{V}_{at,el} + \hat{V}_{at,at}$, where the potential energies involve interactions among electrons and ions and possibly spin–orbit coupling energies $\hat{H}_{SO} = \sum_{at} \hat{h}_{SO}^{(at)}$ [12], as detailed in Chapter 5 on diabatic collisions. The potential energies $\hat{V}_{at,at}$ and $\hat{V}_{at,el}$ are functions of $\boldsymbol{Q}$ and contain long-range Coulomb interactions and short-range repulsive interactions from closed electronic shells. Here the atomic variables are further classified as quantal or classical-like and the potential energies are further expanded in accordance, with $\hat{V}_{at,el} = \hat{V}_{at,el}^{(qu)} + \hat{V}_{at,el}^{(cl)}$ for electrons interacting with quantal or classical atomic charges and dependent on variables $\boldsymbol{q}$ and $\boldsymbol{r}$, respectively. Also, $\hat{V}_{at,at}$ is decomposed into pairs for (*qu,qu*), (*qu,cl*), and (*cl,cl*) components.

The dependence of operators on quantal atomic variables can be further expanded in a basis set of J internal rovibrational states $\phi_j(\boldsymbol{q})$. The combined electronic-nuclear basis set of $K.J = L$ functions $\{\Phi_k(\boldsymbol{X}; \boldsymbol{Q})\phi_j(\boldsymbol{q}) = \Upsilon_\lambda(\boldsymbol{q}, \boldsymbol{X})\}$, with $\lambda = (k, j)$ labelling an electronic-nuclear channel state, provides a density function expansion

$$\Gamma(\xi, \xi', t) = \sum_\lambda^L \sum_{\lambda'}^L \Upsilon_\lambda(\boldsymbol{q}, \boldsymbol{X})\Gamma_{\lambda\lambda'}(\boldsymbol{r}, \boldsymbol{r}', t)\Upsilon_{\lambda'}(\boldsymbol{q}', \boldsymbol{X}')$$

and introduces the $L \times L$ matrix $\boldsymbol{\Gamma}(\boldsymbol{r}, \boldsymbol{r}', t) = [\Gamma_{\lambda\lambda'}(\boldsymbol{r}, \boldsymbol{r}', t)]$ with elements dependent only on classical-like variables. The set $\{\Upsilon_\lambda(\boldsymbol{q}, \boldsymbol{X})\}$ can be combined to form new electronic-nuclear states $\Psi_\mu(\boldsymbol{q}, \boldsymbol{X})$ that diagonalize $\boldsymbol{\Gamma}(\boldsymbol{r}, \boldsymbol{r}', t)$ so that

$$\Gamma(\xi, \xi', t) = \sum_{\mu}^{L} \Psi_\mu(\boldsymbol{q}, \boldsymbol{X}) \Gamma^{(\mu)}(\boldsymbol{r}, \boldsymbol{r}', t) \Psi_\mu(\boldsymbol{q}', \boldsymbol{X}').$$

Further treatment of a classical-like dynamics can be done introducing eikonal phases and amplitudes for the density function by extension of the eikonal representation for a wavefunction in Chapter 5. Details of a similar treatment involving an expansion of $\Gamma(\xi, \xi', t)$ in a basis set of amplitude functions $\chi_n(\xi, t)$ with a common eikonal action $S(\boldsymbol{r}, t)$ leads to

$$\Gamma(\xi, \xi', t) = \left[\sum_{n} w_n \chi_n(\xi, t) \chi_n(\xi', t) \right] \exp\{ i[S(\boldsymbol{r}, t) - S(\boldsymbol{r}', t)]/\hbar \}$$

with equations for action and amplitude functions found in Chapter 5 and [36].

More generally, letting

$$\Gamma^{(\mu)}(\boldsymbol{r}, \boldsymbol{r}', t) = \int d\Lambda \, \mathrm{a}(\Lambda) A^{(\mu)}(\boldsymbol{r}, \boldsymbol{r}', t; \Lambda) \exp[i\sigma(\boldsymbol{r}, \boldsymbol{r}', t; \Lambda)/\hbar]$$

in an eikonal representation, given as a sum of eikonal path functions times a set of amplitudes $\mathrm{a}(\Lambda)$ with parameters Λ chosen to satisfy boundary conditions, in terms of an eikonal path action $\sigma(\boldsymbol{r}, \boldsymbol{r}', t)$ and eikonal path amplitudes $A^{(\mu)}(\boldsymbol{r}, \boldsymbol{r}', t)$, the density function is

$$\Gamma(\xi, \xi', t) = \int d\Lambda \, \mathrm{a}(\Lambda) \sigma(\xi, \xi', t; \Lambda) \exp[i\sigma(\boldsymbol{r}, \boldsymbol{r}', t; \Lambda)/\hbar]$$

$$\sigma(\xi, \xi', t; \Lambda) = \sum_{\mu}^{L} \Psi_\mu(\boldsymbol{q}, \boldsymbol{X}) A^{(\mu)}(\boldsymbol{r}, \boldsymbol{r}', t; \Lambda) \Psi_\mu(\boldsymbol{q}', \boldsymbol{X}').$$

Replacing this in the L–vN equation for $\Gamma(\xi, \xi', t)$ with the Hamiltonian $\hat{H}_Q = \hat{K}_{el} + \hat{V}_{el,el} + \hat{V}_{at,el} + \hat{V}_{at,at}$ and taking the trace over all quantal variables $(\boldsymbol{q}, \boldsymbol{X})$ allows construction of a partial differential equation for $\sigma(\boldsymbol{r}, \boldsymbol{r}', t; \Lambda)$, containing a chosen effective potential energy $\mathcal{V}(\boldsymbol{r})$. This generalizes our previous Hamilton–Jacobi equation for the action function in a wavefunction, extended here for the density path function. It can be solved in the classical-like limit of short wavelengths for paths between $\boldsymbol{r}$ and $\boldsymbol{r}'$, along the lines of the treatment for $S(\boldsymbol{r}, t)$ in Chapter 5, and can be used to calculate transition rates $R_{\beta\alpha}$.

7.4.3.2 Wigner-Transform Treatment

An alternative general procedure introduces the Wigner transform [39, 40] of the density function and leads to a treatment in a phase space for variables $\boldsymbol{r} = (\vec{R}, \vec{r}_1, \vec{r}_2, \dots, \vec{r}_N)$ and related momenta $\boldsymbol{p} = (\vec{P}, \vec{p}_1, \vec{p}_2, \dots, \vec{p}_N)$. This provides a quantum-classical Liouvillian dynamics described in [34] and related reviews. The treatment here for the dynamics of excited molecules follows our work in [41] with some notation changes consistent with present chapters.

The density operator $\widehat{\Gamma}(t)$ can be defined in the joint space of quantal and classical-like states $|\boldsymbol{q}, \boldsymbol{X}\rangle \otimes |\boldsymbol{r}\rangle$ and written as

$$\widehat{\Gamma}(t) = \int d^N\boldsymbol{r} \int d^N\boldsymbol{r}' \; |\boldsymbol{r}\rangle \, \widehat{\Gamma}(\boldsymbol{r}, \boldsymbol{r}', t) \, \langle \boldsymbol{r}'|$$

where $\widehat{\Gamma}(\boldsymbol{r}, \boldsymbol{r}', t) = \langle \boldsymbol{r}|\widehat{\Gamma}(t)|\boldsymbol{r}'\rangle$ is an operator in the space of quantal variables and a function of classical-like variables. Next it can be written as a partial Wigner transform (or PWT), useful for nonadiabatic molecular dynamics [42, 43], introducing new variables $(\boldsymbol{r}+\boldsymbol{r}')/2 = \overline{\boldsymbol{r}}$ and $\boldsymbol{r}-\boldsymbol{r}' = \boldsymbol{s}$, new functions $\langle \overline{\boldsymbol{r}} - \boldsymbol{s}/2|\widehat{\Gamma}(t)|\overline{\boldsymbol{r}} + \boldsymbol{s}/2\rangle$, and their PWT integral

$$\widehat{\Gamma}_W(\overline{\boldsymbol{p}}, \overline{\boldsymbol{r}}, t) = (2\pi\hbar)^{-N} \int d^N\boldsymbol{s} \, \exp(i\,\overline{\boldsymbol{p}}\,\boldsymbol{s}/\hbar)\langle \overline{\boldsymbol{r}} - \boldsymbol{s}/2|\widehat{\Gamma}(t)|\overline{\boldsymbol{r}} + \boldsymbol{s}/2\rangle$$

where the PWT $\widehat{\Gamma}_W(\overline{\boldsymbol{p}}, \overline{\boldsymbol{r}}, t)$ can be interpreted as a quasi-density function of phase-space momentum and position variables $(\overline{\boldsymbol{p}}, \overline{\boldsymbol{r}})$. Normalization of the density operator follows from the trace over quantal variables and

$$tr(\widehat{\Gamma}_W) = tr_{qu}\left[\int d^N\,\overline{\boldsymbol{r}} \int d^N\overline{\boldsymbol{p}}\,\widehat{\Gamma}_W(\overline{\boldsymbol{p}}, \overline{\boldsymbol{r}}, t)\right] = 1.$$

The quasi-classical phase-space density is given by $(\overline{\boldsymbol{p}}, \overline{\boldsymbol{r}}, t) = tr_{qu}[\widehat{\Gamma}_W(\overline{\boldsymbol{p}}, \overline{\boldsymbol{r}}, t)]$, and the averaged quantal density operator is $\widehat{\Gamma}_W^{(qu)} = \int d^N\,\overline{\boldsymbol{r}} \int d^N\,\overline{\boldsymbol{p}}\,\widehat{\Gamma}_W(\overline{\boldsymbol{p}}, \overline{\boldsymbol{r}}, t)$, dependent only on quantal variables $(\boldsymbol{q}, \boldsymbol{X})$. The PWT of a physical operator $\widehat{A}$ is

$$\widehat{\mathrm{A}}_W(\overline{\boldsymbol{p}}, \overline{\boldsymbol{r}}, t) = \int d^N\boldsymbol{s} \, \exp(i\,\overline{\boldsymbol{p}}\,\boldsymbol{s}/\hbar)\langle \overline{\boldsymbol{r}} - \boldsymbol{s}/2|\widehat{\mathrm{A}}(t)|\overline{\boldsymbol{r}} + \boldsymbol{s}/2\rangle$$

with expectation value

$$\langle \widehat{\mathrm{A}}(t)\rangle = tr(\widehat{\Gamma}_W\widehat{\mathrm{A}}_W) = tr_{qu} \int d^N\overline{\boldsymbol{r}} \int d^N\overline{\boldsymbol{p}}\,\widehat{\Gamma}_W(\overline{\boldsymbol{p}}, \overline{\boldsymbol{r}}, t)\widehat{\mathrm{A}}_W(\overline{\boldsymbol{p}}, \overline{\boldsymbol{r}}, t).$$

The equation of motion for $\widehat{\Gamma}_W$ follows from the L–vN equation taking the PWT of both sides. This involves the transform of products of operators, which can be constructed in a compact way as

$$(\widehat{A}\widehat{B})_W = \widehat{\mathrm{A}}_W \exp(-i\hbar\overleftrightarrow{\Lambda}/2)\widehat{\mathrm{B}}_W$$

$$\overleftrightarrow{\Lambda} = \frac{\overleftrightarrow{\partial}}{\partial\overline{\boldsymbol{p}}} \cdot \frac{\vec{\partial}}{\partial\overline{\boldsymbol{r}}} - \frac{\overleftrightarrow{\partial}}{\partial\overline{\boldsymbol{r}}} \cdot \frac{\vec{\partial}}{\partial\overline{\boldsymbol{p}}}$$

with $\overleftrightarrow{\Lambda}$ a bi-directional differential operator acting to the left and right, as shown by the top arrows over differentials. The PWT of the L–vN equation then gives

$$i\hbar\,\partial\widehat{\Gamma}_W/\partial t = \widehat{\mathrm{H}}_W(t)\exp(-i\hbar\overleftrightarrow{\Lambda}/2)\widehat{\Gamma}_W(t) - \widehat{\Gamma}_W(t)\exp(-i\hbar\overleftrightarrow{\Lambda}/2)\widehat{\mathrm{H}}_W(t)$$

to be solved with initial conditions $\widehat{\Gamma}_W(t_{in}) = \widehat{\Gamma}_W^{(in)}(\overline{\boldsymbol{p}}, \overline{\boldsymbol{r}})$ chosen to describe an initial classical distribution in phase space. Here the Hamiltonian can be decomposed into quantal (qu), classical-like (cl), and their coupling (qc) terms $\widehat{\mathrm{H}}_W = \widehat{\mathrm{H}}_W^{(qu)} + \widehat{\mathrm{H}}_W^{(cl)} + \widehat{\mathrm{H}}_W^{(qc)}$, with $\widehat{\mathrm{H}}_W^{(cl)} = \overline{\boldsymbol{p}}^2/(2m) + \widehat{V}_{at,at}^{(cl,cl)}(\overline{\boldsymbol{r}}) = \mathrm{H}^{(cl)}(\overline{\mathbf{p}}, \overline{\boldsymbol{r}})$. Furthermore, the term $\widehat{\mathrm{H}}_W^{(qu)}$ commutes with the differential operator $\overleftrightarrow{\Lambda}$ in the above equation.

A quantum-classical treatment of the dynamics follows by taking the limit of short wavelengths along the paths between $\boldsymbol{r}$ and $\boldsymbol{r}'$ in the above equation, expanding the exponential operator and keeping only the lowest order term in $\hbar\overleftrightarrow{\Lambda}/2$. This gives

$$\frac{\partial \widehat{\Gamma}_W}{\partial t} = -\frac{i}{\hbar}\left[\widehat{\mathrm{H}}_W^{(qu)} + \widehat{\mathrm{H}}_W^{(qc)}, \widehat{\Gamma}_W\right] + \left\{\mathrm{H}^{(cl)}, \widehat{\Gamma}_W\right\} + \frac{1}{2}\left(\left\{\widehat{\mathrm{H}}_W^{(qc)}, \widehat{\Gamma}_W\right\} - \left\{\widehat{\Gamma}_W, \widehat{\mathrm{H}}_W^{(qc)}\right\}\right)$$

in terms of Poisson curly brackets as in $\{\widehat{A}, \widehat{B}\} = \frac{\partial \widehat{A}}{\partial \overline{\boldsymbol{r}}} \cdot \frac{\partial \widehat{B}}{\partial \overline{\mathbf{p}}} - \frac{\partial \widehat{A}}{\partial \overline{\mathbf{p}}} \cdot \frac{\partial \widehat{B}}{\partial \overline{\boldsymbol{r}}}$, where it is necessary to keep the order of operator products. The first two terms to the right describe expected quantal and classical rates, while the last term is a nonintuitive addition. The equation for the quasi-classical phase-space density $\gamma(\overline{\mathbf{p}}, \overline{\boldsymbol{r}}, t)$ follows by taking the trace over quantal variables, as

$$\frac{\partial \gamma}{\partial t} = \left\{\mathrm{H}^{(cl)}, \gamma\right\} + tr_{qu}\left(\left\{\widehat{\mathrm{H}}_W^{(qc)}, \widehat{\Gamma}_W\right\}\right)$$

which must be solved coupled to the equation for the phase space averaged quantal density operator $\widehat{\Gamma}_W^{(qu)}(\boldsymbol{q}, \boldsymbol{X}) = \int d^N\overline{\boldsymbol{r}} \int d^N\overline{\mathbf{p}}\, \widehat{\Gamma}_W(\overline{\mathbf{p}}, \overline{\boldsymbol{r}}, t)$, obtained from the corresponding averages of the two sides of the equation for $\widehat{\Gamma}_W$. The $\widehat{\Gamma}_W^{(qu)}$, dependent only on quantal variables $(\boldsymbol{q}, \boldsymbol{X})$, satisfies the averaged equation

$$\frac{\partial \widehat{\Gamma}_W^{(qu)}}{\partial t} = -\frac{i}{\hbar}\left[\widehat{\mathrm{H}}_W^{(qu)} + \widehat{\mathrm{H}}_W^{(qc)}, \widehat{\Gamma}_W^{(qu)}\right] + \left\{\mathrm{H}^{(cl)}, \widehat{\Gamma}_W^{(qu)}\right\}.$$

Differential cross sections $d\sigma_{\beta\alpha}/d\Omega$ can be obtained from the expectation value of the flux operator $\vec{J} = \vec{J}^{(qu)} + \vec{J}^{(cl)}$, in its PWT form,

$$\widehat{\vec{J}_W}(\overline{\mathbf{p}}, \overline{\boldsymbol{r}}) = \frac{1}{2M}\left[\frac{\hbar}{i}\frac{\partial}{\partial \boldsymbol{q}}, \delta(\widehat{\boldsymbol{q}} - \boldsymbol{q}_s)\right]_+ + \frac{1}{m}\overline{\mathbf{p}}\delta(\overline{\boldsymbol{r}} - \overline{\boldsymbol{r}}_s)$$

shown here at scattering exit location $(\boldsymbol{q}_s, \overline{\boldsymbol{r}}_s)$ and omitting electronic variables $\boldsymbol{X}$ for bound electrons. Taking the trace over all variables of the density operator $\widehat{\Gamma}_W^{(\alpha)}$ starting in state α, times the flux operator term $\widehat{\vec{J}_W^{(\beta)}}(\overline{\mathbf{p}}, \overline{\boldsymbol{r}}) = |\Upsilon_\beta\rangle\left\langle \Upsilon_\beta \middle| \widehat{\vec{J}_W} \middle| \Upsilon_\beta \right\rangle \Upsilon_\beta|$ for a final state β, gives a state-to-state flux

$$\vec{J}_{\beta\alpha}(t) = tr\left(\widehat{\Gamma}_W^{(\alpha)}(t)\widehat{\vec{J}_W^{(\beta)}}\right) = tr_{qu}\int d^N\overline{\boldsymbol{r}} \int d^N\overline{\mathbf{p}}\, \widehat{\Gamma}_W^{(\alpha)}(\overline{\mathbf{p}}, \overline{\boldsymbol{r}}, t)\widehat{\vec{J}_W^{(\beta)}}(\overline{\mathbf{p}}, \overline{\boldsymbol{r}}, t)$$

and a cross section in the limit $t \to \infty$ and $R \to \infty$ with the radial outgoing velocity $dR/dt = v_R^{(\beta)}$, obtained from $d\sigma_{\beta\alpha}/d\Omega = \vec{J}_{\beta\alpha}.\vec{n}\, R^2/|\vec{J}_\alpha^0|$ in terms of the incoming flux $\vec{J}_\alpha^0$.

Numerical integration of the coupled differential equations for the phase space density $\gamma(\overline{\boldsymbol{p}},\overline{\boldsymbol{r}},t)$ and quantal density operator $\widehat{\Gamma}_W^{(qu)}(\boldsymbol{q},\boldsymbol{X},t)$ requires consideration of time scales for classical-like and quantal degrees of freedom. During a collision, the rate of change over time of smoothly varying classical-like degrees of freedom $(\overline{\boldsymbol{p}},\overline{\boldsymbol{r}})$ is usually much smaller than the rate for rapidly oscillating quantal $(\boldsymbol{q},\boldsymbol{X})$ degrees of freedom, except possibly in transitions between quantal states at phase-space locations like a conical intersection or a potential energy crossing seam, where quasi-classical momenta undergo sharp change. The quasi-classical degrees of freedom can be numerically advanced over large time intervals $t_{n-1}\leq t\leq t_n$, except in the sharp transition regions where the interval must be reduced.

A way to account for the self-consistent coupling of the two densities is to advance time in increments Δt as in $t_0\leq t\leq t_1=t_0+\Delta t$, starting with the quantal density operator at phase-space location $(\overline{\boldsymbol{p}},\overline{\boldsymbol{r}})_0$ at t_0, letting it first relax for fixed phase-space variables until time t_1 as $\widehat{\Gamma}_W^{(qu,0)}(\boldsymbol{q},\boldsymbol{X},t)$. It is followed by calculation of $tr_{qu}\left(\left\{\widehat{\mathrm{H}}_W^{(qc)},\widehat{\Gamma}_W^{(0)}\right\}\right)$ and then by integrating the equation for $\gamma(\overline{\boldsymbol{p}},\overline{\boldsymbol{r}},t)$ within the time interval $t_0\leq t\leq t_1$ or equivalently calculating trajectories $\overline{\boldsymbol{p}},\overline{\boldsymbol{r}}$ over time. This provides perturbation terms due to changes in $\mathrm{H}^{(cl)}$ and $\widehat{\mathrm{H}}_W^{(qc)}$ in the equation for $\widehat{\Gamma}_W^{(qu)}(\boldsymbol{q},\boldsymbol{X},t)$. Finally, perturbation corrections are included in calculation of a new, driven quantum density operator, $\widehat{\Gamma}_W^{(qu)}(\boldsymbol{q},\boldsymbol{X},t)$. This "relax-and-drive" procedure is similar to the one described in Chapter 5 for electronically diabatic collisions [36].

Two convenient approximations for molecular scattering are (i) the mentioned short-wavelength approximation and (ii) the introduction of an effective (qc) potential $V'(\overline{\boldsymbol{r}})$ approximating $\left\{\widehat{\mathrm{H}}_W^{(qc)},\widehat{\Gamma}_W^{(qu)}\right\}\cong\left\{V'(\overline{\boldsymbol{r}}),\widehat{\Gamma}_W^{(qu)}\right\}$, obtained from an average over the classical momentum. Then the potential energy can be chosen as $V^{(cl)}(\overline{\boldsymbol{r}})=\widehat{V}_{at,at}^{(cl,cl)}(\overline{\boldsymbol{r}})+V'(\overline{\boldsymbol{r}})$, so that $\overline{\boldsymbol{p}}^2/(2m)+\widehat{V}^{(cl)}(\overline{\boldsymbol{r}})=\mathcal{H}^{(cl)}(\overline{\boldsymbol{p}},\overline{\boldsymbol{r}})$ is a new classical Hamiltonian. The equation for the classical phase space density is now $\partial\gamma/\partial t=\{\mathcal{H}^{(cl)},\gamma\}$ and it can be solved for a bundle of trajectories satisfying the Hamiltonian equations

$$\frac{d\overline{\boldsymbol{r}}}{dt}=\frac{\partial\mathcal{H}^{(cl)}}{\partial\overline{\boldsymbol{p}}},\quad\frac{d\overline{\boldsymbol{p}}}{dt}=-\frac{\partial\mathcal{H}^{(cl)}}{\partial\overline{\boldsymbol{r}}}$$

with initial conditions $(\overline{\boldsymbol{p}},\overline{\boldsymbol{r}})(t_{in})=(\overline{\boldsymbol{p}},\overline{\boldsymbol{r}})^{(in)}$. From this it follows that $\partial\widehat{\Gamma}_W^{(qu)}/\partial t-\left\{\mathcal{H}^{(cl)},\widehat{\Gamma}_W^{(qu)}\right\}=d\widehat{\Gamma}_W^{(qu)}/dt$ is a total time derivative.

Computational implementation involves transforming the operator expressions into matrices by expanding them in a basis set of electronic states $\Phi_k(\boldsymbol{X};\boldsymbol{q},\overline{\boldsymbol{r}})=\langle\boldsymbol{X}|\Phi_k(\boldsymbol{q},\overline{\boldsymbol{r}})\rangle$ for fixed atomic positions, collected in a row matrix of ket states $|\boldsymbol{\Phi}(\boldsymbol{q},\overline{\boldsymbol{r}})\rangle=\left[|\Phi_1(\boldsymbol{q},\overline{\boldsymbol{r}})\rangle,\ldots,|\Phi_N(\boldsymbol{q},\overline{\boldsymbol{r}})\rangle\right]$. The quantal $N\times N$ density matrix $\boldsymbol{\Gamma}(t)$ in

$\hat{\Gamma}_W^{(qu)}(t) = |\mathbf{\Phi}\rangle\, \mathbf{\Gamma}(t)\, \langle\mathbf{\Phi}|$ satisfies the equation

$$\frac{d\mathbf{\Gamma}}{dt} = -\frac{i}{\hbar}\left[\mathbf{H}_W^{(qu)} + \mathbf{H}_W^{(qc)} - i\hbar\left\langle \mathbf{\Phi}\middle|\frac{d\mathbf{\Phi}}{dt}\right\rangle, \mathbf{\Gamma}\right]$$

to be solved with initial conditions $\mathbf{\Gamma}(t_{in}) = \mathbf{\Gamma}^{(in)}$. It is coupled to the above Hamiltonian equations for phase-space trajectories and must be solved for time intervals adapted to the details of potential energy surfaces and crossings resulting from the choice of the electronic basis set.

Here we illustrate the treatment with an example on the photodissociation dynamics of NaI into $Na^+ + I^-$ [44]. A two-state model was adopted from [45] for the neutral and ion-pair species, with potential energy functions obtained for electronically diabatic states, where $\langle \mathbf{\Phi} \,|\, d\mathbf{\Phi}/dt \rangle = 0$. It displays a bonding energy well for the ions, a repulsive potential energy for the neutral state and a crossing of potential energies for neutral and ionic states, as expected for similar systems [12, 30]. This is a challenging model for quantum-classical treatments because a given trajectory can be trapped in the well, can be repulsed, or can transfer between charge states. The Wigner-transform treatment, propagating a bundle of trajectories, is found to nevertheless produce the right answers for scattering populations vs time.

In some detail, diabatic potential energy matrix elements $H_{11}(R)$, $H_{22}(R)$, and $H_{12}(R)$ were calculated and parametrized as functions of the interatomic radial distance R. They are given in au's [44] with the diabatic coupling energy $H_{12}(R) = 2.02 \times 10^{-3} \exp[-0.194(R - R_x)^2]$. Diabatic potential energies are shown in Figure 7.1, for neutral state 1 and ionic state 2 crossing at a distance R_x.

The model dynamics involves only one classical-like degree of freedom, with $(\overline{\boldsymbol{p}}, \overline{\boldsymbol{r}}) = R(t), P(t)$ as position and momentum variables and $P(t) = M dR/dt$ for relative reduced mass M. Initial conditions are constructed for an initial state due to photoabsorption of a femtosecond light pulse, promoting an initial vibrational bound state of NaI in the ionic well with quantum number $v = 0$ to the repulsive neutral potential.

The initial distribution in phase space at time $t_{in} = t_0$ is in atomic units, for $\hbar = 1$,

$$\Gamma_{11}^{(in)}(P, R) = \pi^{-1} \exp\left[-(R - R_0)^2/\sigma^2 - \sigma^2(P - P_0)^2\right]$$

with $R_0 = 5.047\, au(L)$, $P_0 = 0.0$, $\sigma = 0.12462\, au$, while $\Gamma_{12}^{(in)} = \Gamma_{21}^{(in)} = \Gamma_{22}^{(in)} = 0$.

Coupled equations to be solved are, with $\mathbf{H}_W^{(qu)} + \mathbf{H}_W^{(qc)} = \boldsymbol{H}(R)$,

$$\frac{dP}{dt} = -\frac{\partial \mathcal{H}^{(cl)}}{\partial R}, \quad \frac{dR}{dt} = \frac{\partial \mathcal{H}^{(cl)}}{\partial P}$$

$$\frac{d\mathbf{\Gamma}}{dt} = -\frac{i}{\hbar}[\boldsymbol{H}, \mathbf{\Gamma}]$$

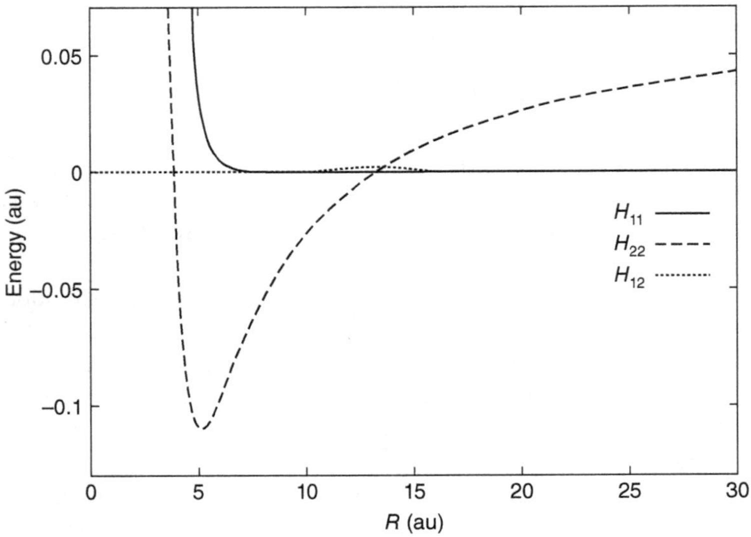

Figure 7.1 Diabatic potential energy matrix elements $H_{11}(R)$, $H_{22}(R)$ for NaI (state 1), and $Na^{+} + I^{-}$ (state 2) and diabatic energy coupling $H_{12}(R)$. Source: Adapted from [45]. The diabatic potential energy functions cross at $R_x = 14.02$ au.

with $\mathcal{H}^{(cl)} = P^2/(2M) + V(R) + V'(R)$, where $V(R) = H_{11}(R)$ and dV'/dR is constructed as a Hellmann–Feynman average force [12], in

$$\frac{dV'}{dR} = tr_{qu}\left[\widehat{\Gamma}_W^{(qu)}(P,R)\left(\frac{\partial\widehat{\mathrm{H}}_W^{(qc)}}{\partial R} - \frac{\partial H_{11}}{\partial R}\right)\right]$$

which accounts for changing forces in the region of the potentials crossing. The coupled equations have been solved with a version of the "relax-and-drive" procedure named here EP-QCLE (for effective potential-quantum classical Liouville equation). The equations were also solved with the accurate but more computationally demanding SO-FFT (or split-operator fast Fourier transform) procedure [46], and results were compared with EP-QCLE for trajectories and quantal populations.

Results for populations of quantal states are shown in Figure 7.2 vs time in au's, for the ionic state, and also for the neutral state broken up into its population in an inner (bound motion) region with $0 \leq R \leq R_x$ and outer (free motion) region with $R > R_x$. Results from the EP-QCLE calculations agree very well with the more accurate results. Figure 7.3 shows the final phase-space distribution of the trajectory bundle, which started with the $\Gamma_{11}^{(in)}(P,R)$ distribution, where $P_{in} = 0$. The final distribution displays the physical features expected in this dynamics, with a region of nearly closed orbits corresponding to motion in the ionic well, and a region of

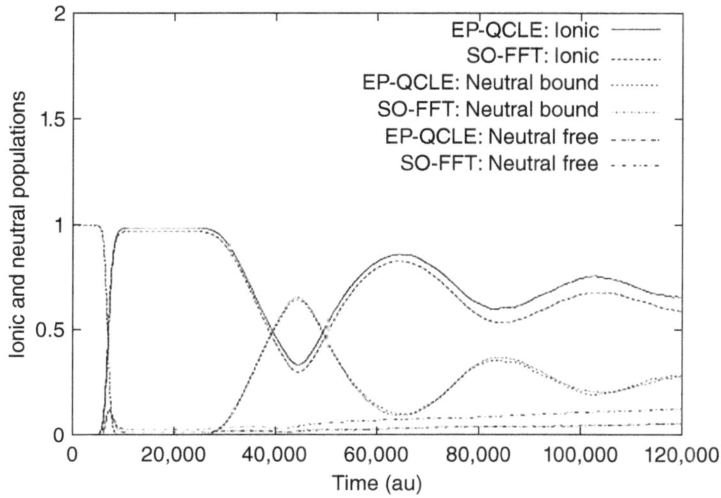

Figure 7.2 Ionic and neutral state populations vs dissociation time after photoexcitation. Neutral populations are separately shown for regions of bound and free-like motions. Results are compared for the quantal-classical Wigner-transformed density solution of the Liouville–von Neuman equation (EP-QCLE) and for an accurate numerical solution of the coupled classical-quantal partial differential equations (SO-FFT). Source: From [44], with permission of Elsevier.

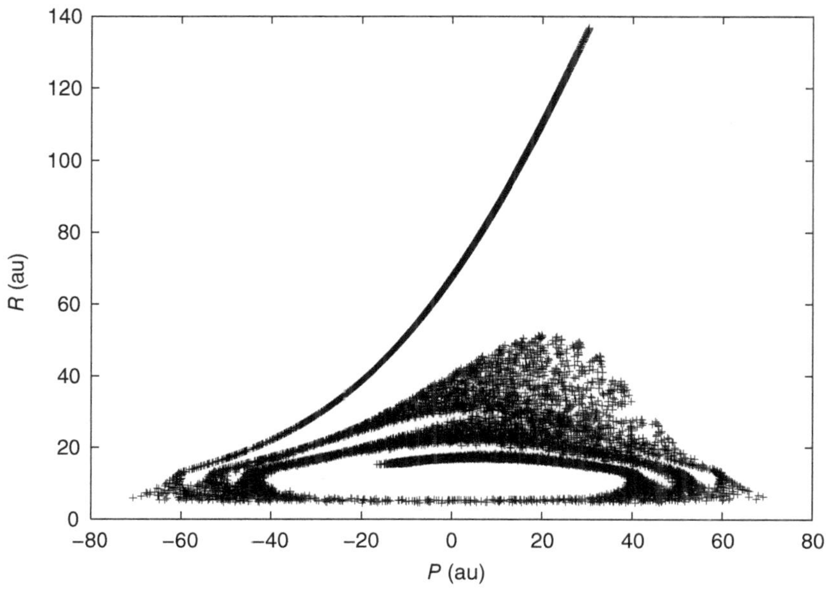

Figure 7.3 Final distribution of phase space trajectories showing regions of nearly orbiting motion in the ionic well and regions of nearly free motion on the potential energy function for the neutral atoms in dissociating NaI. Source: From [44]. With permission of Elsevier.

nearly free motion to large R for the neutral atoms break-up. It is remarkable that even though an individual trajectory may not follow a physical event, the whole bundle of trajectories does reconstruct the correct events.

7.5 Density Operator Treatments for Reactive Collisions

The density operator treatments presented in Section 7.4 are applicable to reactive molecular collisions and provide concepts and computational methods useful for studies of quantum coherence and for calculation of product yields in reactions. This is an active and developing area of research, that is being done with many related approaches. Some recent related reviews and overviews, with methods similar to the ones in this book, are briefly considered in what follows, but the large number of related publications prevents a thorough consideration.

The calculation of rates in reaction dynamics and photodynamics, from propagators and density matrices, has been detailed in reviews [46–50], using numerical integration procedures or expansions in Gaussian basis sets. Several aspects of quantal-classical treatments with density matrices involving multiple electronic surfaces have been covered in [42, 51–54] dealing with nonadiabatic phenomena and aspects of potential energy surface crossing. Coherence of quantal states and their decoherence decay during collisions have been detailed using density matrices in [55, 56]. A methodology closely related to density matrix theory for molecular dynamics is that of path integrals for the dynamics of many-atom systems. Some reviews can be found in [57–59].

References

1 N. F. Mott and H. S. W. Massey, *The Theory of Atomic Collisions*, Oxford, England: Clarendon Press, 1965.

2 B. Lipman and J. Schwinger, "Variational principles for scattering processes. I," *Phys. Rev.*, vol. 79, p. 469, 1950.

3 M. L. Goldberger and K. M. Watson, *Collision Theory*, New York: Wiley, 1964.

4 R. G. Newton, *Scattering Theory of Waves and Particles* (2nd edition), New York: Springer-Verlag, 1982.

5 A. Messiah, *Quantum Mechanics*, Volume 1, Amsterdam: North-Holland Publishing Company, 1962.

6 A. Messiah, *Quantum Mechanics*, Volume 2, Amsterdam: North-Holland Publishing Company, 1962.

7 D. A. Micha, "Effective hamiltonian methods for molecular collisions," *Advances in Quantum Chemistry*, vol. 8, p. 211, 1974.
8 G. Arfken, *Mathematical Methods for Physicists* (2nd edition), New York: Academic Press, 1970.
9 H. Feshbach, "A unified theory of nuclear reactions. II," *Ann. Phys.*, vol. 19, p. 287, 1962.
10 P. O. Lowdin, *Linear Algebra for Quantum Theory*, New York: John Wiley & Sons, 1998.
11 P. J. Kuntz, "Features of potential energy surfaces and their effects on collisions," in *Dynamics of Molecular Collisions Part B*, (Editor: W. H. Miller), New York: Plenum Press, 1976, pp. 53–120.
12 D. A. Micha, *Molecular Interactions: Concepts and Methods*, Hoboken, NJ: John Wiley & Sons, 2020.
13 D. A. Micha, "Collision dynamics of three interacting atoms: The Faddeev equations," *J. Chem. Phys.*, vol. 57, p. 2184, 1972.
14 D. A. Micha, "General theory of reactive scattering: A many-atom approach," in *Theory of Chemical Reaction Dynamics*, Volume II, *Chapter 3*, (Editor: M. Baer), Boca Raton, FL: CRC Press, 1985, pp. 182–218.
15 L. D. Faddeev, *Mathematical Aspects of the Three-Body Problem in Quantum Scattering Theory*, New York: Davey, 1965.
16 E. O. Alt, P. Grassberger and W. Sandhas, "Reduction of the three-particle collision problem to multichannel two-particle Lipmann-Schwinger equations," *Nucl. Phys.*, vol. B2, p. 167, 1967.
17 D. A. Micha, "Rearrangement in molecular collisions: A many-body approach," in *Theory of Chemical Reaction Dynamics*, Volume II, (Editor: M. Baer), Boca Raton, FL: CRC Press, 1985, pp. 181–218.
18 J. A. Olson and D. A. Micha, "Transition operators for atom-atom potentials: The Hilbert-Schmidt expansion," *J. Chem. Phys.*, vol. 68, p. 4352, 1978.
19 Z. C. Kuruoglu and D. A. Micha, "Diatomic transition operators: Results of L2 basis expansions," *J. Chem. Phys.*, vol. 72, p. 3327, 1980.
20 D. A. Micha and L. H. Beard, "Numerical calculation of diatomic transition operators: The variable-phase and -amplitude method," *Chem. Phys. Lett.*, vol. 53, p. 329, 1978.
21 J.-M. Yuan and D. A. Micha, "Collision dynamics of three interacting atoms: Stripping reactions of $Ar^+ + H_2$ and of $K + I_2$," *J. Chem. Phys.*, vol. 64, p. 1032, 1976.
22 L. H. Bear and D. A. Micha, "Collision dynamics of three interacting atoms: Energy transfer and dissociation in collinear motions," *J. Chem. Phys.*, vol. 73, p. 1193, 1980.
23 Z. C. Kuruoglu and D. A. Micha, "Collision dynamics of three interacting atoms: The Faddeev Equations in a diabatic electronic basis," *J. Chem. Phys.*, vol. 79, p. 6115, 1983.

24 Z. C. Kuruoglu and D. A. Micha, "Calculation of resonances in the $H+H_2$ reaction using the Faddeev-AGS method," *Int. J. Quantum Chem.*, vol. Symp. 23, p. 103, 1989.

25 D. A. Micha and E. F. Vilallonga, "The collisional time-correlation function approach to molecular energy transfer," in *Advances in Chemical Physics*, Volume 84, (Editors: I. Progogine and S. A. Rice), New York: John Wiley & Sons, 1993, pp. 1–72.

26 E. F. Vilallonga and D. A. Micha, "The calculation of time-correlation functions for molecular collisions," *Phys. Rep.*, vol. 212, pp. 329–390, 1992.

27 D. A. Micha, "Optical models in molecular collision theory," in *Dynamics of Molecular Collisions, Part A*, (Editor: W. H. Miller), New York: Plenum Press, 1976, pp. 81–129.

28 J. von Neuman, *Mathematical Foundations of Quantum Mechanics (Translation)*, Princeton, NJ: Princeton University Press, 1955.

29 D. A. McQuarrie, *Statistical Mechanics*, New York: Harper & Row, 1973.

30 V. May and O. Kuhn, *Charge and Energy Transfer Dynamics in Molecular Systems* (2nd edition), Weinheim, Germany: Wiley-VCH, 2004.

31 G. C. Schatz and M. A. Ratner, *Quantum Mechanics in Chemistry*, Englewood Cliffs, NJ: Prentice-Hall, Inc., 1993.

32 D. A. Micha and I. Burghardt (Editors), *Quantum Dynamics of Complex Molecular Systems*, Berlin, Germany: Springer-Verlag, 2007.

33 E. Wigner, "On the quantum correction for thermodynamic equilibrium," *Phys. Rev.*, vol. 40, p. 749, 1932.

34 G. Ciccotti, D. F. Coker and R. Kapral, "Quantum statistical dynamics with trajectories," in *Quantum Dynamics of Complex Molecular Systems*, (Editors: D. A. Micha and I. Burghardt), Berlin: Springer-Verlag, 2007, p. 275.

35 D. A. Micha, A. Leathers and B. Thorndyke, "Density matrix treatment of electronically excited molecular systems: Applications to gaseous and adsorbate dynamics," in *Quantum Dynamics of Complex Molecular Systems*, (Editors: D. A. Micha and I. Burghardt), Berlin: Springer-Verlag, 2007, p. 165.

36 D. A. Micha, "Density matrix treatment of electronic rearrangement," *Adv. Quantum Chem.*, vol. 35, p. 317, 1999.

37 D. A. Micha, "From few-atom to many-atom quantum dynamics," *Adv. Quantum Chem.*, vol. 41, p. 139, 2002.

38 D. A. Micha, "Quantum partitioning methods for few-atom and many-atom dynamics," *Adv. Quantum Chem.*, vol. 74, p. 107, 2017.

39 M. Hillery, R. F. O'Connell, M. O. Scully and E. P. Wigner, "Distribution functions in physics: Fundamentals," *Phys. Rep.*, vol. 106, p. 121, 1984.

40 S. Mukamel, *Principles of Nonlinear Optical Spectroscopy*, Oxford, England: Oxford University Press, 1995.

41 D. A. Micha and B. Thorndyke, "The quantum-classical density operator for electronically excited molecular systems," *Adv. Quantum Chem.*, vol. 47, p. 294, 2004.

42 C. C. Martens and J.-Y. Fang, "Semiclassical-limit molecular dynamics on multiple electronic surfaces," *J. Chem. Phys.*, vol. 106, p. 4918, 1997.

43 R. Kapral and G. Ciccotti, "Mixed quantum-classical dynamics," *J. Chem. Phys.*, vol. 110, p. 8919, 1999.

44 B. Thorndyke and D. A. Micha, "Photodissociation dynamics from quantum-classical density matrix calculations," *Chem. Phys. Lett.*, vol. 403, p. 280, 2005.

45 V. Engels and H. Metiu, "A quantum mechanical study of predissociation dynamics of NaI excited by a femtosecond laser pulse," *J. Chem. Phys.*, vol. 90, p. 6116, 1989.

46 M. D. Feit, J. A. Fleck and A. Steiger, "Solution of the Schroedinger equation by an spectral method," *J. Comput. Phys.*, vol. 47, p. 412, 1982.

47 R. Kosloff, "Propagation methods for quantum molecular dynamics," *Annu. Rev. Phys. Chem.*, vol. 45, p. 145, 1994.

48 S. Tretiak and S. Mukamel, "Density matrix analysis and simulation of electronic excitations in conjugated and aggregated molecules," *Chem. Rev.*, vol. 102, p. 3171, 2002.

49 H. B. Schlegel, J. M. Millam, S. S. Iyengar, G. A. Voth, A. D. Daniels, G. E. Scuseria and M. J. Frisch, "Ab initio molecular dynamics: Propagating the density matrix with Gaussian orbitals," *J. Chem. Phys.*, vol. 114, p. 9758, 2001.

50 B. F. E. Curchod and T. J. Martinez, "Ab initio nonadiabatic quantum molecular dynamics," *Chem. Rev.*, vol. 118, p. 3305, 2018.

51 M. Santer, U. Manthe and G. Stock, "Quantum-classical Liouville description of multidimensional nonadiabatic molecular dynamics," *J. Chem. Phys.*, vol. 114, p. 2001, 2001.

52 M. Thoss and H. Wang, "Semiclassical description of molecular dynamics based on initial value methods," *Annu. Rev. Phys. Chem.*, vol. 55, p. 299, 2004.

53 X. Li, J. C. Tully, H. B. Schlegel and M. Frisch, "Ab initio Ehrenfest dynamics," *J. Chem. Phys.*, vol. 123, 084106, 2005.

54 J. E. Subotnik, "Augmented Ehrenfest dynamics yields a rate for surface hopping," *J. Chem. Phys.*, vol. 132, 134112, 2010.

55 C. Zhu, A. W. Jasper and D. G. Truhlar, "Non-Born–Oppenheimer trajectories with self-consistent decay of mixing," *J. Chem. Phys.*, vol. 120, p. 5543, 2004.

56 Y. Shu and D. G. Truhlar, "Decoherence and its role in electronically non-adiabatic dynamics," *J. Chem. Theory Comput.*, vol. 19, p. 380, 2023.

57 N. Makri and W. H. Miller, "Monte Carlo path integration for the real time propagator," *J. Chem. Phys.*, vol. 89, p. 2170, 1988.

58 W. H. Miller, "Electronically nonadiabatic dynamics via semiclassical initial value methods," *J. Phys. Chem. A*, vol. 113, p. 1405, 2009.

59 S. Kundu and N. Makri, "Intramolecular vibrations in excitation energy transfer: Insight from real-time path integral calculations," *Annu. Rev. Phys. Chem.*, vol. 73, p. 349, 2022.

Index

f

g

h

i

j

k

l

m

n

o

p

www.ingramcontent.com/pod-product-compliance
Lightning Source LLC
Chambersburg PA
CBHW060702050426
42133CB00033B/2
9780470383735